EXPERIMENTAL DESIGN AND ANALYSIS

MARVIN LENTNER
Virginia Polytechnic Institute
and State University

THOMAS BISHOP
Battelle Columbus Laboratories

VALLEY BOOK COMPANY
P.O. Box 884
Blacksburg, VA 24060

Printed in the United States of America

10 9 8 7 6 5 4 3 2 1

ISBN 0-9616255-0-3

This book is intended as both a text and a reference. There is ample material included for a one semester course. At Virginia Tech (currently on the quarter system,) we cover most of the material in Chapters 3 through 11, plus some of the material of the remaining chapters, as time permits. The students, primarily undergraduate majors in statistics and graduate students from other disciplines, have had two quarters of statistical methods including a good coverage of completely randomized designs with a brief exposure to mean separation procedures and factorials.

The material in this book is in the mainstream of applied statistics. Many of the examples and problems include data from actual consulting projects. Theoretical development of the design and analysis of experiments is beyond the scope of this book. We state results of established theory and indicate how it applies to the topic being discussed. Chapter 13 contains theoretical aspects of general incomplete block designs. The complexity here is due more to the notation than to the mathematics involved.

Experimentation is a key component of research. In the initial phases of developing a research program, the various aspects of *Experimental Design* should be incorporated. The initial research proposal should address such issues as: Which experimental setup to use, which factors to study, the manner of obtaining the observations, and so on. Aspects of *Experimental Analysis* are used in preparing final reports of the work, in the dissemination of the research results. Extracting relevant information from the data, constructing valid estimates and verifying the existence of meaningful differences are some to these aspects.

This book deals with both the design and analysis aspects of experimentation. We discuss various points to consider in selecting an appropriate experimental design. All classical, commonly used designs are discussed. For the most part, we have limited our coverage to the equal numbers (balanced) cases. Originally we intended to include material on the unequal numbers (unbalanced) cases but the manuscript quickly became too unwieldy. On a number of occasions, though, we do mention the problems of unbalanced designs and indicate possible solutions to such situations.

For reproduction of statistical tabulations, we acknowledge the permissions given by: The Rand Corporation to include a table of random digits (Table A.1); to the Biometrika Trustees to include tables of percentage points of Student T, χ^2, Hartley's maximum F, and the studentized range distributions (Tables A.6, A.7, A.8, and A.11); to Dr. H.L. Harter and the Biometric Society to include a table of Duncan's NMRT values (Table A.12); to Dr. C.W. Dunnett, the Biometric Society, and the American Statistical Association to include a table of values for Dunnett's procedure (Table A.13); and to Dr. O. J. Dunn and the American Statistical Association to include a table of values for Bonferroni's T distributions (Table A.14.)

We are indebted to numerous persons for invaluable comments and suggestions which lead to the final version of this book. First, we would like to express appreciation for the comments and encouragement received from former students and instructors who used earlier versions of this material over the past six years. Particular thanks are due to Professor Francis Giesbrecht, North Carolina State University, and Professor Jill S. Hampton, Radford University, for their detailed review of an earlier version; to Mr. Michael Rozum, Virginia Tech, for his comments on reading an earlier version; and to Professor Eric Smith, Virginia Tech, for his comments and discussions on parts of an earlier version. We also would like to thank Monte Lentner for his assistance with the artwork.

Finally, we would like to express thanks to our families for their patience, understanding, and encouragement while we adhered to the experimental design that resulted in this book.

Marvin Lentner   Thomas Bishop

CONTENTS

viii   Contents

x   Contents

TO THE READER

Each section is identified by a number separated from the chapter number by a period. Thus, Section 5.6 means the sixth section of Chapter 5. Subdivisions of a section (subsections) are identified by the inclusion of a third number; thus, the number 1.5.3 means the third subdivision of the general Section 1.5.

Equations, formulas, and other displayed material are numbered consecutively in each chapter and identified by two numbers appearing at the left margin, in parentheses. Thus, (6.24) refers to the twenty-fourth identified item of Chapter 6. Examples also are numbered consecutively by chapter and each ends with the symbol, ✳.

References are given in alphabetical order at the end of a chapter and numbered in square brackets. Thus, [2] appearing in the text refers to the second reference at the end of the chapter in which it appears.

Problems appear at the end of each chapter. Answers to odd-numbered problems are given at the back of the book.

## Chapter 1

BASIC CONCEPTS OF EXPERIMENTAL RESEARCH

1.1   Introduction

During the past several decades, the boundaries of scientific research activity have been greatly extended. Scientific research methodology is used in such disciplines as agriculture, biology, engineering, chemistry, transportation, business, economics, wildlife preservation, sociology, history, and various environmental issues -- indeed every discipline you can name.

Although we are entertaining a broad definition of the terms *scientist* and *scientific research activity*, those research activities which can be legitimately classified as *scientific* all possess a common characteristic. The information obtained in the course of such research activity is derived from experiments which are objectively planned and designed according to the principles of sound logical reasoning. These principles form the cornerstones of the scientific approach to research. They include among others: precise and defendable definitions, clear statements of objectives, clear statements of assumptions, development of appropriate models, appropriate data collection and analysis techniques, and objective reporting of the results.

Today's research environment is dynamic and multidisciplinary in nature. The important research problems are usually complex, and their solutions often require the combined effort of scientists and researchers whose experiences are wide ranging and include several disciplines. The logical issues associated with today's research problems are often subtle and elusive, and contemporary reseachers must be trained to deal with them. In light of the complexity of current research, the role of sound, fundamental logical reasoning and the planning of experiments cannot be overstated.

As we have noted, the experiment is the common link among the various research activities. The experiment is the vehicle for rationally obtaining new information which can be used to develop new theories and challenge old ones. Because the experiment is the key element in the scientific process, it is essential that researchers have a general knowledge about designing, conducting, and analyzing experiments. The next section of this chapter is devoted to a brief discussion of the basic steps required in conducting an experiment which will provide data amenable to statistical analyses. When properly used, statistical analyses lead to scientifically sound conclusions. It is for this reason that *statistics* are rapidly becoming a universal language of research.

The remainder of this chapter is devoted to basic ideas of experimental analyses. It is assumed that the reader is familiar with elementary concepts of classical statistics, including an exposure to linear regression. In particular we take as known ideas: Randomness, Probability, Population, Sample, Parameter, Statistic, Unbiased, Hypothesis Testing and Interval Estimation. For those who wish to review any of these ideas, most elementary statistics books can be consulted; for example, those by Freund [2] or Lentner [3].

1.2 The Experimental Program

Experiments are conducted to obtain information about populations of interest. Information gained from such research activities may be used in a variety of ways, for example:

(i) in drawing inferences about parameters of interest,
(ii) in making decisions about formulated hypotheses, and
(iii) for planning future research.

One of the major aspects of experimental designs is concerned with efficiency; that is, conducting the experiment which yields the most information for the least cost (as measured by number of observations, time of completion, amount of materials required.) In many cases, optimal designs do not exist or are not unique, and the choice of an efficient experimental plan is left to the discretion of the researcher. It is for this reason that the experimenter should have a broad background in the design of experiments so the choice of an efficient design can be made wisely.

The information obtained from planned experiments is used inductively. That is, generalizations are made about a population from information contained in a random sample of that particular population. In all nontrivial experiments, the effects under investigation are influenced by forces beyond the control of the experiments. Because of these outside forces, inferences and decisions projected from a sample to the population are sometimes erroneous. Proper statistical analyses provide the tools for quantifying the chances of obtaining erroneous results. This point is often misunderstood. Statistics never prove anything. Rather, proper statistical designs and their analyses allow for controlling and quantifying the proportion of erroneous results. This gives the experimenter some quantifiable degree of belief in the observed results. Throughout this book we will consider various techniques and experimental designs useful in reducing the risks and errors of the analytical process.

Experiments which have been improperly designed or blindly performed usually result in irrelevant or unwanted information. To insure the attainment of relevant and precise information, a scientifically performed experiment is required. Such experiments may help in leading a researcher to a deeper understanding of the phenomenon under investigation.

It is imperative that researchers take into account all information at their disposal, such as previous experiences, results of related research, known theories of the discipline, and so on. To insure that all information is effectively used, the following checklist of steps should be used in planning and conducting a scientific experiment.

1. Formulation of the research plan. Included here is precise statement of the problem to be solved and the objectives of the research. What does the researcher plan to study and why?

2. Choice of factors to be used. After the problem has been formulated, a decision about pertinent factors can be made. All important factors must be included but too many factors may result in an unwieldy experiment. Can any factors be omitted without seriously affecting the planned research? The range and specific values of each factor must be considered. To reflect the impact of a factor, a sufficient number of its values must be used.

3. Choice of variables to be measured. A variable should be measured if it provides information about the phenomenon under investigation. Important variables are retained while less important variables may be discarded at the analysis stage. Costs of measuring each variable must be weighed against its importance.

4. Choice of the inference space. This may be defined briefly as the set of populations to which the inferences may be applied. The inference space, once decided upon, will have an influence on the size of the experiment (the number of populations to sample) as well as the applicability of the experimental results.

5. Selection of experimental material. The type and amount of material required may depend upon the objectives, the factors, the inference space, the researcher's budget, availability, and so on.

6. Choice of an experimental design. This is a primary topic of this book and refers to the manner in which the factors are assigned to the experimental material. Should this be

a completely randomized allocation or some other type of random assignment?

7. Formulation of a model. It is customary to give a mathematical model to describe the observations anticipated under the experimental plan. The model should include all factors and components to provide an adequate representation of the observations.

8. Collection of the data. Extreme care should be exercised to insure that the entire experimental plan (particularly the randomization) is carried out correctly and that the observations are properly and accurately recorded. When deemed necessary, a system of "checks-and-balances" should be used.

9. Analysis of the data. Depending upon the model and the design, necessary computations must be performed. Complex or large experiments may require a computer to perform some of these computations.

10. Conclusion and interpretations. This step is extremely important. The practical implications of the experimental results must be presented clearly and objectively together with any qualifying remarks regarding applicability of the results. Limitations of the study should be noted.

The first four steps constitute the experimental set-up phase, the next three steps represent the design phase, and the last three steps represent the analysis phase.

In any experiment, certain *random variables* are required to address important aspects of the phenomenon under investigation. The variables selected for measurement (in step 3 above) might be used directly. Yield per plot, growth of a plant, and the number of defects per manufactured unit are measured variables which might serve as random variables. If not used directly, measured variables might be converted (by taking logarithms, square roots, and so on) or combined (by taking ratios, products, and so on) to produce the desired random variables. For example, an important random variable for a nutrition scientist is the copper/zinc ratio of individuals, a function of two measured variables, copper content and zinc content.

A *random variable* may be any function of experimental observations, provided that the function produces numerical values which follow a probability distribution. Thus, each experimental observation provides one *realization*(that is, one specific value) of any random variable. And, except in trivial cases, the realizations of a random variable exhibit variation. To measure the magnitude of variation for a random sample of n realizations of a variable y, we usually calculate a quantity called the *sample variance* defined by

(1.1) $$s^2 = \frac{1}{n-1} \sum_{i=1}^{n} (y_i - \bar{y})^2$$

where

(1.2) $$\bar{y} = \frac{1}{n} \sum_{i=1}^{n} y_i$$

is the *sample mean*.

1.3   Some Important Normal Theory

The normal distribution is one of the most frequently encountered distributions in statistical inferences. Its role in experimental analyses is equally prominent. Because this book deals only with normal based analyses, we will state, without proof, and discuss three important theorems regarding normal distributions. We let $N(\mu, \sigma^2)$ denote a normal distribution with mean μ and variance σ^2.

Theorem 1.1

If our variable of interest, y, has the $N(\mu, \sigma^2)$ distribution, then in random samples of size n the sample means have the $N(\mu, \sigma^2/n)$ distribution.

This theorem tells us how the sample means are distributed when sampling from a normal population. Because sample means are used extensively for making inferences in experimental research, Theorem 1.1 is particularly useful. Included as part of Theorem 1.1 are the mean value and standard deviation of sample means; namely,

(1.3) $\mu_{\bar{y}} = \mu$ and $\sigma_{\bar{y}} = \sqrt{\sigma^2/n}$

The sample means have the same mean value as the population from which the samples were drawn but the variance of the sample means is the population variance reduced by the sample size.

Example 1.1

Suppose variable y has the N(24,50) distribution. Then in samples of size 10, the sample mean $\bar{y}$ has a normal distribution with mean value of μ = 24 and standard deviation of $\sigma_{\bar{y}} = \sqrt{\sigma^2/n} = \sqrt{5}$; that is, $\bar{y}$ has the N(24,5) distribution. ✳

   In   addition to sample means,   we have occasion to use   other functions of random variables.   The next theorem deals with   the distribution of linear combinations of normal variables.

Theorem 1.2

   Let y_1, y_2, ..., y_k be any random variables having normal distributions and let $a_1, a_2,$ ..., a_k be constants, not all zero. Then, the linear combination

(1.4)$$L = \sum_{i=1}^{k} a_i y_i$$

   has a normal distribution with mean value of

(1.5)$$\mu_L = \sum_{i=1}^{k} a_i \mu_i$$

   and a variance of

(1.6)$$\sigma_L^2 = \sum_{i=1}^{k} a_i^2 \sigma_i^2 + \sum_{i \neq j}^{k} a_i a_j \sigma_{ij}$$

   where μ_i is the mean value of y_i, σ_i^2 is the variance of y_i, and σ_{ij} is the covariance between y_i and y_j.

Several points are worth making about Theorem 1.2.   First of all, if the y_i's are mutually independent, all σ_{ij} are zero and the variance of L has a simple form

$$(1.7) \qquad \sigma_L^2 = \sum_{i=1}^{k} a_i^2 \sigma_i^2$$

Secondly, the sample mean is a linear combination of the n sample observations, which are independent under random sampling. Thus, the distribution of the sample mean is specified by Theorem 1.2 as well as by Theorem 1.1, when random sampling from a normal distribution.

Finally, the y's used in Theorem 1.2 need not be individual observations; the y's could be means or other quantities so long as each has a normal distribution. We will indicate what form the y's take as we apply Theorem 1.2 to given situations.

Example 1.2

Let y_1, y_2, y_3, and y_4 have, respectively, $N(5,20)$, $N(10,15)$, $N(8,30)$, and $N(12,25)$ distributions, with all y's mutually independent. Then, the linear sum

$$L_1 = y_1 + y_2 + y_3$$

has a normal distribution with mean value of

$$\mu_{L_1} = 5 + 10 + 8 = 23$$

and variance of

$$\sigma_{L_1}^2 = 20 + 15 + 30 = 65.$$

Likewise, the linear combination

$$L_2 = y_1 + y_2 - 2y_3 + y_4$$

has a normal distribution with mean value of

$$\mu_{L_2} = 5 + 10 - 2(8) + 12 = 11$$

and a variance of

$$\sigma_{L_2}^2 = 20 + 15 + 4(30) + 25 = 180. \; ※$$

The two previous theorems dealt with situations where the underlying populations are normal. These theorems have limited application for it is generally considered that few, if any, random variables have exact normal distributions. The assumption of normality is not critical, and may be relaxed, when making inferences about the mean of any distribution so long as the sample size is sufficiently large. This important concept is an immediate consequence of the Central Limit Theorem, one of the major theorems in statistical analysis. We state one of the more common versions.

Theorem 1.3 (Central Limit Theorem)

Let the variable y have any distribution with finite mean μ and variance σ^2. Then, in random samples of size n from this distribution, the standardized sample mean $\sqrt{n}(\overline{y} - \mu)/\sigma$ has a distribution which approaches the N(0,1) distribution as n approaches infinity.

As a result of the Central Limit Theorem, the standard normal (or Z) distribution can be used for inferences about the mean μ of any random variable so long as n is large. The degree of approximation in the inference depends upon the population departure from normality and the sample size. For continuous variables having symmetric distributions, samples of size 20 or more are generally sufficient. Lengths (or diameters) of components, growths (heights) of plants, and weights of fruit per plant are examples of continuous variables which may have symmetric distributions.

For discrete random variables, sample sizes of several hundred may be required. Count variables, such as the number of insects per plot, the number of paint defects per auto, and the number of industrial accidents per week are examples of discrete variables which generally require large sample sizes for good results from application of the Central Limit Theorem. Smaller sample sizes may suffice for specific discrete distributions. For example, a binomial distribution is symmetric when the proportion p equals 0.5. It is fairly well-known that application of the Central Limit Theorem gives good results for samples as small as n = 10 when the proportion p is near 0.5. Proportions near zero or unity would require considerably larger sample sizes.

In most practical situations, application of the Central Limit Theorem involves an estimate of the population variance, σ^2. Using the sample variance, s^2, as the estimator of σ^2 gives a random variable $\sqrt{n}(\overline{y} - \mu)/s$ which has an *approximate* Student's t distribution when based on a *large* sample size. As expected, larger sample sizes are required when using the t distribution instead of the Z distribution for making inferences about a mean.

We now give an example to illustrate the role of the Central Limit Theorem in statistical inference when the variance σ^2 is unknown and estimated by the sample variance, s^2.

Example 1.3

A vegetable grower wishes to estimate the average weight, in grams (g), per fruit for a particular variety of tomatoes.   He records the average weight per fruit for n = 50 tomato plants of this year's crop and obtains

$$\bar{y} = 4237(g) \quad \text{and} \quad s = 22.5(g)$$

Applying the Central Limit Theorem, we can take the average weight, $\bar{y}$, to be an approximately normal variable. So using a Student's t distribution with 49 degrees of freedom, an approximate 99% CI for the average weight per fruit is

$$L,U = \bar{y} \pm (s/\sqrt{n})t_{\alpha/2}$$
$$= 84.74 \pm 3.18(2.68)$$
$$= 84.74 \pm 8.52 \rightarrow (76.22; \ 93.26)$$

The vegetable grower can be 99% confident that this variety of tomatoes produces fruit having an average weight between 76.2 and 93.3 grams. ✖

1.4   Definitions and Terminology of Experimental Research

An experiment is a planned scientific inquiry designed to invest- igate one or more populations.   A sociologist considers several socio-economic classes and asks a random sample of people from each class how much money is spent per month on entertainment. An engineer might measure the oxidation rate of aluminum rods having different coatings and subjected to different humidities.   In the first case, each socio-economic class would define a population; in the latter case, a population is defined by each coating- humidity combination.   The various conditions (processes, tech- niques, operations, and so on) which distinguish the populations of interest are called *treatments*. Every treatment uniquely defines a population. Thus, each socio-economic class would be a treatment in the first experiment while each combination of coating and environment would be a treatment in the engineer's experiment.

A *control* is a term used for a "standard treatment" (or maybe a placebo) included in an experiment so that there is a reference value to which other treatments may be compared. In some experiments a control is a "dummy" treatment where nothing is applied to the experimental material (for example, no fertilizer, no heat applied, no insecticide) and in some experiments a standard or "accepted" treatment may be designated as a control.

When several aspects are studied in a single experiment, such as the coatings and humidities above, each is usually called a *factor*. The different categories within a factor are called the *levels* of the factor. If the engineer uses 5%, 15% and 25% humidities, he would be using 3 levels of the humidity factor.

In many experiments, the researcher can exercise control over the particular levels of the factors appearing in the experiment. The sociologist can decide to obtain information on only two specific socio-economic classes. The engineer, within limitations of his experimental equipment, can decide which humidities and which coatings to investigate. For some experiments though, the researcher has little or no control over the levels appearing in the experiment. Such is the case of a plant scientist studying yields at different amounts of annual rainfall, at different seasonal temperatures, and so on.

Each treatment is assigned to certain experimental material. An *experimental unit*, denoted by EU, is the smallest entity receiving a single treatment, provided that two such entities could receive different treatments. For the sociologist an EU would likely be an individual person whose entertainment is determined. For the engineer, each aluminum rod would likely be an EU. But a whole group of items could be an EU. This would be the case for an animal scientist who feeds a single ration (a treatment) to a pen of cattle---the pen would be the EU because individual cattle fail to satisfy the second requirement of an EU. And if an educator tried different teaching methods on various third grade classes, each class would be an EU.

In many experiments it may be impractical to measure or observe the entire EU. For example, if a tree is an EU, one would not want to remove all leaves from a tree just to obtain calcium content. The nature of the treatments and EU also might make such an observation impractical. An agronomist studying the rate of application of limestone may use trucks to apply the limestone on large fields; yields may be obtained from small segments of each field. Each subset of an EU that is measured or observed is called a *response unit* or *sampling unit*. Several response units may be selected at random from each EU for observation. If the entire EU is observed, the response unit is the EU, of course.

In the most basic experimental situations, each population is the set of conceptual measurements from EU receiving the same treatment. These treatment populations, as they are called, often are differentiated on the basis of their mean values and variances. In many experimental situations, it is assumed that the treatment populations have roughly the same variance. When this assumption is reasonable, the analyses and inferences about population means are considerably simpler than the situations where unequal variances occur.

From this brief discussion it is apparent that experimental observations might exhibit variation (differences) for any number of reasons: the particular treatments being investigated, the nature of the experimental material, environmental conditions, different trainings and abilities of workers used, and so on. Any worthwhile analysis of an experimental situation should account for as much variation as possible. To do this, it first is necessary to recognize or identify the potential sources of variation likely to be present in the experiment. Then the impact or effect of each source of variation must be accounted for in the experimental analysis. This subdivision of "total variability" is the essence of the analysis of variance procedure which we now define formally.

Definition 1.1

Analysis of variance (ANOVA) is the process of subdividing the total variability of experimental observations into portions attributable to recognized sources of variation.

As we shall see later, it is actually the total sum of squares which gets partitioned into portions that can be meaningfully identified. The results obtained in an analysis of variance generally are summarized in tabular form, in an ANOVA table.

From an ANOVA of our experimental data, we estimate the (common) variance of the treatment populations by a quantity usually called *experimental error* (Exp. Error, for short.) As will be apparent later, Exp. Error includes variation due to all unaccounted sources of variation. Major elements that contribute to the Exp. Error variability are: (i) naturally inherent differences, (ii) lack of uniformity in repeating experimental setups, and (iii) other extraneous elements. There is very little that can be done about the impact of naturally inherent variation, short of defining new populations. Refinement of experimental techniques may be necessary if "lack of uniformity" is excessive. Finally, the "other extraneous variation" includes such things as measurement errors, computational errors, and human errors, as well as variation due to "an incomplete" model.

By the last we mean a model which should contain additional terms
and as a result does not remove an adequate amount of variation
from the residual components.   Experimental design looks at this
very problem by considering alternative models and experimental
plans that provide better recognition of important sources of
variation and lead to "cleaner residual components".   In summary
then, an effort to reduce Exp. Error may include refinement of
experimental techniques, use of more homogeneous experimental
materials, use of different experimental models, and so on.
    To estimate the variance for any treatment population would
require observations from at least 2 EU.   Of course the estimate
gets better as the number of EU increases.   And under the basic
assumption of equal variances for the treatment populations, we
can use the pooling concept (first encountered in two-population
analyses) to obtain a better, more appropriate value of Exp.
Error.
    A set of EU upon which each treatment appears once is called
a *replication* of the treatments, and throughout the experiment we
might say that "the i-th treatment is replicated r_i times".
While some replication is necessary for calculation of Exp.
Error, additional replication may be necessary as smaller and
smaller differences of treatment parameters are to be detected.
The ability of an experiment to detect population differences is
sometimes referred to as the *sensitivity* of the experiment.   The
chance of detecting smaller differences is increased as the
sensitivity is increased, and this occurs when Exp. Error is
decreased.
    The size of an experiment, as measured by the number of
replications, is highly dependent upon the desired sensitivity.
In turn, sensitivity depends upon the size of the differences to
be detected and the size of Exp. Error.   The type and number of
treatments also have some effect on the number of replications
needed.
    Accuracy and *precision* are two additional terms that require
some elaboration.   There seems to be various definitions of these
terms in existence so we consider a common definition of each.
Accuracy refers to the closeness of an estimate to the true
value.   Thus, accuracy is a measure related to bias, which for a
parameter θ and its estimator $\hat{\theta}$, is defined by

$$\text{Bias in } \hat{\theta} = E[\hat{\theta}] - \theta$$

where the capital letter E denotes expected value.
    Precision, on the other hand, denotes the repeatability of
measurements, and therefore is related to Exp. Error. The following display gives the 4 possibilities.

Accurate

	Yes	No
Precise Yes	Both accurate and precise	Precise but not accurate
No	Accurate but not precise	Neither accurate nor precise

Think of a person who repeatedly tosses a dart at a target as in Figure 1.1 below

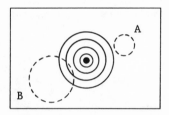

Figure 1.1   Dart Target

If a person tosses the dart and all strikes are dispersed within the smallest circle around the bullseye, his tossing is both accurate and precise. If all tosses land in the small area labelled A, his tossing would be precise but not accurate. Tosses landing within the largest circle containing the bullseye would be accurate but not precise. Finally, tosses landing within the dotted region labelled B would be neither accurate nor precise.

For any set of treatments, there is usually a number of different experimental plans available for the researcher to use. The experimental layout, including the arrangement of the EU and the allocation of the treatments to the EU, will be called an *experimental design*, or simply a design. Some of the frequently occurring designs will be discussed in the ensuing chapters. In the comparison of treatments, a scientist always uses a design. Hopefully, care and thought were used in selecting a design rather than accepting whatever happened without planning.

One of the prime objectives of experimental research is to ascertain whether or not different treatment populations have the same parameters, in particular, the same means. To detect significant differences among treatment means, the experimenter should use the most sensitive experimental design; that is, the design having the smallest Exp. Error.

1.5   Linear Models

The purpose of experimental research is to study the behavior of some variable of interest, such as the income of a person, the yield of a crop, the weight gain of certain animals, the baking time of cakes, the learning ability of students, and so on. A convenient way to describe the behavior of a variable of interest is by providing a model, an expression or equation which explains the response of the variable in terms of its component parts. Naturally, the simplest model is desired. It is for this reason that considerable attention is given to linear models. Such was the case in the study of regression where linear models were used to explain the relationship between a dependent variable, y, and one or more independent (auxiliary) x variables. The adjective *linear* is used, in ANOVA as well as in regression models, to indicate that the parameters appear in the equation in a linear manner.

In the following subsections, linear models will be discussed for normal populations. First we will discuss the model for one normal population. Then we will consider the model for two normal populations and show how this can be extended easily to any number of normal populations. The argument given for these normal populations can be applied to any non-normal situation, and in most cases, the same model would be obtained.

1.5.1   Model for One Normal Population

When sampling from one normal population, where the variable of interest is denoted by y, the experimental data may appear as in Figure 1.2 below where the sample observations are indicated by * on the horizontal axis. For illustration, we assume that $\mu = 20$.

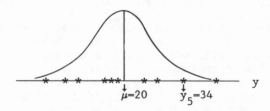

Figure 1.2   Sample from $N(20, \sigma^2)$

Each response, each observed value y_i, can be represented as the linear sum of two components:

$$\mu = \text{mean value of } y$$

(1.8)

$$\epsilon_i = y_i - \mu = \text{deviation of } y_i \text{ from the} \\ \text{mean value}$$

Thus, the observation y_5 can be expressed as

$$y_5 = 34 \\ = 20 + 14 = \mu + \epsilon_5$$

And, in general

(1.9) $$y_i = \mu + \epsilon_i$$

where any ϵ_i may be positive, negative, or zero and all ϵ_i are such that their expected (average) value is zero.

   Note that, on the right hand side of (1.9), μ is a constant while ϵ_i is a random component which explains all extraneous variation of the experimental and analytical procedure. One might say, and rightfully so, that ϵ_i is a "catch-all" term.

1.5.2 Model For Two Normal Populations

We assume that a sample of size r_1 is drawn from one normal population, say $N(\mu_1, \sigma_2^2)$, and a sample of size r_2 is drawn from a second normal population, say $N(\mu_2, \sigma_2^2)$. The observations from the first population, $y_{11}, y_{12}, \ldots, y_{1r_1}$ are indicated on the horizontal axis below by * and those from the second population, $y_{21}, y_{22}, \ldots, y_{2r_2}$ are indicated by o. For illustration, we will assign $\mu_1 = 20$, $\mu_2 = 42$, $r_1 = 12$ and $r_2 = 10$.

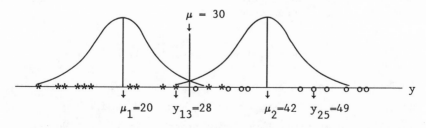

Figure 1.3 Samples from Two Normal Populations

One observation from each population has been identified.   These are y_{13} = 28 and y_{25} = 49.   Also shown in Figure 1.3 is a point labeled μ = 30.   This is an overall mean value defined by

(1.10)
$$\mu = \frac{r_1\mu_1 + r_2\mu_2}{r_1 + r_2}$$

which is a weighted mean of μ_1 and μ_2. The value of defining μ is that it gives us a reference point to which we can compare μ_1 and μ_2. Consequently, we could have taken any value on the horizontal axis as a reference point --- just so we have one. The use of μ, as   defined   in   (1.10),   as   the   reference   point   has   certain advantages in interpretations.

With the above terminology, we note that an observation from the first population, say y_{1j}, can be written as the sum of 3 components:

μ = overall mean value of y

(1.11) τ_1 = a deviation from the
                    overall mean

            ϵ_{1j} = a deviation of the response
                    from its mean value

Here μ = 30 and τ_1 = 20 - 30 = -10. Now let us look at y_{13},   the identified response from the first population:

$$y_{13} = 28$$
$$= 30 + (-10) + (+8) = \mu + \tau_1 + \epsilon_{13}$$

And, a typical observation from the first population could be explained by the linear model

(1.12) $y_{1j} = \mu + \tau_1 + \epsilon_{1j}$ for j = 1, 2, ..., r_1

Similarly, an observation from the second population, say y_{2j}, can be written as the sum of μ, $\tau_2 = \mu_2 - \mu$, and $\epsilon_{2j} = y_{2j} - \mu_2$, where these components have definitions as in (1.11). We see that

$r_2 = 42 - 30 = 12$. Then, the response y_{25} can be written as

$$y_{25} = 49$$
$$= 30 + (+12) + (+7) = \mu + r_2 + \epsilon_{25}$$

Likewise, a typical observation from the second population could be explained by the linear model

(1.13) $y_{2j} = \mu + r_2 + \epsilon_{2j}$ for $j = 1, 2, \ldots, r_2$

The models in (1.12) and (1.13) differ only in the subscripts 1 and 2. Therefore, they may be combined into a single expression as follows:

(1.14) $y_{ij} = \mu + r_i + \epsilon_{ij}$ for $i = 1, 2$
                                                              $j = 1, 2, \ldots, r_i$

1.5.3   Model for More than 2 Normal Populations

If we took a random sample from each of k normal populations, we can easily extend the ideas of the previous section to arrive at the linear model. First, the reference point would be taken as the overall mean.

(1.15) $$\mu = \frac{r_1\mu_1 + r_2\mu_2 + \cdots + r_k\mu_k}{r_1 + r_2 + \cdots + r_k}$$

$$= \frac{1}{n} \sum_{i=1}^{k} r_i\mu_i \quad \text{where } n = \sum_{i=1}^{k} r_i$$

Then, for each of the k populations, we would have a model as in (1.12) and (1.13). These k "individual population" models could be combined into a single expression:

(1.16) $y_{ij} = \mu + r_i + \epsilon_{ij}$ for $i = 1, 2, \ldots, k$
                                                              $j = 1, 2, \ldots, r_i$

which is the same as the model in (1.14) with the range on the subscript i extended to k.

1.6   Characterization of Model Components

Linear models were given in the preceding sections for various
populations.   These were the simplest models that one could
consider.   For two or more populations, each model contained (i)
an overall mean, a reference point, (ii) a treatment component,
τ, and (iii) a residual component, ϵ.

Each component in an analysis of variance model explains a
specific aspect of the experiment.   Stated in another way, every
facet of the experimental procedure that has a significant,
identified impact on the observed responses can exhibit this
influence only through a component in the model.   For example,
the treatment component, τ_i, explains the effect of the various
treatments by measuring the location of the population means
relative to the reference point.   The only way the effect of the
treatments can be assessed, or linked to the responses, is
through the presence of a treatment component in the model.

Each component in a model provides a measure of variation
due to an identified source.   Thus, the treatment component, τ_i,
would explain some, but usually not all, of the differences in
the observed responses.   As we have indicated earlier, ANOVA is
the process of measuring the identified sources of variation; the
model components are the quantities needed for the process.

As experiments become more complicated, more components must
be included in the model.   Some components may be needed to
measure variations due to arrangements or groupings of the EU,
others to measure variations due to the treatment factors, and
others to measure variations due to samplings of the EU.   These
3 major classes of components address the major design aspects of
scientific research; namely,

    (i)   experimental design
    (ii)   treatment design
    (iii) response or sampling design.

This terminology is used by others, though not extensively.
Federer [1] made specific reference to treatment design.   More
recently, Urquhart [4] discussed all three categories in his
expository article.

The components of the experimental design aspect account for
relationships among EU, generally addressed by considering groups
or arrangements of similar experimental material.   The experimental design components indicate the manner in which the treatments
are associated with the EU.   As we will see later, certain EU
can be grouped so that each unit of a given group is expected to
produce the same effect.   All treatments should appear in such a
group so that group effects may be removed from the response.

The components of the treatment design account for relationships among the various treatments. If several treatment factors are present, differences in the responses are expected with respect to the various levels of each factor---components are included in the model to account for possible effects of the different levels.

The components of the response design account for residual variation and for differences present because of samplings and subsamplings of the basic units, the EU. In some experiments, recorded observations are obtained from response units selected within each EU. In a majority of the experiments the response units and EU are one and the same, in which case the model contains only one component for the response design. But when several subsets of each EU are randomly selected for observation, rather than obtaining a single observation for the entire EU, there would be a component included in the model to reflect differences among the observed subsamples of the EU. Several stages of subsampling may be used: each EU may be partially sampled, each of these subsamples may be partially sampled, and so on. An additional component would be included in the model for each stage of subsampling.

By and large, the experimental design components represent the uninteresting portion of the model and the research. These components are included when the EU are grouped and arranged according to differences known to exist prior to experimentation. Such groupings are necessary, however, so that the corresponding variations are removed from Exp. Error. Failure to do so would result in contaminated variances of treatment components.

The treatment design components are of primary importance. Their presence allows us to make inferences about the treatment populations. Generally the type of inferences that a researcher makes is dependent on the type of treatments appearing in the experiment. In very broad terms, the treatments define either unstructured or structured populations.

In experiments having *unstructured* treatments, the objective is to compare a set of unrelated treatment. For example, a home economist might test cake mixes sold by six different companies. An experiment would be designed to determine if cakes baked from the six mixes are different, as measured by texture, taste, appearance, and so on. If differences were detected among the means under the 6 mixes, it may be of interest to pin-point the differences; this may be attempted by one of the mean separation techniques. A later chapter deals with this topic.

Structured treatments, those defining related populations, may be classified in the following ways:

(i)   those related by specific variable structures
(ii) those having a factorial structure

(iii)   those having a gradient structure
(iv)   those having a grouping structure.

The treatments belonging to the first class may be described as those whose population means are related by some type of regression function.   In other words, the means have a specific, defined relationship with each other.   Included in this case are situations where the treatment means have a complicated (and maybe unknown) structure which can be approximated very well by a regression function, such as a linear regression line.

Factorial treatments are formed from combinations of two or more different factors of interest.   Experiments having factorial treatments are quite common and we will devote considerable time to them (see Chapter 5.)

Treatments having a *gradient* structure are those for which incremental differences are present.   Thus, an engineer might use steel having 1%, 2% and 3% carbon.   A biologist might use 10%, 20% and 30% relative humidities during the larval stage of a certain insect.

Treatments having *group* structure may be described simply as two or more sets (groups) of related treatments.   Often, the relationship from set to set is weak or non-existent.   As an example, consider an experiment to study milk production under three grain-supplemented rations and four rations without grain. An engineer might study wind resistance for different models of automobiles and use groups of "compacts", "mid-sized", and "full-sized" automobiles.

1.7   The Basic Experimental Designs

Most experimental designs can be considered as modifications or extensions of a few basic designs.   New designs are developed chiefly as a matter of necessity to handle new or more complex research situations.   Which experimental designs should be placed in the basic category is mostly a matter of preference so we consider the following three as basic experimental designs:

(a)   the completely randomized (CR) designs
(b)   the randomized complete block (RCB) designs
(c)   the Latin square (LS) designs.

A chapter will be devoted to each of these designs.   Numerous modifications and extensions will be discussed also.

1.8   Fixed, Mixed, and Random Effects

If an experiment was conducted using all treatments of interest
to the researcher, the inferences would be restricted to the set
of treatments used.   For this type of situation, the inference
set is "fixed" and the linear model would be termed a *fixed
effects* model, or simply a fixed model. The fixed effects model
is by far the most commonly used model; most experimenmts are
designed to investigate a specific set of treatments.

In other situations, a researcher may wish to make inferences
about a set of treatments larger than those appearing in the
experiment.   The treatments used in such an experiment represent
a random sample of all treatments about which inferences are
desired.   The linear model for experiments of this type would be
termed a *random effects* model. If an experimenter wanted to make
inferences about an interval of temperatures but could use only
ten temperatures in an experiment, temperature would be a random
effects factor.   Day or environmental factors frequently are
considered to have random effects unless the experimenter chooses
specific levels at which the experiment is to be conducted.

When a linear model contains components other than a single
treatment component, it is termed a:

    (i) fixed model if all components represent fixed effects,
   (ii) random model if all components represent random effects,
  (iii) mixed model if some components represent fixed effects
       and some represent random effects.

The overall mean, μ, and the random component, ϵ, do not play a
role in determining whether a model is fixed, mixed, or random.

1.9   Combining Estimates

In later analyses, particularly those dealing with incomplete
blocks, we sometimes obtain two independent estimates of the
treatment parameters.   How can we combine such estimates to
obtain a single, "better" estimate of the parameter? There is a
general theory available for making such combinations.   We state
the general theorem without proof.

Theorem 1.4

Let $\hat{\theta}_1$ and $\hat{\theta}_2$ be independent estimators of a parameter θ,
and let σ_1^2 and σ_2^2 be respective variances of the estimators.
Then, the minimum variance estimator of θ is given by

$$(1.17) \qquad \theta^* = \frac{w_1 \hat{\theta}_1 + w_2 \hat{\theta}_2}{w_1 + w_2} \qquad \text{where } w_i = 1/\sigma_i^2$$

We note from (1.17) that the combined estimator of θ is a weighted mean, where the weight are reciprocals of the variances of the individual estimators. In effect, an individual estimator with a larger variance is given lesser influence in the combining process. This makes sense because larger variance implies less stability (low reliability) of an estimator which in turn implies "less faith should be placed in such an estimator."

Example 1.4

As a simple illustration of Theorem 1.4, suppose a person obtains a random sample of size n_1 = 20 from a $N(\mu, 120)$ distribution and obtains $\bar{y}_1$ = 48. A friend obtains a second, independent sample of size n_2 = 12 from the same distribution and obtains $\bar{y}_2$ = 40. The two mean estimates have variance of

$$\sigma_{\bar{y}_1}^2 = 120/20 = 6 \quad \text{and} \quad \sigma_{\bar{y}_2}^2 = 120/12 = 10$$

This indicates we should place more trust in the estimate from sample 1 (not surprising because $\bar{y}_1$ is based on more observations than $\bar{y}_2$ is.) The minimum variance estimator of μ is the combined estimator

$$\mu^* = \frac{48/6 + 40/10}{1/6 + 1/10} = 45. \ \blacksquare$$

1.10 Hypothesis Testing vs. Significance Testing

Classical hypothesis testing can be described as a two-decision process: The null hypothesis, H_0, is either rejected or not rejected. These actions result because the researcher fixes the Type I error rate, α, which leads to a definite rejection region for H_0. The decisions are based on the test statistic; that is, whether or not its value falls in the rejection region.

Many researchers feel that this is too rigid a process, especially when one observes a value of the test statistic "close to the boundary values" of the rejection region. Also, selecting a value of α may be somewhat arbitrary, usually there is some flexibility in the value chosen. Different values of α may lead to opposite decisions.

Generally the results of scientific research are not reported in this two-decision manner of classical hypothesis testing. Rather, the observed value of the test statistic is reported together with an indication of the significance level. Some computer packages calculate and present a single value for this level. Often, a range of values must be reported due to the inadequacy of available tables. A reader can then make his/her own judgment on whether the reported results are "significant" for the intended purpose.

The term *significance testing* is applied to hypothesis testings where the "observed significance level" is reported in lieu of a classical decision. The observed(reported) significance level is equal to the size of the rejection region *that would have been used* for a classical test when the boundary is set at the observed value of the test statistic. Common notation for the observed significance level is P. An example will illustrate the general idea.

Example 1.5

Let us use the sample from the first normal distribution of the previous example. Suppose we wish to test

$$H_0: \mu = 44 \quad \text{vs.} \quad H_1: \mu > 44$$

The rejection region for a classical test would be the upper tail area of the Z distribution. The value of our test statistic is

$$Z = (48 - 44)/\sqrt{6} = 1.63$$

To obtain the observed significance level, we need only determine the area of the Z distribution in the upper tail, above 1.63. From the standard normal tables, Table A.5, this value is found to be P = 0.052. ▓

There is debate about how one should measure the observed significance level of tests having two sided critical regions. Fortunately, all ANOVA tests (of no treatment effects) have one-sided critical regions so this is not a problem of great concern in this book. A few isolated tests of individual treatment contrasts (to be discussed later) may have two-sided critical regions.

1.11   Least Squares Estimation

One of the oldest and most widely used procedure for estimating
parameters of a model is the method of least squares (LS).   Two
famous scientists, Gauss (1777-1855) and Legendre (1752-1833),
are credited with independently developing the procedure.   Gauss
published his version in 1809 but evidence indicates he used it
around 1803; Legendre published his version in 1805.

The method of least squares is applicable to simple models,
such as those in Section 1.5, or extremely complex, nonlinear
models.   In any case, least squares estimates depend not only on
the proposed model but also on the assumptions satisfied by the
model components, particularly the ϵ's.   If no assumptions are
made about the ϵ's, the LS estimators generally contain variances
and covariances of the ϵ's. The estimates cannot be calculated
unless these variability measures are known.   As the ϵ's satisfy
more assumptions, the LS estimators generally have simpler forms.
They are simplest when the ϵ's are independent and have equal
variances---in this case, the estimators do not contain any
variances and covariances of the ϵ's. Most of our LS estimators
will be generated under these broad assumptions.

Use of LS estimation does not require specific assumptions
about the underlying distribution of the ϵ's.   In particular,
normality of the ϵ's is not required. But as usual, inferences
in the form of tests of hypotheses or confidence intervals would
require some type of underlying distribution of the ϵ's.

We first give a statement of the basic principle upon which
least squares estimation is accomplished.

Definition 1.2   The Least Squares Principle

Least squares estimates of model parameters are those
values which minimize the sum of squares of residuals
(of the experimental observations) from the fitted model.

Without getting specific, let us represent an experimental model
by

$$(1.18) \qquad y_i = \text{"parameters"} + \epsilon_i$$

A fitted model would be obtained by replacing the parameters with
estimates calculated from the y_i, the experimental observations.

The residuals of the observations, called e_i, can be expressed as

$$(1.19) \qquad y_i = \text{"estimated parameters"} + e_i$$

so we see that the sum of squares to be minimized under the Least Squares principle is

$$(1.20) \qquad \sum_{i=1}^{n} e_i^2 = \sum_{i=1}^{n}(y_i - \text{estimated parameters})^2$$

Generally for each given model, a system of equations must be solved to obtain the estimated parameters which minimize this sum of squares. As a matter of convention, these equations are called the *normal equations,* but they have nothing whatsoever to do with normal distributions (recall the earlier statement about the normality assumption.)

Depending on the complexity of the assumed model in (1.18), obtaining and solving the normal equations may be a tedious process. [In many cases, the normal equations may be obtained by techniques of the calculus: Taking derivatives in (1.20) with respect to each estimated parameters, setting these derivatives equal to zero, and solving.]

Example 1.6

The linear model for a random sample of observations from a single population is, from (1.9):

$$y_i = \mu + \epsilon_i \qquad \text{for } i = 1, 2, \ldots, n$$

Residuals from the fitted model are given by

$$y_i = \hat{\mu} + e_i \qquad \text{for } i = 1, 2, \ldots, n$$

For the LS estimation of μ, the sum of squares to be minimized is

$$\sum_i e_i^2 = \sum_i (y_i - \hat{\mu})^2$$

The single "normal equation" to estimate the parameter μ, in this case, can be shown to be

$$\sum_i (y_i - \hat{\mu}) = 0$$

Solution of this last equation gives the LS estimator of the parameter μ; namely,

$$\hat{\mu} = \frac{1}{n} \sum_i y_i = \bar{y} \quad \text{\Maltese}$$

We stated that normality of the ϵ's is not a requirement for LS estimation. But if the ϵ's do have a normal distribution, we could estimate parameters by using either normal theory or the LS principle. Are the estimates the same or different? It is not too difficult to show that, when the ϵ's have a normal distribution, estimates of the model parameters are identical under normal theory and LS.

Example 1.7

In the previous example we obtained $\hat{\mu} = \bar{y}$ as a least squares estimator of μ. This is also the estimator of the mean, μ, of a normal distribution (that is, obtained under normal theory.) We also note that the sum of squares of minimized residuals is

$$\sum_i e_i^2 = \sum_i (y_i - \bar{y})^2$$

after inserting the LS estimator. But this is precisely the sum of squares used in calculating the estimate of variance (of ϵ's) under normal theory. So in sampling from a single normal distribution, the estimates of μ and σ^2 are identical under both normal theory and LS estimation. ▓

1.12 Robustness of Statistical Procedures

Every statistical procedure requires specific assumptions to provide valid inferences. Validity refers to the correctness of the reported Type I and II error rates. In other words, if a person makes a test at the 5% level of significance, the decision made by the tester is valid if the Type I error rate is actually 0.05 (and not 0.049 or 0.062 or some other value unequal to 0.05.)

A few procedures give valid inferences when the single assumption of random sampling is satisfied. Most procedures require several assumptions. For example, testing the equality of two means with a t distribution requires four assumptions: random samples, equal population variances, independence of the two samples, and normal populations. Other statistical procedures may require other assumptions for giving valid inferences.

When any statistical procedure is applied to an experimental plan, some of the required assumptions may not be completely satisfied. If not, there is concern about validity of the results obtained under the procedure. Logically, one would expect a given procedure to give poorer quality inferences when applied to populations where the assumptions are violated to a greater

degree.   But this is not always true.   Some procedures give good
quality inferences when some of the assumptions are not satisfied
completely.   We give a definition to address this concept.

Definition 1.3

A statistical procedure is said to be *robust* if the validity
of its inferences is relatively unaffected by departures
from the underlying assumptions.

Robustness   is not a condition that can be quantified in practice
(because departures from the underlying assumptions are unknown.)
Nevertheless,   most statistical procedures have been analyzed by
theoreticians for robustness.   Some procedures may be robust with
respect to violations of one assumption but not with respect to
violations of other assumptions.   We know, by virtue of the
Central Limit Theorem, that a Z test about a single population
mean is robust with respect to nonnormality provided that the
population variance is known and a sufficiently large sample size
is   used.   Likewise,   a Student's t test about the means of two
normal distributions is fairly robust with respect to departures
from equal population variances so long as equal sample sizes are
used,   see Welch [5].   Finally, ANOVA tests are quite robust over
a wide spectrum of experimental conditions.   We will discuss this
more fully in Section 2.15.

1.13 Preliminary Tests

Every statistical procedure requires certain assumptions for
valid inferences. For example, using the Student's t distribution
for inference about the difference of two means requires 4
assumptions: independent random samples drawn at random from 2
normal populations having equal variances. Random sampling and
independence usually are easy to discern. Normality and equality
of variances may be questionable and may require statistical
tests for verification.   These tests, made to verify the validity
of the t test, are called *preliminary tests*.
    Preliminary tests are ones made prior to the test of
interest,   and   are   necessary   for assessing its   validity.
Therefore,   preliminary tests are conditional tests; that is, the
test of interest is valid given that all preliminary tests
indicate that required assumptions hold.
    Two   points need to be made about preliminary tests.   First,
preliminary tests often are not very robust. The test of equality

of variances, mentioned in the first paragraph above, depends heavily on normality, more so than the Student's t test of equal means(the test of interest.) Second, each preliminary test alters the significance level of any following (dependent) test, in particular, the test of interest. Thus, for any test following a preliminary test, a chosen or observed significance level of 5% may be actually 3% or 6%. There is no way to assess the amount by which a significance level is altered by preliminary tests---many factors are involved: the number of tests, their robustness, and so on.

1.14 Dot Notation

Totals are used extensively in experimental analyses. Each mean value requires a total. And most sums of squares are calculated from totals. Therefore, it will be advantageous to introduce *dot notation*, a shorthand notation for totals. Specifically, a *dot* is used as a subscript on a variable to indicate a summation. We begin with a defintion.

Definition 1.4

Each dot appearing as a subscript on a variable denotes the sum of that variable over the entire range of the index previously occupying the position of the dot.

For a sample of n observations indexed by a single subscript, say $y_1, y_2, \ldots, y_n$, the total, using dot notation, is

$$(1.21) \qquad y_{.} = \sum_{i=1}^{n} y_i$$

One might ask, "Why not eliminate the dot for further simplification?" We could for this simple type of data, but generally cannot for data having two or more subscripts. Suppose our data are described by

$$(1.22) \qquad y_{ij} \quad \text{for } i = 1, 2, \ldots, a \\ j = 1, 2, \ldots, b$$

Then, we might calculate totals such as

$$y_{i.} = \sum_{j=1}^{b} y_{ij}$$

(1.23)
$$y_{.j} = \sum_{i=1}^{a} y_{ij}$$

$$y_{..} = \sum_{i=1}^{a} \sum_{j=1}^{b} y_{ij} = \sum_{i=1}^{a} y_{i.} = \sum_{j=1}^{b} y_{.j}$$

As one would discover, eliminating the dots for this type of data would make it impossible to tell $y_{1.}$ and $y_{.1}$ apart, $y_{2.}$ and $y_{.2}$ apart, and so on.

It is unfortunate that some of the commonly used quantities, such as, $\sum_{i} y_i^2$ and $\sum_{i}(y_i - \bar{y})^2$, do not have representations in dot notation. But by adopting the convention that "the dots take precedence," many functions of totals can be abbreviated by dot notation. For example,

$$y_.^2 = (\sum_{i} y_i)^2 \qquad \text{not} \quad \sum_{i} y_i^2$$

(1.24)
$$\sqrt{y_.} = \sqrt{\sum_{i} y_i} \qquad \text{not} \quad \sum_{i} \sqrt{y_i}$$

$$y_{i.}^2 = (\sum_{j} y_{ij})^2 \qquad \text{not} \quad \sum_{j} y_{ij}^2$$

Example 1.8

Suppose the owner of a business has recorded sales figures, y, (in dollars), for a two year period using 4 indices:

$$y_{ijkm} \quad \begin{array}{ll} \text{for } i = 1, 2 & \text{(years)} \\ j = 1, 2, \ldots, 12 & \text{(months)} \\ k = 1, 2, 3, 4 & \text{(weeks)} \\ m = 1, 2, \ldots, 6 & \text{(days)} \end{array}$$

Then, a few of the many totals are:

$$y_{111.} = \sum_{m} y_{111m}$$

which represents the total sales for Year 1, Month 1 (January), Week 1,

$$y_{24..} = \sum_{k} \sum_{m} y_{24km}$$

which represents the total sales for Year 2, Month 4 (April),

$$y_{.2..} = \sum_{i} \sum_{k} \sum_{m} y_{i2km}$$

which represents the total sales for Month 2 (February), and

$$y_{1...} = \sum_{j} \sum_{k} \sum_{m} y_{1jkm}$$

which represents the total sales for Year 1. ※

PROBLEMS

1. Random samples are drawn from a normal distribution with mean $\mu = 30$ and variance of $\sigma^2 = 80$. Give the distribution of the sample means

   (a) in samples of size n = 10
   (b) in samples of size n = 15.

2. Random samples are drawn from a normal distribution with mean $\mu = 125$ and variance of $\sigma^2 = 270$. Give the distribution of the sample means

   (a) in samples of size n = 12
   (b) in samples of size n = 25.

3. Random samples of size 16 are selected from a population whose elements have a N(20, 40) distribution. Give the standard deviation of $\bar{y}$.

4. Random samples of size 24 are selected from a population whose elements have a N(160, 300) distribution. Give the standard deviation of $\bar{y}$.

5. Suppose y_1, y_2, y_3, y_4, and y_5 have respective normal distributions: $N(10, 20)$, $N(15, 25)$, $N(18, 20)$, $N(15, 10)$, and $N(20, 30)$, with all y's mutually independent. Give the mean and variance for each of the following linear combinations:

(a) $L_1 = y_1 + y_3 + y_4$

(b) $L_2 = y_1 + y_3 - 2y_5$

(c) $L_3 = y_1 + y_2 + y_3 + y_4 + y_5$

(d) $L_4 = y_2 - y_3 + y_4 - y_5$

6. Suppose that y_1, y_2, and y_3 each have a $N(15, 40)$ distribution while y_4, y_5, and y_6 each have a $N(12, 30)$ distribution, with all y's mutually independent. Give the mean and variance for each of the following linear combinations:

(a) $L_1 = y_1 + y_2 + y_3$

(b) $L_2 = y_4 + y_5 + y_6$

(c) $L_3 = L_1 + L_2$

(d) $L_4 = L_1 - L_2$

(e) $L_5 = y_1 - y_2 + y_3 - y_4 + y_5 - y_6$

7. Let random variables y_1, y_2, and y_3 have respective normal distributions: $N(10, 15)$, $N(12, 20)$, and $N(16, 25)$. Suppose the covariances between pairs of y's are: $\sigma_{12} = 10$, $\sigma_{13} = -8$, and $\sigma_{23} = 4$. Find the mean and variances for each of the following linear combinations:

(a) $L_1 = y_1 + y_2 + y_3$

(b) $L_2 = y_1 - 2y_2 + y_3$

(c) $L_3 = y_1 - y_3$

8. Repeat Problem 7 if y_1, y_2, and y_3 have respective normal distributions: $N(12, 10)$, $N(15, 18)$, and $N(10, 24)$, and with covariances: $\sigma_{12} = 4$, $\sigma_{13} = -10$, and $\sigma_{23} = -5$.

REFERENCES

[1]   Federer, W. T., *Experimental Design*, Macmillan, New York,
       1955.
[2]   Freund, J. E., *Modern Elementary Statistics*, 5th. ed.,
       Prentice-Hall, Englewood Cliffs, NJ, 1979.
[3]   Lentner, M., *Introduction to Applied Statistics*, 2nd.ed.,
       Blacksburg, Va., 1984.
[4]   Urquhart, N. S., The Anatomy of a Study. *HortScience*, Vol.
       16, 1981.
[5]   Welch, B. L., The Significance of the Difference between Two
       Means When the Population Variances Are Unequal. *Biometrika*,
       Vol. 29, 1938.

## Chapter 2

COMPLETELY RANDOMIZED DESIGNS
(CR Designs)

2.1   Introduction

Now we consider experiments that are conducted to investigate a set of t treatments. For simplicity of discussions in the next few sections, we assume unstructured treatments(see Section 1.6); structured treatments will be considered later. One of the treatments may be a control, a placebo or standard treatment, included to provide a reference for comparison.

A total of n experimental units (EU) are available for use in the experiment. The EU are as homogeneous as possible; that is, no source of variation can be recognized among them under any grouping or arrangement. Stated in another way, there is no basis for grouping the EU as there is in some of the more complex designs studied in later chapters. This does not rule out the existence of variation among the EU; we simply have no relevant information about it. The reason for insisting upon homogeneous EU is that when treatments can be applied to uniform material, observed differences will contain no effects attributable to the EU.

2.2   The Experimental Plan

Each treatment is to be randomly assigned to several EU in an unrestricted manner†. Suppose the i-th treatment appears r_i times in the experiment; we usually describe this as r_i replications of the i-th treatment. The analysis of a CR design is accomplished in a straightforward manner with equal or unequal replications of treatments. But from the standpoint of efficiency, all treatments should occur with the same frequency; that is, all $r_i = r$.

---

†Except that no treatment may be assigned to an EU which has previously been assigned a treatment. This trivial restriction must hold for any design.

Example 2.1

Suppose a graduate student in agronomy is required to investigate the yield of two new varieties of soybeans as part of his M.S. program.   Additionally, he is to compare these two new varieties to a standard variety which has been widely used for a number of years.   Available for his use is an area of land 50'x60' on the departmental research farm. There are t = 3 treatments: the new varieties of soybeans and a control (the standard variety).   If EU of size 10'x10' are desired, the area can be subdivided into 30 plots which would be the EU.   Without further information, such as results of soil tests, there is no reason to believe that certain EU are better for growing soybeans than others;   that is, there is no reason to believe that the EU are non-homogeneous with respect to any factor that might influence the growing of soybeans.   For equal replications of treatments,   each variety would be randomly assigned to 10 EU. ※

We will use the term "basic CR design" to refer to CR designs having t unstructured treatments, r_i EU for the i-th treatment, and a single observation recorded from each EU.   More general CR designs will be considered later, including those with structured treatments and subsampling of EU.

To conduct a basic CR design,   each treatment is assigned randomly to a number of EU.   Random numbers are useful in making these assignments. Many computer packages, such as SAS [26] and BMDP [12], will generate random numbers, or they may be read from existing tables.   Table A.1 contains random numbers reproduced from the Rand tables [25].

Suppose a researcher is planning to conduct a basic CR design with equal replications.   Thus, n = rt homogeneous EU are to receive the t treatments equally often.   Then random numbers from 1 thru n are needed; the first r specify the EU to receive the first treatment, the next r numbers for the second treatment, and so on.

If two-digit random numbers are needed, for example, one reads pairs of digits from Table A.1---reading down columns, across rows, on diagonals, or on any arbitrary path that does not repeat itself.   An arbitrary starting point is selected by the user. Duplications and numbers greater than n are omitted because each number from 1 thru n must appear precisely once in the random sequence.   We illustrate this process in the following example.

Example 2.2

The agronomy student of the previous example has 30 available plots, numbered from 1 to 30, say. To avoid confusion, suppose the 3 treatments are labeled A, B, C. Table A.1 will be used to obtain a random sequence for the integers 1 to 30. Using an arbitrary starting point of row 21 and column 33, and reading down columns 33 and 34 (then columns 35, 36, columns 37, 38, and so on, if necessary), the following sequence is obtained:

11, 3, 14, 6, 25, 30, 10, 29, 15, 16, 20, 23, 22, 21, 4, 17,

24, 1, 26, 27, 9, 2, 28, 5, 19, 12, 13, 7, 8, 18

Treatment A is assigned to the EU having the first 10 numbers in the sequence, treatment B the next 10 numbers, and treatment C the last 10 numbers. The design setup would have the following association of plot numbers and treatments:

1 B	2 C	3 A	4 B	5 C	6 A
7 C	8 C	9 C	10 A	11 A	12 C
13 C	14 A	15 A	16 A	17 B	18 C
19 C	20 B	21 B	22 B	23 B	24 B
25 A	26 B	27 B	28 C	29 A	30 A

With unequal replications, one uses unequal segments of the sequence of random numbers in making the treatment assignments. Remember that unequal replications are not recommended unless there is a good reason, such as wanting more precision on certain treatments.

2.3   Some Design Considerations

One of the requirements for using a CR design is that the EU be homogeneous. In what type of experimental situations does such a requirement hold? In some experiments, there may be so many design factors involved that perfect homogeneity would not be anticipated. Can the researcher do anything to enhance the degree of homogeneity?

For many greenhouse studies, the researcher might obtain a given quantity of soil. Rocks, particles, and foreign matter may be sifted out.   Then the remaining soil is thoroughly mixed and placed in appropriate containers.   Depending upon the amount of mixing,   the soil   in the containers (the EU) should be fairly homogeneous.

Likewise, a biologist could make up fairly homogeneous petri dishes of agar (the EU) if he started with one large batch of well-mixed agar.   A home economist could mix a large batch of cake mix to form a set of homogeneous cakes, EU for testing different ovens or temperatures, perhaps.   An engineer might "bake" a large silicon crystal to be sliced into homogeneous chips (the EU).

Finally, the techniques of cloning show promise of developing homogeneous EU, at least in the plant sciences. Already available is a process of generating thousands of ferns in a relatively short period of time.

While we have indicated situations where researchers are able to construct homogeneous EU, what about the multitude of situations where EU are already in existence, or where construction of homogeneous EU would be impractical or impossible? Such would be the case of a scientist in a behavioral science or education discipline.   An educator, trying to compare different teaching methods, undoubtedly would use different classes of students as the EU.   Can homogeneous classes be obtained?   Probably not. Maybe all fourth grade classes in one large school, or in several schools in a large city, are the most homogeneous EU that can be obtained.   Special classes, such as those for students with learning disabilities or for gifted students, would introduce an element of heterogeneity.   A single class could not be used repeatedly for all treatments because there might be "carryover effects" from the earlier to later uses of the class.   Similar problems might be encountered by a sociologist, psychologist, economist, and others who might need to use a class of students or group of people as an EU.

Whether an experimenter uses existing EU or constructs them in no way guarantees that homogeneity is preserved throughout the experiment.   It won't do much good to painstakingly construct homogeneous containers of soil if they are placed in a greenhouse in such a way that some EU will be exposed to a heat source while other EU will not.   The home economist might unknowingly destroy the homogeneity aspect so carefully introduced by mixing one large batch initially, if some of the cakes are baked immediately while others stand for several hours before facilities become available. Clearly then, the researcher must consider the entire experimental plan. When a CR design is used, all non-treatment factors should be kept as constant as possible.   This might include: Having one person apply the treatments to the plots, having repetitive techniques standardized, having one laboratory

do all analyses, and so on.   Things to avoid might include:
Plots with a history of flooding or drought; experimental setups
with limited access, such as greenhouses or laboratories used
excessively by other personnel; a prolonged timetable, especially
if the EU have a potential of changing through time; and so on.
     In summary, a researcher wishing to use a CR design should
assess all factors or sources which might lead to nonhomogeneous
EU, and should take whatever steps are necessary to achieve or
preserve homogeneity of EU.   When nonhomogeneity becomes an issue
that cannot be resolved adequately, the researcher must consider
experimental designs other than the CR.   Later chapters of this
book deal with alternative designs.

2.4 Advantages and Disadvantages of CR Designs

The CR designs are the simplest of all experimental designs and
have a number of advantages. Important advantages include:

  (1) Flexibility. Subject to available resources, there is no
      restriction on either the number of treatments or on the
      number of replications of the treatments.
  (2) Ease of analysis. Of all designs, the CR designs are
      fairly easy to analyze, even with unequal replications of
      treatments.
  (3) Maximum degrees of freedom for estimation of Exp. Error.
      This is important only for small experiments or for
      experiments where extraneous variation is great.

The chief disadvantage of a CR design is that it is relatively
inefficient when another experimental design is more appropriate.
This stems from the fact that all unrecognized variation (which
if recognized would dictate the use of another design) and all
extraneous variation are included in Exp. Error (see comments in
Section 1.6 about experimental design components).

2.5   Linear Model and Least Squares Estimates

The t treatments used in a CR experiment define a set of t
treatment populations having means: μ_1, μ_2, $\cdots$ μ_t, often called
the treatment means.   Then, the linear model for a CR design is:

(2.1) $y_{ij} = \mu + \tau_i + \epsilon_{ij}$; $\begin{aligned} i &= 1, 2, \ldots, t \\ j &= 1, 2, \ldots, r_i \end{aligned}$

where

y_{ij} = observation from the j-th EU of the i-th treatment

μ = overall mean (reference point)

$\quad = \frac{1}{n} \Sigma r_i \mu_i$ with $n = \underset{i}{\Sigma} r_i$

(2.2)

$\tau_i = \mu_i - \mu$ = i-th treatment effect

$\epsilon_{ij} = y_{ij} - \mu_i$ = deviation of the (ij)-th observation from the i-th treatment mean, called the residual component

The least squares estimation of parameters is accomplished by minimizing the residuals from the fitted model. For a CR model (2.1), the quantity to be minimized is

(2.3) $Q = \underset{ij}{\Sigma\Sigma} e_{ij}^2 = \underset{ij}{\Sigma\Sigma}(y_{ij} - \hat{\mu} - \hat{\tau}_i)^2$

where values of $\hat{\mu}$ and the $\hat{\tau}$'s are chosen so that Q is minimal. Here, the normal equations form a system of t+1 equations in t+1 unknowns [using the calculus, as indicated in Section 1.11]:

$-2\underset{ij}{\Sigma\Sigma}(y_{ij} - \hat{\mu} - \hat{\tau}_i) = 0$

(2.4)

$-2\underset{j}{\Sigma}(y_{ij} - \hat{\mu} - \hat{\tau}_i) = 0$ for i = 1, 2, ---, t

or after simplification

$n\,\hat{\mu} + \underset{i}{\Sigma}\, r_i\hat{\tau}_i = y_{..}$

(2.5)

$r_i\mu_i + r_i\hat{\tau}_i = y_{i.}$ for i = 1, 2, ---, t

Note that the system (2.5) represents the normal equations for a basic CR design. Unfortunately, the last t equations add to the first equation which implies a singular system; that is, a unique solution does not exist. The system can be solved though if we impose a constraint on the estimates, $\hat{\mu}$ and the $\hat{\tau}$'s. For all of the "important inferences," it doesn't matter what constraint is imposed on the system so we opt for a simple one. A visual examination of (2.5) reveals that a convenient choice is

(2.6) $\sum_i r_i \hat{\mu}_i = 0$

Then, the least squares estimates *under this constraint* are:

(2.7)
$$\hat{\mu} = \bar{y}_{..}$$
$$\hat{\tau}_i = \bar{y}_{i.} - \bar{y}_{..} \quad \text{for } i = 1, 2, \cdots, t$$

Two general points are worth noting:

(1) For all experimental design studies, the system of normal equations will be singular. (This generally is not so in regression studies.)

(2) The specific estimates of the model parameters will depend upon the imposed constraint. But for balanced designs, the ANOVA table, the overall ANOVA test, and estimates of *parametric contrasts* (see Section 4.2) will be the same no matter what constraint is imposed. For unbalanced designs, one must consider "estimable functions," a concept beyond the scope of this book.

2.6 Assumptions for Estimation and Inferences in a CR Design

Consider the following list of assumptions for components of the basic CR model (2.1):

(i) μ is a fixed constant, common to all observations.

(ii) The ϵ_{ij} are normally and independently distributed with zero mean and variance σ_ϵ^2 for all populations (treatment groups.)

(2.8)

(iii) In the fixed effects model, the τ's are fixed constants which satisfy $\sum_i r_i \tau_i = 0$.

In the random effects case, the τ's are normally and independently distributed with zero mean and variance σ_τ^2, and the τ's and ϵ's are distributed independently of each other.

We emphasize that assumption (ii) specifies both normality and homogeneous variances for all treatment populations.

Calculation of the least squares estimates indicated in the previous section and construction of the ANOVA table (given in the next section) require all assumptions except normality. For valid inferences, tests and CI based on the ANOVA table, all assumptions, including normality, are required. As usual, any violation of the required assumptions would result in approximate estimates and inferences. When the assumptions are seriously violated, the researcher should consider analyses other than classical ANOVA; maybe nonparametric analyses, see Conover [10] or Hollander and Wolfe [19], or data analyses, see Hoaglin, et. al. [18] or Mosteller and Tukey [23].

2.7   Analysis for a CR Design

We continue to assume unstructured treatments but allow for equal or unequal replications of treatments. First we give a table to show how the observed responses might appear after recording.

Table 2.1   Format for Recording CR Data

Observation	1	2	---	j	---	r_i	Treatment Total	Mean
Treatment 1	y_{11}	y_{12}	---	y_{1j}	---	y_{1r_1}	$y_{1.}$	$\bar{y}_{1.}$
2	y_{21}	y_{22}	---	y_{2j}	---	y_{2r_2}	$y_{2.}$	$\bar{y}_{2.}$
.								
.								
i	y_{i1}	y_{i2}	---	y_{ij}	---	y_{ir_i}	$y_{i.}$	$\bar{y}_{i.}$
.								
.								
t	y_{t1}	y_{t2}	---	y_{tj}	---	y_{tr_t}	$y_{t.}$	$\bar{y}_{t.}$
Overall total and mean							$y_{..}$	$\bar{y}_{..}$

Note: $y_{i.} = \sum_j y_{ij}$ and $\bar{y}_{i.} = y_{i.}/r_i$ for i = 1, 2, ---, t

As we have indicated earlier, the ANOVA is a procedure for identifying and measuring specific variation components in the experimental data. The use of the CR model (2.1) implies that we expect differences due to the various treatments, as measured by the τ's, and differences within each treatment population, as measured by the ϵ's. We therefore need to measure these two sources of variation in the observed data.

The total variation in the experimental data may be measured by the total sum of squares which can be expressed as

$$(2.9) \qquad SSY = \sum_{ij}(y_{ij} - \bar{y}..)^2 = \sum_{ij} y_{ij}^2 - C$$

where C is an "overall correction term" given by

$$(2.10) \qquad C = y_.^2./n$$

The total sum of squares for y will be partitioned into two parts: one explaining variation among the treatment means, the other explaining variation among measurements within treatments, often called "within-population variability." This partitioning is most easily accomplished by adding and subtracting $\bar{y}_{i.}$ within the parentheses in (2.9). Applying rules of summations provides the following simplification

$$(2.11) \qquad \begin{aligned} SSY &= \sum_{ij}(y_{ij} - \bar{y}_{i.} + \bar{y}_{i.} - \bar{y}..)^2 \\ &= \sum_{ij}(y_{ij} - \bar{y}_{i.})^2 + \sum_{ij}(\bar{y}_{i.} - \bar{y}..)^2 \end{aligned}$$

The second term measures variation of the treatment means, $\bar{y}_{i.}$, from the overall mean and consequently measures "among treatment" variation. The first term measures variation of each observation from its treatment mean, hence measures the "within population" variation. The two sums of squares can now be identified as

$$(2.12) \qquad \begin{aligned} SST &= \text{treatment SS} \\ &= \sum_{ij}(\bar{y}_{i.} - \bar{y}..)^2 = \sum_{i}(y_{i.}^2/r_i) - C \end{aligned}$$

$$\text{SSE} = \text{Exp. Error SS}$$

(2.13)
$$= \sum_{ij}(y_{ij} - \bar{y}_{i.})^2 = \text{SSY} - \text{SST}$$

The calculation forms appearing in (2.12) and (2.13) are easily derived using rules of summations and algebra.

Recall that the true treatment effect $\tau_i = \mu_i - \mu$ has a sample estimate of $\hat{\tau}_i = \bar{y}_{i.} - \bar{y}..$ so the treatment sum of squares can be expressed as

(2.14)
$$\text{SST} = \sum_{ij} \hat{\tau}_i^2 = \sum_i r_i \hat{\tau}_i^2$$

In other words, the treatment variation is estimated from the observed treatment effects. When $r_i = r$ for all treatments (equal replications) some of the above expressions simplify, but only slightly. The following ANOVA table summarizes the results of this section.

Table 2.2   ANOVA for CR Designs

Source	df	SS	MS	EMS Fixed	EMS Random*
Treatments (among treatment means)	t-1	SST	MST	$\sigma_\epsilon^2 + \dfrac{1}{t-1}\sum_i r_i \tau_i^2$	$\sigma_\epsilon^2 + r_o \sigma_\tau^2$
Exp. Error (within treatments)	n-t	SSE	MSE	σ_ϵ^2	σ_ϵ^2
Total	n-1	SSY			

*The coefficient is $r_o = (n - \sum_i r_i^2/n)/(t-1)$

The term 'Source' is an abbreviation for the phrase 'Source of Variation'. The column headed MS is the mean squares column; any mean square is simply a SS divided by its degrees of freedom. The EMS column gives the expected mean squares; the expected (average) value of the mean square under the model (2.1) and assumptions (2.8). While it is not obvious at this time, the EMS are essential for the derivation of exact tests, which are made by comparing two mean squares having the same expectation under the null hypothesis. Hypothesis testing is discussed in the next section.

With equal replication of treatments, all $r_i = r$, the EMS for treatments in Table 2.2 simplify to

$$\sigma_\epsilon^2 + \frac{r}{t-1} \sum_i \tau_i^2 \quad \text{and} \quad \sigma_\epsilon^2 + r\sigma_\tau^2$$

To provide some unification of fixed and random effects components, let us introduce a notation for "finite (or pseudo) variances." Any sum of squares of fixed effects parameters divided by corresponding degrees of freedom will be denoted in the remainder of this book by a κ^2 with a subscript to indicate the particular factor or effect involved. In this case

$$(2.15) \qquad \kappa_\tau^2 = \frac{1}{t-1} \sum_i \tau_i^2$$

whereby the two EMS can be written as

$$(2.16) \qquad \sigma_\epsilon^2 + r\kappa_\tau^2 \quad \text{and} \quad \sigma_\epsilon^2 + r\sigma_\tau^2$$

when all treatments are equally replicated. Thus, a κ^2 will indicate a fixed effects component while a σ^2 will indicate a random effects component.

Example 2.3

Suppose the agronomy student, Example 2.1, received the following results from his experiment. Variety 1 is a control, a standard treatment. Assume the plants were thinned to a uniform stand per plot. The yields, in pounds per plot, were:

Variety	1	2	3
	6.6	5.6	6.9
	6.4	5.2	7.1
	5.9	5.3	6.4
	6.6	5.1	6.7
	6.2	5.7	6.5
	6.7	5.6	6.6
	6.3	5.6	6.6
	6.5	6.3	6.6
	6.5	5.0	6.8
	6.8	5.4	6.8
Totals: $y_{i.}$	64.5	54.8	67.0 → y.. = 186.3

$$C = (186.3)^2/30 = 1156.9$$

$$SSY = (6.6)^2 + (6.4)^2 + \ldots + (6.8)^2 - C$$

$$= 1167.5 - 1156.9 = 10.6$$

$$SST = [(64.5)^2 + (54.8)^2 + (67)^2]/10 - C$$

$$= 1165.2 - 1156.9 = 8.3$$

$$SSE = 10.6 - 8.3 = 2.3$$

Under the fixed effects model, the ANOVA table has the form shown below.

ANOVA for Yields of New Soybean Varieties

Source	df	SS	MS	EMS
Varieties	2	8.3	4.15	$\sigma_\epsilon^2 + 5 \sum_i \tau_i^2$
Exp. Error	27	2.3	0.09	σ_ϵ^2
Total	29	10.6		※

Example 2.4

A large department store decided to train a new group of sales representatives by four training programs. From a group of 24 new applicants, six were randomly assigned to each program. After 3 weeks of training and two weeks on the job, the sales commissions (15% of sales) were determined for each applicant. Two weekly sales figures were incomplete due to illness of the trainees. There is interest in determining whether average weekly sales differ for the four training programs. Therefore, each program could generate a (treatment) population of weekly sales values.

Program	Weekly Commissions						Totals	Means
1	231	209	226	214	230	218	1328	221.33
2	160	183	210	179	191		932	184.60
3	251	246	238	227	240		1202	240.40
4	195	188	204	192	210	197	1186	196.67

$C = (4639)^2/22 = 978,196.4$

$SSY = (231)^2 + (209)^2 + \ldots + (197)^2 - C$
$= 990,097 - 978,196.4 = 11,900.6$

$SST = (1328)^2/6 + (923)^2/5 + (1202)^2/5 + (1186)^2/6 - C$
$= 987,709.9 - 978,196.4 = 9,513.5$

$SSE = 11,900.6 - 9,513.5 = 2,387.1$

Under the fixed model, the ANOVA table is presented below

ANOVA of Weekly Sales of Four Trainees

Source	df	SS	MS	EMS
Trainees	3	9,513.5	3,171.2	$\sigma_\epsilon^2 + \dfrac{1}{3} \sum_i r_i \tau_i^2$
Exp. Error	18	2,387.1	132.6	σ_ϵ^2
Total	21	11,900.6		※

2.8   Inference Under the Fixed Effects Model

Recall that the i-th treatment defines a population of potential values assumed to have a $N(\mu_i, \sigma_\epsilon^2)$ distribution. Consequently, there is interest in making inferences about the parameters of these t treatment distributions. The population means, the μ_i, have been described earlier as treatment means.

Possible inference about the treatment means may include one or more of the following:

   (i) a test of simultaneous equality of all means
   (ii) tests or CI about specific treatment comparisons
   (iii) tests or CI about pairs of treatment means.

In the next subsection the first category will be discussed. Inferences of the last two categories will be covered in later chapters.

2.8.1   Testing the Equality of Treatment Means

Here we consider an overall test of the simultaneous equality of all treatment means. The null hypothesis for such a test would be

(2.17) $H_0: \mu_1 = \mu_2 = \cdots = \mu_t$

which, for the CR design, would be equivalent to

H_0: All treatment means are equal.

These null hypotheses state that there is no variation among the population means, the μ_i. But in any experiment, we expect some variation among the sample treatment means. How much of the observed variation is due to sampling and chance? Or, at what point should we say that the variation is so great that we no longer consider it due to chance but due instead to differences among the treatment means? This depends, to a large extent, on the size of the within-population variance, σ_ϵ^2.

The observed treatment variation is based on the treatment sum of squares, SST. Statistical theory tells us to compare the related mean square, MST, to the Exp. Error mean square, MSE. Under the assumptions (2.3) and the null hypothesis (2.17), the ratio of these two mean squares has an F distribution with t-1 and n-t degrees of freedom. Thus, to test the null hypothesis (2.17), the test statistic is

(2.18)   F = MST/MSE

an F variable with t-1 and n-t degrees of freedom.

Note that the equality of all μ_i implies all $\tau_i = \mu_i - \mu$ are zero. As a result of this implication, several observations can be made:

(i) The hypothesis (2.17) is equivalent to the hypothesis

$H_0: \tau_i = 0$, for all i.

(ii) Under the null hypothesis, treatment and residual mean squares have the same expectation, as seen from Table 2.2 (because $\Sigma r_i \tau_i^2 = 0$ under the null hypothesis). Thus the ratio of these two MS forms our test statistic. See the comment following Table 2.2 about the usefulness of EMS.

(iii) If H_0 is not true, because not all τ_i are zero, MST has a larger expected value than MSE [$\Sigma r_i \tau_i^2$ must be positive if some τ_i are not zero]. Consequently, significantly different treatments are indicated only when the value of our F statistic is sufficiently greater than 1.

In summary, then, the test of equality of treatment means in a CR design has null and alternative hypotheses of

(2.19)

$$H_0: \text{All } \tau_i = 0.$$

$$H_1: \text{At least two } \tau_i \text{ are not zero.}$$

uses a test statistic of

(2.20) $F = MST/MSE$

which has the F distribution with t-1 and n-t degrees of freedom. The critical region for this test is the upper tail area of this distribution. Therefore, the *observed* significance level is given by the area corresponding to values of this F distribution which exceed "the value of F calculated from the expression in (2.20)."

Example 2.5

The agronomy graduate student (see also Example 2.3), would make a test of equality of variety means as follows:

H_0: The 3 variety means are equal.

H_1: The 3 variety means differ.

Test statistic: $F = MST/MSE$ with 2 and 27 df.

$F = 4.15/0.09 = 46.1$

The observed significance level is $P < 0.005$.

Interpretation: There is sufficient evidence at the 0.5% level of significance to conclude that the 3 soybean varieties have different mean yields. ⊠

Example 2.6

A test of equal mean weekly sales for the four training programs, Example 2.4, is:

$$H_0: \mu_1 = \mu_2 = \mu_3 = \mu_4$$

H_1:   At least two μ_i differ.

Test statistic: F = MST/MSE   with 3 and 18 df.

$$F = 3,171.2/132.6 = 23.9$$

The observed significance level is   P < 0.005.

Interpretation: There is sufficient evidence at the 0.5% level of significance to conclude that the mean weekly sales is not the same for the four training programs. ▓

2.8.2   Differences of Pairs of Means

When the ANOVA test provides evidence of significantly different treatment means, a table of differences of all pairs of means may be given along with the corresponding standard deviations of these differences.   Without attaching any significances, the information in such a table might be used by a researcher to identify general similarities and differences of treatment means. The difference of two treatment means, say $\mu_i - \mu_h$, where $i \neq h$, is estimated unbiasedly by the difference of the corresponding sample means, $\bar{y}_{i.} - \bar{y}_{h.}$; with estimated standard deviation given by

(2.21) $$s_{\bar{y}_{i.} - \bar{y}_{h.}} = \sqrt{MSE(1/r_i + 1/r_h)}$$

Example 2.7

In the previous example we saw evidence of differences among the mean weekly sales for the four training programs. A table of mean differences and corresponding standard deviations would be:

i	h	$\bar{y}_{i.} - \bar{y}_{h.}$	$s_{\bar{y}_{i.} - \bar{y}_{h.}}$
1	2	$ 36.73	6.97
1	3	-19.07	6.97
1	4	23.66	6.65
2	3	-55.80	7.28
2	4	-13.07	6.97
3	4	42.73	6.97

The first standard deviation listed above is

$$6.97 = \sqrt{132.6(1/5+1/6)}$$

From the values given we note that training programs 2 and 4 have the smallest mean differences and all other pairs of training programs have at least a two standard deviation difference in mean sales with programs 2 and 3 having the greatest difference.※

2.9   Observed F Values Less Than Unity

From statistical theory, we know that F ratios calculated in ANOVA tests of equal means are expected to be at least 1. As we indicated earlier in observations (ii) and (iii), MST is expected to be at least as large as MSE under the CR model (2.1) and the assumptions in (2.8). In practice, however, an F ratio less than unity may be observed. Two reasons may be proposed for this: (i) random sampling fluctuations or (ii) an incorrect linear model. Due to random sampling fluctuations, there is always a chance that MST is an under-estimate (of its expected value) while MSE is an over-estimate. This is more likely to happen when H_0 is true. The use of an incorrect model may generate an F value less than 1 because the Exp. Error MS contains variation which the correct model would remove. Some researchers use the criterion: The model is incorrect if a calculated value of F is less than a *lower* percentage point, for some pre-assigned value of α.

2.10   Inference Under the Random Model

The random effects models do not occur as often in practice (as the fixed effects models) because the researcher usually designs an experiment to investigate all treatments of interest. We will discuss the basic inferences when a random model is encountered. Reference here is to the CR designs but the general idea applies to other designs.

Under a random effects model, the treatments appearing in the experiment are not the only ones of interest. Consequently, we are interested not so much in the particular treatment effects occurring in our experiment but rather in knowing something about the variability in the distribution of all treatment effects. Recall from assumptions (2.8) that for random treatment effects, the τ's have a $N(0, \sigma_\tau^2)$ distribution. Therefore, the hypothesis of interest is

(2.22) H_0: $\sigma_\tau^2 = 0$

The statistic for testing this null hypothesis is the same F ratio as used for testing in the fixed-effects model. Likewise, the observed significance level is obtained from the upper tail area of the given F distribution.

Estimation of the two variance components, σ_ϵ^2 and σ_τ^2, may be of interest. Point estimates of these parameters are

(2.23) $\hat{\sigma}_\epsilon^2 = \text{MSE}$

(2.24) $\hat{\sigma}_\tau^2 = (\text{MST-MSE})/r_o$

where r_o is defined in Table 2.2. Interval estimation of $\hat{\sigma}_\epsilon^2$ and $\hat{\sigma}_\tau^2$ is accomplished through the chi-squared distribution. These intervals are

(2.25) $\dfrac{\text{n-t}}{a} \hat{\sigma}_\epsilon^2 < \sigma_\epsilon^2 < \dfrac{\text{n-t}}{b} \hat{\sigma}_\epsilon^2$

where a and b are, respectively, upper and lower $\alpha/2$ percentage points of a χ^2 variable with n-t degrees of freedom, and

(2.26) $\dfrac{\nu}{c} \hat{\sigma}_\tau^2 < \sigma_\tau^2 < \dfrac{\nu}{d} \hat{\sigma}_\tau^2$

where c and d are, respectively, upper and lower $\alpha/2$ percentage points of a χ^2 variable with degrees of freedom approximated by

(2.27) $\nu = \dfrac{(\text{MST} - \text{MSE})^2}{\dfrac{(\text{MST})^2}{\text{t-1}} + \dfrac{(\text{MSE})^2}{\text{n-t}}}$

The CI for σ_τ^2 is an approximation; there is no exact procedure for this interval estimation. Note that the estimated degrees of freedom given by (2.27) generally are not integers which requires interpolation in the χ^2 tables, such as Table A.7.

Example 2.8

To illustrate the concepts and inferences of a random effects model, let us assume that the agronomy student, (Example 2.3),

wishes   to   make   inferences   about   a   large   set   of   varieties   from
which   the three used in the experiment were chosen.   To test the
null hypothesis H_0: $\sigma_\tau^2 = 0$, one would calculate an F value

$$F = MST/MSE = 4.15/0.09 = 46.1$$

which gives an observed significance level of $P < 0.005$. We would
conclude that $\sigma_\tau^2 > 0$, most likely.

Estimates of the variance components are

$$\hat\sigma_\epsilon^2 = MSE = 0.09$$

$$\hat\sigma_\tau^2 = (MST - MSE)/r_o = (4.15 - 0.09)/10 = 0.41.$$

A 95% CI for the variance component σ_ϵ^2 is:

$$\frac{27(0.09)}{43.2} < \sigma_\epsilon^2 < \frac{27(0.09)}{14.6}$$

$$0.056 < \sigma_\epsilon^2 < 0.166$$

With degrees of freedom given by (2.20)

$$\nu = \frac{(4.15 - 0.09)^2}{(4.15)^2/2 + (0.09)^2/27} = \frac{16.48}{8.6} = 1.91$$

the 95% approximate CI for σ_τ^2 is

$$\frac{1.91(0.41)}{7.17} < \sigma_\tau^2 < \frac{1.91(0.41)}{0.046}$$

$$0.109 < \sigma_\tau^2 < 17.02$$

The denominators,   7.17 and 0.046, were obtained by interpolating
in the Table A.7, using 1.91 df. ▓

It   is   worth   noting   that the use of only a   few   treatments
(such   as the use of only 3 as in the last example) generally   is
inadequate   in an experiment dealing with   random   effects.   Any
estimated   variance   with so few degrees of freedom is of   little
value.

2.11   Subsampling in CR Designs

Here we consider the analysis of an experiment conducted in the completely randomized setting where the entire EU is not measured or observed. Such might be the case if EU are

(i) large fields of cropland where different rates of liquid fertilizers are applied,
(ii) batches of steel having different percentages of carbon, and
(iii) hulls of ships upon which different paints are applied.

Not only might large EU be necessary from an economic standpoint, their use might better represent the real-world status. If farmers customarily apply liquid fertilizer with large machinery, using such machinery in a small scale experiment with 25 square-foot plots probably would be very impractical, provided it could even be done without contaminating surrounding plots. A more reasonable approach would be to use larger plots or fields from which small areas could be sampled.

If precisely one subset is measured or analyzed on each EU, no matter how large the EU, we have a basic CR design considered in the first part of this chapter. We therefore consider a more typical situation where several subsets, called sampling units, are selected at random on each EU. An observation is recorded for each selected sampling unit. Generally one can expect more accurate and precise results as the number of sampling units increase per EU, this gain being more pronounced as the variable of interest exhibits more variation within the EU.

Large EU make subsampling a necessity. In other experiments, the researcher may introduce subsampling in order to study the within EU variability. Knowledge of this variation may be of value in future research experiments.

2.11.1   The Equal Numbers Case

We assume r EU per treatment and s subsampling units per EU. A total of rst observations will be recorded then. As might be expected, the required calculations take more time than in a basic CR design but they are straight-forward. The total sum of squares is based on rst observations (instead of only rt). As we will see later, one additional sum of squares must be calculated, to measure variation among responses within the EU. Hopefully, the additional time required for calculations when subsampling is present is offset by other considerations, such as time saved by not measuring entire EU, accuracy of the results, and so on.

Except in extremely rare cases, the s observations obtained
on any EU will not be identical. Differences among these observations represent variation that was not present in the basic
designs without subsampling. To account for this subsampling
variation, our linear model must contain a component to identify
and measure this new variation. These subsampling components
measure a second type of "error variation", an amount attributed
only to observations within the EU. The linear model to explain
the rst experimental observations obtained in a CR subsampling
experiment has the form

(2.28) $y_{ijk} = \mu + \tau_i + \epsilon_{ij} + \delta_{ijk}$; $\begin{aligned} i &= 1, 2, \ldots, t \\ j &= 1, 2, \ldots, r \\ k &= 1, 2, \ldots, s \end{aligned}$

where

y_{ijk} = observation for the k-th subsample on
the j-th EU of the i-th treatment

μ = overall mean, a constant

(2.29) $\tau_i = \mu_{i..} - \mu$ = i-th treatment effect

ϵ_{ij} = random component explaining variation
among EU on the same treatment

δ_{ijk} = random component explaining variation
within EU, among subsamples of the same EU

For valid statistical inferences, we assume that

(2.30)
ϵ's are i.i.d. $N(0, \sigma^2_\epsilon)$

δ's are i.i.d. $N(0, \sigma^2_\delta)$

with the ϵ's and δ's distributed independently of each other.
The treatment components, the τ's, may be either fixed or random
depending on their inference space. If they are random, they
are assumed to be distributed independently of all other random
components.
    To calculate the sums of squares, we need the raw data, the
total for each EU, and the treatment totals. These totals are:

$$y_{ij.} = \sum_k y_{ijk} = \text{total for the j-th EU of the i-th treatment}$$

(2.31)

$$y_{i..} = \sum_j y_{ij.} = \text{total for the i-th treatment}$$

$$y... = \sum_i y_{i..} = \text{overall total}$$

The sum of squares calculations are

$$C = y_{...}^2 / rts = \text{general correction term}$$

(2.32)   SST = Treatment SS

$$= \sum_{ijk} (\bar{y}_{i..} - \bar{y}...)^2$$

$$= \frac{1}{rs} \sum_i y_{i..}^2 - C$$

(2.33)   SSE = Exp. Error SS

$$= \sum_{ijk} (\bar{y}_{ij.} - \bar{y}...)^2$$

$$= \frac{1}{s} \sum_{ij} y_{ij.}^2 - \frac{1}{rs} \sum_i y_{i..}^2$$

(2.34)   SSS = sampling error SS

$$= \sum_{ijk} (y_{ijk} - \bar{y}_{ij.})^2$$

$$= \sum_{ijk} y_{ijk}^2 - \frac{1}{s} \sum_{ij} y_{ij.}^2$$

(2.35)   SSY = Total SS

$$= \sum_{ijk} (y_{ijk} - \bar{y}...)^2$$

$$= \sum_{ijk} y_{ijk}^2 - C$$

We now summarize the above in a general ANOVA table.

Table 2.3   ANOVA for Subsampling in a CR Design
(Equal Numbers, Fixed Effects)

Source	df	SS	MS	EMS
Treatments	t-1	SST	MST	$\sigma_\delta^2 + s\sigma_\epsilon^2 + rs\kappa_\tau^2$
Exp.Error	t(r-1)	SSE	MSE	$\sigma_\delta^2 + s\sigma_\epsilon^2$
Sampling Error	tr(s-1)	SSS	MSS	σ_δ^2
Total	trs-1	SSY		

Example 2.9

An engineer is studying three electroplating processes. Eighteen specimens, all uniform in size and thickness, were obtained for the experiment. Six specimens were randomly assigned to each process. At the completion of the electroplating each specimen was measured for thickness at three randomly selected locations. The results were, in mm:

		Process	
Specimen	I	II	III
1	0.52,0.49,0.49	0.54,0.51,0.53	0.56,0.57,0.57
2	0.53,0.51,0.50	0.54,0.56,0.54	0.58,0.55,0.57
3	0.52,0.49,0.51	0.53,0.55,0.52	0.57,0.58,0.57
4	0.51,0.53,0.52	0.52,0.53,0.53	0.56,0.58,0.56
5	0.54,0.52,0.50	0.54,0.54,0.53	0.55,0.57,0.58
6	0.52,0.51,0.52	0.52,0.51,0.54	0.57,0.54,0.56

The specimen and process totals are:

		Process	
Specimen	I	II	III
1	1.50	1.58	1.70
2	1.54	1.64	1.70
3	1.52	1.60	1.73
4	1.56	1.58	1.70
5	1.56	1.61	1.70
6	1.55	1.57	1.67
Total	9.23	9.58	10.20

$$y... = 29.01$$

The   necessary sums of squares are calculated easily from the raw data and totals:

Total: $SSY = (0.52)^2 + (0.49)^2 + \ldots + (0.56)^2 - (29.01)^2/54$

$= 15.6205 - 15.5848 = 0.0357$

Process: $SSP = [(9.23)^2 + (9.58)^2 + (10.20)^2]/18 - 15.5848$

$= 15.6116 - 15.5848 = 0.0268$

Exp. Error: $SSE = [(1.5)^2 + (1.58)^2 + \ldots + (1.67)^2]/3 - 15.6116$

$= 15.6143 - 15.6116 = 0.0027$

Sampling: $SSS = 15.6205 - 15.6143 = 0.0062$

The   ANOVA table for this experiment,   assuming   fixed   treatment effects, is:

ANOVA for Electroplating Processes

Source	df	SS	MS	EMS
Processes	2	0.0268	0.0134	$\sigma_\delta^2 + 3\sigma_\epsilon^2 + 18\kappa_\tau^2$
Exp.Error	15	0.0027	0.0002	$\sigma_\delta^2 + 3\sigma_\epsilon^2$
Sampling Error	36	0.0062	0.0002	σ_δ^2
Total	53	0.0357		✷

From   the EMS shown in Table 2.3,   it is apparent that a test of no treatment effects uses an F statistic given by

(2.36) $F = MST/MSE$

with (t-1) and t(r-1) degrees of freedom.   It is important to remember that the sampling error mean square is not used in the F test of treatment effects.

Example 2.10

A test of no difference in mean thickness of electroplating for the three processes, Example 2.9, would be

$$H_0: \tau_i = 0, \text{ for all } i$$

$$H_1: \text{Some } \tau_i \text{ are not zero}$$

From (2.31) the value of the test statistic is

$$F = 0.0134/0.0002 = 67$$

with 2 and 15 degrees of freedom.   The observed significance level is $P < 0.005$. ✳

2.11.2   The Unequal Numbers Cases

Missing values, or unequal numbers may occur at any of the stages of the design. Three different possibilities exist:

(i) unequal numbers of EU per treatment, but a constant number of sampling units for each EU,

(ii) an equal number of EU per treatment, but an unequal numbers of sampling units per EU, and

(iii) unequal numbers of EU per treatment and unequal numbers of sampling units per EU.

The first case listed is the most common of the three because the experimenter generally can control the number of observations taken per EU.

The totals given in (2.31) are appropriate for the unequal numbers cases. (Here the ranges of summations may be variable whereas they were constant in the equal numbers case.) Then, the sums of squares can be obtained from the first expression (the definitional formula) given in each of (2.32) thru (2.34). Only the divisors are changed in the second expression(the calculation formula) given in each of these equations, to reflect the unequal numbers. The total sum of squares is unchanged from the expressions given in (2.35).

Letting n_{ij} denote the number of sampling units measured on the (i,j)-th EU, the general correction term would be

$$(2.37) \qquad\qquad C = y^2.../n..$$

where $n.. = \underset{ij}{\Sigma\Sigma} \ n_{ij}$ represents the overall number of observations.

The remaining calculation formulas are:

(2.38)   SST = Treatment SS

$$= \sum_i (y_{i..}^2 / n_{i.}) - C$$

where $n_{i.} = \sum_j n_{ij}$ represents the number of observations on the i-th treatment.

(2.39)   SSE = Exp. Error SS

$$= \sum_{ij} (y_{ij.}^2 / n_{ij}) - \sum_i (y_{i..}^2 / n_{i.})$$

(2.40)   SSS = Sampling Error SS

$$= \sum_{ijk} y_{ijk}^2 - \sum_{ij} (y_{ij.}^2 / n_{ij})$$

With these calculations we can now present the ANOVA table for the general unequal numbers case.

Table 2.4   ANOVA for Subsampling in CR Design
                  (Unequal Numbers, Fixed Treatments)

Source	df	SS	MS	EMS
Treatments	t-1	SST	MST	$\sigma_\delta^2 + c_1 \sigma_\epsilon^2 + c_2 \kappa_\tau^2$
Exp. Error	r.-t	SSE	MSE	$\sigma_\delta^2 + c_3 \sigma_\epsilon^2$
Samp. Error	n..-r.	SSS	MSS	σ_δ^2
Total	n..-1	SSY		

where $r. = \sum_i r_i$ = number of EU, and the coefficients of the EMS components will be discussed next. For case (i):

(2.41)
$$c_1 = c_3 = s$$
$$c_2 = [n.. - s \sum_i r_i^2/r.]/(t-1)$$

For cases (ii) and (iii):

(2.42)
$$c_1 = [\sum\sum_{ij}(n_{ij}^2/n_{i.}) - \sum\sum_{ij}(n_{ij}^2/n..)]/(t-1)$$

$$c_2 = [n.. - \sum_i(n_{i.}^2/n..)]/(t-1)$$

$$c_3 = [n.. - \sum\sum_{ij}(n_{ij}^2/n_{i.})]/(t-1)$$

From (2.41) and the EMS in Table 2.4 it is clear that a test of
no treatment effects is made in case (i) as in the equal numbers
case, using an F statistic given by

(2.43) $F = MST/MSE$

with $(t-1)$ and $(r.-t)$ degrees of freedom.

The coefficients c_1 and c_3 appearing in (2.42) generally
are unequal.   Hence, there is no exact test of treatment effects
for cases (ii) and (iii).   One can see from the EMS in Table 2.4
that, when $c_1 \neq c_3$, no two mean squares have the same expectation
under a hypothesis of zero treatment effects.   An approximate
test procedure will be discussed in the next section.

Example 2.11

A turfgrass specialist investigated four root growth stimulators.
Each was applied to eight plots of established turfgrass on a
fringe area of a golf course.   Unfortunately six plots were
destroyed by workers who had to dig up a faulty sprinkler system.
Two equal-sized cores (samples) were randomly selected from each
of the final 26 plots, analyzed for root weight, with results, in
grams:

| | | Stimulator | | |
Plot	1	2	3	4
1	4.3,4.8 (9.1)	3.9,4.1 (8.0)	3.8,3.4 (7.2)	3.7,3.9 (7.6)
2	4.0,4.3 (8.3)	3.8,3.3 (7.1)	3.6,3.1 (6.7)	3.5,3.4 (6.9)
3	4.3,3.8 (8.1)	3.7,3.4 (7.1)	3.5,3.7 (7.2)	3.6,4.1 (7.7)
4	4.2,4.7 (8.9)	3.6,3.2 (6.8)	3.2,3.7 (6.9)	3.8,3.5 (7.3)
5	4.3,4.5 (8.8)	3.0,3.7 (6.7)	3.4,3.0 (6.4)	3.6,3.9 (7.5)
6	3.9,4.2 (8.1)		3.2,3.6 (6.8)	4.2,4.0 (8.2)
7	4.4,4.0 (8.4)		3.3,3.5 (6.8)	
8	4.6,4.1 (8.7)			
$y_{i..}$	68.4	35.7	48.0	45.2 → 197.3 = y...
r_i	8	5	7	6

The numbers in parentheses are $y_{ij.}$, the plot (the EU) totals. The numbers designated by r_i represent the number of EU per treatment. Then, using (2.38) thru (2.40), the sums of squares calculations are

$$C = (197.3)^2/52 = 748.60$$

$$SSY = [(4.3)^2 + (4.8)^2 + \ldots + (4.0)^2] - C$$
$$= 758.41 - 748.60 = 9.81$$

$$SST = \frac{(68.4)^2}{16} + \frac{(35.7)^2}{10} + \frac{(48.0)^2}{14} + \frac{(45.2)^2}{12} - C$$
$$= 754.68 - 748.60 = 6.08$$

$$SSE = [(9.1)^2 + (8.3)^2 + \ldots + (8.2)^2]/2 - 754.68$$

$$= 1512.83/2 - 754.68 = 1.74$$

$$SSS = 758.41 - 756.42 = 1.99$$

And from (2.41), the coefficients for the EMS components are

$$c_1 = c_3 = 2$$

$$c_2 = [52 - 2(64+25+49+36)/26]/3$$
$$= (52 - 13.38)/3 = 12.87$$

ANOVA for Root Growth Stimulators

Source	df	SS	MS	EMS
Stimulators	3	6.08	2.030	$\sigma_\delta^2 + 2\sigma_\epsilon^2 + 12.87\kappa_\tau^2$
Exp. Error	22	1.74	0.079	$\sigma_\delta^2 + 2\sigma_\epsilon^2$
Sampling	26	1.99	0.077	σ_δ^2
Total	51	9.81		

A test of no stimulator effects would have hypotheses

$$H_0: \text{All } \tau_i = 0$$

$$H_1: \text{Some } \tau_i \neq 0$$

Because this experiment belongs to the category we have labeled case (i), we can make an exact test with the F distribution, provided the assumptions in (2.32) hold. The value of the test statistic is

$$F = 2.03/0.079 = 25.70$$

which has 3 and 22 degrees of freedom. The observed significance level is $P < 0.005$. ※

2.12 Satterthwaites' Procedure

In certain experiments, the analysis of variance does not provide exact tests for some of the model parameters. We saw this in the previous section when we discussed subsampling in the CR designs. Two of the three unequal numbers cases did not have exact tests. And various other experimental situations will be encountered where exact F tests do not exist.

Satterthwaite [27] has described an approximate procedure that may be used when the ANOVA does not provide an exact test. Basically, the problem exists because no two mean squares have the same expectation under a null hypothesis.   This occurred in the previous section because of unequal coefficients of an EMS component.   In principle, Satterthwaites' procedure is quite simple: Just build a synthetic MS so that two mean squares have the same expectation under the null hypothesis.   But one has options: Either a numerator or a denominator mean square can be synthesized , or both.

Let MS_1, MS_2, ..., MS_m denote the mean squares appearing in the ANOVA table.   To build a synthetic MS, one constructs a linear combination of the MS_i, say

(2.44) $M = a_1 MS_1 + a_2 MS_2 + \ldots + a_m MS_m$

where a_1, a_2, ..., a_m are appropriate constants (some of which may be zero) so that M has the desired expectation.   Forming a ratio with the synthetic MS gives an approximate F statistic. The degrees of freedom for M are approximated by

(2.45) $df = \dfrac{M^2}{\Sigma(a_i MS_i)^2/df_i}$

where df_i represents the degrees of freedom on MS_i.

Usually in practice, a denominator MS is synthesized to have expectation of the numerator MS under H_0.   In so doing, some of the a_i may be negative; this could result in a negative synthetic mean square, M. To eliminate a negative M, Cochran [9] proposed a modification of Satterthwaites' procedure where both numerator and denominator MS are synthesized, but using only non-negative constants in constructing the linear combinations.   Note that Cochran's proposal would require the calculation of both degrees of freedom, approximately by (2.45).

Example 2.12

Refer to the EMS appearing in Table 2.4.   Should we encounter an experiment belonging to either case (ii) or (iii), we would have

$c_1 \neq c_3$. Then to investigate treatment effects, we would need to make an approximate test. Using Satterthwaites' procedure, we consider building a synthetic Exp. Error MS having expectation equal to $\sigma_\delta^2 + c_1 \sigma_\epsilon^2$ (to match the expectation of the Treatment MS under the null hypothesis of zero treatment effects.) With constants

$$a_1 = 0$$

$$a_2 = c_1/c_3$$

$$a_3 = 1 - c_1/c_3$$

construct a synthetic MS

$$M = a_1 MS_1 + a_2 MS_2 + a_3 MS_3$$

$$= (c_1/c_3)MSE + (1-c_1/c_3)MSS$$

From the EMS given in Table 2.4, we note that M has expectation

$$(c_1/c_3)(\sigma_\delta^2 + c_3\sigma_\epsilon^2) + (1 - c_1/c_3)\sigma_\delta^2 = \sigma_\delta^2 + c_1\sigma_\epsilon^2$$

as desired.

The degrees of freedom for M are approximated by

$$df = \frac{M^2}{[(c_1/c_3)MSE]^2/(r.-t) + [(1-c_1/c_3)MSS]^2/(n..-r.)} \text{ ※}$$

2.13 A Test of Equal Population Variances

An assumption required for valid inferences under classical ANOVA is that the residual components, the ϵ's, are independently distributed with equal variances for all populations, see (2.8). Empirical evidence suggests that mild departures from equality of variances can be tolerated so long as the replications (sample sizes) are equal, or nearly so, see Welch[29] and [30]. As the variances differ by larger amounts, the error rates may deviate appreciably from the indicated values.

In some experimental situations, there may be a suspicion of unequal variances. Here one may wish to make the test indicated below. In other situations, biological experiments in particular, the researcher is fairly certain of unequal variances. Such is the case with many "biological growth responses" which tend to become more variable with the age of organisms, time elapsed, and so on.   Here one may consider a transformation of the data,   see Section 2.14.

Testing the equality of variances for two normal populations is no problem;   an exact F test exists,   see Freund [15] or Lentner [22]. But there is no exact test for testing the equality of 3 or more population variances,   even if one assumes underlying normal distributions.   Several approximate tests have been proposed for testing

$$(2.46) \qquad H_0: \; \sigma_1^2 = \sigma_2^2 = \text{---} = \sigma_k^2$$

when   random sampling from $k > 2$ independent normal populations. Unfortunately,   none of these tests is very robust; that is, each test is very sensitive to departures from normality.

Hartley [16] has proposed a test procedure using a "pseudo-F" statistic equal to the ratio of the largest observed variance   to the smallest; namely,

$$(2.47) \qquad F^* = \max\{s_i^2\}/\min\{s_i^2\}$$

This expression also is called Hartley's maximum F.   Percentage points of the F* distributions are given in Table A.8 for various values   of   k and ν degrees of freedom,   those on   each   observed variance.   Note that Table A.8   has been   constructed under   the assumption of equal degrees of freedom on each observed variance.

The critical region for testing the null hypothesis in (2.46) against   an alternative of "different variances" is the   upper α area   of the appropriate F* distribution.   In other   words,   the hypothesis in (2.46) is rejected if

$$(2.48) \qquad F^* > F^*_\alpha$$

for a Type I error rate of α.

If the degrees of freedom are unequal, but not too different, a conservative test of (2.44) may be made using F*,   but with the parameters ν equal to "the minimum degrees of freedom" on any s_i^2.

Example 2.13

We   will   use   the trainee data of Example   2.4   to   illustrate Hartley's max. F test of

$$H_0: \; \sigma_1^2 = \sigma_2^2 = \sigma_3^2 = \sigma_4^2$$

The four population (training program) variances are estimated by

$$s_1^2 = 407.33/5 = 81.47$$

$$s_2^2 = 1325.2/4 = 331.3$$

$$s_3^2 = 329.2/4 \;\; = \;\; 82.3$$

$$s_4^2 = 325.33/5 = 65.1$$

For a conservative 5% test of equal variances for the training programs, Hartley's maximum F statistic is

$$F^* = s_2^2/s_4^2 = 331.3/65.1 = 5.1$$

The upper 5% tabulated point of F^*, with $k = 4$ and $\nu = 4$, is $F^*_{0.05} = 20.6$. Thus, equality of variances of the 4 training programs is not rejected at the 5% level of significance. ▓

2.14   Transformations of Data

We have mentioned the robustness of the ANOVA tests with respect to departures from normality and mild departures from equal population variances. When the departures from equal variances become severe, we seek possible remedies so that better quality inferences may be obtained. An often used technique is to apply some transformation to the data to stabilize the variances. The foremost problem is finding the most appropriate transformation.

The development and application of variance-stabilizing transformations has been reported by many researchers, for example, Bartlett [1], [2], Bartlett and Kendall [3], Beall [4] and Curtiss [11].

The appropriate transformation can be derived mathematically if one knows the relationship between the means and variances (or standard deviation) of the populations, see Bartlett [2] or Box and Cox [7], for example. One may wish to plot the observed standard deviations against the means to detect the possible relationship. This is especially useful for biological data where the populations are ordered by a gradient, such as time, age, and so on.

We give a list of the commonly occurring relationships, situations producing such data, and the transformation which will stabilize the variances of such data.

Table 2.5   Variance Stabilizing Transformations

Variance Proportional to	Type of Data	Transformation
μ	Count data (e.g., Poisson)	$\sqrt{y}$ $\sqrt{y+0.5}$ if small integers are present
μ^2	No specific type	$\log (y)$ $\log(y+1)$ if values near 0 are present
$\mu(1 - \mu)$	Percentages Proportions (e. g., Binomial)	$\text{Arcsin}\sqrt{y}$
μ^4	Business, Economic	$1/y$ $1/(y+1)$ if values near 0 are present

One might suspect that a transformation applied to a set of experimental data to stabilize population variances might destroy other needed properties, such as normality and independence. It turns out, though, that non-normality and unequal variances tend to go hand-in-hand so that a transformation to stabilize the variances also corrects some non-normality in the data, [2]. The impact of a transformation on independence of residual components (the ϵ's) is more difficult to assess, as we have indicated before. There is empirical evidence that transformations do help in achieving additivity of factors in more complex models (blocks, factorials, and others to be studied.) On the basis that all other properties generally are corrected simultaneously by an appropriate transformation, one would expect independence to be helped rather than destroyed.

When data are transformed to correct violation of assumptions, the "usual analyses" are performed on the transformed data. Assumptions must hold, therefore, on the transformed scale. After all statistical decisions are made, one usually transforms the results back to the original scale.

2.15   Robustness of ANOVA Inferences

In most real world experiments it is generally agreed that one or more assumptions are violated to some degree. The violations in

a given research setting affect the validity of inferences   under
some   statistical procedures more severly than under others.   We
are interested primarily in validity of inferences under ANOVA as
this is the basic tool of analyzing designed experiments.   Thus,
our   concern   is   whether or not the   underlying   assumptions   in
ANOVA,   such   as those in (2.8),   are violated to the   extent   of
appreciably changing the error rates.

The   validity of inferences from ANOVA has been discussed   in
three classical papers by Bartlett[2], Cochran [8], and Eisenhart
[14]. Additionally, validity of ANOVA has been investigated under
very   general randomization theory which requires minimal assumptions. Empirical results were reported by Eden and Yates [13] and
Hey [17]; theoretical results were reported by Pitman [24], Welch
[28],   and Kempthorne [20] and [21].   All evidence indicates that
the   classical   ANOVA F test (established under normality)   is   a
very good approximation of the general randomization test, especially in fixed effects cases.

To   summarize these reported results,   the classical ANOVA   F
test is very robust with respect to non-normality,   [8] and [17].
And   when   there   are equal replications,   mild   departures   from
homogeneous variances have little impact on validity of the ANOVA
test, [30] and Box [5] and [6].   Unequal variances have a greater
impact on estimates of treatment effects and their variances than
on tests [8]. Variance stabilizing transformations are therefore,
more important for estimation than testing purposes.   In cases of
severe   inequalities,   they are important for both estimation and
testing, of course.

The   above   remarks hold for validity of ANOVA inferences   in
general,   not just for CR designs.   In fact,   many of the   cited
references   dealt   with more complex designs.   An assumption   of
"additivity of effects" enters the picture when a researcher uses
a design other than the basic CR design,   unless such   additivity
is   accounted   for by inclusion of appropriate model   components.
When   nonadditivity is unaccounted for in the model,   a   frequent
situation,   the effects of nonadditivity combine with the residual
components,   the ϵ's.   This tends to inflate Exp. Error variation
but   some   cancellation is expected under proper   randomization.
The   greatest impact occurs when the nonadditivity effects   overshadow   the residual components;   that is,   when the variation of
the ϵ's is small,   [8].   Appropriate transformations may   offer
relief   of   nonadditivity if these effects are   believed   to   be
severe.

PROBLEMS

1.   An   agronomy student subdivided a plot of land into 10' x 10'
     subplots and labeled them as shown below

1	2	3	4	5	6
7	8	9	10	11	12
13	14	15	16	17	18
19	20	21	22	23	24
25	26	27	28	29	30

(a) Use the table of random numbers, Table A.1, to randomly assign a set of 5 treatments equally to these subplots.

(b) Repeat part (a) if treatment 1 is to occur in the experiment twice as often as each other treatment (still on the same 30 subplots.)

2. (a) Use Table A.1 to make a random assignment of 6 treatments to 24 plots numbered 101 thru 124.

(b) Repeat part(a) if treatments 1 and 2 are each to occur in the experiment twice as often as each of the other 4 treatments (still on the same 24 plots.)

3. A graduate student in horticulture obtained permission to use a small parcel of land at the University's branch station for his experiment. His proposed research is to study the effect of 6 different rates (dosages) of liquid fertilizer on the yield of potatoes. Each rate is to be applied to 5 plots of potatoes. Each plot was to receive its designated rate of application every two weeks. When the student measured his parcel, he discovered it was 125 feet by 20 feet---large enough for 126 rows; each row 20 feet long with 12 inches between rows. He obtained the necessary seed potatoes (a high quality certified seed) and planted the 126 rows. He now comes to you for help.

(a) How should he set up his experimental units(EU)? Be specific. Assume that the parcel is small enough to be uniform throughout. You may use a diagram to supplement your explanation. Hint: "Buffer rows" may be required between EU so fertilizer applied to one plot does not contaminate neighboring plots.

(b) An accepted practice in experiments of this type is to use the yield from 8 feet of row(s) per EU. How should the observations be obtained? Be specific. Again, you may find a diagram useful.

(c) Give the model for his statistical analysis based on your answers to parts (a) and (b) above.

4. A researcher in Forestry received a grant to investigate five newly developed varieties of pine trees. He carefully and painstakingly planned an elaborate greenhouse experiment to determine which variety has the greatest annual growth. After 4 years, he statistically analyzed the results of his experiment and recommended variety B as the best (for maximum annual growth). His results were so overwhelming that 10 lumber companies planted the recommended variety on half of their replacement acreage. After 4 years, 8 of the companies complained that variety B pine trees were only 75% as tall as "an old standby variety". Several principles of good research were violated in the experimentation and recommendation.

(a) Give one of the violations and briefly explain how the pitfall could have been avoided.

(b) Repeat part (a) for another violation.

5. The following ANOVA resulted from an experiment dealing with the strength of steel having different percentages of carbon.

ANOVA for Steel Strength, Different Percentages of Carbon

Source	df	SS	MS	EMS
Percentages		120		
Exp. Error	24	144		
Total	29	264		

(a) Give the linear model for this experiment. Explain terms used.

(b) Complete the above table. Assume equal replications and fixed treatment effects.

(c) Test the equality of mean strengths for the different percentages.

(d) Given the estimated standard deviation of $\bar{y}_{i.} - \bar{y}_{h.}$

(e) If $\bar{y}_{1.} = 13$ and $\bar{y}_{2.} = 11$, set and interpret a 95% CI for the mean difference $\mu_1 - \mu_2$

6. An animal scientist recorded the information in the following table from an experiment dealing with wool length measured on 12 sheep randomly selected from each of 5 breeds. Assume the fixed model.

ANOVA of Wool Length

Source	df	SS	MS	EMS
Breeds			4.23	
Exp. Error		29.7		

(a) Give the linear model for this experiment. Explain terms used.
(b) Complete the above ANOVA. Use numbers wherever possible.
(c) Test the equality of mean wool length for the 5 breeds.
(d) Give the estimated standard deviation of the difference of observed mean wool lengths of two different breeds.
(e) If the observed lengths for breeds 2 and 4 were 2.63 and 3.14, respectively, construct a 90% CI for $\mu_2 - \mu_4$.

7. A plant science researcher recorded the transpiration of tobacco plants of 5 different varieties. The following information is presented:

$$r_1 = 10 \qquad r_2 = r_3 = 12 \qquad r_4 = r_5 = 9$$

$$\bar{y}_{1.} = 8.7, \quad \bar{y}_{2.} = 9.3, \quad \bar{y}_{3.} = 7.4, \quad \bar{y}_{4.} = 11.1, \quad \bar{y}_{5.} = 9.6$$

ANOVA for Transpiration of Tobacco Plants

Source	df	SS	MS	EMS
Varieties		75.7		
Exp. Error		333.7		

(a) Complete the ANOVA table under the fixed model.
(b) Test for equality of mean transpirations for the 5 varieties.
(c) Set and interpret a 99% C.I. for $\mu_1 - \mu_4$.
(d) What assumptions are necessary for part (a) to be valid?
(e) What assumptions are necessary for part (c) to be valid?

8. The federal tax department is reviewing the laws governing withholding for federal income tax. They would like to know if the average amount refunded is about the same for the single, married filing jointly and married filing separately groups. A random sample of returns for each group was obtained giving:

	Single	Married Jointly	Married Separately
r_i:	100	200	50
$\bar{y}_{i.}$:	186	523	214

SSY = 71,291,000

(a) Give the linear model and assumptions for valid tests and confidence intervals.
(b) Set up the ANOVA under this model and assumptions.
(c) Give the estimated standard deviation for the difference of observed mean refunds for single individuals and those married filing separately.
(d) Set and interpret a 95% CI for the difference of mean refunds for the two "married" categories.

9. A nutrition specialist is measuring the amount of fatty acid present in blood when four different diets are used. Initially forty subjects (uniform with respect to the fatty acid) were obtained and 10 were randomly assigned to each diet. After a 3 week period, the following was recorded:

Diet (i):	1	2	3	4
Mean, $\bar{y}_{i.}$:	21.3	28.6	25.7	20.4
Variance, s_i^2:	8.40	13.08	16.54	9.62

ANOVA of Fatty Acid under Four Diets

Source	df	MS	EMS
Diets		147.70	
Exp. Error		11.91	

Assume normality and independence of the observations.

(a) Complete the ANOVA under the fixed model.
(b) Test equality of mean fatty acid under the 4 diets. Interpret.
(c) Give the estimated standard deviation of the difference of two observed treatment means.

10. In a greenhouse experiment, 5 types of light were used:
     natural, white, green, blue, and yellow. Eight tobacco
     plants, each 10 mm tall, were subjected to each type of
     light with the following means and variances of stem growths
     (in mm.)

	Natural	White	Yellow	Blue	Green
Mean:	4.09	4.30	3.55	3.01	3.19
Variance:	0.027	0.083	0.066	0.030	0.073

A partial ANOVA table was found to be:

ANOVA for Growth of Tobacco Plants: Different Light

Source	df	SS	MS	EMS
Treatments		9.93		
Exp. Error		1.95		

(a) Complete the ANOVA under the fixed model.
(b) Test the equality of mean stem growth under the 5 light
     types. Interpret.
(c) Give the estimated standard deviation of the difference
     of two observed treatment means.

11. Test equality of variances for the fatty acid measurements
     of the 4 dietary populations of Problem 9. Let $\alpha = 0.05$.

12. Test equality of variances for the stem growths of tobacco
     plants subjected to the five types of light, Problem 10. Let
     $\alpha = 0.05$.

13. A city considered several new water filtration systems.
     There were 4 systems available; A, B, C, and D. The costs
     of these are approximately equal; so the primary concern is
     the amount of impurities (measured in ppm) left in the water
     after it has been filtered. The four systems were hooked,
     individually by parallel piping, to one of the city's water
     tanks for testing. Over a period of one week, gallons of
     filtered water were randomly selected from each system's
     output, and the amount of impurities was determined. The
     results were:

		Method	
A	B	C	D
15.1	9.3	7.3	10.1
16.4	8.7	8.8	9.8
16.7	10.5	8.4	11.3
15.9	8.3	7.9	10.3
14.8	9.4	7.5	9.6
15.3	11.2	9.1	10.9
17.1	10.4	8.5	8.9
15.5	10.5	7.9	10.2
14.7	9.3	8.5	9.4
16.3	8.9	7.7	10.1

(a) What are the EU for this experiment?
(b) Give the linear model for this experiment. Define all terms used.
(c) Set up the ANOVA table under the fixed model.
(d) Test the hypothesis that the mean amount of impurities remaining in the water after filtration is the same for all 4 systems.
(e) What assumptions are required for part (c)? part (d)?
(f) Give a table of mean differences and their standard deviations.

14. The following are ascorbic acid contents for 3 varieties of ripe peaches grown in a certain locality. All measurements were converted to mg/100 g.

Variety 1		Variety 2		Variety 3	
5.34	5.50	7.12	6.80	6.28	6.40
5.58	5.42	6.89	6.91	6.01	6.12
5.26	5.47	6.93	6.76	6.27	6.24
5.47	5.71	6.82	6.97	6.15	6.31
5.39	5.62	7.06	6.88	6.38	6.37

(a) Give the linear model for this experiment. Explain each term appearing in the model. Assume fixed effects.
(b) Give assumptions necessary for valid inferences under this model.
(c) Set up the ANOVA table under this model.
(d) Test that the average ascorbic acid content is the same for the 3 varieties.
(e) Give a table of mean differences and their standard deviations.
(f) Set a 95% CI for $\mu_2 - \mu_1$

15. The warm-up time of color TV's was determined for 5 different models. Four TV's were randomly selected from the stockpile of each model giving the following warm-up times, in seconds:

I	II	III	IV	V
22	22	12	14	17
25	25	10	11	15
17	27	13	16	17
19	32	9	14	20

(a) Give the appropriate model and assumptions. Assume fixed effects.
(b) Set up the ANOVA table under the fixed model.
(c) Test the equality of mean warm-up times for the five models.
(d) Give a table of observed treatment mean differences and their standard deviations.

16. A horticulturist has been asked to test a new fungicide, Nurelle, for its effect on azaleas. He selects 49 uniform plants from a particular variety growing in a greenhouse. A standard fungicide, Truban, also will be used in the experiment for comparative purposes. All plants will be inoculated with a fungus two weeks after the experiment begins. For some plants the fungicide will be applied before the inoculation; for others, it will be applied after inoculation. The seven treatments to be used are defined as follows:

#	Treatment	
1	Truban	
2	1 oz. Nurelle	
3	3 oz. Nurelle	3 days before inoculation
4	5 oz. Nurelle	
5	1 oz. Nurelle	
6	3 oz. Nurelle	7 days after inoculation
7	5 oz. Nurelle	

(Courtesy of Dr. R. Lindstrom, Horticulture Dept., Va. Tech.)

The data recorded were fresh weights after three weeks. The weights from the seven plants of each treatment were:

			Treatment			
1	2	3	4	5	6	7
43	28	30	47	37	36	34
37	35	41	43	41	35	44
36	45	40	44	40	37	37
42	45	44	37	26	30	47
38	38	33	52	23	44	44
31	32	37	39	45	37	43
34	45	46	42	34	40	42

(a) Set up the ANOVA for this experiment. Assume fixed treatment effects. (For now ignore the treatment structure.)

(b) Test overall equality of mean growths under the seven treatments. Interpret.

(c) Give the estimated standard deviation of the difference of 2 observed treatment means.

17. A psychologist studied the effect of sound on the behavior pattern of monkeys. Sound was supplied at 4 different intensity levels. Six monkeys were randomly assigned to each sound level and placed in individual cages with food. When a monkey touched his food, the sound was supplied and the time (in seconds) until the monkey reached for the food again was recorded:

Intensity	1	2	3	4
	6.1	7.3	10.5	13.7
	7.3	9.1	12.4	14.8
	5.4	7.6	10.6	13.6
	6.3	8.4	13.0	14.4
	6.6	9.3		14.9
	7.2			

Unequal replications resulted from sound system failures.

(a) Set up the ANOVA under the fixed model.

(b) Test equality of mean times for the 4 sound intensities.

(c) Give a table of observed mean differences and their standard deviations.

(d) Set and interpret a 95% CI for the difference in mean time under intensities 1 and 4.

18. A government testing agency is checking five types of industrial filters for their ability to remove pollutants from industrial exhausts. A number of filters of each type is

tested under uniform exhaust conditions giving the following amounts of pollution remaining in the filtered air:

Type:	1	2	3	4	5
	21	14	26	18	7
	15	15	18	25	5
	10	20	14	22	9
	12		21		13
	9				10
	17				

Unequal replications resulted from failure or improper operation of exhaust set-ups.

(a) Set up the ANOVA under the fixed model.
(b) Test equality of mean amounts of pollution under the different types of filters.
(c) Give a table of observed mean differences and their standard deviations.
(d) Set and interpret a 95% CI for the mean amount of pollution under filter types 1 and 3.

19. A human factors engineer is studying the reaction times of a certain stimulus presented under four different conditions (arrangements of a control panel, the treatments). A total of 48 volunteers were obtained and 12 were randomly assigned to each condition. The stimulus was presented at random, on four occasions, to the volunteer and the reaction time on each occasion was recorded. [Note: the stimulus is given under only one condition to any one volunteer.]

ANOVA of Reaction Times to Stimulus

Source	df	SS	MS	EMS
Condition			816.4	
Volunteers per condition			361.5	
Repeats per volunteer			83.1	
Total	191			

(a) Give the linear model for this study, indicate what each term measures, and indicate ranges on subscripts. Assume these are the only conditions of interest.

(b) Complete the ANOVA under this model.
(c) Test that the mean reaction times are equal for the four
     conditions.
(d) Give the estimated standard deviation of the difference
     of observed mean reaction times for any two conditions.

20. A range scientist is studying the growth of fescue grass
    under three management systems.   Twelve plots were available
    for each system.   All plots were clipped to the same height
    at the beginning of the experiment.   Upon completion of the
    experiment, the growth (as measured by the height of grass)
    was measured at two randomly selected sites within each plot.
    The following partial ANOVA was obtained.

ANOVA for Growth of Fescue Grass

Source	df	SS	MS
Management Systems		101.3	
Plots(MS)	33	455.4	
Sites(P)			5.4

Total

(a) Complete the ANOVA table. Omit the EMS.
(b) Do the management systems appear to be equivalent with
     respect to mean growth?
(c) Give the standard deviation of the difference of mean
     growth for two management systems.
(d) If the observed mean growth for systems 1 and 2 were,
     respectively, 14.4 cm. and 17.3 cm., set and interpret a
     99% confidence interval for the difference in true mean
     growth for these 2 systems.

21. An engineer investigated four methods of electroplating wire
    stock.   A total of 40 uniform foot long sections were elec-
    troplated, 10 by each method. The number of flaws was counted
    per wire giving:

				Method				
1		2		3		4	
0	0	1	1	3	4	3	5
1	0	2	1	1	2	6	3
1	1	1	2	2	3	2	5
0	0	2	1	2	1	4	4
2	1	3	2	3	2	2	4

(a)   Transform the data by $\sqrt{y+1}$ and give the ANOVA table.
(b)   Test equality of mean number of flaws for the 4   methods of electroplating.

22. A plant   scientist studied 3 systems of rotation in trying to control   for   nematodes.   Twelve plots were used   with   each system.   After several years of growing the same varieties of crops on the plots,   soil samples were taken,   and the number of nematodes per sample were

	Rotation System				
1		2		3	
3	8	31	10	9	7
5	6	15	19	7	14
2	5	12	11	3	8
6	4	18	20	6	5
3	6	22	15	8	9
5	3	12	10	7	6

(a) Apply the square root transformation and set up the ANOVA table.
(b) Test equality of mean number of nematodes for the three rotation systems.

23.   An   entomologist performed an experiment to   study   bollworm control   by   5 hormones.   Bollworm larvae for   one   hybrid strain   were raised under controlled laboratory   conditions. At 15 days of age, 50 uniform larvae were each injected with a   fixed   dose of one hormone,   randomly selected   from   the five.   After 24 hours,   the number of deaths and   survivors was   recorded.   Each   hormone   was used on 4 groups of   50 larvae.   The percentages of deaths were:

		Hormone		
1	2	3	4	5
18	28	12	26	48
30	32	8	32	44
36	38	20	34	56
16	38	24	42	50

(a) Use Table A.15 to perform an arcsin   transformation prior to setting up the ANOVA table.
(b) Test equality of percentage deaths due to the 5 hormones.

24. A range scientist recorded the percentage ground cover on plots seeded with 4 drought resistant varieties of legume. Ten plots were seeded with each variety giving

Variety							
1		2		3		4	
10	10	10	5	15	20	10	15
15	15	5	10	20	20	15	5
25	15	5	10	15	20	15	10
10	20	10	5	15	20	10	10
5	20	0	5	20	15	10	15

(a) Use Table A.15 to perform an arcsin transformation prior to setting up the ANOVA table.
(b) Test equality of percentage cover for the 4 varieties.

REFERENCES

[1] Bartlett, M.S., The square-root transformation in the analysis of variance. *Suppl. Jour. Roy. Stat. Soc.*, Vol 3, 1936.
[2] Bartlett, M. S., The use of transformations. *Biometrics*, Vol. 3, 1947.
[3] Bartlett, M.S., and M.G. Kendall, The statistical analysis of variance-heterogeneity and the logarithmic transformation. *Suppl. Jour. Roy. Stat. Soc.*, Vol 8, 1946.
[4] Beall, G., The transformation from entomological field experiments so that the analysis of variance becomes applicable. *Biometrika*, Vol. 32, 1942.
[5] Box, G.E.P. Some theorems on quadratic forms applied in the study of analysis of variance problems, I: Effect of inequality of variance in the one-way classification. *Annals Math Stat*, Vol. 25, 1954.
[6] Box, G.E.P. Some theorems on quadratic forms applied in the study of analysis of variance problems, II: Effects of inequality of variance and of correlation between errors in the two-way classification. *Annals Math Stat*, Vol. 25, 1954
[7] Box, G.E.P., and D.R. Cox, An analysis of transformations. *Jour. Roy. Stat. Soc.*, Series B, Vol. 26, 1964.
[8] Cochran, W.G., Some consequences when the assumptions for the analysis of variance are not satisfied. *Biometrics*, Vol. 3, 1947.
[9] Cochran, W.G., Testing a linear relation among variances. *Biometrics*, Vol. 13, 1957.
[10] Conover, W.J., *Practical Nonparametric Statistics*, 2nd. ed., Wiley, 1980.

[11] Curtiss, J. H., On transformations used in the analysis of variance. *Annals Math. Stat.*, Vol. 14, 1943.

[12] Dixon, W.J. (Ed.), BMDP Statistical Software, 1985 Printing, Univ. of California Press, Berkeley, 1985.

[13] Eden, T., and F. Yates, On the validity of Fisher's Z-test when applied to an actual sample of non-normal data. *Jour. Agr. Sci.*, Vol.23, 1933.

[14] Eisenhart, C., The assumptions underlying the analysis of variance. *Biometrics*, Vol. 3, 1947.

[15] Freund, J. E., *Modern Elementary Statistics*, 5th. ed., Prentice-Hall, New York, 1979.

[16] Hartley, H. O., The maximum F-ratio as a short-cut test for heterogeneity of variance. *Biometrika*, Vol.37, 1950.

[17] Hey, G. B., A new method of experimental sampling illustrated on certain non-normal populations. *Biometrika*, Vol. 30, 1938.

[18] Hoaglin, D. C., F. Mosteller and J. W. Tukey (Eds.), *Understanding Robust and Exploratory Data Analysis*, Wiley, 1983.

[19] Hollander, M. and D. Wolfe, *Nonparametric Statistical Methods*. Wiley, 1973.

[20] Kempthorne, O., *The Design and Analysis of Experiments*, Wiley, New York, 1952.

[21] Kempthorne, O., The Randomization Theory of Experimental Inference. *Jour. Amer. Stat. Assoc.* Vol. 50, 1955.

[22] Lentner, M., *Introduction to Applied Statistics*, 2nd. ed., Blacksburg, VA, 1984.

[23] Mosteller, F. and J. W. Tukey, *Data Analysis and Regression*, Addison-Wesley, 1977.

[24] Pitman, E. J. G., Significance tests which may be applied to samples from any population. III. The analysis of variance test. *Biometrika*, Vol. 29, 1937.

[25] RAND Corporation, *A Million Random Digits with 100,000 Normal Deviates*, The Free Press, New York, 1955.

[26] SAS Institute Inc. *SAS® User's Guide: Statistics, Version 5 Edition.* Cary, NC: SAS Institute Inc., 1985. 956 pp.

[27] Satterthwaite, F. E., An approximate distribution of estimates of variance components. *Biometrics Bulletin*, Vol. 2, 1946.

[28] Welch, B. L., On the Z-test in randomised blocks and Latin squares. *Biometrika*, Vol. 29, 1937.

[29] Welch, B.L., The significance of the difference between two means when the population variances are unequal. *Biometrika*, Vol. 29, 1938.

[30] Welch, B.L., The generalization of 'Student's' problem when several different population variances are involved. *Biometrika*, Vol. 34, 1947.

## Chapter 3

MEAN SEPARATION PROCEDURES

3.1   Introduction

We restrict attention in this chapter to unstructured treatments
having fixed effects. Structured treatments are discussed in the
next chapter with factorial treatments presented in a separate,
later chapter.

Recall that the basic ANOVA test addresses the hypothesis of
simultaneous equality of all treatment means.   Other tests or
inferences about treatment means are possible but usually one
must calculate numerical quantities not supplied directly by the
ANOVA table.

When the ANOVA test fails to detect significant differences
among the treatment means, the experiment simply does not provide
enough evidence to claim differences among the treatments. So
without additional information, a nonsignificant test leaves the
researcher with a general conclusion that the treatment populations have essentially equivalent mean values. From a practical
standpoint, no additional inferences about treatment means can
ordinarily be made.   We might point out, however, that specific
comparisons, pre-planned before the experiment was conducted, may
be of interest.   The techniques of the next chapter may be more
appropriate for investigating such pre-planned comparisons.

When an ANOVA F test reveals significant differences among
the treatment means, it does not in any way indicate which means
differ or the magnitude of the differences.   Thus, detection of
significant differences of specific means will require some kind
of post-anova analysis.   One simple procedure would be to give a
table of estimated differences of all pairs of treatment means
and the corresponding standard deviations.   From such a table, a
reader could make his own judgment about which pairs of means are
significantly different.   This idea was introduced in Section
2.8.2.

Note that a table of mean differences and their standard deviations displays only the magnitude of differences; no significance would be attached to any of them. For a few isolated mean differences, the experimenter might wish to set a CI or make a test. Inference about a pre-planned comparison of the difference of two means, say $\mu_i - \mu_h$, can be accomplished from a t variable

$$(3.1) \qquad t = \frac{(\bar{y}_i - \bar{y}_h) - (\mu_i - \mu_h)}{\sqrt{MSE(1/r_i + 1/r_h)}}$$

having degrees of freedom equal to those of the Exp. Error mean square, MSE, from the ANOVA.

At this point, one might ask the following question, "Why not make t tests to judge significances among all pairs of mean differences?" The answer is that such t tests would not be mutually independent because they all are based on a single set of data (dependent due to the common value of MSE in their denominator.) Consequently it would be difficult to determine Type I and II error rates for the collection of such tests. Considerable research in statistics has dealt with modifications and extensions of the basic t tests in an effort to improve upon or to pinpoint the error rates. But definitive error rates seem to be the exception rather than the rule. When error rates can be specified, they frequently are not the ones of most interest.

Because the number of post-anova techniques is so extensive and because only some of the techniques have determinable error rates of interest, the subject of post-anova analyses is quite controversial. In addition, personal preferences also contribute to the controversy. Nevertheless, there is considerable demand for post-anova analyses, including a requisite for publication in some scientific journals. Without indicating preferences, we will discuss a few procedures in each of several broad classes. Chew [1] and Miller [9] may be consulted for information on additional procedures.

One class of post-anova techniques may be labeled broadly as mean separation procedures. Their function is to determine which specific means or sub-sets of means differ. The list of mean separation procedures is fairly extensive so we do not intend to give an exhaustive coverage.

One category of mean separation procedures consists of the multiple comparisons and multiple range tests. These are designed to judge significances among all pairs of mean differences. The multiple comparisons use a single critical value while multiple range procedures use two or more critical values. The "least significant difference" and "Tukey's method" are two multiple comparison procedures to be discussed. The "Student-Newman-Keuls method" and "Duncan's New Multiple Range Test "are two multiple range procedures to be discussed. Note that for k means, there are k(k-1)/2 paired comparisons investigated by any procedure in this category.

Another category of mean separation procedures investigates only the paired comparisons involving a control. Thus, in an experiment dealing with a control and k-1 other means, one would make only k-1 paired comparisons. This number, k-1, is a small fraction of all possible comparisons indicated in the preceding paragraph, unless k is small. A widely used procedure in this category is "Dunnett's procedure" which we shall discuss.

3.2   Error Rates of Mean Separation Procedures

Let us first state the Type I and Type II errors in the context of mean separation procedures. A Type I error would occur by judging a pair of means significantly different when in reality the means are equal. A Type II error would occur when a pair of means is actually different but not detected as being different. As usual, we would like both error rates to be mimimal.

For any statistical procedure, it is well-known that Type I and Type II error rates are inversely related. As one decreases, the other increases. To reduce both error rates to small values, one generally needs very large sample sizes (replications.) Thus, we seek procedures which strike some balance between the Type I and Type II error rates, unless one of the errors is more important than the other.

One can now begin to see the complexities of post-anova procedures and the reason why so many different procedures have been proposed. They have evolved because of (i) the inability to specify certain error rates (see the preceding section), (ii) the desire to balance the two error rates, (iii) the need to control only one error rate for specific problems, and so on. Continued research and new proposals for post-anova analysis indicate that researchers are still wrestling with the problem.

We will describe Type I error rates as they are applied to different mean separation procedures. These include comparison-wise, experimentwise, family, per-experiment, and rangewise error rates. The first 2 error rates are fairly straightforward. A family error rate would refer to some prescribed subset of comparisons of means (a specific family of comparisons.) The term "rangewise" will be used to represent a collection of "p-mean significance levels", to be defined later. These and other error terms, as well as related quantities, are discussed by Miller [9].

Consider all repetitions of a given experimental situation. For all comparisons judged by a mean separation procedure in these repetitions, a certain fraction are incorrectly judged significant. This fraction represents the *comparisonwise error rate*, defined by the ratio

$$(3.2) \quad \alpha_C = \frac{\text{Number of comparisons incorrectly judged significant}}{\text{Total number of comparisons}}$$

where the numerator and denominator are each enumerated for all repetitions of a given experimental situation and mean separation procedure. Thus α_C represents a long range frequency of making wrong decisions among the set of all comparisons made. An exact value of α_C cannot be specified for all procedures.

Instead of an error rate for the total number of comparisons made in all repetitions of an experiment (such as α_C), most researchers would prefer an error rate for only the comparisons made within the experiment (that is, within the single repetition the has been conducted.) This error rate, a *within experiment* error rate, is defined as

$$(3.3) \quad \alpha_W = \frac{\text{Number of comparisons incorrectly judged significant within a single experiment}}{\text{Total number of comparisons within a single experiment}}$$

Because of the dependencies stated in Section 3.1, α_W cannot be specified for most post-anova procedures except in terms of an upper bound. Over all repetitions of a given experimental situation, α_W is a random variable whose average (expected) value is called the *per-experiment* error rate, denoted by α_{PE}.

To better understand α_{PE} and to set the stage for later defining the rangewise error rate, we need to consider the notion of a "composite error rate." Suppose we could make a total of p independent tests of comparisons among a set of means. If each comparison is tested using a Type I error rate of α, then it is fairly easy to show that the chance (probability) of making one or more incorrect decisions among these p tests is

$$(3.4) \quad \alpha_p = 1 - (1-\alpha)^{p-1}$$

It is very important to remember that α_p, which we may call a *composite error rate*, is an error rate for p *independent* tests.

Example 3.1

If sets of p independent tests were made, each test at a Type I error rate of $\alpha = 0.05$, the following composite error rates of would prevail for the sets indicated.

p	α_p	p	α_p
1	.05	8	.3366
2	.0975	9	.3698
3	.1426	10	.4013
4	.1855	15	.5367
5	.2262	20	.6594
6	.2649	25	.7226
7	.3017		

Thus, for sets of 6 independent tests there is more than a 25% composite error rate; for 8 independent tests, slightly more than a 33% composite error rate; and so on. These values indicate that the composite error rates increase rapidly as the number of tests increase. ❉

It is known that α_p (3.4) serves as an *upper bound* for the composite error rate of *dependent* tests. Thus, an upper bound of α_{PE} can be obtained from (3.4). Across all repetitions of an experiment, the values of α_p fluctuate about the long range error rate of α_{PE}. Hopefully, many repetitions have small values of α_p with only a few having large values, but there is no way to assess whether this is so or not. As one sees from the values in Example 3.1, α_p could increase drastically though as the number of comparisons increases.

Another error rate that is easy to determine for many mean separation procedures is the so-called *experimentwise* error rate given by the ratio

$$(3.5) \quad \alpha_E = \frac{\text{Number of experiments in which comparisons are incorrectly judged significant}}{\text{Total number of experiments}}$$

where the "total number of experiments" would be all possible repetitions of a specific experimental situation. Note that an experiment is included in the numerator of (3.5) if any number of incorrect comparisons are detected. In this sense, experiments with one or fifty incorrect decisions are considered "equally undesirable", and each would be counted once in the numerator.

Decisions made by multiple comparison and multiple range tests are not independent but (3.4) can be used to obtain an upper bound on their composite error rates. For certain specific subsets of p means, the composite error rate will be called a p-mean significance level, which may or may not be given by (3.4). The specific subsets will be defined shortly.

The development of multiple range tests, and the accompanying rangewise error rates, is based on the following logic. The p-mean significance levels (and the composite error rate) may increase dramatically as the number of comparisons increases, see Example 3.1. This is intuitive: There are more opportunities for making erroneous decisions in larger families of comparisons. To have a "better control" of the p-mean significance level then, the total number of comparisons is not made as one single set but rather as a series of smaller subsets. Each subset should have a smaller p-mean significance level, and different subsets might have different p-mean significance levels.

When a multiple range procedure is applied to k means, the totality of k(k-1)/2 paired comparisons are divided into k-1 subsets according to the positions they occupy when all means are ordered. All adjacent pairs of means are placed in a "subset of order 2". All pairs of means separated by one other mean (and hence span 3 ordered positions) are placed in a "subset of order 3". In general, all pairs of means which span p positions in the ordered array are placed in the "subset of order p." A multiple range procedure has a p-mean significance level of α_p for the subset of order p; the value of α_p may or may not be given by (3.4). The collection of α_p (for p=2,3,...,k) form the *rangewise error rates* for a given procedure. As we discuss each procedure, we will indicate the rangewise error rates.

Example 3.2

Suppose k = 5 means denoted by A,B,C,D,E are ordered, giving the array

$$B < D < A < E < C$$

Then, the set of 5(4)/2 = 10 paired comparisons would be divided
into the following subsets of order p for various multiple range
procedures.

Order p	Comparisons			
2	B,D	D,A	A,E	E,C
3	B,A	D,E	A,C	
4	B,E	D,C		
5	B,C			

There would be rangewise error rates of α_1, α_2, α_3, and α_4. ✴

3.3 The Least Significant Difference (LSD)

For any set of k means, the LSD procedure consists of making
ordinary t tests of all k(k-1)/2 paired differences. The paired
differences are comparisons of the form $\mu_i - \mu_h$ in a CR design.
For this design then, the LSD would test

(3.6) $H_0: \mu_i = \mu_h$

individually for all i and h, with i $\neq$ h.

Known also as Fisher's LSD [5], this procedure is used only
when the basic ANOVA test shows significantly different means.
This insistence provides a better control of the error rates.
The term "protected LSD" is sometimes attached to this procedure.
An "unprotected LSD" does not require an ANOVA test of equal
treatment means; that is, an "unprotected LSD" would be made
whether or not the ANOVA test indicated significant treatment
differences. We will not consider the unprotected LSD.

Because the t tests of all possible hypotheses in (3.6) are
not independent, the α used to provide the critical value is
applicable to each individual comparison, not to the entire
collection of comparisons within any one experiment. The LSD
procedure has a comparisonwise error rate $\alpha_c = \alpha$ over all
repetitions of the experiments. For a single experiment, an
upper bound for the composite error rate is α_p given by (3.4)
with p = k(k-1)/2. See also Example 3.1.

Until we state to the contrary, we assume equal replication
of all treatments. Further we assume each null hypothesis in

(3.6) is tested against a two-sided alternative. The t statistic used in testing the null hypothesis (3.6) is the difference of the corresponding sample means divided by the standard deviation of this difference, given earlier in (2.21). For brevity, let us call this standard deviation $(s_{\bar{d}})$. To reject H_0, as stated in (3.6), and conclude that μ_i and μ_h are significantly different, the absolute difference of the observed means $|\bar{y}_i - \bar{y}_h|$ must exceed

$$(3.7) \qquad LSD = (s_{\bar{d}}) t_{\alpha/2}$$

The right hand side of (3.7) sometimes is called "the least significant difference," the critical difference for any pair of means. The quantity $t_{\alpha/2}$ is the upper $\alpha/2$ percentage point of a t variable with degrees of freedom equal to those of MSE.

Let us give another expression for the right hand side of (3.7), one that is more convenient for modification when other procedures are considered. The tabular values of t distributions are related to another collection of tabulated values, called *studentized range values*, in the following manner

$$(3.8) \qquad t_{\nu;\alpha/2} = r(2,\alpha,\nu)/\sqrt{2}$$

where $r(2,\alpha,\nu)$ is the studentized range value of order 2, level α, and for ν degrees of freedom. Table A.11 gives studentized range values. Then, we may rewrite (3.7) as

$$(3.9) \qquad LSD = (s_{\bar{d}})r(2,\alpha,\nu)/\sqrt{2}$$

where $r(2,\alpha,\nu)$ denotes the studentized range value of order 2, error rate α, and for ν degrees of freedom.

Each absolute difference $|\bar{y}_i - \bar{y}_h|$ is compared to LSD as calculated equivalently from either (3.7) or (3.9). For those differences which exceed LSD, the corresponding population means are judged to be significantly different by the LSD procedure.

The data in the next example will be used to illustrate and compare the different mean separation procedures introduced in the remainder of this chapter.

Example 3.3

Suppose the following information is available on the dry weight of a legume grown on reclaimed mine land. The treatments were different rates of fertilization. Each mean is based on six plots in a CR design.

$$\bar{y}_{1.} = 30.5 \qquad \bar{y}_{2.} = 30.3 \qquad \bar{y}_{3.} = 35.7 \qquad \bar{y}_{4.} = 33.0$$

MSE = 4.76 with ν = 20 df.   Then

$$s_{\bar{d}} = \sqrt{2(4.76)/6} = 1.26.$$

For a 5% LSD procedure, we find the studentized range value from Table A.11 to be 2.95 so that

$$LSD = 1.26(2.95)/\sqrt{2} = 2.63$$

The table of mean differences is conveniently formed as follows:

i	$\bar{y}_{i.}$	$\bar{y}_{i.}$ - 30.3	$\bar{y}_{i.}$ - 30.5	$\bar{y}_{i.}$ - 33.0
3	35.7	5.4*	5.2*	2.7*
4	33.0	2.7*	2.5	-
1	30.5	0.2	-	-
2	30.3	-	-	-

The asterisk denotes significance according to the LSD procedure. Thus, we conclude that the mean of treatment 3 differs from each of the other treatment means (row 1) and treatment mean 4 differs from treatment mean 2 (row 2), by a 5% LSD analysis. ✱

Forming the table of mean differences as we did in Example 3.3 makes the decision process easier to accomplish, especially for multiple range tests. First write down the treatment means in decreasing order. In the next column, the smallest mean is subtracted from each mean. Then in the next column, the second smallest mean is subtracted from each mean, and so on. Only the upper part of the table need be completed as the lower entries are either zero or negatives of the upper entries. When the table of differences is constructed in this manner, the lowest diagonal provides the subset of differences of order 2, the next lowest diagonal provides the subset of differences of order 3, and so on.

The studentized range table has been constructed under equal replications. The error rates depend upon this equality. Thus, when the LSD procedure is used in an experiment having unequal replication of treatments, approximate error rates would result. The degree of approximation is a function of the replication numbers. This is to say, minor deviations from equality would have lesser effects on the error rates than would more severe departures. The formulas given earlier are appropriate for equal or unequal replications. With equal replications, the standard deviation in (3.9) is constant for all pairs of means, and as a simplification, the factor $\sqrt{2}$ may be cancelled. With unequal replications though, a new standard deviation and therefore a new LSD value would need to be calculated for each new combination of unequal r's.

3.4   Tukey's Range Procedure (The HSD)

Tukey [11] believed that the LSD procedure was too liberal and should be modified by using a studentized range value of order k equal to the total number of means being compared rather than order 2 which LSD uses. Basically this implies that LSD assumes all pairs of means are adjacent whereas Tukey's procedure assumes every pair of means occurs at the end positions of the ordered array. This assumption of extreme ordering demands a large critical difference and leads to a more conservative procedure than LSD. For this reason some have labeled Tukey's method the Honestly Significant Difference or HSD method.

Tukey's method has an experimentwise error rate of size α, the value used in obtaining a tabular point. Thus, if $\alpha = 0.05$ is used in obtaining the studentized range value, the researcher would make erroneous decisions in 5% of the experiments (of the given research setting).

Tukey's procedure is the same as the LSD procedure except for the critical difference. Now the observed difference of any pair of means is compared to Tukey's critical difference given by

$$(3.10) \qquad TK = (s_{\bar{d}})r(k,\alpha,\nu)/\sqrt{2}$$

where $r(k,\alpha,\nu)$ is a studentized range value of order k, the number of means being compared. The second parameter, α, needed to get a studentized range value (Table A.11) is the experiment-wise error rate for Tukey's method. The third parameter, ν, is the degrees of freedom of MSE used in calculating the standard deviation appearing in (3.10).

Example 3.4

We now apply Tukey's procedure to the experimental data given in the previous example. With k = 4, α = 0.05, and ν = 20, the tabular range value is r(4,0.05,20)= 3.96. Then, the critical difference is, by (3.10)

$$TK = 1.26(3.96)/\sqrt{2} = 3.52$$

The table of mean differences is the same as before. We see that Tukey's procedure judges only two pairs of treatment means significantly different; namely (2,3) and (1,3). ※

For unequal replications of treatments, Tukey's procedure is used in an approximate sense, as discussed at the end of the previous section.

3.5 Student-Newman-Keuls (S-N-K) Procedure

For several reasons, the LSD and Tukey's procedures seem to be at opposite ends of the spectrum: The LSD assumes all pairs of means are ordered side-by-side (order 2) while Tukey's assumes all pairs of means have extreme ordering (order k). And, applied to the same sets of experimental data, the LSD procedure will usually judge more pairs of means significantly different than will Tukey's procedure. This disparity of significances is due to the standardized range values used by each procedure.

The use of multiple table values, and therefore multiple critical differences, was proposed independently by Newman [10] and Keuls [7]. The name Student likewise is associated with this procedure and accordingly we shall call it the S-N-K procedure. In Section 3.2 we discussed the idea of dividing the set of paired comparisons into subsets of order p. The S-N-K procedure has a different critical value for each subset. Specifically, the observed difference of each pair of means in the subset of order p is compared to the critical difference given by

(3.11) $SNK(p) = (s_{\bar{d}})r(p,\alpha,\nu)/\sqrt{2}$ for p = 2, 3, $\cdots$, k

where $r(p,\alpha,\nu)$ is a studentized range value of order p. Here, k is the total number of means being compared and ν are the degrees of freedom on MSE used in calculating the standard deviation on the right hand side of (3.11).

The S-N-K procedure, being a multiple range procedure, has rangewise error rates. The comparisons made for each subset of order p have a p-mean significance level of $\alpha_p = \alpha$, where α is the value used to obtain the various range values, $r(p, \alpha, \nu)$.

Example 3.5

We now apply the S-N-K procedure to the means for the legume experiment. From Example 3.3, $(s_{\bar{d}}) = 1.26$ with $\nu = 20$ degrees of freedom. With 4 means being compared, critical differences are needed for orders p = 2,3,4. Using $\alpha = 0.05$ and Table A.11, we obtain

$$SNK(2) = 1.26(2.95)/\sqrt{2} = 2.63$$

$$SNK(3) = 1.26(3.58)/\sqrt{2} = 3.19$$

$$SNK(4) = 1.26(3.96)/\sqrt{2} = 3.52$$

The table of mean differences given earlier is reproduced here:

		h	
i	2	1	4
3	5.4*	5.2*	2.7*
4	2.7	2.5	-
1	0.2	-	-
2	-	-	-

We compare the 3 differences on the lowest diagonal with SNK(2); the 2 differences on the second lowest diagonal with SNK(3); and finally the single difference on the top diagonal with SNK(4). We see that the S-N-K procedure judges significance for 3 pairs of means; namely, (3,2), (3,1), and (3,4). ▒

Unequal replications are handled in the S-N-K procedure as in the two previous procedures. The modifications may be more extensive here due to the different subsets of order p.

3.6   Duncan's New Multiple Range Test (NMRT)

The NMRT procedure, developed by Duncan [2] and [3], is similar to the S-N-K procedure in that it uses several range values, one for each subset of means of order p. The NMRT differs from S-N-K by using p-mean significance levels as given by (3.4) to obtain the range values for the different subsets (S-N-K used the same α for all subsets). These variable α_p are called special protection levels by Duncan and are intended to provide protection against finding false significant differences.   Because of the variables α_p's, a special table is required for the NMRT.   Duncan's new multiple range values are given in Table A.12.

To perform the NMRT, the critical differences are calculated and applied to the subsets of paired differences. Except for the special range values, the NMRT is applied in the same manner as the S-N-K procedure.   The critical differences for the NMRT for the subsets of order p = 2, 3, ---, k are given by

$$(3.12) \qquad D(p) = (s_{\bar{d}})r^*(p,\alpha_p,\nu)/\sqrt{2}$$

where $r^*(p,\alpha_p,\nu)$ denote the special range value of order p, "error rate" α_p and ν degrees of freedom. The special protection level is calculated from (3.4) and is already incorporated in the NMRT tables, such as Table A.12.

Example 3.6

We illustrate the NMRT for the legume experiment of Example 3.3. With $(s_{\bar{d}})$ = 1.26, ν = 20 and α = 5%, the critical differences are

p:	2	3	4
$r^*(p,0.05,20)$:	2.95	3.10	3.18
D(p):	2.63	2.76	2.83

The table of mean differences is

		h	
i	2	1	4
3	5.4*	5.2*	2.7*
4	2.7	2.5	-
1	0.8	-	-
2	-	-	-

Applying D(2) to the 3 differences on the lowest diagonal, D(3) to the 2 differences on the next lowest diagonal, and D(4) to the single difference (on the top diagonal), we obtain significances as indicated by asterisks. Thus, by a 5% NMRT procedure, the following pairs of means are judged significantly different: (3,2), (3,1), and (3,4). ✻

When treatments are replicated unequally, the NMRT procedure is used, in an approximate sense, as discussed for the three previous procedures. This modification was first given by Kramer [8], and verified to be a good approximation when comparing k = 3 treatment means.

3.7 Dunnett's Procedure: Comparing Treatments to a Control

Dunnett [4] introduced a procedure that is appropriate for comparing each treatment with a control or standard treatment. Only the paired comparisons involving the control treatment are tested. The error rate is on the experimentwise basis with α_E = α. Special values are required for Dunnett's procedure, these are given in Table A.13. Tabulations are given for both one and two-sided alternatives.

Each treatment mean is tested against the mean of the control by comparing the corresponding mean difference with a single critical difference specified by

$$(3.13) \qquad D = (s_{\overline{d}})d(k,\alpha,\nu)$$

where d(k,α,ν) is the special Dunnett's value from Table A.13 for making k comparisons with the control, α experimentwise error rate and ν degrees of freedom. The tabulated values in Table A.13 are based on equal replications but may be used for unequal replications of treatments, subject to approximations indicated earlier for other procedures.

Example 3.7
Let us apply a two-sided Dunnett's procedure to the variety yield data of Example 2.2 where variety 1 was a control. From earlier analyses we have MSE = 0.09 with 27 degrees of freedom,

$$\overline{y}_{1.} = 6.45, \quad \overline{y}_{2.} = 5.48, \quad \text{and } \overline{y}_{3.} = 6.70$$

(each mean based on 10 observations).

For a 5% error rate, Dunnett's critical difference is

$$D = \sqrt{2(0.09)/10}(2.33) = (0.134)2.33 = 0.31$$

The two comparisons with the control provide mean differences of:

$$\bar{y}_{2.} - \bar{y}_{1.} = 6.45 - 5.48 = 0.97$$

$$\bar{y}_{3.} - \bar{y}_{1.} = 6.45 - 6.70 = 0.25$$

Only the magnitude of the first difference exceeds the critical difference so we conclude that only the mean under variety 2 differs from the mean under the control (standard) variety. ▧

3.8 Selecting the Best Treatment and Simultaneous Confidence Intervals

For a number of experimental situations, the primary task of the researcher is to determine the "best" treatment. We shall first consider the case where the "best" treatment is the one having the largest mean value. An engineer may want to determine which mixture (of ingredients) produces the strongest concrete. An agronomist may want to determine the best yielding variety, from a group of newly formed varieties. A medical researcher may want to determine which drug is most effective in curing a particular disease.

In addition to knowing the best treatment, the researcher might like to know if any of the "other" treatments could serve as substitutes for the best one. Comparisons of the "best" and "each other" treatment mean are relevant for making such assessments. Usually there is little or no interest in any other paired comparisons of treatment means. The desired comparisons are achieved by constructing a set of *simultaneous confidence intervals*(SCI). For the current objective then, we wish to give a set of SCI for paired comparisons of the best with each other treatment mean. Let $\mu*$ denote the "best"(largest) treatment mean, then we are interested in differences of the form $\mu* - \mu_i$. We shall consider only *upper bound* SCI for these differences; that is, intervals which specify a difference no greater than the upper limit.

Hsu [6] has discussed a procedure for pinpointing the "best" treatment mean while also providing a set of SCI for paired differences involving the unknown best treatment. Specifically, the procedure gives

(i) a collection C of treatments, one of which is from the population having the "largest" mean, and

(ii) SCI for differences between the largest population   mean and each other population mean.

The   procedure has an assurance (probability) of $1 - \alpha$ that both (i) and (ii) are true for a given experiment.

From the above discussion, one notes a similarity of Hsu's method with Dunnett's comparisons involving a   control.   Indeed, if the control is the "best" treatment, the two procedures are equivalent in the sense that the treatments not in the collection C are those significantly different from the control by Dunnett's method.   As further evidence of the similarity, the same tabular values are used in both methods. We now describe Hsu's procedure.

To construct the collection containing the best treatment, the following steps are taken:

(i) corresponding to each treatment mean, $\bar{y}_{i.}$, calculate

(3.14) $$M_i = \max_{h \neq i} (\bar{y}_{h.})$$

= the largest treatment mean, not including the i-th one

and

(3.15) $$K_i = M_i - (s_{\bar{d}})d(k,\alpha,\nu)$$

where $d(k,\alpha,\nu)$ is the *one-sided* tabular value from Table A.13 with k=t-1, level α and ν degrees of freedom (on MSE.)

(ii) the i-th treatment is placed in the collection C containing the best treatment if

(3.16) $$\bar{y}_{i.} > K_i$$

To construct the SCI for the paired differences with the largest mean, we calculate the intervals $(0,D_i)$, where D_i is the greater of 0 and

(3.17) $$M_i - \bar{y}_{i.} + (s_{\bar{d}})d(k,\alpha,\nu)$$

Example 3.8

We return to the legume data of Example 3.3 to illustrate Hsu's method. From earlier work, we have

$$s_{\bar{d}} = 1.26 \quad \text{and} \quad \nu = 20$$

and for 95% assurance, the tabular value from Table A.13 is $d(3,0.05,20) = 2.03$ giving

$$(s_{\bar{d}})d(k,\alpha,\nu) = 1.26(2.03) = 2.6$$

Then, using (3.14) and (3.15)

i	$\bar{y}_{i.}$	M_i	$K_i = M_i - 2.6$
1	30.5	35.7	33.1
2	30.3	35.7	33.1
3	35.7	33.0	30.4
4	33.0	35.7	33.1

According to (3.16), the only treatment to be placed in the collection C is treatment #3.

To construct the SCI, we use (3.17) and calculate

i	$M_i - \bar{y}_{i.} + 2.6$	D_i
1	35.7 - 30.5 + 2.6 = 7.8	7.8
2	35.7 - 30.3 + 2.6 = 8.0	8.0
3	33.0 - 35.7 + 2.6 = -0.1	0
4	35.7 - 33.0 + 2.6 = 5.3	5.3

giving SCI for paired differences with the largest variety mean:

$$(0, 7.8), \quad (0, 8.0), \quad (0, 0), \quad (0, 5.3)$$

for varieties 1, 2, 3, and 4, respectively. ✱

We now consider cases where the "best" treatment is the one having the smallest mean value: (1) the environmental conditions giving the least plant damage, (2) the filtering procedure leaving the fewest impurities, (3) the diet providing the least cholesterol, and so on.   Letting $\mu**$ denote the smallest mean, the above procedure is easily modified.

To construct the collection C containing the "best" treatment (having the smallest mean), the following steps are taken:

(i) corresponding to each treatment mean, $\bar{y}_{i.}$, calculate

(3.18) $m_i = \min_{h \neq i} (\bar{y}_{h.})$

= the smallest treatment mean, not including the i-th one

and

(3.19) $k_i = m_i + (s_{\bar{d}})d(k,\alpha,\nu)$

where $d(k,\alpha,\nu)$ is the *one-sided* tabular value from Table A.13 with k=t-1, level α and ν degrees of freedom (on MSE.)

(ii) the i-th treatment is placed in the collection C containing the "best" treatment if

(3.20) $\bar{y}_{i.} < k_i$

To construct the SCI for the paired differences with the smallest mean, $\mu_i - \mu**$, we calculate the intervals $(0, D_i)$, where D_i is the greater of 0 and

(3.21) $\bar{y}_{i.} - m_i + (s_{\bar{d}})d(k,\alpha,\nu)$

PROBLEMS

1. In a CR design, a person recorded the following information:

   (i)   Each treatment had 11 observations

   (ii) $\bar{y}_{1.} = 14$ $\bar{y}_{2.} = 20$ $\bar{y}_{3.} = 12$ $\bar{y}_{4.} = 5$

   (iii) SS for treatments was 200; SS for Exp. Error was 440 SS for total was 640

   (iv) The ANOVA test of equal treatment means showed significance with $P < 0.01$.

What differences of pairs of treatment means would be judged significantly different at the 5% level by

(a)   Fisher's LSD?
(b)   Tukey's procedure?
(c)   Duncan's procedure?
(d)   Dunnett's procedure if treatment 3 is the control?

2. From an ANOVA, the following information is available:

MSE = 28.08 with 16 degrees of freedom
MST = 221.2 with 3 degrees of freedom.

Assume equal replication of treatments having means:

$$\bar{y}_{1.} = 22 \qquad \bar{y}_{2.} = 32 \qquad \bar{y}_{3.} = 16 \qquad \bar{y}_{4.} = 25$$

What differences of pairs of treatment means would be judged significantly different at the 1% level by

(a)   Fisher's LSD?
(b)   Tukey's procedure?
(c)   Duncan's procedure?
(d)   Dunnett's procedure if treatment # 1 is the control?

3. A study of growth regulators was undertaken to evaluate their effect on the diameter of a particular variety of apples. Each regulator was applied to six randomly selected trees in an orchard.   At harvest, ten apples were randomly selected from each tree and their average diameter recorded.   For the five treatments, the mean diameter were

| A | 6.86 (cm.) | B | 5.64 | C | 6.10 |
| D | 6.38 | E | 6.91 | | |

From the ANOVA, the Exp. Error mean square was MSE = 0.365.

What pairs of means would be judged significantly different, at the 5% level

(a)   by Duncan's NMRT?
(b)   by Tukey's procedure?
(c)   by Dunnett's procedure?   Assume treatment A is the control.

4. A consulting client brings the following partial ANOVA from which he has rejected the hypothesis of equal mean yields for the seven varieties.   The client now wants help in deciding which means are "significantly different".

Source	df	MS
Varieties	6	367.0
Exp. Error	35	79.6

Each variety occurred on six plots with the following means:

A	49.6		E	71.3
B	71.2		F	58.1
C	67.6		G	61.0
D	61.5			

Judge all pairs of variety means for significance at the   5% level by

(a) Tukey's procedure
(b) Duncan's procedure
(c) Dunnett's procedure if treatment G is the control.

5. Refer to Problem 3 above and use Hsu's method to specify a subset of growth regulators containing the population having the greatest mean diameter.   Also give the SCI of the other treatment means with the greatest.   Use a coverage assurance of 0.95.

6. Refer to Problem 4 above and use Hsu's method to specify a subset of the varieties containing the population having the greatest mean yield. Also give the SCI of the other treatment means with the greatest. Use a coverage assurance of 0.95.

7. A nutrition specialist is measuring the amount of a fatty acid present in blood when four different diets are used.   Forty subjects (uniform with respect to the fatty acid) were obtained and 10 were randomly assigned to each diet. After a 3 week period, the following was recorded:

	Diet (i)			
	1	2	3	4
Mean $\bar{y}_{i.}$	21.3	28.6	25.7	20.4

From the ANOVA, Exp. Error MS was 11.91 with 36 degrees of freedom. (See also Problem 9, Chapter 2.)

(a) Use Dunnett's method to judge significant differences of treatment means from the control(assumed to be Diet # 1) Let $\alpha = 0.05$.

(b)   Use Hsu's method to specify a subset containing the smallest fatty acid mean and the set of SCI of the other treatment means with the smallest. Use a coverage assurance of 0.95

8. In a greenhouse study, a researcher studied the effect of different types of light on the growth of tobacco plants, (see also Problem 10, Chapter 2). From the ANOVA, MSE = 0.48 with 45 degrees of freedom.   The mean stem growths (in mm.) were:

Natural	4.09	Blue	3.01
White	4.30	Green	3.19
Yellow	3.55		

Use Dunnett's method to judge significance of differences of treatment means with the control (natural light.) Let α = 0.05.

9. The amounts of impurities remaining in commercial grade alcohol was measured for each of 8 samples taken randomly under each of 4 filtration methods.   The means, in mg/100cc, were:

		Method		
1	2	3	4
14.875	15.988	15.862	16.05

The Exp. Error MS from the ANOVA was 0.19.   The ANOVA showed significant differences among treatment means with P < 0.01.

Analyze for significant differences of pairs of treatment means by

(a) Tukey's procedure with α = 0.05.
(b) Duncan's NMRT with α = 0.05.
(c) Dunnett's procedure with α = 0.05.   Assume treatment 1 is the control (standard) treatment.

10. An engineer, studying the strength of steel wire stock, made with different percentages of carbon, recorded the following treatment means, each based on 5 pieces of wire:

$$\bar{y}_{1.} = 193 \quad \bar{y}_{2.} = 214 \quad \bar{y}_{3.} = 236$$

$$\bar{y}_{4.} = 241 \quad \bar{y}_{5.} = 252 \quad \bar{y}_{6.} = 248$$

The Exp. Error MS from the ANOVA was 458.6.   The ANOVA showed significant differences among treatment means with P < 0.025.

Analyze for significant differences of pairs of means by

(a) Tukey's procedure with $\alpha = 0.05$.
(b) Duncan's NMRT with $\alpha = 0.05$.
(c) Dunnett's procedure with $\alpha = 0.05$. Assume treatment 1 is the control (standard) treatment.

11. Refer to Problem 9 above and use Hsu's method to specify a subset of filtration methods containing the population with the least mean impurities. Also give the SCI of the other treatment means with the least. Use a coverage assurance of 0.95.

12. Refer to Problem 10 above and use Hsu's method to specify a subset of carbon percentages containing the population with the greatest strength. Also give the SCI of the other treatment means with the greatest. Use a coverage assurance of 0.95.

13. A horticulture graduate student investigated the yield of a variety of sweet potatoes under 4 levels of fertilizer. The results were:

$$\bar{y}_{1.} = 136 \qquad \bar{y}_{2.} = 188 \qquad \bar{y}_{3.} = 156 \qquad \bar{y}_{4.} = 163$$

Each mean is based on the results of 8 plots. The Exp. Error MS from the ANOVA was 643. The ANOVA test of equal means was significant with $P < 0.01$.

Analyze for significant differences of pairs of means by

(a) Fisher's LSD procedure with $\alpha = 0.05$.
(b) Dunnett's procedure with $\alpha = 0.05$ if treatment 1 is the control.

14. The animal scientist, Problem 6, Chapter 2, recorded the following mean wool lengths for the 5 breeds of sheep:

$$\bar{y}_{1.} = 1.94; \ \bar{y}_{2.} = 2.63; \ \bar{y}_{3.} = 3.52; \ \bar{y}_{4.} = 3.14; \ \bar{y}_{5.} = 2.71$$

MSE $= 0.54$ with 55 degrees of freedom

Analyze for significant differences of pairs of means by Duncan's NMRT. Let $\alpha = 0.05$.

15. The following was given in Problem 7, Chapter 2, for the transpiration of tobacco plants of 5 varieties:

Variety (i)	1	2	3	4	5
r_i	10	12	9	12	9
$\bar{y}_{i.}$	8.7	9.3	7.4	11.1	9.6

MSE = 7.1 based on 47 df.

Judge differences of all pairs of means for significances by Duncan's NMRT. Let $\alpha = 0.05$.

16. The following was given in Problem 8, Chapter 2, for refunds of federal tax returns of three groups of filers:

	Single	Married,Jointly	Married,Separately
r_i	100	200	50
$\bar{y}_{i.}$	186	523	214

SSY = 71,291,000

Judge differences of all pairs of means for significances by Duncan's NMRT.

17. The following was given for water filtration systems in Problem 13, Chapter 2:

System(i)	1	2	3	4
$\bar{y}_{i.}$	15.78	9.65	8.16	10.06

MSE = 0.601.

Analyze differences of pairs of treatment means by

(a) Fisher's LSD procedure with $\alpha = 0.05$.
(b) the S-N-K procedure with $\alpha = 0.05$.

18. Refer to Problem 14, Chapter 2 and analyze differences of pairs of treatment means by

(a) Fisher's LSD procedure with $\alpha = 0.05$.
(b) the S-N-K procedure with $\alpha = 0.05$.

19. The following information is from Problem 17, Chapter 2, for the effect of sound intensities on the behavior patterns of monkeys:

Intensity (i)	1	2	3	4
r_i	6	5	4	5
$\bar{y}_{i.}$	6.48	8.34	11.62	14.28

MSE = 0.75 based on 16 df.

Analyze differences of pairs of means for significance by

(a) the LSD procedure with $\alpha = 0.01$.
(b) Duncan's NMRT procedure with $\alpha = 0.01$.

20. Refer to Problem 18, Chapter 2, for data on an emission study of exhaust systems. Analyze differences of all pairs of means for significance by

(a) the LSD procedure with $\alpha = 0.01$.
(b) Duncan's NMRT procedure with $\alpha = 0.01$.

21. Refer to Problem 19 above and use Hsu's method to specify a subset of intensities containing the population having the greatest mean time.   Also give SCI of the other treatment means with the greatest. Use a coverage assurance of 0.95.

22. Refer to Problem 20 above and use Hsu's method to specify a subset of exhaust systems containing the population having the least mean pollutants in the air.   Also give SCI of the other treatment means with the least. Use a coverage assur- ance of 0.95.

23. The human factors engineer Problem 19, Chapter 2, recorded the following mean reaction times:

Condition (i)	1	2	3	4
$\bar{y}_{i..}$	17.72	22.50	15.42	24.03

MS for Volunteers (Condition) = 361.5
MS for Repeats (Volunteer) = 83.1
r = 12 volunteers per condition
s = 4 repeats per volunteer

Analyze differences of all pairs of condition means for significance by

(a) Tukey's procedure with $\alpha = 0.05$.
(b) Duncan's NMRT with $\alpha = 0.05$.

24. The range scientist, Problem 20, Chapter 2, recorded the following mean growths (in cm):

Management System	1	2	3
$\bar{y}_{i..}$	14.4	17.3	15.7

MS for Plots (System) = 13.8
MS for Sites (Plot) = 5.4
r = 12 plots per system
s = 2 sites per plot

Analyze differences of all pairs of system means for significance by

(a) Tukey's procedure with α = 0.05.
(b) Duncan's NMRT with α = 0.05.

25. Refer to Problem 21, Chapter 2, which contains data on electroplating wire stock. Analyze differences of all pairs of method means for significance by the LSD procedure with α = 0.05. Hint: Use the transformed data.

26. Refer to Problem 22, Chapter 2, which contains data on rotation systems for control of nematodes. Analyze differences of all pairs of system means by the LSD procedure with α = 0.05. Hint: Use the transformed data.

27. Refer to Problem 23, Chapter 2, which contains data on percentage deaths of bollworms under 5 hormones. Analyze differences of all pairs of hormone means by Duncan's NMRT with α = 0.05. Hint: Use the transformed data.

28. Refer to Problem 24, Chapter 2, which contains data on percentage ground cover under 4 varieties of a legume. Analyze differences of all pairs of variety means by Duncan's NMRT with α = 0.05. Hint: Use the transformed data.

REFERENCES

[1] Chew, V., "Comparisons Among Treatment Means in Analysis of Variance," U. S. D. A., Technical Bulletin, ARS/H/6, 1977.
[2] Duncan, D. B., A Significance Test for Differences between Ranked Treatments in an Analysis of Variance," *Virginia Jour. Sci.*, Vol. 2, 1951.

[3] Duncan, D. B., Multiple range and multiple F Tests, *Biometrics*, Vol. 11, 1955.

[4] Dunnett, C. W., A Multiple comparisons procedure for comparing several treatments with a control, *Jour. Amer. Stat. Assoc.*, Vol. 50, 1955.

[5] Fisher, R. A., *The Design of Experiments*, Oliver and Boyd, Ltd., Edinburgh and London, 1935.

[6] Hsu, J. C., Simultaneous confidence intervals for all distances from the 'best', *Annals of Stat.*, Vol. 9, 1981.

[7] Keuls, M., The use of the 'studentized range' in connection with the analysis of variance, *Euphytica*, Vol. 1, 1952.

[8] Kramer, C. Y., Extension of multiple range tests to group means with unequal numbers of replications, *Biometrics*, Vol. 12, 1956.

[9] Miller, R. G., Jr., *Simultaneous Statistical Inference*, 2nd. ed., Springer-Verlag, New York, 1981.

[10] Newman, D., The distribution of the range in samples from a normal population, expressed in terms of an independent estimate of standard deviation, *Biometrika*, Vol. 31, 1939.

[11] Tukey, J. W., Comparing individual means in the analysis of variance, *Biometrics*, Vol. 5, 1949.

## Chapter 4

INFERENCES ABOUT STRUCTURED MEANS

4.1   Introduction

In this chapter we consider various inferences which are suitable for structured means; that is, means having some relationship that can be identified.   Major focus will be given to gradient and group structures;   factorial structures will be covered in detail in the next chapter.

The mean separation procedures of Chapter 3, discussed for unstructured(unrelated) means, might be applied in some instances to structured means.   It depends upon the objectives that the researcher wishes to address.   As we will see though, many of the procedures discussed in this chapter look at special aspects of the treatments,   aspects which the mean separation procedures are unable to address.   Therefore,   these new procedures are usually more appropriate than mean separation procedures for post-anova inferences of structured means.   Needless to say,   procedures which require a particular treatment structure are not applicable to unstructured means;   this will become apparent as we continue our discussion.

Before we present the various procedures,   let us briefly review some of the treatment structures,   see Section 1.6. If the treatments differ only by increasing (or decreasing) amounts of a factor,   they have a gradient structure.   Thus, an engineer might use steel having 1%,   2% and 3% carbon.   A biologist might use temperatures of 10°C, 15°C, and 20°C during the larval stage of a certain insect.   For the engineer,   the treatments differ only in the percentages of carbon present. Those for the biologist differ only in the amounts of heat applied.   While we have illustrated equally-spaced increments,   this is by no means necessary.   Equal spacing,   whenever possible, does provide certain simplifications as we will show shortly.

Treatments having group structure may be described simply as two or more sets of related treatments.   The relationship from set to set may be weak or non-existent.   As an example, suppose a home economist is studying various properties of cakes baked by 7 commercial mixes (box mixes from supermarket shelves).   Suppose 3 of the cake mixes call for the addition of buttermilk, 2 call for addition of whole milk and the remaining 2 call for addition of water. The seven mixes form 3 groups of treatments. In addition to making comparisons within each group, the home economist may want to make comparisons of (i) one treatment in one group with one treatment in a second group, (ii) all treatments in one group with all treatments in a second group, and so on. Notice that some of these comparisons could be made with mean separation procedures while others (the last one listed above, for example) cannot.

Experiments dealing with group structured treatments are quite common. An agronomist might study yields under groups of liquid and granular fertilizers. A business researcher might study investment characteristics for the 3 groups: National banks, state banks, and savings and loan companies. An engineer might study energy provided by groups of liquid and solid-state fuels.

There are some experiments whose treatments are related but not in a clear-cut manner. Some elements of group and gradient structure may be present but not so totally. For example, suppose a standard item (a tool, a manufacturing machine, an insecticide, and so on) has been widely used but a number of alternatives are now available from various competitors. Each alternative item undoubtedly has certain desirable and undesirable features. Some features may be common to several alternatives. Conceivably, different features could lead to different groupings among the treatments, and in turn, to different sets of comparisons. The following example gives an illustration of this.

Example 4.1

Suppose a textile manufacturer is considering four new machines which might be replacements for their standard machine. The machines are described as follows:

Machine	Features
S	(standard) roller bearings, manual oiling
A	roller bearings, automatic oiling
B	brass bushings, manual oiling
C	roller bearings, sealed
D	nylon bushing, automatic oiling

From a study of the features, we could set up 3 groups on the basis of the "Type of bearing/bushing", ignoring the oiling system, giving (S,A,C), B, and D. But if we grouped according to "oiling system" only, we would get 3 groups (S,B), (A,D), and C. So we see that the treatments have some structure though not so clear-cut that groups can be uniquely or easily defined. ✠

Finally, it should be apparent that an experiment could have treatments with a combination of structures. As a simple example, the agronomist mentioned earlier might have a gradient structure on each group of liquid and granular fertilizers. Various other combinations will be encountered in later examples and problems.

4.2   Contrasts of Means

Mean separation procedures dealt with very basic comparisons of means; namely, the difference of a pair. With unstructured treatments there really isn't any other meaningful comparisons that could be considered. Now with structured means we are not restricted to the basic paired comparisons, we may construct more general comparisons as dictated by our treatment structure and research objectives.

We begin with a definition. For simplicity of notation, suppose we are interested in a set of k means denoted simply by $\mu_1, \mu_2, \ldots, \mu_k$.

Definition 4.1

A contrast is a linear combination of means whose coefficients sum to zero. Thus, for constants $c_1, c_2, \ldots, c_k$, not all zero,

(4.1) $$\theta = \sum_i c_i \mu_i$$

is a contrast if $\sum_i c_i = 0$.

It should be apparent that the right hand side of (4.1) will generate numerous comparisons. With only two nonzero coefficients equal to +1 and -1, the contrast is a paired difference of means. All paired differences of means can be generated if all possible assignments of +1 and -1 are made among $c_1, c_2, \ldots, c_k$. But this is only the beginning, for we can consider such contrasts as

$$\theta_1 = \mu_1 + \mu_2 - 2\mu_3$$

$$\theta_2 = \mu_1 + 2\mu_2 - 3\mu_4$$

$$\theta_3 = \mu_2 + \mu_3 - \mu_4 - \mu_6$$

(4.2)

$$\theta_4 = 3\mu_1 - \mu_2 - \mu_3 - \mu_4$$

$$\theta_5 = 3\mu_1 + 3\mu_2 - 2\mu_3 - 2\mu_4 - 2\mu_5$$

$$\theta_6 = \mu_1 - 0.6\mu_2 - 0.4\mu_5$$

to list a few.   From these contrasts one sees that the potential is unlimited.   Which contrasts are considered in any experimental situation   depend on the structure of the means and the   research objectives.

It is known for statistical theory that each contrast has one degree   of freedom.   This is so because any contrast represents a single (linear) aspect of the means,   a "one-dimensional piece of information   about   the total set of means." More   will   be   said about this later.

To   gain   some insight into the nature of contrasts,   let   us consider   a test of hypothesis of a contrast.   Usually the   null hypothesis states that   the contrast has a zero value.   Thus,   a test about θ_4 given in (4.2) would be

(4.3) $H_0: 3\mu_1 - \mu_2 - \mu_3 - \mu_4 = 0$

which can be rewritten in the following equivalent form

(4.4) $H_0: \mu_1 = (\mu_2 + \mu_3 + \mu_4)/3$

From this last hypothesis one sees that a researcher would use θ_4 to   compare the mean under treatment 1 with "the average   of   the means under treatments 2, 3, and 4." This contrast is of interest when   treatment   1 is a control---the researcher wants to see   if the control treatment has the same effect as the other treatments taken collectively.

For   the contrast θ_5 given in (4.2),   an analysis of H_0 would imply a comparison of

(4.5) $\dfrac{1}{2}(\mu_1 + \mu_2) = \dfrac{1}{3}(\mu_3 + \mu_4 + \mu_5)$

and would be appropriate for comparing "overall averages" of two groups, one group composed of treatments 1 and 2, the second group composed of treatments 3, 4 and 5.

Example 4.2

Let us return to the textile machines described in Example 4.1. For the sake of discussion, suppose that the variable of interest is hours of maintenance-free operation and that information can be obtained from several machines of each type. Some contrasts of interest might be the following:

$$\theta_1 = 4\mu_S - \mu_A - \mu_B - \mu_C - \mu_D$$

$$\theta_2 = \mu_S - \mu_A$$

$$\theta_3 = 2\mu_S + 2\mu_A + 2\mu_C - 3\mu_B - 3\mu_D$$

Contrast θ_1 compares the standard machine with "the rest"; that is, the average maintenance time for the standard machine versus the average maintenance time of all other machines. Contrast θ_2 compares the mean for the standard machine with the mean for machines of type A --- to see if the automatic oiling system has any effect. This contrast has the most meaning when the type A and S machines are identical except for the oiling system. And contrast θ_3 is a comparison of the overall mean for machines with roller bearings with the overall mean for machines not having roller bearings. This contrast is of interest if the oiling system is believed to have no effect (or a very minimal effect relative to the type of bearings). ❋

Example 4.3

A home economist is interested in determining the effect of baking temperature on the moisture of cakes. She plans to use a High and a Low temperature in two different brands of oven, say A and B. There are four treatments: AH, AL, BH, BL, with means μ_1, μ_2, μ_3, and μ_4, respectively.

Suppose she is particularly interested in comparing the results of High and Low temperatures in the Brand A oven (a new model, perhaps). Thus, she is interested in a contrast

$$\theta_1 = \mu_1 - \mu_2.$$

Likewise, for comparing Brand A and Brand B ovens, ignoring temperatures, she would be interested in a contrast

$$\theta_2 = \frac{1}{2}[(\mu_1 + \mu_2) - (\mu_3 + \mu_4)].$$ ✶

Hypothesis testing of contrasts has been mentioned already. Confidence intervals also can be constructed for contrasts. We assume the "usual ANOVA assumptions" so that an estimate of Exp. Error variability, MSE, is available from an ANOVA table. An unbiased estimator of a contrast $\theta_1 = \sum_i c_{i1}\mu_i$ is given by the same linear combination of sample means; namely,

(4.6) $C_1 = \sum_i c_{i1}\bar{y}_i.$

Both C_1 and θ_1 will be called contrasts (one of sample means, the other of population means). With the independence of the $\bar{y}_i.$ and assumption of equal variances, the contrast C_1 has an estimated variance of

(4.7) $s_{C_1}^2 = \sum_i c_{i1}^2 (MSE/r_i) = MSE \sum_i (c_{i1}^2/r_i)$

where r_i is the number of observations upon which $\bar{y}_i.$ is based.

Under the assumption of normality, each $\bar{y}_i.$ is a normal variable (Theorem 1.1). Then by Theorem 1.2, any contrast of the $\bar{y}_i.$ is likewise a normal variable because any contrast is a linear combination of means. The contrast C_1 can be converted to a Student t variable

(4.8) $t_\nu = (C_1 - \theta_1)/s_{C_1}$

where the degrees of freedom are ν, those of MSE. This t variable can be used for making a test or setting a CI about θ_1.

A two-sided test about a contrast θ_1 can be made using the t variable in (4.8) or using an F variable because of the relationship $t_\nu^2 = F_{1,\nu}$. For the hypothesis $H_0: \theta_1 = 0$, the F statistic may be obtained from (4.8) upon replacing θ_1 by zero (under H_0) and squaring:

(4.9) $F_{1,\nu} = C_1^2/s_{C_1}^2$

$$= C_1^2/\text{MSE} \sum_i (c_{i1}^2/r_i)$$

In the context of ANOVA tests, an F statistic is the ratio of 2 mean squares. In the current setting we must have

(4.10)   F = [MS for contrast C_1]/MSE = [SS for C_1]/MSE

the last expression resulting from the fact that any contrast has one degree of freedom. Matching terms of (4.9) and (4.10) leads to the following general expression for the sum of squares of a contrast:

(4.11) $SS[C_1] = C_1^2/\sum_i (c_{i1}^2/r_i)$

Note that expression (4.11) gives a simple way of calculating the sum of squares for a contrast. One merely needs the estimated value, C_1, the coefficients (the c_{i1}), and r_i, the replication numbers.

Example 4.4

Suppose the home economist of the previous example baked r = 10 cakes under each treatment, made a moisture determination on each, and obtained the following treatment means

AL	5.6	BL	5.0
AH	4.4	BH	4.2

From an ANOVA, residual variation was MSE = 1.5 with 36 degrees of freedom.
       We will set a 95% CI for $\theta_1 = \mu_1 - \mu_2$, a comparison of Low and High temperatures for Brand A ovens. An estimate is

$$C_1 = \bar{y}_1. - \bar{y}_2. = 5.6 - 4.4 = 1.2$$

with an estimated standard deviation, using (4.7), equal to

$$s_{C_1} = \sqrt{1.5(1/10+1/10)} = 0.55$$

The 97.5 percentile of t_{36} is 2.03; the confidence limits are

$$1.2 \pm 0.55(2.03) = 1.2 \pm 1.12$$

One can be 95% confident that the difference in mean moisture for these two temperatures is between 0.08 and 2.32.
    A test about equal mean moistures for the two brands of oven (see contrast θ_2 in the previous example) would be a test of

$$H_0: \frac{1}{2}(\mu_1 + \mu_2) = \frac{1}{2}(\mu_3 + \mu_4)$$

From the contrast estimate

$$C_1 = (\bar{y}_{1.} + \bar{y}_{2.} - \bar{y}_{3.} - \bar{y}_{4.})/2$$
$$= (5.6 + 4.4 - 5.0 - 4.2)/2 = 0.4$$

we obtain sums of squares of

$$SS[C_1] = (0.4)^2/[1/2^2(10) + 1/2^2(10) + 1/2^2(10) + 1/2^2(10)]$$
$$= 0.16/0.1 = 1.6$$

The calculated value of F is, by (4.10)

$$F_{1,36} = 1.6/1.5 = 1.07$$

which gives an observed significance level of $P > 0.10$.   Hence, there appears to be little evidence that the two brands of ovens give different overall means of moisture. ▓

4.3   Orthogonal Contrasts

There are infinitely many contrasts that one can construct for any k means.   This is easy to see for we can multiply each given contrast by any nonzero constant. The simple contrast $\theta = \mu_1 - \mu_2$ can be multiplied by nonzero constants to give " new contrasts"

$$2\theta = 2\mu_1 - 2\mu_2$$

$$0.5\theta = 0.5\mu_1 - 0.5\mu_2$$

$$8.61\theta = 8.61\mu_1 - 8.61\mu_2$$

$$217\theta = 217\mu_1 - 217\mu_2$$

and so on. Clearly the values of the estimates change as we change multipliers, and it can be shown that the confidence limits change accordingly by the same constant multiple.

But what about tests of hypotheses? It shouldn't really matter whether we test

$$H_0: \theta = \mu_1 - \mu_2 = 0$$

or

$$H_0: b\theta = b\mu_1 - b\mu_2 = 0$$

for any constant $b \neq 0$. For all of these tests to be equivalent, the F statistics would have to be identical (the degrees of freedom and critical values certainly are). And identical test statistics imply that "contrasts which differ only by a constant multiple have the same sum of squares". Let us show this. If we begin with a general contrast $C = \sum_i c_i \bar{y}_i.$ and multiply by a constant b, we get

(4.12) $$C* = bC = \sum_i (bc_i)\bar{y}_i. = \sum_i c_i^* \bar{y}_i.$$

then the sum of squares for C* is, by (4.11)

(4.13) $$SS[C*] = (C*)^2/\sum_i \{(c_i^*)^2/r_i\}$$

$$= (bC)^2/\sum_i \{(bc_i)^2/r_i\}$$

$$= C^2/\sum_i (c_i^2/r_i) = SS[C]$$

This is a very important and useful property of contrasts for it gives us considerable flexibility in calculating sums of squares as well as eliminating the need to consider "contrasts related by constant multiples". We note a certain ambiguity among contrasts which differ only by constant multiples. For this reason, some people prefer to define contrasts with an additional condition imposed; namely, $\Sigma c_i^2 = 1$. This provides some sort of uniqueness of contrasts but might result in "unpleasant" coefficients.

Example 4.5

In the previous example we calculated the sum of squares for the the contrast $C_2 = (\bar{y}_1. + \bar{y}_2. - \bar{y}_3. - \bar{y}_4.)/2$ when testing

$$H_0: \frac{1}{2}(\mu_1 + \mu_2) = \frac{1}{2}(\mu_3 + \mu_4)$$

We found $SS[C_2] = 1.6$. Let us calculate the sum of squares for

$$C^* = 2C_2 = \bar{y}_1. + \bar{y}_2. - \bar{y}_3. - \bar{y}_4.$$

$$= 5.6 + 4.4 - 5.0 - 4.2 = 0.8$$

From (4.11), $SS[C^*] = \dfrac{(0.8)^2}{1/10 + 1/10 + 1/10 + 1/10}$

$$= 0.64/0.4 = 1.6 = SS[C].\ \text{\rlap{\char"25}}$$

While the above idea has eliminated the need to consider contrasts that differ only by constant multiples, there still are many contrasts which might be considered. An indication of the many patterns appears in (4.2). Most of the contrasts that we write down will contain "overlapping information" to some degree. Ideally we would like each contrast to address a different and "unrelated" aspect of the treatments, in other words, to provide non-overlapping information. We will now consider contrasts that are unrelated in a very specific manner. To set the stage for a definition, let us denote two general contrasts by

(4.14) $C_p = \sum_i c_{ip} \bar{y}_i.$ and $C_q = \sum_i c_{iq} \bar{y}_i.$

Definition 4.2

Two contrasts, C_p and C_q, are said to be *orthogonal* if

(4.15) $$\sum_i c_{ip} c_{iq}/r_i = 0$$

or when all r_i are equal, if

(4.16) $$\sum_i c_{ip} c_{iq} = 0.$$

Example 4.6

Two contrasts considered by the home economist, Example 4.4, were

$$C_1 = \bar{y}_{1.} - \bar{y}_{2.} \quad \text{and} \quad C_2 = (\bar{y}_{1.} + \bar{y}_{2.} - \bar{y}_{3.} - \bar{y}_{4.})/2.$$

Because each $r_i = 10$, we may use (4.16) to check whether C_1 and C_2 are orthogonal.

$$\sum_i c_{i1} c_{i2} = (1)(1) + (-1)(1) + (0)(-1) + (0)(-1) = 0$$

Thus, C_1 and C_2 are orthogonal.   We can see that C_1 and C_2 provide   non-overlapping information because C_1 is a   comparison within Brand A ovens while C_2 is a comparison between Brand A and Brand B ovens.   Neither comparison provides any information about the other. ▓

One of the major uses of orthogonal contrasts is to subdivide treatment   sums of squares,   say SST,   into components which   are additive(whose sums of squares add to SST.) Assuming that SST has t-1   degrees of freedom,   we would need t-1 orthogonal   contrasts for   the required additivity,   because each contrast has 1 degree of freedom. We now define additive sets of contrasts.

Definition 4.3

Any set of t-1 orthogonal contrasts whose sums of squares add to   the   treatment sum of squares will be called   a   *complete orthogonal set* of treatment contrasts.

As Definition 4.3 implies, there is not a "unique" complete orthogonal sets of contrasts among the treatment means. In fact, when treatments have 2 or more degrees of freedom, there are always 2 or more complete orthogonal sets of treatment contrasts.

Example 4.7

Let us return to the home economics experiment again. There are four treatments whereby a complete orthogonal set of treatment contrasts will consist of 3 contrasts. Earlier we defined two orthogonal contrasts

$$c_1 = \bar{y}_{1.} - \bar{y}_{2.}$$

$$c_2 = (\bar{y}_{1.} + \bar{y}_{2.} - \bar{y}_{3.} - \bar{y}_{4.})/2$$

and it is fairly easy to show, using (4.6) and (4.16), that the third contrast for a complete orthogonal set is

$$c_3 = \bar{y}_{3.} - \bar{y}_{4.}$$

From the treatment totals given in Example 4.4, the treatment SS is calculated as

$$SST = [(56)^2 + (44)^2 + (50)^2 + (42)^2]/10 - (192)^2/40$$

$$= 933.6 - 921.6 = 12.$$

Earlier we found $SS[c_1] = 1.6$. For c_2 and c_3 the sums of squares are

$$SS[c_2] = (5.6 - 4.4)^2/[1/10 + 1/10]$$

$$= 1.44/0.2 = 7.2$$

$$SS[c_3] = (5.0 - 4.2)^2/[1/10 + 1/10]$$

$$= 0.64/0.2 = 3.2$$

and we see that $SST = 1.6 + 7.2 + 3.2$. ⁂

4.4.1   Some Comments About Orthogonal Contrasts

Quite often there are several complete orthogonal sets which might be constructed. The contrasts in some of the complete sets may be more meaningful than those in other sets. In fact when we insist that certain contrasts be members of a complete orthogonal set, some contrasts needed "to complete the set" may have obscure meaning. This happens mostly with large sets or when structure of the means is considered "weak", as with the textile machines of Example 4.1. Contrasts of two groups provide the most meaningful information when the groups differ by only one factor.

Example 4.8

Let us consider a complete set of orthogonal contrasts for the textile machines of Example 4.1. At the outset, suppose the researcher insists on contrasts of "the standard machine versus the rest" and "roller bearings versus non-roller bearings, on new machines". Then one complete orthogonal set is:

| | | Coefficients for Machines | | | |
Contrast	S	A	B	C	D
1	4	-1	-1	-1	-1
2	0	1	-1	1	-1
3	0	1	0	-1	0
4	0	0	1	0	-1

The third contrast may be described as "roller, automatic oiling versus roller, manual oiling" which represents a comparison worth considering. But the fourth contrast is difficult to interpret; it is simply a "paired comparison".

Note also that the usefulness of the second contrast might be questioned because the two groups being compared have several differing factors. ✳

Orthogonal contrasts have two minor advantages over arbitrary sets of contrasts, including paired differences of the mean separation procedures. Sums of squares for orthogonal contrasts are additive; that is, sums of squares for a complete orthogonal set of treatment contrasts add to the treatment sum of squares. And, the composite error rate for a set of orthogonal contrasts should be lower than for an arbitrary set of contrasts because the number of orthogonal contrasts is usually smaller than the number of contrasts considered in an arbitrary set. But the exact value of the composite error rate cannot be specified even for tests of orthogonal contrasts because these tests are not independent. All F statistics of orthogonal contrasts have the same denominator (namely MSE) which causes the dependency among the tests. Therefore, any appreciable advantage of orthogonal

contrasts must come from considering fewer contrasts. See Section
3.2 for a discussion of error rates; these error rates apply to
both orthogonal and nonorthogonal contrasts.

In view of the previous remarks, one shouldn't feel compelled
to consider only orthogonal contrasts. In fact, in quite a few
experiments, nonorthogonal contrasts may be required to address
the research objectives. In other experiments some orthogonal
and some nonorthogonal contrasts might be utilized jointly to
address research objectives. The composite error rate undoubtedly
increases with the addition of nonorthogonal contrasts, but with
only a few, the effect should not be too great. The danger with
nonorthogonal contrasts, however, is that they may be addressing
the same unusual feature of the data. Insisting on orthogonality
provides some sort of protection against this "duplication."

Example 4.9

Suppose the home economist (see Example 4.3) is interested in
comparing "oven A versus oven B", "low versus high temperature,
in oven A" and "oven A versus oven B, at the high temperature".
We have

Contrast	AH	Coefficients for AL	BH	BL
1	+1	+1	-1	-1
2	+1	-1	0	0
3	+1	0	-1	0

Contrasts 1 and 2 are orthogonal to each other but the third one
is not orthogonal to either of the first two. But if these
three contrasts represent aspects of most interest to the home
economist, there is no reason why these three contrasts shouldn't
be considered. ✖

4.4.2 A Special Set of Orthogonal Contrasts

Any sum of squares can be calculated from a set of orthogonal
contrasts. In many cases, a specific set of contrasts would be
used, as dictated by some underlying structure of the data. For
certain cases, the researcher has some flexibility in choosing
the contrasts to use. This would be true for certain incomplete
block designs and factorial experiments where the primary goal is
the calculation of specific sums of squares, not in the interpretation of individual contrasts.

In the remainder of our discussion here, we suppose that a complete orthogonal set of contrasts is required. When only a subset (of a complete set) is required, the following procedure would be modified accordingly. For simplicity let the desired sum of squares have m degrees of freedom, and be based on m+1 means. Then, one may easily construct the coefficients for a complete set of orthogonal contrasts as shown in Table 4.1.

Table 4.1   Orthogonal Contrast Coefficients

	1	2	3	Means 4	5	...	m	m+1
Contrast								
1	+1	-1	0	0	0	...	0	0
2	+1	+1	-2	0		...	0	0
3	+1	+1	+1	-3	0	...	0	0
.								
.								
.								
m-1	+1	+1	+1	+1	+1	...	-(m-1)	0
m	+1	+1	+1	+1	+1	...	+1	-m

Thus, for 2 degrees of freedom, the first 2 rows (and 3 columns) would be used. For 3 degrees of freedom, the first 3 rows (and 4 columns.) And so on.

Example 4.10

For comparative purposes, let us calculate the treatment sum of squares for the home economics experiment, using Table 4.1. From Example 4.7, the 4 means (each based on 10 observations) were: 5.6, 4.4, 5.0, and 4.2. Using the first 3 contrasts from Table 4.1 and (4.11) gives:

$$C_1 = 5.6 - 4.4 = 1.2 \rightarrow SS[C_1] = (1.2)^2/(0.1 + 0.1) = 7.2$$

$$C_2 = 5.6 + 4.4 - 10.0 = 0 \rightarrow SS[C_2] = 0$$

$$C_3 = 5.6 + 4.4 + 5.0 - 12.6 = -2.4 \rightarrow SS[C_3] = 5.76/1.2 = 4.8$$

and

$$SS[C_1] + SS[C_2] + SS[C_3] = 7.2 + 0 + 4.8 = 12.0 = SST. ▓$$

4.5   Simultaneous Confidence Intervals (SCI)

A special type of simultaneous confidence intervals was discussed in Section 3.8, those for paired differences involving the "best" mean.   Now we consider SCI for general sets of contrasts which, as before, need not be orthogonal.   Three different procedures for generating SCI will be considered: the studentized range, Scheffe's, and Bonferroni's procedures. For each we will indicate the error rate.

Each of these three procedures could be used to generate SCI for all paired differences of means.   These will be considered as we discuss each procedure.

4.5.1   Studentized Range SCI

As the name suggests, this procedure is based upon the student-ized range distributions. Table A.11 provides tabular points of these distributions for equal replications.   Consequently, this studentized range procedure should not be applied to experiments with unequal replications, unless there is only slight departures from equality.

Let $\theta = \sum_i c_i \mu_i$ denote any contrast of k means of interest

and $C = \sum_i c_i \bar{y}_{i.}$ be its estimator. Then, the studentized range SCI

of level α for θ is given by the limits

$$(4.17) \qquad \sum_i c_i \bar{y}_{i.} \pm r(k,\alpha,\nu)\sqrt{MSE/r_i} \sum_i |c_i|/2$$

where $r(k,\alpha,\nu)$ is the studentized range value of rank k, error rate α, and ν degrees of freedom of MSE. The error rate α is on the experimentwise basis; that is, in a proportion α of the experiments *some* SCI do not contain the corresponding parametric function.   Note that for paired differences, say $\mu_i - \mu_h$, the quantity $\sum_i |c_i|/2$ reduces to unity.

Example 4.11

For the legume data, Example 3.3, let us set 5% studentized range SCI for the following contrasts:

$$\theta_1 = \mu_1 - \mu_2$$

$$\theta_2 = \mu_1 + \mu_2 - 2\mu_4$$

$$\theta_3 = \mu_1 + \mu_2 - 3\mu_3 + \mu_4$$

We have

$$C_1 = 30.5 - 30.3 = 0.2$$

$$C_2 = 30.5 + 30.3 - 2(33.0) = -5.2$$

$$C_3 = 30.5 + 30.3 - 3(35.7) + 33.0 = -13.3$$

$$MSE = 4.76, \quad \nu = 20, \quad \text{and } r = 6$$

From Table A.6, $r(4,0.05,20) = 3.96$. The three intervals have limits:

$$0.2 \pm 3.96\sqrt{4.76/6}(1 + 1)/2 = 0.2 \pm 3.53$$

$$-5.2 \pm 3.96(0.89)(1 + 1 + 2)/2 = -5.2 \pm 7.05$$

$$-13.3 \pm 3.96(0.89)(3) = -13.3 \pm 10.58$$

Many other SCI could have been constructed; we have done only 3 of infinitely many. ※

4.5.2   Scheffe's SCI

The SCI proposed by Scheffe [2] are based upon percentage points of the F distribution rather than the studentized range distribution.   For the collection of all SCI constructed at level α, the quantity $1 - \alpha$ is a joint confidence coefficient and refers to the chance(probability) that all intervals simultaneously contain the corresponding parametric function. Thus, α is an experiment-wise error rate relative to the set of all possible contrasts among the means.   A major difference between Scheffe's SCI and the studentized range SCI is that Scheffe's is an exact procedure for both equal and unequal replications of treatments.

   For the parametric function $\theta = \Sigma\, c_i\mu_i$, a contrast of k means, the corresponding SCI has limits   i

(4.18) $$\Sigma_i c_i \bar{y}_i. \pm \sqrt{(k-1)F_\alpha MSE \,\Sigma_i (c_i^2/r_i)}$$

where F_α is the upper α percentage point of F with degrees of freedom $(k-1)$ and ν, those of treatments and MSE, respectively.

Example 4.12

For comparative purposes we shall construct the Scheffe's SCI for the 3 parametric functions of the previous example. Recall that

$$C_1 = 0.2 \qquad C_2 = -5.2 \qquad C_3 = -13.3$$

$$MSE = 4.76 \qquad \nu = 20 \qquad r = 6$$

$$F_{0.05} = 3.10$$

The three Scheffe's SCI have limits

$$0.2 \pm \sqrt{3(3.10)4.76(2/6)} = 0.2 \pm 3.84$$

$$-5.2 \pm \sqrt{3(3.10)4.76(6/6)} = -5.2 \pm 6.65$$

$$-13.3 \pm \sqrt{3(3.10)4.76(12/6)} = 13.3 \pm 9.41 \text{ ❋}$$

Scheffe's procedure is recommended for general "data snooping" among means when no pre-planned comparisons are specified. But when the researcher is interested only in SCI for all possible pairs of means, the studentized range procedure is better in the sense of giving shorter intervals. This is easily verified by comparing the intervals obtained in Examples 4.11 and 4.12. The lengths of intervals for paired differences are seen to be 2(3.53) and 2(3.84). Equal lengths should not be expected because different distributions are used in formulating the intervals. With unequal replication of the means, Scheffe's procedure is exact and preferred.

4.5.3 Bonferroni's SCI

Bonferroni's SCI procedure was designed for situations where only a fixed number of SCI are desired. Specifically, let m denote the number of contrasts of importance to the researcher. If m is small, we usually can improve considerably on the lengths of the corresponding SCI (that is, shorter intervals than we would obtain under the two previous procedures).

For Bonferroni's procedure, the error rate α refers to the specific set of m contrasts under consideration; the quantity $1-\alpha$ refers to the chance (probability) that the m SCI simultaneously contain the true values of the parametric functions of interest. Thus, α is a per-experiment error rate for the specified set of m contrasts of interest.

For a fixed set of m SCI to be constructed by Bonferroni's procedure, an error rate of α/m is assigned to each interval. Then the error rate for the collection of m SCI is at most α, and the confidence coefficient is at least $1-\alpha$.

Bonferroni's procedure uses percentage points of Student t distributions. One may interpolate in the regular t tables to get percentage points for the desired error rates or use Table A.14. For a contrast $\theta = \Sigma_i c_i \mu_i$, its estimator is $C = \Sigma_i c_i \bar{y}_i$. and limits for the Bonferroni SCI are

(4.19) $$\Sigma_i c_i \bar{y}_i. \pm t_{\nu;\alpha/2m}\sqrt{MSE \; \Sigma_i c_i^2/r_i}$$

where $t_{\nu;\alpha/2m}$ is the upper $\alpha/2m$ percentage point of the Student's t distribution having ν degrees of freedom, those of MSE.

Example 4.13

Let us define the three parametric contrasts appearing in the two previous examples as the only ones of interest. Then, for an overall error rate of $\alpha = 0.05$, we use an error rate of $\alpha* = 0.05/3 = 0.0167$ for each interval. From earlier work, we have:

MSE $= 4.76$ $\nu = 20$ $C_1 = 0.2$ $C_2 = -5.2$ $C_3 = -13.3$

Table A.14 gives $t_{20;0.0083} = 2.61$ so we obtain respective limits for the SCI of:

$$0.2 \pm 2.61\sqrt{4.76(2/6)} = 0.2 \pm 3.29$$

$$-5.2 \pm 2.61\sqrt{4.76(6/6)} = -5.2 \pm 5.69$$

$$-13.3 \pm 2.61\sqrt{4.76(12/6)} = -13.3 \pm 8.05$$

Note that these intervals are shorter than the corresponding intervals constructed by either of the other SCI procedures. ▓

As we have indicated earlier, and as the last example just verified, Bonferroni SCI are shorter than those obtained by the studentized range or Scheffe's procedure when m is small. But as m gets larger, the Bonferroni SCI get prohibitively large. As the number of contrasts increases, so do the chances of erroneously finding significances. Bonferroni's SCI attempts to guard against such errors by providing wider intervals. This implies that the Bonferroni procedure would be impractical for setting SCI for all paired differences of means because $m = t(t-1)/2$. With $t = 10$ treatments, there would be $m = 45$ contrasts of pairs of means.

4.6   Trend Contrasts for Means Having Gradient Structure

When our treatments have a gradient structure (Section 4.1), a special set of orthogonal contrasts can be considered which allow the researcher to investigate various polynomial trends among the means. The trend analysis, as some people call it, amounts to fitting polynomial regressions to the treatment means using the "gradient levels" as the independent variable. The question is: "What degree polynomial should be fit?" The researcher may have some knowledge or theoretical basis for using a particular polynomial model. If not, one would sequentially investigate the linear trend, then the quadratic, then the cubic, and so on. For most experiments, linear, quadratic, and cubic trends generally are sufficient. For k treatment levels, the highest degree that can be fitted is k-1.

4.6.1   Equally Spaced, Equally Replicated Treatments

Usually a researcher must make choices of what levels of factors to use, whether they should be equally spaced or not, whether the levels should be equally replicated or not. Rarely is there a single best choice. Research objectives, costs, time and so on must be considered.

Here we assume equal replication of treatments whose levels are equally spaced. The next sub-section covers the general case where unequal replications and/or unequally spaced levels are present.

The importance of each polynomial component is measured by a specific (orthogonal) contrast. In the equal replication, equal spacing case, these contrasts are constructed very easily from orthogonal polynomial coefficients. Table A.9 gives coefficients for 10 or fewer treatments. A more extensive tabulation of these coefficients is given in Anderson and Houseman [1].

We denote the orthogonal polynomial coefficients by ξ_{ij}, where $i = 1, 2, \ldots, k$ refers to the treatment designation while $j = 1, 2, \ldots, (k-1)$ refers to the degree of the polynomial. The coefficients are applied to treatment means to obtain the desired trend contrasts. Specifically, the contrasts and their corresponding sums of squares are

$$(4.20) \qquad C_j = \sum_i \xi_{ij} \bar{y}_i.$$

and

$$(4.21) \qquad SS[C_j] = C_j^2 / \sum_i (\xi_{ij}^2 / r)$$

The sum of squares of the coefficients, $\Sigma_i \xi_{ij}^2$, are provided also in Table A.9 and in the more extensive table [1].

Example 4.14

A CR design was conducted to investigate four equally spaced treatments. There were 6 EU per treatment which provided the following means:

$$\bar{y}_{1.} = 17.43 \qquad \bar{y}_{2.} = 16.63$$

$$\bar{y}_{3.} = 22.6 \qquad \bar{y}_{4.} = 19.93$$

With four treatments, we can investigate linear, quadratic, and cubic trends (only). A plot of the treatment means would suggest linear and cubic trends.

Using Table A.9 for the orthogonal polynomial coefficients, we calculate the following contrasts for measuring the linear, quadratic, and cubic trends:

$$C_L = \sum_i \xi_{i1}\bar{y}_{i.} = -3(17.43) - 16.63 + 22.6 + 3(19.93) = 13.5$$

$$C_Q = \sum_i \xi_{i2}\bar{y}_{i.} = 17.43 - 16.63 - 22.6 + 19.93 = -1.9$$

$$C_C = \sum_i \xi_{i3}\bar{y}_{i.} = -17.43 + 3(16.63) - 3(22.6) + 19.93 = -15.5$$

The sums of squares for these trend components are:

$$SS[C_L] = C_L^2/\Sigma_i(\xi_{i1}^2/r) = (13.5)^2/(20/6) = 54.675$$

$$SS[C_Q] = (-1.9)^2/(4/6) = 5.415$$

$$SS[C_C] = (-15.5)^2/(20/6) = 72.075 \; \text{※}$$

The special contrasts and sums of squares in (4.20) and (4.21) are used to judge significance of the various polynomial trends among the treatment means. This is accomplished with the usual F tests for contrasts; namely,

$$(4.22) \qquad F_{1,\nu} = SS[C_j]/MSE$$

where MSE is the Exp. Error mean square with ν degrees of freedom from the ANOVA.

Example 4.15

For the experiment described in the previous example, the ANOVA gave MSE = 3.65 with 20 degrees of freedom. To test the 3 trends, the observed values of the F statistics are, respectively:

$$F = 54.675/3.65 = 14.98$$

$$F = 5.415/3.65 = 1.48$$

$$F = 72.075/3.65 = 19.75$$

giving observed significance levels, respectively, of $P < 0.005$, $P > 0.1$, and $P < 0.005$. We see that the linear and cubic trends are quite significant. ✱

4.6.2   Gradient Treatments with General Spacing and Replication

If we conduct an experiment with treatments having a gradient structure but spaced and/or replicated unequally, orthogonal polynomial coefficients of Table A.9 cannot be used for a trend analysis. Procedures are available for calculating the contrast coefficients in the general case but these are fairly complicated and would need to be done anew for each different spacing and replication pattern.

Instead of calculating the polynomial coefficients in an experiment where unequal spacings or unequal replications are present, a trend analysis can be accomplished equivalently by fitting a series of regression models. Most computer packages have the capability to perform such an analysis. One begins with a model containing only a linear term, then sequentially adding a quadratic term, then a cubic term, and so on. The "sequential SS due to adding a quadratic term (to a model previously containing only a linear term)" would be the sum of squares measuring the quadratic trend of the treatment means.

4.6.3   Some Trend Patterns

When treatments have a gradient structure, a plot of the means can be quite valuable in providing general information about various trends, unusual responses, and so on. The tests described in the previous section would be used to verify which polynomial trends are significant. A plot of the means ordinarily would be included as part of any final report or data presentation, and the significant trends might be indicated on such a plot.

In Figure 4.1, we present some plots of means and indicate

the most prominent trend(s).   In each case we assume   treatments have   equally-spaced   levels.   We give one possible plot   of   the means   having the indicated trend;   other possibilities exist   in each case.   In practice, the actual significance of the indicated trend would be assessed by a statistical test, of course.

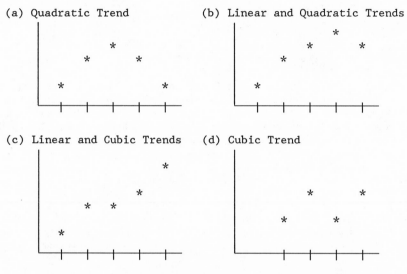

Figure 4.1   Prominent Trends of Plotted Means

PROBLEMS

1. A research worker brings you the following coefficients for a set   of contrasts of interest.   Assume equal replications of all treatments.

Contrasts	Treatment					
	A	B	C	D	E	F
1	3	0	3	-2	-2	-2
2	1	0	-1	0	-1	1
3	0	4	0	-4	0	0
4	0	0	0	2	-1	-1

(a)   Are contrasts 1 and 2 orthogonal?   Justify your answer.
(b)   Are contrasts 1 and 3 orthogonal?   Justify your answer.
(c)   Give a fifth contrast which is orthogonal to contrast 4. Your   new contrast cannot be one of the first three, but may be orthogonal to any of the first 3 contrasts.

2. A research worker brings you the following coefficients for contrasts of interest. Assume equal replication of all treatments.

Contrast	Treatment					
	A	B	C	D	E	F
1	3	0	3	-2	-2	-2
2	1	0	-1	0	-1	1
3	0	4	0	-4	0	0
4	0	0	0	2	-1	-1

(a) Are contrasts 1 and 2 orthogonal? Justify your answer.
(b) Are contrasts 1 and 3 orthogonal? Justify your answer.
(c) Give a fifth contrast which is orthogonal to contrast 4. Your new contrast cannot be one of the first three but it may be orthogonal to any or all of the first 3 given contrasts.

3. Given below is a table of coefficients for a set of contrasts of interest.

Contrast	Treatment					
	A	B	C	D	E	F
1	1	-1	0	0	0	0
2	2	2	0	-3	0	-1
3	1	1	0	0	-2	0
4	1	1	1	-1	-1	-1
5	-2	2	-2	0	-1	1
6	-2	-2	0	0	4	0
7	0	0	1	0	1	-2

Which pairs of contrasts are orthogonal and which are not? Assume equal replications.

4. Repeat Problem 3 with a replacement of the following:

Contrast	A	B	C	D	E	F
3	0	0	-2	-1	0	3
5	-2	2	0	1	0	-1
6	0	0	1	0	1	-2

5. A biochemist conducts an experiment to study the effectiveness of two similar antibiotic drugs, D_1 and D_2, and a control C. The design was a CR with r = 6 EU per treatment.

(a) Give the coefficients for a meaningful, complete set of orthogonal contrasts.

(b) State in words what each of your contrasts is comparing (or measuring).

(c) Let μ_1, μ_2 and μ_3 be the respective means for drugs D_1, D_2 and control. Give the null hypothesis tested by each of your contrasts.

6. A plant pathologist studied 4 herbicides that   are described as follows:

Herbicide	Description
1	2% active ingredients, liquid form
2	2% active ingredients, powder form
3	5% active ingredients, liquid form
4	5% active ingredients, powder form

Assume a CR design with r = 8 EU per treatment.

(a) Give the coefficients for a meaningful, complete set of orthogonal contrasts.

(b) State in words what each of your contrasts is comparing (or measuring).

(c) Let μ_1, μ_2, μ_3, and μ_4 be the respective means of the 4 treatments.   Give the null hypothesis tested by each of your contrasts.

7. Repeat Problem 5 if $r_1 = r_3 = 5$ and $r_2 = 6$.

8. Repeat Problem 6 if $r_1 = r_4 = 8$ and $r_2 = r_3 = 6$.

9. In a CR design, the 5 observed treatment means, each based on 8 observations, were:

Treatment	1	2	3	4	5
$\bar{y}_{i.}$	5.4	6.1	4.3	7.0	4.5

Apply the following "weird set" of orthogonal polynomial coefficients to this data to show that the treatment sum of squares can be calculated even from a set of orthogonal contrasts which are almost impossible to interpret in any meaningful fashion.

Contrast	Coefficients				
1	1	0	-3	0	2
2	-9	1	1	1	6
3	0	1	0	-1	0
4	3	-7	5	-7	6

10. Apply the special set of orthogonal polynomial coefficients (Section 4.4.2) to the data of the previous example to show that the treatment sum of squares can be calculated from such a set of contrasts.

11. Suppose the biochemist, Problem 5 above, recorded the following means:

Treatment	D_1	D_2	C
r_i	6	6	6
$\bar{y}_{i.}$	14.6	20.3	30.7

(a) Calculate the sum of squares for each of your contrasts specified in Problem 5(a).
(b) If the Exp. Error MS from the ANOVA was 135.4, test each hypothesis given in Problem 5(c).

12. The plant pathologist, Problem 6 above, recorded the following mean yields (in grams) after plots received an initial application of one herbicide:

Herbicide	1	2	3	4
$\bar{y}_{i.}$	14.3	12.6	16.4	13.1

(a) Calculate the sum of squares for each of your contrasts specified in Problem 6(a).
(b) If the Exp. Error MS from the ANOVA was 4.88, test each hypothesis given in Problem 6(c).

13. Repeat Problem 11 using the replication numbers given in Problem 7.

14. Repeat Problem 12 using the replication numbers given in Problem 8.

15. A biologist studied the effects of different colors of light on the growth of bacteria. The treatments (colors) were

designated A,B,C, and D.   An equal amount of bacteria was
placed on each of 20 petri dishes containing a certain growth
medium.   Five of the dishes were randomly assigned to each
type of light and after 30 hours, the following growth means
(in cm.) were obtained:

Light	A	B	C	D
$\bar{y}_{i.}$	9.4	6.8	5.8	8.2

The ANOVA test of equal growth means showed significant
differences with $P < 0.01$. What post-ANOVA analysis would be
done for each of the following specification of treatments?
After explaining why your choice is the most appropriate,
perform your post-ANOVA analysis.   The Exp.Error MS from the
ANOVA was 0.725 with 16 df.

(a)   Suppose the 4 treatments were:

     A   White light for 30 hours
     B   Blue light for 30 hours
     C   Green light for 30 hours
     D   Yellow light for 30 hours

(b)   Suppose the 4 treatments were:

     A   No light, total darkness
     B   5 hours Blue light, rest total darkness
     C   7 hours Green light, rest total darkness
     D   5 hours Blue light, then 7 hours Green light, rest
       total darkness

(c)   Suppose the 4 treatments were:

     A   Natural light for 30 hours
     B   Blue light for 30 hours
     C   Green light for 30 hours
     D   Red light for 30 hours

16. Repeat Problem 15 if the 4 treatments were

(a)   A   No light, total darkness for 30 hours
     B   Blue light for 30 hours
     C   Green light for 30 hours
     D   Both Blue and Green light for 30 hours

(b)   A   Natural light (and darkness) for 30 hours
     B   Blue light for 30 hours
     C   Green light for 30 hours
     D   Blue light for 15 hours, Green light for 15 hours

   (c) A No light, total darkness for 30 hours
        B Blue light for 10 hours, total darkness for 20 hours
        C Blue light for 20 hours, total darkness for 10 hours
        D Blue light for 30 hours.

17. The merchant's association of a university town asks for your help in designing an experiment to gather information about money spent by the university students (that is, spent within the town). In preliminary discussions, you discover that there is an interest in knowing of differences in the amount spent when the students are first divided into three groups:

      (i)   freshmen, sophmores
     (ii)   juniors, seniors
   (iii)   graduate students

with each of these groups divided into

      (a)   members of fraternities/sororities
      (b)   non-members of fraternities/sororities

In other words, the merchant's association has six (6) groups in mind. They have heard about orthogonal contrasts and insist on considering some. Give four orthogonal contrasts which are meaningful for the merchant's association. State what each is measuring and give the coefficients. Assume equal replications.

18. A graduate student in finance planned an experiment to study the amount of loans written by two types of institutions

     (i)   Savings and Loan Associations
    (ii)   Banks

for each of three sizes of cities

     (a)   small (<25,000)
     (b)   medium (25,000 - 100,000)
     (c)   large (> 100,000)

Give four orthogonal contrasts which are meaningful for the student. State what each is measuring and give the coefficients. Assume equal replications of the groups.

19. Research scientists of a paint company studied the weathering ability of two new paints when applied to soft and hard wood boards. Two soft woods (pine and poplar) and two hard woods (maple and oak) were used. Ten boards, each 2 feet long, were obtained of each type of wood. Five boards of each type were painted with paint I, five with paint II. (The design

was a CR.)   At the end of the test, each board was scored (on a 1-10 scale) giving the following means:

A	pine, paint I	9.2
B	pine, paint II	7.4
C	poplar, paint I	8.6
D	poplar, paint II	8.2
E	maple, paint I	9.0
F	maple, paint II	8.4
G	oak, paint I	7.6
H	oak, paint II	7.4

(a) Calculate the treatment sum of squares.
(b) Give a set of 4 meaningful orthogonal contrasts among the 8 treatment means.   Describe what each contrast is measuring.

20. Refer to the paint experiment of the previous problem.   For each of the 4 contrasts given in part (b):

(a) Calculate the sum of squares.
(b) Test the appropriate hypothesis.   From the ANOVA, the Exp. Error MS was 0.62.

21. A horticulturist was asked to test a new fungicide, Nurelle, for its effect on azaleas.   He selected 35 uniform plants from a particular variety growing in a greenhouse.   A stand-ard fungicide, Truban, also was used in the experiment for comparative purposes.   All plants were inoculated with a fungus two weeks after the experiment began.   For some plants the fungicide was applied at the time of inoculation; for others, it was applied after inoculation.   The seven treat-ments were:

#	Treatment	
1	Truban	
2	1 oz. Nurelle	
3	3 oz. Nurelle	at time of inoculation
4	5 oz. Nurelle	
5	1 oz. Nurelle	
6	3 oz. Nurelle	7 days after inoculation
7	5 oz. Nurelle	

The design was a CR with r = 7 plants per treatment.

(Courtesy of Dr. R. Lindstrom, Horticulture Dept., Va. Tech.)

(a) Give a complete set of meaningful   orthogonal   contrasts
among the means of these treatments.
(b) Give   the null hypothesis that would be tested by each of
your contrasts.

22. Refer to the azalea experiment of the previous   problem.   The
data   recorded   were fresh weights after 4 weeks.   The   mean
weights of the 7 plants were:

Treatment	1	2	3	4	5	6	7
$\bar{y}_{i.}$	37.3	38.3	38.7	43.4	35.1	37.0	41.6

(a) Calculate the sum of squares for each contrast given   in
Problem 19 (a).
(b) Test each hypothesis given in Problem 19(b).   From   the
ANOVA, the Exp. Error MS was 30.7.

23. A biologist investigated the   retardation of   growth of a
particular bacteria when heat was applied.   Four amounts of
heat were used: 90°F,   105°F,   120°F,   and   135°F.   Eight
cultures of   the bacteria were assigned to each   temperature
and after a fixed period of time,   the weight loss was determined.   From the ANOVA,   MSE = 4.07.   The mean weight losses
were (in grams):

90°F	105°F	120°F	135°F
8.3	6.4	9.1	9.8

Do a trend analysis on the mean weight losses.

REFERENCES

[1] Anderson, R. L.,   and E. E. Houseman, Tables of Orthogonal
Polynomial Values Extended to N = 104.   Research Bulletin
297, Ag. Expt. Station, Iowa State University, Ames, 1963.
[2] Scheffe, H.,   A method for judging all contrasts in the
analysis of variance. *Biometrika*, Vol. 40, 1953.

## Chapter 5

FACTORIAL EXPERIMENTS

5.1 Introduction

As stated in Section 1.6, the term *factorial* refers to a class of experiments where the treatments have a well-defined structure. A set of factorial treatments consists of all combinations of all levels of two or more factors. Each *treatment combination* must contain one level of every factor.

Example 5.1

Suppose a chemist is performing an experiment with 3 forms of a catalyst (Factor A) each used at 4 pressures (Factor B). Denote levels of A and B by a_1, a_2, a_3 and b_1, b_2, b_3, b_4, respectively. Then, a factorial experiment to investigate these factors has t = 12 treatment combinations given by

$$a_1 b_1 \qquad a_1 b_2 \qquad a_1 b_3 \qquad a_1 b_4$$

$$a_2 b_1 \qquad a_2 b_2 \qquad a_2 b_3 \qquad a_2 b_4$$

$$a_3 b_1 \qquad a_3 b_2 \qquad a_3 b_3 \qquad a_3 b_4 \; ※$$

We may think of a factorial experiment as a combination of two or more smaller-scaled experiments. In the above example, the chemist could perform two experiments ... one investigating the forms of the catalyst, another investigating pressures. Not only would this probably be an inefficient way to investigate

these factors,   the two single factor experiments would not allow
one to investigate any dependencies of the factors.   This is an
important aspect of factorials which we shall discuss shortly.
    One potential problem of factorial experiments is the number
of treatments.   In a factorial dealing with factors A, B, C, and
D, having levels a, b, c, and d, respectively, there are t = abcd
different treatments. Thus, with many factors and/or many levels,
the number of treatments can get prohibitively large. This might
present problems in finding uniform experimental material.   Later
chapters will discuss ways of dealing with this problem.
    The term factorial refers to the treatment design,   not the
experimental design.   Factorial treatments may be investigated in
most experimental designs,   in particular in each of the basic
experimental designs as well as more complicated designs.   The
term factorial treatment design is used sometimes.

5.2   The 2x2 Factorial Experiments

Suppose we first consider the simplest factorial--one dealing
with two factors each at two levels.   When a factor has only two
levels, they frequently are called low and high levels, and may
be denoted respectively by a_1 and a_2 for factor A and by b_1 and
b_2 for factor B.   For the 2x2 factorial,   the 4 treatments would
be designated as in Table 5.1.

Table 5.1   Treatments of a 2x2 Factorial

		Factor B	
		low	high
Factor A	low	a_1b_1	a_1b_2
	high	a_2b_1	a_2b_2

If the treatments occur equally often,   the researcher would need
4r EU, where r is the number of times each treatment occurs.
    The table of treatment totals will be called the AB   subclass
table.   If the experimental observations are denoted by y_{ijk},
where subscripts i and j refer to levels of factors A and B,
respectively,   the AB subclass table appears as in Table 5.2
below.

Table 5.2   The AB Subclass Table and Factor Totals

		\multicolumn{5}{c}{Levels of Factor B}	
		1	2	--- j ---	b	A totals
Levels of Factor A	1	$y_{11.}$	$y_{12.}$	--- $y_{1j.}$ ---	$y_{1b.}$	$y_{1..}$
	2	$y_{21.}$	$y_{22.}$	--- $y_{2j.}$ ---	$y_{2b.}$	$y_{2..}$
	.					
	i	$y_{i1.}$	$y_{i2.}$	--- $y_{ij.}$ ---	$y_{ib.}$	$y_{i..}$
	.					
	a	$y_{a1.}$	$y_{a2.}$	--- $y_{aj.}$ ---	$y_{ab.}$	$y_{a..}$
B totals		$y_{.1.}$	$y_{.2.}$	--- $y_{.j.}$ ---	$y_{.b.}$	$y_{...}$

Example 5.2

An engineer performed an experiment to study the thickness of electroplating when two different Amps (amperages, a measure of electrical current) were used in combination with two different temperatures of the solution. Uniform rods were randomly assigned (one each) to the plating tanks, identical except for temperature and amps. Six rods were electroplated by each amp-temperature combination from which the following AT subclass was obtained. Each cell entry represents the total thickness of electroplating, in mm, when the six rods were measured:

B Temperature

		Low	High	Amp. totals
A Amps	Low	19.8 3.3	23.4	43.2
	High	28.2 4.7	18.6	46.8
Temp. totals		48.0	42.0	90.0 = y...

Note that the overall total can be obtained by summing either set of factor totals (48.0 + 42.0, or 43.2 + 46.8). ✵

Referring to Table 5.2 or the subclass table in the last example, it is readily apparent that many treatment comparisons can be made. Not all of these comparisons are orthogonal which is to be expected. Let us look at some of these comparisons and the information that they reveal.

First consider the comparison in the first column of a 2x2 table. This is the low level of B (j=1) and the comparison of these means

$$(5.1) \qquad C_1 = \bar{y}_{21.} - \bar{y}_{11.}$$

measures the effect of changing levels of factor A while keeping factor B constant. This quantity will be called the *simple effect of factor A at the low level of factor B*. Thus,

$$C_1 = 4.7 - 3.3 = 1.4$$

measures the difference in the mean electroplating when changing from low to high amps while keeping the temperature at the low level. Therefore, at this low temperature, the simple effect suggests that increased amperage results in increased thickness of the electroplating.

We can make a similar comparison at the high level of B (j=2), within the second column of a 2x2 table:

$$(5.2) \qquad C_2 = \bar{y}_{22.} - \bar{y}_{12.}$$

the *simple effect of A at the high level of B*. Now we see that

$$C_2 = 3.1 - 3.9 = -0.8$$

indicating that increased amperage results in decreased thickness of mean electroplating when high temperature is used, a reversal of the trend observed at the low temperature.

An overall effect of factor A is obtained from a comparison of the A totals; that is,

$$(5.3) \qquad C_3 = \bar{y}_{2..} - \bar{y}_{1..}$$

which we shall call the *main effect of factor A*. Because C_3 is based upon marginal means, it reflects the effect of factor A combined across both levels of factor B. In our experiment,

$$C_3 = 3.9 - 3.6 = 0.3$$

which indicates (for the combined levels of temperature) that increased amperage results in increased mean thickness of the electroplating.

At this time, let us look at some equivalent expressions for the overall effect of factor A. We have

(5.4)
$$C_3 = \bar{y}_{2..} - \bar{y}_{1..}$$

$$= \frac{1}{2}(\bar{y}_{21.} + \bar{y}_{22.}) - \frac{1}{2}(\bar{y}_{11.} + \bar{y}_{12.})$$

$$= \frac{1}{2}(\bar{y}_{21.} - \bar{y}_{11.}) + \frac{1}{2}(\bar{y}_{22.} - \bar{y}_{12.})$$

$$= \frac{1}{2}(C_1 + C_2)$$

In other words, the main effect of A can be measured from the average of the simple effects of A. From the last expression in (5.4), it is clear that C_3 is not orthogonal to either C_1 or C_2

Simple and main effects of factor B can be defined in a similar fashion. The rows of the subclass in Table 5.1 provide this information. The comparison

(5.5)
$$C_4 = \bar{y}_{12.} - \bar{y}_{11.}$$

gives the *simple effect of B at the low level of A*. From the data in Example 5.2, we have

$$C_4 = 3.9 - 3.3 = 0.6$$

which indicates that when amps are maintained at the low level, an increase from low to high temperature results in an increased mean thickness of the electroplating.

At the high level of A, the comparison

(5.6)
$$C_5 = \bar{y}_{22.} - \bar{y}_{21.}$$

gives the second simple effect of B. For our example

$$C_5 = 3.1 - 4.7 = -1.6$$

which indicates that when amps are maintained at the high level, an increase from low to high temperature results in a decreased mean thickness of the electroplating.

Finally, the *overall or main effect of B* is measured by the comparison

$$(5.7) \qquad C_6 = \bar{y}_{.2.} - \bar{y}_{.1.}$$

For the data in Example 5.2, we find the value of this contrast to be

$$C_6 = 3.5 - 4.0 = -0.5$$

which indicates (for combined levels of amperage) that increasing temperature from low to the high level results in a decreased mean thickness of the electroplating.

Equivalent expressions for comparison C_6 are

$$(5.8) \qquad C_6 = \frac{1}{2}(\bar{y}_{12.} + \bar{y}_{22.}) - \frac{1}{2}(\bar{y}_{11.} + \bar{y}_{21.})$$

$$= \frac{1}{2}(\bar{y}_{12.} - \bar{y}_{11.}) + \frac{1}{2}(\bar{y}_{22.} - \bar{y}_{21.})$$

$$= \frac{1}{2}(C_4 + C_5)$$

From the last expression, we can say that (i) the main effect of B is equal to the average of the simple effects of B and (ii) comparison C_6 is not orthogonal to either C_4 or C_5. But it is easy to show that C_6 and C_3 are orthogonal.

Example 5.3

For the electroplating data of the previous example, a plot of the treatment means appears in the following graph.

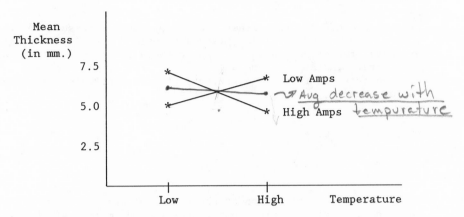

A second graph could have been drawn with the roles of Temperature and Amps interchanged. Mean profile lines would then be obtained for Low and High Temperatures. ▓

In the electroplating experiment we have been discussing, we found that the two simple effects of factor A(amps) were unequal, and so were the two simple effects of factor B (temperature). This illustrates the non-additive nature of factors A and B in that experiment: the effect of A is not the same for both levels of B (and the effect of B is not the same for both levels of A). There is a lack of consistency in each factor across the levels of the other factor. This non-additive feature is indicated also by the non-parallelism of the two lines of the graph in Example 5.3. (Had we drawn the second graph indicated in that example, we would have seen non-parallel lines also.) We now define this concept.

Definition 5.1

The lack of additivity of factor effects, or the non-parallelism of mean profiles, is called the *interaction* of the factors. When the factor effects are additive, we say there is no interaction of the factors.

We point out that interaction exists when there is any deviation from parallel profiles. Whether the interaction is significant or not is a separate issue, to be assessed by a statistical test.

In the 2x2 factorial, interaction is present when the simple effects of either factor are unequal. Thus, the difference of the two simple effects (of either factor) provides a measure of the interaction. In the electroplating example, the simple effects of each factor were unequal, hence there is an interaction of temperature and amps. Generally for a 2^2 factorial, interaction can be measured from the difference of simple effects for either factor; namely,

$$(5.9) \qquad C_7 = C_2 - C_1 = C_5 - C_4$$
$$= \bar{y}_{22.} - \bar{y}_{12.} - \bar{y}_{21.} + \bar{y}_{11.}$$

which also is a comparison of the treatment means. Now let us take a look at the three comparisons for the main effects and interaction in the 2^2 factorial. We have:

$$C_3 = -\bar{y}_{11.} - \bar{y}_{12.} + \bar{y}_{21.} + \bar{y}_{22.}$$
$$(5.10) \qquad C_6 = -\bar{y}_{11.} + \bar{y}_{12.} - \bar{y}_{21.} + \bar{y}_{22.}$$
$$C_7 = +\bar{y}_{11.} - \bar{y}_{12.} - \bar{y}_{21.} + \bar{y}_{22.}$$

These three comparisons represent a complete orthogonal set for the four treatments. Therefore, the main effects and interaction form a partitioning of treatment effects. The special structure of factorials always permits a partitioning of treatment effects into portions to measure the main effects and the interaction(s).

Example 5.4

The Amp-Temperature subclass of Example 5.2 is reproduced below.

		Temperature 1	2
	1	19.8	23.4
Amps			
	2	28.2	18.6

Recall that each total is based upon measurements from 6 rods.

The treatment sum of squares is

$$SST = [(19.8)^2 + (28.2)^2 + (23.4)^2 + (18.6)^2]/6 - (90)^2/24$$

$$= 2080.8/6 - 337.5 = 9.3$$

The comparisons to measure the main effects and interaction of Amps and Temperature are, by (5.10)

$$C_A = (-19.8 + 28.2 - 23.4 + 18.6)/6 = 0.6$$

$$C_B = (-19.8 - 28.2 + 23.4 + 18.6)/6 = -1.0$$

$$C_{AB} = (+19.8 - 28.2 - 23.4 + 18.6)/6 = -2.2$$

from which we obtain sums of squares

$$SSA = (0.6)^2/(4/6) = 0.54$$

$$SSB = (-1.0)^2/(4/6) = 1.5$$

$$SSAB = (-2.2)^2/(4/6) = 7.26$$

We see that these three sums of squares add to the treatment sum of squares. ▓

We have used comparisons to calculate the sums of squares of the factorial treatment components. Alternatively, we could have calculated these sums of squares from the main effect totals and the AB subclass totals. For the 2^2 factorial, these alternative sums of squares are special cases of the formulas given in the next section.

5.3.1 General Two Factor Factorials

We now consider factorial experiments having only two factors but any number of levels. Suppose factor A has a ≥ 2 levels and factor B has b ≥ 2 levels resulting in an axb factorial. Such would be the case in the electroplating experiment if the chemist had decided to use more than two temperatures or more than two amps. Another two-factor experiment, a 3x4 factorial appeared in Example 5.1.

To aid in discussing the concepts as they apply to the general two-factor experiments, we will begin by giving a subclass table and a graph for a 3x3 factorial.

Table 5.3   A 3x3 Subclass Table and Factor Totals

		Level of Factor B			
		1	2	3	A totals
Level of Factor A	1	a_1b_1 $y_{11.}$	a_1b_2 $y_{12.}$	a_1b_3 $y_{13.}$	$y_{1..}$
	2	a_2b_1 $y_{21.}$	a_2b_2 $y_{22.}$	a_2b_3 $y_{23.}$	$y_{2..}$
	3	a_3b_1 $y_{31.}$	a_3b_2 $y_{32.}$	a_3b_3 $y_{33.}$	$y_{3..}$
B totals		$y_{.1.}$	$y_{.2.}$	$y_{.3.}$	

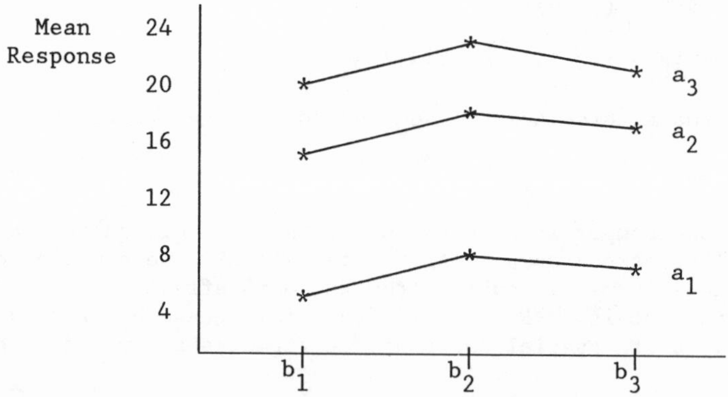

Figure 5.1   Mean Profiles for a 3x3 Factorial

First we point out that Definition 5.1 is used to determine the presence or absence of interaction in factorial experiments. We see from the plot of means in Figure 5.1 that interaction does not exist between factors A and B of this experiment. The mean profiles for the 3 levels of A are parallel. We leave it to the reader to verify that the mean profiles for the 3 levels of B also are parallel when the positions of A and B are interchanged in a graph like Figure 5.1.

Secondly, we note that there may be a large number of simple effects, calculated from adjacent cells of the AB subclass table.

Twelve such comparisons can be formed for the 3x3 subclass in Table 5.3 (and for the subclass in Example 5.1, there are 15 such comparisons). Calculating all simple effects to investigate interaction is inefficient and generally not done except for small factorials.

When a factor has more than two levels, the notion of its main effect is more obscure in relation to the simple effects. At least 2 orthogonal contrasts of simple effects would be required to explain such a "main effect." For this reason simple effects are of lesser importance in factorials beyond the 2x2.

Finally, when A and B have more than 2 levels, calculation of the sums of squares for each main effects and interaction cannot be done by a single comparison as in the 2x2 case. For example, when the factor A has "a" levels, one can form a-1 orthogonal comparisons among the A means. Therefore one would need to pool the sums of squares for a complete orthogonal set of comparisons to get the sum of squares for factor A effects. The sum of squares for A would have a-1 degrees of freedom. Similar pooling of sums of squares would be needed for the main effect B and interaction sums of squares. Instead of using comparisons, the sums of squares for the two main effects and the interaction can be calculated from "main effect totals" and "subclass totals." The general formulas are:

$$(5.11) \qquad SSA = \frac{1}{rb} \sum_i y_{i..}^2 - C$$

$$(5.12) \qquad SSB = \frac{1}{ra} \sum_j y_{.j.}^2 - C$$

$$(5.13) \qquad SSAB = SAB - SSA - SSB$$

where $C = y_{...}^2/rab$ is the overall correction term and SAB is the AB subclass sum of squares given by

$$(5.14) \qquad SAB = \frac{1}{r} \sum\sum_{ij} y_{ij.}^2 - C$$

We will reserve the single capital S for subclass sums of squares. We emphasize that the cell totals from the AB subclass are used in calculating SAB. In the two-factor factorials, these totals are really treatment totals so SAB is actually the treatment sum of squares. We are introducing the subclass notation for use later when generalizing to factorials having any number of factors.

Example 5.5

We will use the general formulas (5.11)-(5.14) to calculate the sums of squares for the main effects and interaction of the electroplating data.

$C = (90)^2/24 = 337.5$

$SAB = [(19.8)^2 + (28.2)^2 + (23.4)^2 + (18.6)^2]/6 - 337.5$

$= 2080.8/6 - 337.5 = 9.3$

$SSA = [(42)^2 + (48)^2]/12 - 337.5 = 339.0 - 337.5 = 1.5$

$SSB = [(43.2)^2 + (46.8)^2]/12 = 337.5 = 338.04 - 337.5 = 0.54$

$SSAB = 9.3 - 1.5 - 0.54 = 7.26$

These same values were obtained for the sums of squares of Amps, Temperature, and their interaction in Example 5.4 when orthogonal contrasts were used. ※

Example 5.6

Consider the data given below as that which might have been obtained in a chemical experiment as described in Example 5.1. Take factor A to be the forms of the catalyst, B to be pressure, and y to be the weight of the resulting product.

Factor B

		b_1	b_2	b_3	b_4
	a_1	11	8	12	9
		12	10	10	11
		9	10	13	10
	a_2	13	14	8	9
Factor A		11	10	12	9
		14	10	10	8
	a_3	9	10	11	7
		9	8	11	11
		9	11	9	6

The overall total is y... = 364 from which we obtain a correction term of $C = (364)^2/36 = 3680.44$. The total sum of squares for y is:

$$SSY = (11)^2 + (12)^2 + \ldots + (11)^2 + (6)^2 - C$$
$$= 3798 - 3680.44 = 117.56$$

The AB subclass and factor totals are

	b_1	b_2	b_3	b_4	A totals: $y_{i..}$
a_1	32	28	35	30	125
a_2	38	34	30	26	128
a_3	27	29	31	24	111
B totals:	97	91	96	80	

The AB subclass (treatment) sum of squares is

$$SAB = [(32)^2 + (28)^2 + \text{---} + (24)^2]/3 - C$$
$$= 11216/3 - 3680.44$$
$$= 3738.67 - 3680.44$$
$$= 58.23$$

The sums of squares for the main effects and interaction of A and B are

$$SSA = [(125)^2 + (128)^2 + (111)^2]/12 - 3680.44$$
$$= 3694.12 - 3680.44$$
$$= 13.73$$

$$SSB = [(97)^2 + (91)^2 + (96)^2 + (80)^2]/9 - C$$
$$= 3700.67 - 3680.44$$
$$= 20.23$$

$$SSAB = 58.23 - 13.73 - 20.23 = 24.27$$

Finally, Exp. Error sum of squares is obtained as the difference of the total and treatment sums of squares:

$$SSE = 117.56 - 58.23 = 59.33$$

The above calculations may be conveniently summarized in the following partial ANOVA table.

Analysis of Variance for Yield of a Chemical Product

Source	df	SS	MS	EMS
Catalysts	2	13.73	6.86	$\sigma^2_\epsilon + 12\kappa^2_A$
Pressures	3	20.23	6.74	$\sigma^2_\epsilon + 9\kappa^2_B$
CxP	6	24.27	4.04	$\sigma^2_\epsilon + 3\kappa^2_{AB}$
Exp. Error	24	59.33	2.47	σ^2_ϵ
Total	35	117.56		✻

5.3.2   The Linear Model for a Two-Factor Factorial

The linear model given in (2.1) for a basic CR design is easily modified for a factorial experiment. Remember that factorial refers to the treatment design so that one needs only replace the τ's by components to reflect the factorial structure. For a two factor factorial in a CR design, the linear model is

$$(5.15) \qquad y_{ijk} = \mu + \alpha_i + \beta_j + (\alpha\beta)_{ij} + \epsilon_{ijk}; \quad \begin{array}{l} i = 1, 2, \cdots, a \\ j = 1, 2, \cdots, b \\ k = 1, 2, \cdots, r \end{array}$$

where

$$\mu = \text{overall mean}$$

$$\alpha_i = \mu_{i.} - \mu$$
$$= \text{effect due to the i-th level of factor A}$$

$$(5.16) \qquad \beta_j = \mu_{.j} - \mu$$
$$= \text{effect due to the j-th level of factor B}$$

$$(\alpha\beta)_{ij} = \mu_{ij} - \mu_{i.} - \mu_{.j} + \mu$$
$$\begin{aligned} = \; & \text{component to measure the interaction} \\ & \text{resulting when the i-th level of A and} \\ & \text{j-th level of B are combined} \end{aligned}$$

Here, μ_{ij} denotes the mean in the (i,j)-th cell of the AB subclass, $\mu_{i.}$ and $\mu_{.j}$ denote marginal means of this subclass, and μ denotes the overall mean.

A couple of points are worth making at this time. First, we could write

(5.17) $$\tau_{ij} = \alpha_i + \beta_j + (\alpha\beta)_{ij}$$

whereby the expression in (5.15) is identical to the CR model (2.1) except for an additional subscript. This again illustrates that factorial refers to treatment design and not to experimental or response design. Secondly, we may rewrite the interaction component in (5.16) as follows

(5.18) $$(\alpha\beta)_{ij} = (\mu_{ij} - \mu) - (\mu_{i.} - \mu) - (\mu_{.j} - \mu)$$
$$= (\mu_{ij} - \mu) - \alpha_i - \beta_j$$

In the last form, we note that $(\alpha\beta)_{ij}$ measures the discrepancy in the (i,j)-th cell over and above the amount contributed by the sum of the 2 main effects. This illustrates that the interaction components measure non-additivity of factors.

5.4.1 General Factorial Experiments

Calculation of sums of squares in a factorial experiment having three or more factors is accomplished without undue difficulty by extending the results given in the general two factor case of Section 5.3. Of course, the number of calculations increases greatly as the number of factors increases. A sum of squares must be calculated for each main effect, each two factor interaction, each three factor interaction, and so on.

Suppose we have a 3 factor factorial, factors A, B, and C with respective levels a, b, and c. There are t = abc treatments. Besides 3 main effects, there are 3 two-factor interactions and 1 three-factor interaction. A subclass table is needed to calculate the sum of squares of each interaction; here, we would need 4 subclasses: AB, AC, BC and ABC. The three factor subclass table would be the treatment table, as the three factors define the treatments being investigated. It is true in general that the highest order subclass is the factorial treatment table.

The sum of squares for each main effect is easily calculated from the corresponding factor level totals. In the above three factor example, let y_{ijkm} represent the individual observations with subscripts denoting, respectively, factors A, B, C and replications. Then, for example, the sum of squares for main effect B would be

(5.19) $$SSB = \frac{1}{acr} \sum_j y^2_{.j..} - C$$

where $C = y^2.../abcr$ is the general correction term.

The sum of squares for any interaction can be calculated by a simple algorithm. First find the corresponding subclass sum of squares; then subtract the sums of squares of all lower order interactions and main effects of the subclass (that is, sums of squares of all combinations of subclass letters.) For example, suppose we want the sum of squares for the ABC interaction. We first find SABC, the sum of squares for the ABC subclass. From SABC, we subtract the sums of squares for the interactions AB, AC, and BC as well as main effects A, B, and C. Thus,

(5.20)   SSABC = SABC - SSAB - SSAC - SSBC - SSA - SSB - SSC

And for the AC interaction

(5.21)   SSAC = SAC - SSA - SSC

Example 5.7

For a 3x4x2 factorial on factors A, B, and C, the following represents the treatment table, the ABC subclass. Assume each treatment total is based on r = 5 observations.

			B		
C	A	1	2	3	4
1	1	10	6	8	7
1	2	9	8	7	7
1	3	12	11	8	10
2	1	19	16	12	18
2	2	14	15	15	15
2	3	16	14	13	18

The   table entries are y_{ijk},   where the subscripts i, j,   and k
refer to the levels of factors A, B, and C, respectively.
  The sums of squares for the ABC subclass table is really the
treatment sum of squares

$$SST = SABC = [(10)^2 + (6)^2 + \ldots + (13)^2 + (18)^2]/5 - (288)^2/120$$
$$= 3822/5 - 691.2 = 764.4 - 691.2 = 73.2$$

We now form the two-factor subclasses:

| | | Levels of B | | | |
Levels of A	1	2	3	4	A totals
1	29	22	20	25	96
2	23	23	22	22	90
3	28	25	21	28	102
B totals	80	70	63	75	

| | | Levels of B | | | |
Levels of C	1	2	3	4	C totals
1	31	25	23	24	103
2	49	45	40	51	185

| | Levels of C | |
Levels of A	1	2
1	31	65
2	31	59
3	41	61

The main effect and two-factor subclass sums of squares are

$$SAB = [(29)^2 + (22)^2 + \ldots + (28)^2]/5(2) - C$$
$$= 7010/10 - 691.2 = 12.0$$

$$SAC = [(31)^2 + (65)^2 + \ldots + (61)^2]/5(4) - C$$
$$= 15030/20 - 691.2 = 60.3$$

$SBC = [(31)^2 + (25)^2 + \ldots + (51)^2]/5(3) - C$
$= 11318/15 - 691.2 = 63.33$

$SSA = [(96)^2 + (90)^2 + (102)^2]/5(4)(2) - C$
$= 27720/40 - 691.2 = 1.8$

$SSB = [(80)^2 + (70)^2 + (63)^2 + (75)^2]/5(3)(2) - C$
$= 20894/30 - 691.2 = 5.27$

$SSC = [(103)^2 + (185)^2]/5(3)(4) - C$
$= 44834/60 - 691.2 = 56.03$

The interaction sums of squares are

$SSAB = SAB - SSA - SSB$
$= 12.0 - 1.8 - 5.27 = 4.93$

$SSAC = SAC - SSA - SSC$
$= 60.3 - 1.8 - 56.03 = 2.47$

$SSBC = SBC - SSB - SSC$
$= 63.33 - 5.27 - 56.03 = 2.03$

$SSABC = SABC - SSAB - SSAC - SSBC - SSA - SSB - SSC$
$= 73.2 - 4.93 - 2.47 - 2.03 - 1.8 - 5.27 - 56.03$
$= 0.67$

The sums of squares to measure the main effects and interactions are summarized in the following partial ANOVA table.

Partial ANOVA Table for 3x4x2 Factorial

Source	df	SS
A	2	1.80
B	3	5.27
AB	6	4.93
C	1	56.03
AC	2	2.47
BC	3	2.03
ABC	6	0.67
(Treatments)	(23)	(73.20)

We have shown the information only for treatments and their factorial partitioning. We would need the raw data to provide the total and Exp. Error components of the ANOVA table. ✳

5.4.2 Linear Models for a General Factorial

The linear model for a factorial experiment having more than two factors simply includes more components, one component for each main effect and interaction.   Thus for a three-factor   factorial in a CR design, the linear model would be

$$(5.22) \quad y_{ijkm} = \mu + \alpha_i + \beta_j + (\alpha\beta)_{ij} + \gamma_k + (\alpha\gamma)_{ik}$$
$$+ (\beta\gamma)_{jk} + (\alpha\beta\gamma)_{ijk} + \epsilon_{ijkm}$$

where all components are defined as in the two-factor experiments except for $(\alpha\beta\gamma)_{ijk}$ which measures the interaction of the three factors.   This component may be expressed as

$$(5.23) \quad (\alpha\beta\gamma)_{ijk} = \mu_{ijk} - \mu_{ij.} - \mu_{i.k} - \mu_{.jk} + \mu_{i..} + \mu_{.j.} + \mu_{..k} - \mu$$

5.5 Advantages and Disadvantages of Factorial Experiments

A few of the desirable and undesirable features of factorial experiments may be apparent at this time. We already know that additional sums of squares calculations are required. Instead of a single treatment sum of squares, a factorial analysis requires sums of squares for all main effects and interactions.   Even though these additional sums of squares calculations must be made, they are straightforward and follow the pattern discussed earlier.   To compensate for this additional work though, we hope to get "better, more meaningful" results from the experiment.
    Some of the important advantages of factorial experiments are:

  (i) Flexibility --- We may use any number of factors and any
      number of levels for each, subject only to available
      resources or having an experiment of a manageable size.
 (ii) Factorial treatments can be used in most experimental
      designs.
(iii) Interactions of factors can be investigated.
 (iv) In the absence of interactions, a factorial experiment
      is equivalent to single factor experiments conducted
      simultaneously.

The essence of the last advantage is that when no interactions exist, a factorial experiment is a very efficient utilization of resources; single factor experiments would be conducted at the same time and on the same experimental units. Unfortunately a complete absence of interactions is quite unlikely, especially when the number of factors is large.

Disadvantages of factorial experiments include:

  (i) More complex analysis if missing values are present.

 (ii) Because the treatments consist of several factors, exp-
      erimental results may be more difficult to interpret.
      This is especially true when interactions are   present,
      but using a model that ignores such interactions likely
      would lead to erroneous conclusions. Inferences will be
      covered in a later section.

(iii) Because at least two replications are needed to estimate
      Exp. Error variation, the size of the experiment may get
      prohibitively large if the number of factors or levels
      is large.

5.6   Assumptions in Factorial Experiments

Assumptions about the error components remain unchanged from the CR designs. Here we consider assumptions only about the factorial treatment parameters. The basic ideas concerning fixed and random effects apply here (see Section 1.8).

For a two-factor factorial with both factors having fixed effects, the assumptions are

$$\sum_i \alpha_i = 0, \quad \sum_j \beta_j = 0$$

(5.24)

$$\sum_i (\alpha\beta)_{ij} = 0 \quad \text{and} \quad \sum_j (\alpha\beta)_{ij} = 0$$

If both factors have random effects the assumptions are

$$\alpha's \text{ are i.i.d. } N(0, \sigma_A^2)$$

(5.25) $$\beta's \text{ are i.i.d. } N(0, \sigma_B^2)$$

$$(\alpha\beta)'s \text{ are i.i.d. } N(0, \sigma_{AB}^2)$$

The $\alpha's$, $\beta's$, and $(\alpha\beta)'s$ are distributed independently of each other and independently of any other random components in the model.

In the two-factor factorials, there are two mixed models for the treatment effects. The assumptions for these are basically a blend of the appropriate items from (5.24) and (5.25). Thus, for the mixed model with A effects fixed and B effects random, the assumptions are

$$\sum_i \alpha_i = 0, \quad \sum_i (\alpha\beta)_{ij}$$

(5.26) β's are i.i.d. $N(0, \sigma_B^2)$

$(\alpha\beta)$'s have a $N(0, \sigma_{AB}^2)$ distribution.

The β's and $(\alpha\beta)$'s are distributed independently of each other and independently of any other random components in the model.

Notice that the interaction components are randomly distributed because of the randomness of factor B. This is true in general: Any interaction component is randomly distributed if at least one of its component factors is random.

The assumptions stated above for the two factor models can be extended, in the obvious and straightforward manner, to any factorial experiment having three or more factors. There are increasingly more treatment components in the model as more and more factors are included in the experiment. And, the number of possible mixed models likewise increases as the number of factors increases.

5.7 Inferences For Fixed Effects Models

We will discuss aspects of analyses for factorial experiments in which all factors have fixed effects. The next section will be devoted to experiments having one or more factors with random effects.

5.7.1 Estimation of Means and Effects

We first consider estimation of treatment parameters appearing in a factorial model. In a two-factor factorial, for example, we see from (5.16) that estimates of main effect and interaction components would be

$$\hat{\alpha}_i = \bar{y}_{i..} - \bar{y}_{...}$$

(5.27) $$\hat{\beta}_j = \bar{y}_{.j.} - \bar{y}_{...}$$

$$\widehat{(\alpha\beta)}_{ij} = \bar{y}_{ij.} - \bar{y}_{i..} - \bar{y}_{.j.} + \bar{y}_{...}$$

Quite often the researcher is interested in the difference of certain means; for example, the difference in mean response at two levels of factor A. As we note from (5.27), differences of mean estimates are related to the estimates of model parameters.

For the difference of main effect means in a two factor factorial, consider $\alpha_i - \alpha_h = \mu_{i.} - \mu_{h.}$ and $\beta_j - \beta_h = \mu_{.j} - \mu_{.h}$ which have estimates of

(5.28) $\hat{\alpha}_i - \hat{\alpha}_h = \bar{y}_{i..} - \bar{y}_{h..}$

(5.29) $\hat{\beta}_j - \hat{\beta}_h = \bar{y}_{.j.} - \bar{y}_{.h.}$

These differences have estimated standard deviations of:

(5.30) $s_{\bar{y}_{i..} - \bar{y}_{h..}} = \sqrt{2MSE/rb}$

(5.31) $s_{\bar{y}_{.j.} - \bar{y}_{.h.}} = \sqrt{2MSE/ra}$

Differences of certain subclass means also may be of interest. Generally there is interest only in the difference of means for the same levels of all factors except one. In an AB subclass these differences would be of the form $\mu_{ij} - \mu_{ih}$, for the same level of factor A, and $\mu_{ij} - \mu_{hj}$, for the same level of factor B. These represent differences of means in a single row or column of the AB subclass. Estimates of these differences are

(5.32) $\hat{\mu}_{ij} - \hat{\mu}_{ih} = \bar{y}_{ij.} - \bar{y}_{ih.}$

(5.33) $\hat{\mu}_{ij} - \hat{\mu}_{hj} = \bar{y}_{ij.} - \bar{y}_{hj.}$

with estimated standard deviation $\sqrt{2MSE/r}$ for either difference.

Under the usual ANOVA assumptions, CI can be constructed through the Student-t distribution; for example, for $\alpha_i - \alpha_h$, the confidence limits would be

(5.34) $(\bar{y}_{i..} - \bar{y}_{h..}) \pm (s_{\bar{y}_{i..} - \bar{y}_{h..}}) t_{\alpha/2}$

where $t_{\alpha/2}$ is the upper $\alpha/2$ percentage point of the Student's t distribution with degrees of freedom equal to those of MSE.

Example 5.8

Using the information in the AB subclass of Example 5.6, we obtain estimates for the three forms of catalyst

$$\hat{\alpha}_1 = \bar{y}_{i..} - \bar{y}_{...} = 125/12 - 364/36$$
$$= 10.417 - 10.111 = 0.306$$

$$\hat{\alpha}_2 = 128/12 - 10.111 = 10.667 - 10.111 = 0.556$$

$$\hat{\alpha}_3 = 111/12 - 10.111 = 9.250 - 10.111 = -0.861$$

As they should, these three estimates sum to zero within rounding error. The difference $\alpha_1 - \alpha_3$ would be estimated by $\bar{y}_{1..} - \bar{y}_{3..} = 10.417 - 9.250 = 1.167$ with estimated standard deviation

$$s_{\bar{y}_{1..}-\bar{y}_{3..}} = \sqrt{2(2.472)/12} = 0.642$$

A 95% CI for $\alpha_1 - \alpha_3 = \mu_{1.} - \mu_{3.}$ would be

$$1.167 \pm (0.642)(2.064) = 1.167 \pm 1.325 \rightarrow (-0.158;2.492)$$

Because the interval encompasses zero, no difference in the means for the first and third levels of catalyst is a distinct possibility. ▓

5.7.2 Expected Mean Squares

In Section 2.6, we stated that the expected mean squares are essential for setting up the proper test statistics. When all factorial treatment components are assumed to have fixed effects, the expected mean squares have a simple form. As we will see, exact tests are possible for all main effects and interactions in the fixed effects case. (This is not true for all mixed or random models, as we will see in Section 5.8)

When all factors have fixed effects, the EMS of any main effect or interaction has two components: σ^2_ϵ and a *source component*. Apart from a coefficient, the source component is a κ^2 symbol with subscripts equal to the source symbols, the letters identifying that source of variation of the ANOVA. For main effect A, the source component is

(5.35) $\kappa_A^2 = \sum_i \alpha_i^2/(a-1)$

and for the ABC interaction, the source component is

(5.36) $\kappa_{ABC}^2 = \sum_{ijk}\sum\sum(\alpha\beta\gamma)_{ijk}^2/(a-1)(b-1)(c-1)$

The coefficient of any source component is determined by the following simple rule: If a letter doesn't appear as a subscript, its range multiplies the component. In addition the number of replications, r, multiplies the component.

A partial ANOVA is given below for a three-factor factorial conducted in a CR design, all factors assumed fixed.

Table 5.4   Partial ANOVA for Three-Factor Factorial

Source	df	MS	EMS
A	a-1	MSA	$\sigma_\epsilon^2 + rbc\kappa_A^2$
B	b-1	MSB	$\sigma_\epsilon^2 + rac\kappa_B^2$
AB	(a-1)(b-1)	MSAB	$\sigma_\epsilon^2 + rc\kappa_{AB}^2$
C	c-1	MSC	$\sigma_\epsilon^2 + rab\kappa_C^2$
AC	(a-1)(c-1)	MSAC	$\sigma_\epsilon^2 + rb\kappa_{AC}^2$
BC	(b-1)(c-1)	MSBC	$\sigma_\epsilon^2 + ra\kappa_{BC}^2$
ABC	(a-1)(b-1)(c-1)	MSABC	$\sigma_\epsilon^2 + r\kappa_{ABC}^2$
Exp.Error	abc(r-1)	MSE	σ_ϵ^2
Total	rabc-1		

Example 5.9

A 3x4x2 factorial was conducted in a CR design with 5 EU per treatment. All treatment components were assumed to have fixed effects. The partial ANOVA is:

Source	df	EMS
A	2	$\sigma_\epsilon^2 + 40\kappa_A^2$
B	3	$\sigma_\epsilon^2 + 30\kappa_B^2$
AB	6	$\sigma_\epsilon^2 + 10\kappa_{AB}^2$
C	1	$\sigma_\epsilon^2 + 60\kappa_C^2$
AC	2	$\sigma_\epsilon^2 + 20\kappa_{AC}^2$
BC	3	$\sigma_\epsilon^2 + 15\kappa_{BC}^2$
ABC	6	$\sigma_\epsilon^2 + 5\kappa_{ABC}^2$
Exp. Error	96	σ_ϵ^2
Total	119	✳

Example 5.10

The complete ANOVA table for the chemical experiment analyzed in Example 5.6, under the fixed effects model, is:

ANOVA for Chemical Yield

Source	df	SS	MS	EMS
A	2	13.73	6.865	$\sigma_\epsilon^2 + 12\kappa_A^2$
B	3	20.23	6.743	$\sigma_\epsilon^2 + 9\kappa_B^2$
AB	6	24.27	4.045	$\sigma_\epsilon^2 + 3\kappa_{AB}^2$
Exp.Error	24	59.33	2.472	σ_ϵ^2
Total	35	117.56		✳

5.7.3 Tests of Main Effects and Interactions

Once the standard ANOVA table has been completed, certain tests of hypotheses can be made about the factorial parameters. From such a completed ANOVA, tests about main effects and interactions are possible. Inferences about paired differences and contrasts are discussed later.

Because higher order interactions might mask lower order interactions or main effects, it is advisable to first test for significance of the interactions --- beginning with the highest order, then the next lower order, and so on. When all factors are represented among significant high order interactions, tests of lower order interactions and main effects are useful only to indicate what masking has occurred. For example, in a 4-factor experiment, the significance of the ABC and BCD interactions would imply that all four factors are important (and necessary) in explaining the response. The tests of 2-factor interactions and main effects would be made only to determine whether or not the ABC and BCD interactions have produced any masking. We will give a graphic illustration and further discuss this idea in the latter part of this subsection.

To make a test that there is no interaction of factors A, B, and C, the null hypothesis is

$$(5.37) \qquad H_0: \ (\alpha\beta\gamma)_{ijk} = 0 \text{ for all } i,j,k$$

Under this hypothesis, we note from (5.36) that $\kappa^2_{ABC} = 0$ whereby the mean square for ABC in Table 5.4 has the same expected value as the mean square for Exp. Error. This implies that the ratio of these two mean squares would be the F statistic to test the hypothesis of no ABC interaction; that is,

$$(5.38) \qquad F = MSABC/MSE$$

with degrees of freedom of $(a-1)(b-1)(c-1)$ in the numerator and those of MSE in the denominator.

To make a test that there is no interaction of factors A and B, the null hypothesis is

$$(5.39) \qquad H_0: \ (\alpha\beta)_{ij} = 0 \text{ for all } i,j$$

Under this hypothesis, we note that $\kappa^2_{AB} = 0$ whereby the mean square for AB in Table 5.4 has the same expected value as the mean square for Exp. Error. This implies that the ratio of these two mean squares would be the F statistic to test the hypothesis of no AB interaction; that is,

$$(5.40) \qquad F = MSAB/MSE$$

with degrees of freedom of $(a-1)(b-1)$ in the numerator and those of MSE in the denominator. Similar results hold for each of the other two-factor interactions in the fixed effects model.

To make a test that there are no factor A effects, the null hypothesis would be

$$(5.41) \qquad H_0: \alpha_i = 0 \text{ for all } i$$

Under this hypothesis, we note from (5.35) that $\kappa_A^2 = 0$ whereby the mean square for A in Table 5.4 has the same expected value as the mean square for Exp. Error. This implies that the ratio of these two mean squares would be the F statistic to test the hypothesis of no factor A effects; that is,

$$(5.42) \qquad F = MSA/MSE$$

which has degrees of freedom of $(a-1)$ in the numerator and those of MSE in the denominator. Analogous results hold for any other main effects in a fixed effects model.

Before presenting an example of these test procedures, let us take another look at the implications of the null hypotheses for main effects and interactions. The hypothesis "H_0: All $\alpha_i = 0$" is equivalent to one stating that "there is no effect contributed by any level of factor A". But upon replacing α_i by $\mu_i. - \mu$, we obtain an equivalent hypotheses that "the mean of our response variable y is identical for all levels of Factor A."

On the other hand, any hypothesis stating "no interaction" is not equivalent to a simple equality of means. For example, the hypothesis of no AB interaction "H_0: $(\alpha\beta)_{ij} = 0$ for all i,j" is equivalent to

$$H_0: \mu_{ij} - \mu_i. - \mu_{.j} + \mu = 0 \text{ for all } i,j.$$

This does not imply equality of means in the AB subclass but rather that there exists a certain consistency of means in any pair of rows (or columns) of this subclass. The consistency required by this hypothesis is that of parallel mean profiles in the subclass of AB means, that of additivity of effects for factors A and B.

Example 5.11

For the chemical experiment, Examples 5.6 and 5.8, the null hypothesis of no interaction of factor A (forms of catalyst used) and factor B (pressure) would be

$$H_0: (\alpha\beta)_{ij} = 0 \quad \text{for i, 2, 3 and j} = 1, 2, 3, 4$$

which would be tested by

$$F = MSAB/MSE = 4.045/2.472 = 1.64$$

with 6 and 24 degrees of freedom. The observed significance level is $P > 0.10$.

The null hypothesis of no effect due to factor A would be

$$H_0: \alpha_1 = \alpha_2 = \alpha_3 = 0$$

which would be tested by

$$F = MSA/MSE = 6.865/2.472 = 2.78$$

with 2 and 24 degrees of freedom. The observed significance level is $0.05 < P < 0.10$.

The null hypothesis of no effects due to factor B would be

$$H_0: \beta_1 = \beta_2 = \beta_3 = \beta_4 = 0$$

which would be tested by

$$F = MSB/MSE = 6.743/2.472 = 2.73$$

with 3 and 24 degrees of freedom. The observed significance level is $0.05 < P < 0.10$.

The nonsignificance of interaction implies that the observed data offered no evidence against the additivity of effects for the Catalyst and Pressure factors. Additionally, the low significance exhibited by the two main effect tests indicates that neither factor offers very strong information about our response variable, weight of the chemical produced. ✖

In the previous example a significant interaction was not detected (not at the 10% level, at least.) At this time though, we need to consider the interpretations and implications when significant main effects and interactions are detected by ANOVA tests. Suppose in a three factor experiment, ANOVA tests reveal significance only for the ABC interaction and the A main effect. One should not hastily conclude that factors B and C have an

insignificant impact on explaining the response variable.   Quite
the contrary is true, however, because the significance of an ABC
interaction implies that all three factors are interrelated.   One
needs to know the particular levels of each factor before making
meaningful statements about treatment responses. Apparently some
sort of cancellation of effects has occurred whereby factors B
and C are exhibiting no individual effects.   Consequently one
should be careful of inferences about any single factor when
significant interactions include that factor.

To further reinforce the above implications,   let us consider
the three possible situations for a 2x2 factorial:

(1) interaction significant, main effects not significant,
(2) interaction and only one main effect significant, and
(3) interaction and both main effects significant.

Tables of means and corresponding graphs which belong to these
categories are given in the following figure.

Figure 5.2   Subclass Means for 2x2 Factorials

(1)	b_1	b_2		(2)	b_1	b_2		(3)	b_1	b_2
a_1	14	10		a_1	10	16		a_1	13	14
a_2	10	14		a_2	11	15		a_2	18	12

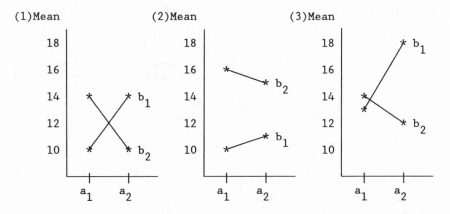

From either the table or graph, we observe in the first situation
that there are no A effects because the A means are each 12.

And, there are no B effects because both B means are also 12.   In this extreme case, the two factors exhibit no effects because of "cancellation" (the criss-crossing of the mean profiles).   The test of neither main effect would show significance but the levels of both factors are obviously necessary in specifying a treatment mean.

In the second situation, we note that the two A means are each 13 while the two B means are 10.5 and 15.5.   There is evidence of B effects but the test of A effects indicates no significance. Even so, knowledge of the particular level of A is required in specifying a treatment mean.

In the third situation, the two A means are 13.5 and 16 while the two B means are 15.5 and 13. Both factors exhibit effects and therefore both factors provide information about the treatment means.   This situation rarely causes problems of interpretation in practice.

Summarizing these three cases, we note that the factors may be important in specifying treatment means even when tests on the main effects show nonsignificance.   As mentioned earlier, when interactions are significant, one must be careful interpreting main effects when tests of them show nonsignificance.

The ideas expounded above carry over to factorials of three or more factors. For example, with three factors, the interpretation of main effects and two-factor interactions must be done with care when there is significant three-factor interaction.

5.7.4   Multiple Comparison and Range Procedures

In Chapter 3 we presented various mean separation procedures for assessing significances of paired mean differences. All of these procedures may be considered for use in factorial experiments. They are most appropriate when the levels of the factors are unstructured.   When the levels of the factors have group or gradient structures, the techniques of Chapter 4 are recommended as they are more appropriate. The next subsection will be devoted to structured factor levels.

We will emphasize the two factor factorials but all techniques of Chapter 3 may be applied to factorial experiments having any number of factors. As indicated earlier, comparisons of means in any subclass are generally most meaningful when the means have the same levels of all factors except one.   Thus, in an ABC subclass, one might compare means belonging to each of several subsets defined by fixing the levels of two factors, say B and C. The number of potential subsets of this type is bc, the number of different combinations of factor B and C levels.

Application of any technique of Chapter 3 to various sets of means in a factorial experiment requires the appropriate standard deviation, degrees of freedom, and value of k. Some useful quantities for mean separation procedures in the two factor factorials are summarized in Table 5.5 below. In all cases listed, the degrees of freedom are those of MSE appearing in the standard deviation.

Table 5.5 Mean Separation Quantities, Two Factor Factorial

Differences of two	Estimator	Standard deviation	k
A means	(5.28)	$\sqrt{2MSE/rb}$	a
B means	(5.29)	$\sqrt{2MSE/ra}$	b
AB means, same level of B	(5.33)	$\sqrt{2MSE/r}$	a
AB means, same level of A	(5.32)	$\sqrt{2MSE/r}$	b

Example 5.12

Refer to the chemical experiment of Example 5.6. Tests of null hypotheses in Example 5.10 indicated nonsignificant interaction but somewhat significant main effects. Therefore, the chemist may wish to investigate differences of the 3 forms of catalyst. [Due to the gradient structure of pressure means, it would be more appropriate to use techniques of the next section to investigate differences of these means.]
  The observed catalyst means are:

$$10.417; \quad 10.667; \quad 9.250$$

For a 1% Tukey's test of differences of catalyst means, the critical difference is found, using the first row of Table 5.5, to be

$$TK = \sqrt{2MSE/rb} \; r(3,0.01,24)/\sqrt{2}$$

$$= \sqrt{2.472/12} \; (4.55) = 2.05$$

We leave it to the reader to verify that this Tukey's test fails to detect significances among pairs of catalyst means. ✖

5.7.5   Contrasts and Gradient Trend Analyses

In factorial experiments, some of the treatment factors may have
group or gradient structures.   For example, the factors of the
chemical experiment, Example 5.6, would have gradient structure
if the levels were increasing amounts of catalyst and increasing
pressures. As indicated earlier, many aspects of these structures
are not addressed by multiple comparison procedures; contrasts
are necessary for addressing these issues.

     Because the contrast procedures of Chapter 4 were presented
in terms of "single factor treatments", they need to be modified
to accommodate factorial treatments. And, whether one considers
contrasts of main effect or interaction components may depend
upon which of the components are significant, see the discussion
following Example 5.10.

     We first consider 2 factor experiments having group structure.
For either set of main effect means, contrasts among and within
groups would be made as the structure and research objectives
warrant.   When there is no significant AB interaction, contrasts
for the individual factors should be sufficient.   On the other
hand, when a significant AB interaction exists, the AB subclass
means contain relevant information about each factor.   Contrasts
of these subclass means would be needed to understand the nature
of the interaction.   Which contrasts are required for this is
difficult to specify. Quite often, the contrasts applied to an
individual factor would be considered for each level of the other
factor.   Thus in an AB subclass, the researcher might apply "the
A contrasts" to each level of B.

Example 5.13

A turfgrass specialist investigated 5 nitrogen fertilizers(factor
A) with each of 6 varieties of turfgrass (factor B). The nitrogen
fertilizers were made by different companies, but 2 were liquid,
and 3 were granular. Each was used in an amount to give the same
percentage of available nitrogen. The turfgrasses were three warm
season and three cool season varieties.   Thus, each factor has a
group structure.

     If there is no fertilizer-variety interaction, all relevant
information is provided by the two sets of main effect means.
The researcher may use a contrast to compare the liquid and
granular subsets, and maybe other contrasts to make comparisons
within the liquid and the granular forms.   Likewise, a contrast
may be used to compare the warm and cool season varieties, with
perhaps other contrasts used to make comparisons within each of
the two seasonal varieties.

If there is a significant fertilizer-variety interaction, the researcher needs to consider means in this two-factor subclass. All contrasts mentioned above for fertilizers (between and within groups) might be considered for each variety. By the same token, all contrasts of varieties mentioned above might be considered for each fertilizer. Without some additional information, it is difficult to identify other contrasts of the AB subclass means which might be relevant. ▧

Now we consider trend analyses for factorial treatments having gradient structure. We limit the discussion to 3x3 factorials, where each factor has equally spaced levels.

Linear and quadratic trends can be investigated for each set of main effect means. These will be denoted by A_L, A_Q and B_L, B_Q and can be calculated from the orthogonal polynomial coefficients of Table A.9. With three levels, the coefficients are (for either factor);

$$\text{Linear :} \quad +1 \quad\quad 0 \quad\quad -1$$
(5.43)
$$\text{Quadratic :} \quad +1 \quad\quad -2 \quad\quad +1$$

The results of Section 4.6.1 may be applied to each set of main effect means by taking "each factor to represent a separate experiment." Values of the trends and their sums of squares are given by (4.20) and (4.21), using either A means based on rb observations or B means based on ra observations. Tests are made using the F statistic of (4.22).

The two trend polynomials (each with one degree of freedom) of either factor represent the two degrees of freedom of that factor. For a 3x3 factorial, the AB interaction has 4 degrees of freedom. We consider a partitioning of the AB interaction SS into a set of 4 meaningful, orthogonal trend components, each with one degree of freedom. We will denote these interaction trends by

(5.44)
$$A_L x B_L \qquad\qquad A_L x B_Q$$
$$A_Q x B_L \qquad\qquad A_Q x B_Q$$

The interpretation of these components will be discussed later. We first show how to calculate their values and sums of squares.

The value of any trend component in (5.44) is calculated from the entries in the AB subclass table. The coefficients for a given interaction trend are obtained by taking products of the corresponding single factor trend coefficients. For the (i,j)-th entry, we simply take the product of the coefficients for the i-th level of A and the j-th level of B (of the appropriate single factor trends). The table below illustrates the coefficients for $A_Q x B_L$.

Table 5.6   Coefficients for $A_Q x B_L$, Equally Spaced Levels

		Coefficients for A_Q		
		+1	-2	+1
Coefficients for B_L	-1	-1	+2	-1
	0	0	0	0
	+1	+1	-2	+1

The coefficients in the cells of Table 5.6 would be applied to the corresponding AB means which, using (4.20) and (4.21), would give a value for the contrast of $A_Q x B_L$ trend and its sum of squares. The other three interaction trends of (5.44) would be handled similarly.

Example 5.14

An engineer investigated the yield of a polymer under three pressures (80,   100,   and 120 pounds per square inch) and   three temperatures (100, 125 and 150°C.) Four runs were made at   each combination of temperature and pressure giving the following means:

Temp.	Pressure 80	Pressure 100	Pressure 120	Temp. Means
100	14.2	15.3	14.7	14.73
125	16.1	18.2	17.5	17.27
150	15.4	15.9	15.1	15.47
Pressure Means:	15.23	16.47	15.77	

The linear and quadratic trends of temperature are measured by

$$T_L = 14.73 - 15.47 = -0.74$$

$$T_Q = 14.73 - 2(17.17) + 15.47 = -4.34$$

with sums of squares, by (4.21),

$$SS[T_L] = (-0.74)^2/(2/12) = 3.29$$

$$SS[T_Q] = (-4.34)^2/(6/12) = 37.67$$

In a similar fashion, one finds the linear and quadratic trends of pressure to be 0.6 and (-20), with sums of squares 2.16 and 8.0, respectively.

Using coefficients formed as in Table 5.6, we calculate the values of the interaction trends:

$$P_L xT_L = 14.2 - 14.7 - 15.4 + 15.1 = -0.8$$

$$P_L xT_Q = 14.2 - 14.7 - 2(16.1) + 2(17.5) + 15.4 - 15.1 = 2.6$$

and analogously

$$P_Q xT_L = -0.4$$

$$P_Q xT_Q = 2.6$$

The sums of squares for these trend components are, by (4.21)

Trend	SS
L x L	$(-0.8)^2/(4/4) = 0.64$
L x Q	$(2.6)^2/(12/4) = 2.25$
Q x L	$(-0.4)^2/(12/4) = 0.05$
Q x Q	$(2.6)^2/(36/4) = 0.75$

Each of the trend components could be tested by an F statistic with 1 and 27 degrees of freedom. ※

At this time let us discuss the interpretation of interaction trend components. Suppose we made a three dimensional plot of

the means in an AB subclass, such as those appearing in Example
5.14. A "Linear x Linear" trend would be indicated if the means
exhibit a linear response in the dimension of each factor. A
"Quadratic x Linear" trend would be indicated if means exhibit a
quadratic response in the dimension of one factor and a linear
response in the dimension of the other factor. And a "Quadratic x
Quadratic" trend would be indicated if means exhibit a quadratic
response in the dimensions of both factors. Any number of the
interaction trends might be significant in a given experiment.
In many cases, the most dominant (significant) trend may actually
overshadow the others. Statistical tests would need to be made to
provide the final judgement on significance of interaction trend
components. When the (overall) AB interaction is nonsignificant,
it would be unusual for any of the interaction trend components
to be significant.

Figure 5.3 below gives a pattern of subclass means that might
be observed when the indicated trend is the most dominant one.
Many other possibilities exist for each case.

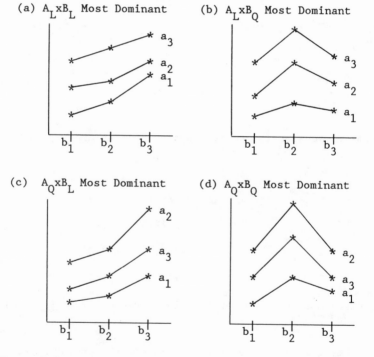

(a) $A_L x B_L$ Most Dominant

(b) $A_L x B_Q$ Most Dominant

(c) $A_Q x B_L$ Most Dominant

(d) $A_Q x B_Q$ Most Dominant

Figure 5.3 Some Dominant Trend Patterns of AB Interaction

Example 5.15

Referring to the trend components calculated in the previous example, we can get some idea of the dominant trends by comparing sums of squares (each would be used in the numerator of an F statistic with a common denominator, MSE.) Thus, we see that the quadratic trend of temperature is the most significant of all. Within pressure, the quadratic trend is more significant than the linear trend. Of the four trend components of the interaction, $P_L x T_Q$ is the most significant. A visual inspection of the means in the pressure-temperature subclass, Example 5.14, indicates some quadratic trend within rows and within columns---but considerably greater within columns (across the temperature levels). The greater significance of T_Q and $P_L x T_Q$ bear this out. ✱

The reader should be able to do a trend analysis for any 2-factor factorial. The only difference from the above results is the number of trend components that might be considered. For any factor, the maximum number corresponds to the degrees of freedom. Thus, a general axb factorial would allow for a maximum of (a-1) polynomial trends for main effect A means, (b-1) for main effect B means, and (a-1)(b-1) for interaction components.

Trend analyses for experiments with three or more factors are accomplished by extending the two factor results. With three factors having equally spaced levels, trend components for main effects and two-factor interactions should present no problem. For the ABC interaction, three dimensional trend components may be considered. These are of the form $A_L x B_L x C_L$, $A_L x B_L x C_Q$, and so on. For each desired trend, the means from the ABC subclass are multiplied by appropriate coefficients. These coefficients are formed in a three-way table from products of the corresponding single factor trend coefficients, in the same manner that Table 5.6 was formed.

5.7.6 Augmented Factorials

In experiments dealing with factorial treatments, a researcher may want to include a control, if such is not represented among the factorial combinations. For example, a person applying 3 levels of nitrogen (say 20, 30, and 40 lbs) in combination with 2 levels of phosphorus (say 5 and 10 lbs) may want to include plots on which no nitrogen or phosphorus is applied. This experiment would have t = 7 treatments: those of the 3x2 factorial plus a control.

In other experiments, a researcher may want to include both a "control" and a "standard" treatment in addition to those of a complete factorial. Here the treatments would consist of the complete factorial set "plus two others."

Consider a slightly different experiment situation where a researcher begins with a complete factorial and adds certain (but not all) combinations of factor levels. For the 3x2 experiment discussed above, one might add a seventh treatment consisting of (50 lbs. nitrogen, 15 lbs. phosphorus). Or, one might add two additional treatments

50 lbs nitrogen, 15 lbs. phosphorus
60 lbs nitrogen, 20 lbs phorphorus.

A control could be added also. As one can see, the number of possibilities is unlimited.

The term *augumented factorial* will refer to any experiment having a complete factorial set of treatments plus one or more additional treatments. How does one analyze the data from an augumented factorial? Can any factorial components (main effects and interactions)be investigated? The answer to the last question is in the affirmative. Let us view the treatments as those of two groups: (i) those of the complete factorial, and (ii) the other treatment(s). Then, the total treatment sum of squares can be partitioned into components by orthogonal contrasts. The special contrasts of Section 4.4.1 can be applied to the complete factorial group to provide information about the main effects and interaction(s). Other contrasts are formed to provide comparisons of the 2 groups (such as: Control vs. all factorial treatments.) The "other treatment(s)" may form several groups instead of one; for example:

(ii)   Control
(iii)   Standard
(iv)   other nonfactorial treatments.

Example 5.16

Suppose the 3x2 nitrogen-phosphorus experiment described earlier was conducted with two additional treatments: a control and a "standard" defined by 50 lbs nitrogen, 20 lbs. phosphorus. There are t = 8 treatments whose overall sums of squares have 7 degrees of freedom. Thus, a set of 7 orthogonal contrasts will partition this sum of squares. Using the following two contrasts

$$C_1 = \text{Standard versus 3x2 Factorial}$$

$$C_2 = \text{Control versus All others}$$

one of several possible treatment partitions is

Source	df
(Treatments)	(7)
Nitrogen	2
Phosphorus	1
N x P	2
c_1	1
c_2	1

Suppose the 8 treatments are defined by

1 $n_1 p_1$	2 $n_1 p_2$	3 $n_2 p_1$	4 $n_2 p_2$
5 $n_3 p_1$	6 $n_3 p_2$	7 Standard	8 Control

Then, using Table 4.1, the coefficients needed to partition the overall treatment sum of squares are:

Contrast	Treatment							
	1	2	3	4	5	6	7	8
Nitrogen	+1	+1	-1	-1	0	0	0	0
	+1	+1	+1	+1	-2	-2	0	0
Phosphorus	+1	-1	+1	-1	+1	-1	0	0
N x P	+1	-1	-1	+1	0	0	0	0
	+1	-1	+1	-1	-2	+2	0	0
c_1	+1	+1	+1	+1	+1	+1	-6	0
c_2	+1	+1	+1	+1	+1	+1	+1	-7

The first contrast really compares the low and middle levels of nitrogen which are represented by the pairs of treatments: (1,2) and (3,4), respectively. This is the reason why the coefficients occur in pairs. The second contrast compares the high level of nitrogen to the two lower levels. These two contrasts are orthogonal, and taken collectively, are one way to explain the overall effect of nitrogen.

The single contrast for phosphorus compares the low and high levels, treatments (1,3,5) and (2,4,6), respectively. Again, the coefficients are the same for each level. Coefficients for the interaction are obtained by multiplying those for main effects (as was done to obtain the interaction trend components, see the previous section.) ▓

5.8   Inferences For Mixed and Random Effects Models

When all levels of all treatment factors occur in the experiment, a fixed effects factorial model is appropriate. Otherwise, a mixed or random effects factorial model would be required. The random model may be considered a special case of the mixed model, where all factors have random effects, and thus need not be discussed separately.

The linear models are identical for the three broad types of factorial effects. Two and three factor models for CR designs appear in (5.15) and (5.22).

Sums of squares calculations needed to construct the regular ANOVA table are the same for all factorial models. And so are the degrees of freedom and mean squares.

Assumptions about factorial treatment components depend upon the type of effects. Earlier in (5.25) and (5.26), we gave the assumptions for the two factor random model and one of the mixed models. It should be clear how one can extend these assumptions to experiments having more than two factors. We emphasize again that all random model components are assumed to be distributed independently of each other. This independence implies that our observations can be explained as a sum of "independently supplied random components." Each random component becomes important in its own right, irrespective of the significance of related inter-actions. This will be apparent again later when we discuss the basic ANOVA tests.

5.8.1 Expected Mean Squares

In ANOVA, the EMS determine the format of the test statistic for many hypotheses. Nearly all of the interesting hypotheses are concerned with treatment aspects so that EMS are particularly important in factorial experiments. But writing the EMS is more complicated for mixed and random effects models than for fixed effects models.

When some factors have fixed effects and others have random effects, the EMS for any factorial effect may have variance components in addition to σ^2_ϵ and a source component. The source component will be a κ^2 component if all source symbols are fixed, otherwise it will be a σ^2 component. Recall that we earlier defined source symbols as the factor letters identifying each particular source of variation in the ANOVA table. So in a 3-factor experiment with factor A random and factors B, C fixed, the source component for the AB interaction would be σ^2_{AB} but the source component for the BC interaction would be κ^2_{BC}.

To aid in generating the EMS components for the mixed and

random effects models, we define the term *random symbols* to be the letters of all factors having random effects. Coefficients of EMS source components were discussed in the previous section; that rule holds for all factorial components. We now concentrate on writing just the components of each source of variation. A few simple rules aid in generating the EMS components of any mixed or random factorial model:

(1) Every EMS contains σ^2_ϵ and a source component.

(2) All EMS components, except σ^2_ϵ, of any given source contain *at least the source symbols* as subscripts. This is true whether the components are σ^2 or κ^2. Therefore, σ^2_{ABC} might appear in the EMS for A but σ^2_{BC} could not.

(3) Add a σ^2 component for every combination of random symbols not appearing in the source symbols. These combinations of random symbols appear as subscripts together with the source symbols.

For more information about these rules and extensions to more complex experiments, see Lentner [4].

Example 5.17

Suppose we consider a 3 factor factorial experiment where factors A and B are fixed and factor C is random. For the factorial treatment sources of variation, we give the EMS components, omitting the coefficients and plus signs for simplicity.

Source	EMS components		
A	σ^2_ϵ	σ^2_{AC}	κ^2_A
B	σ^2_ϵ	σ^2_{BC}	κ^2_B
AB	σ^2_ϵ	σ^2_{ABC}	κ^2_{AB}
C	σ^2_ϵ	σ^2_C	
AC	σ^2_ϵ	σ^2_{AC}	
BC	σ^2_ϵ	σ^2_{BC}	
ABC	σ^2_ϵ	σ^2_{ABC}	

Note that the EMS for each source of variation contains σ^2_ϵ and a source component. For the last four sources of variation all random symbols (only the letter C) are present among the source symbols so no components are added to the two required by the first rule. For the first three sources of variation though, the random symbol C is missing so a σ^2 component is added to each, the subscripts being C together with the source symbols. The source component for each of the last three sources of variation is a σ^2 component because at least one random symbol is present.※

Example 5.18

EMS components are given for a 4 factor factorial where factors A and C are fixed. Coefficients and plus signs are omitted.

Source	EMS components				
A	σ^2_ϵ	σ^2_{ABD}	σ^2_{AB}	σ^2_{AD}	κ^2_A
B	σ^2_ϵ	σ^2_{BD}	σ^2_B		
AB	σ^2_ϵ	σ^2_{ABD}	σ^2_{AB}		
C	σ^2_ϵ	σ^2_{BCD}	σ^2_{BC}	σ^2_{CD}	κ^2_C
AC	σ^2_ϵ	σ^2_{ABCD}	σ^2_{ABC}	σ^2_{ACD}	κ^2_{AC}
BC	σ^2_ϵ	σ^2_{BCD}	σ^2_{BC}		
ABC	σ^2_ϵ	σ^2_{ABCD}	σ^2_{ABC}		
D	σ^2_ϵ	σ^2_{BD}	σ^2_D		
AD	σ^2_ϵ	σ^2_{ABD}	σ^2_{AD}		
BD	σ^2_ϵ	σ^2_{BD}			
ABD	σ^2_ϵ	σ^2_{ABD}			
CD	σ^2_ϵ	σ^2_{BCD}	σ^2_{CD}		
ACD	σ^2_ϵ	σ^2_{ABCD}	σ^2_{ACD}		
BCD	σ^2_ϵ	σ^2_{BCD}			
ABCD	σ^2_ϵ	σ^2_{ABCD}			
Exp. Error	σ^2_ϵ				

Note: for each source of variation, all components except σ_ϵ^2
contain the source symbols among the subscripts.   On the A line,
A is a subscript on the last four components;   on the AC line,   A
and C are subscripts on all components except σ_ϵ^2.

The middle three components on the A line were added by rule
(3).   Random symbols B and D do not appear among the source
symbols of this line---the three combinations: B,   D and BD are
added as subscripts.   On the D line, only one random symbol, B is
missing so only one additional component is added.   And on the BD
line, no random symbols are missing---this EMS contains only
σ_ϵ^2 and a source component. ✱

5.8.2   Estimation of Effects and Means

As we stated in Section 1.9, inferences about a random effects
factor are directed toward a population of parameters not all of
which appear in any one repetition of the experiment.   For a
random factor then, the pertinent inference is directed toward a
variance component for the population of effects;   estimating the
effects corresponding to the levels which appear in one specific
experiment is of little or no interest.   In the random effects
models, all inferences are concerned with variance components.
And in the mixed models, some inferences are directed toward
variance components(for random components) while other inferences
may involve estimation of specific parameters (fixed components).

For the remainder of this section, let us consider a three
factor factorial with factors A and B fixed and factor C random.
Here it may be of interest to estimate effects and means for A,
for B, and for the AB subclass. These estimates may be calculated
as in Section 5.7.1.

Estimating the variances of means and differences of means is
more complicated than in the fixed effects case.   The general
format can be specified, in terms of ANOVA mean squares, for main
effects in a three factor mixed model, such as the one mentioned
above.   The estimated standard deviations are, for the difference
of A means

$$(5.45) \qquad s_{\bar{y}_{i\ldots}-\bar{y}_{h\ldots}} = \sqrt{2MSAC/rbc}$$

and for the difference of B means

$$(5.46) \qquad s_{\bar{y}_{.j..}-\bar{y}_{.h..}} = \sqrt{2MSBC/rac}$$

One notes that the mean squares involved in these two variances
are precisely the ones which would appear in the denominators of
the F statistics used to test the corresponding effects. See the
EMS components in Example 5.16 to verify this.

For a four factor mixed model, the difference of means in a
subclass defined by fixed factors (such as the AB subclass when
factors C and D are random) have variances which are combinations
of mean squares of the ANOVA.   This same complexity occurs   even
for main effect means in any mixed factorial with two or more
random factors. One can readily see this from the EMS components
in Example 5.17 where differences of main effect A (or C) means
would be of interest. We shall not pursue this any further.

5.8.3   Estimation of Variance Components

Whereas estimation of effects is of interest for fixed treatment
parameters, the estimation of variance components is of interest
for random treatment parameters.   Once the EMS are available,
variance components are estimated quite easily by taking differences of appropriate mean squares.

Example 5.19

After supplying the coefficients and plus signs in the EMS of the
previous example, we see that the estimate of the ACD variance
component is

$$\hat{\sigma}^2_{ACD} = (MSACD - MSABCD)/rb$$

and the estimate of the BC variance component is

$$\hat{\sigma}^2_{BC} = (MSBC - MSBCD)/rad \quad \text{✻}$$

5.8.4   Tests of Hypotheses

Tests about main effects and interaction components in a mixed or
random effects factorial are made with reference to the EMS. The
denominator of any given F ratio may not be the Exp. Error mean
square; it might be any one of several interaction mean squares,
depending upon the random and fixed factors.   And as the number
of random factors increases, more tests of treatment components
fall in the approximate category.

Upon examination of the EMS components given in Example 5.17, it is apparent that only a few factorial effects will be tested against the Exp. Error mean square. Only tests of no interactions of BD, ABD, BCD and ABCD will use test statistics having MSE in the denominator. A test of no AB interaction would use a test statistic having the mean square for ABD in the denominator. And we see that there are no exact tests for main effects A nor for the AC interaction. When testing no A effects, H_0: $\kappa_A^2 = 0$, we see that no two EMS are equivalent even when $\kappa_A^2 = 0$. Similarly for null hypotheses of no C effects and no AC interaction, no pair of EMS are equivalent under either of these hypotheses. Approximate tests could be made by Satterthwaite's procedure, discussed in Section 2.12.

For the mixed and random models, some of the implications are different than in the fixed model. Remember that all components having random effects are assumed to be distributed independently of each other. Consequently, the nonsignificance of any σ^2 component would imply that the corresponding random effect has no appreciable value in predicting the responses of the variable y. In Example 5.17, if H_0:$\sigma_B^2 = 0$ was not rejected, we would conclude that factor B does not contribute significantly to explaining the responses. Such an implication doesn't necessarily follow when B has fixed effects, see the discussion in Section 5.7.3 concerning significant interactions.

5.8.5 Contrasts and Post-ANOVA Procedures

Contrasts and mean separation procedures would rarely, if ever, be considered for random treatment components. Variances of random components is of primary concern but usually there is no interest in differences of means for the specific levels which happen to appear in a given experiment.

For the fixed treatment components of a mixed factorial, the researcher might consider contrasts or mean separation procedures depending upon the type of structure among their levels. The procedures discussed in Sections 5.7.4 and 5.7.5 are appropriate, for the most part, for use in the mixed factorials. Certain modifications will need to be made in the variances of means and their differences, and in the corresponding degrees of freedom. We have discussed these variances and problems of estimating them (Section 5.8.2). When variances involve combinations of mean squares, approximate inference procedures must be used. These are beyond the scope of this book; the interested reader is referred to Graybill [1].

PROBLEMS

1. A 3x5 factorial experiment (on factors A and B, respectively) was conducted in a completely randomized design with r = 6 observations per treatment. Both factors have fixed effects. For this experiment:

    (a) give the linear model. Indicate what each term represents and specify ranges of subscripts.
    (b) what assumptions are required for valid tests and CI?
    (c) give a partial ANOVA: sources, df, and EMS.

2. Repeat Problem 1 if both factors have random effects.

3. A 3x2x4 factorial experiment (on factors A,B,and C, respectively) was conducted in a CR design with r = 5 observations per treatment.   Assume all factors have fixed effects.   For this experiment:

    (a) give the linear model. Indicate what each term represents and specify ranges of subscripts.
    (b) what assumptions are required for valid tests and CI?
    (c) give the partial ANOVA: sources, df, and EMS.

4. Repeat Problem 3 if all factors are assumed to have random effects.

5. The following partial ANOVA table was constructed from a CR design having r = 8 EU per treatment:

Source	df	MS	EMS
A	1	32.7	
B	2	18.6	
AB		27.1	
Exp.Error		9.4	
Total			

    (a) Give the linear model which accompanies an ANOVA table such as this. Indicate ranges on subscripts.
    (b) Complete the above table under a fixed effects model.

6. Repeat Problem 5 under a random effects model.

7. The following partial ANOVA table was constructed from a CR design having r = 3 EU per treatment:

Source	df	MS	EMS
A	3	0.0143	
B	2	0.0267	
C	1	0.5535	
AB		0.0442	
AC		0.0810	
BC		0.0421	
ABC		0.0110	
Exp. Error		0.0262	

(a) Give the linear model which accompanies an ANOVA table such as this. Indicate ranges on subscripts.

(b) Complete the above table under a fixed effects model.

8. Repeat Problem 7 if factor A has fixed effects and factors B and C have random effects.

9. In a home economic experiment, two temperatures(factor A) and two lengths of washing time (factor B) gave the following treatment means:

$y_{11\cdot}$ $y_{21\cdot}$ $y_{12\cdot}$ $y_{22\cdot}$

(1)   6.2   a   14.1   b   5.4   ab   7.8

The experiment was conducted as a CR design with r = 5. The variable of interest was y = amount of impurities (in grams) remaining in the material washed. Assume fixed effects for both factors. Give numerical values for estimates of

(a) the simple effects of A   (b) the simple effects of B
(c) the main effect of A   (d) the main effect of B
(e) the interaction of A and B.

10. An agricultural engineer investigated 2 hitch arrangements (factor A) and 2 hitch lengths (factor B) to determine their effect on the force needed to move a stationary object. Six trials were made (on a fixed, hard surface) for each of the 4 combinations, giving means:

(1)   184   a   207   b   230   ab   218

(Courtesy of Dr. J. Perumpral, Ag. Eng. Dept., Va. Tech.)

Assume fixed effects for both factors. Give numerical values for estimates of

(a) the simple effects of A   (b) the simple effects of B
(c) the main effect of A   (d) the main effect of B
(e) the interaction of A and B.

11. Refer to the home economic experiment of Problem 9.

(a) Calculate the sums of squares for the main effects A and

   B and the AxB interaction.
(b)   From the ANOVA, the Exp. Error MS was 29.8.   Test   the
      factorial treatment components for significance.

(12) Refer to the engineering experiment of Problem 10.

   (a)   Calculate the sums of squares for the main effects A and
         B and the AxB interaction.
   (b)   From the ANOVA, the Exp. Error MS was 352.7.   Test the
         factorial treatment components for significance.

13. Consider the table accompanying each plot of treatment   means
    shown below.   Check the most appropriate cell of each   treat-
    ment component--without doing "sums of squares calculations."
    Assume Exp.Error MS is small enough to judge "any difference"
    as   significant.   (Sig = Significant;   NS = Not Significant)
    All means are based on the same number of observations.

(a)

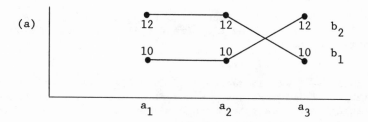

	Sig	NS	Meaningless	Can't tell
Main effect A				
Main effect B				
AxB Interaction				

(b)

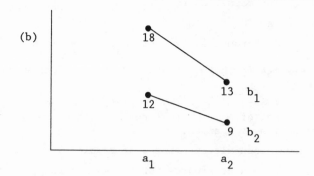

	Sig	NS	Meaningless	Can't tell
Main effect A				
Main effect B				
AxB Interaction				

14. Repeat Problem 13 with the following plots of treatment means. Again assume equal replications.

(a)

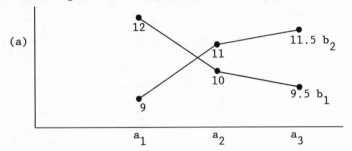

	Sig	NS	Meaningless	Can't tell
Main effect A				
Main effect B				
AxB Interaction				

(b)

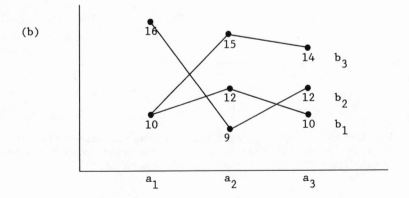

	Sig	NS	Meaningless	Can't tell
Main effect A				
Main effect B				
AxB Interaction				

15. Use the mean plots of Problem 13 above and assume the levels of each factor are equally spaced. For each trend component, check the most appropriate cell of the table.

	Sig	NS	Does not exist.	Can't tell.
(a) A_L				
A_Q				
B_L				
B_Q				
$A_L xB_L$				
$A_L xB_Q$				
$A_Q xB_L$				
$A_Q xB_Q$				

	Sig	NS	Does not exist.	Can't tell.
(b) A_L				
B_L				
$A_L xB_L$				

16. Use the mean plots of Problem 14 above and assume the levels of each factor are equally spaced. For each trend component, check the most appropriate cell of the table.

(a)

	Sig	NS	Does not exist.	Can't tell.
A_L				
A_Q				
B_L				
B_Q				
$A_L \times B_L$				
$A_L \times B_Q$				
$A_Q \times B_L$				
$A_Q \times B_Q$				

(b)

	Sig	NS	Does not exist.	Can't tell.
A_L				
A_Q				
B_L				
B_Q				
$A_L \times B_L$				
$A_L \times B_Q$				
$A_Q \times B_L$				
$A_Q \times B_Q$				

17. To investigate the effect of sulphur and nitrogen on growth of red clover, a plant scientist conducted a greenhouse experiment using a CR design with 4x2 factorial treatments. The levels of sulphur were applied at rates equivalent to 0, 3, 6, and 9 lbs/A, while nitrogen was was either absent or applied at a rate equivalent to 20 lbs/A.

A sufficient number of greenhouse pots were prepared with uniform soil and a uniform stand of red clover so that r = 3

were available for each factorial treatment combination.   The
following dry matter yields (in grams/pot) were obtained:

Nitrogen	Sulphur 0	3	6	9
0	4.48	4.70	5.21	5.88
	4.52	4.65	5.23	5.98
	4.63	4.57	5.38	5.91
20	5.76	7.01	5.88	6.26
	5.64	7.11	5.82	6.26
	5.78	7.02	5.73	6.37

(a) Construct the ANOVA table under a fixed effects model.
(b) Make appropriate tests of treatment components.
(c) For significant treatment components, give tables of mean
    differences and standard deviations.

18. To investigate the effect on potato yields of water loss due
    to transpiration, a horticulturist used shade covers on plots
    at various stages of their growth and development. Plots were
    shaded to reduce solar input(to the plants) by 0, 1/3 or 2/3.
    Each of the 3 shadings were applied to 4 plots for a one
    month period during "early", "middle", and "late" stages of
    growth. The design was an RCB. Yields per plot (in lbs) were:

Block	Shading	Growth Stage Early	Middle	Late
	0	60	65	62
1	1/3	54	57	58
	2/3	41	53	56
	0	53	68	70
2	1/3	46	53	62
	2/3	42	58	54
	0	64	58	54
3	1/3	48	59	63
	2/3	36	50	60
	0	50	61	57
4	1/3	42	52	60
	2/3	50	49	51

(a) Construct the ANOVA table under a fixed effects model.
(b) Make appropriate tests of treatment components.
(c) For significant treatment components, give tables of mean
    differences and standard deviations.

19. A research and development department measured the tensile
    strength of aluminum wire made under all combinations of

    Factor A: Amounts of magnesium, 0, 1, or 2%
    Factor B: Formation, rolled or annealed.

Three specimens were randomly selected from each treatment
group and the following tensile strengths were obtained (in
psi, each value was divided by 1000):

| | % Magnesium | | |
Formation	0	1	2
Rolled	17.8	21.4	30.6
	18.2	20.1	32.1
	16.9	22.3	31.0
Annealed	6.5	10.4	14.6
	5.2	11.7	14.3
	5.9	12.6	15.9

(a) Construct the ANOVA table under a fixed effects model.
(b) Make appropriate tests of treatment components.
(c) For significant treatment components, give tables of mean
    differences and standard deviations.

20. Three different lubricants were investigated, each with and
    without a graphite additive, for their effect on the wear of
    a new steel alloy. Twenty-four specimens were used in the
    experiment, four were randomly assigned to each factorial
    treatment combination. Each specimen was weighed, subjected
    to a 20-hour wear test, and weighed again. The weight losses
    (in mg.) were:

| | | Lubricant | |
Additive	1	2	3
With	35.8, 36.1	29.4, 28.6	30.6, 31.7
	33.2, 34.5	27.5, 30.2	31.4, 30.2
Without	40.5, 42.2	37.5, 36.4	34.4, 33.8
	41.8, 41.0	40.1, 38.6	33.6, 34.9

(a) Construct the ANOVA table under a fixed effects model.
(b) Make appropriate tests of treatment components.
(c) For significant treatment components, give tables of mean differences and standard deviations.

21. Refer to Problem 17 above and perform the most appropriate post-ANOVA analyses for this experiment. Consider results of statistical tests performed earlier. Let $\alpha = 0.05$.

22. Refer to Problem 18 above and perform the most appropriate post-ANOVA analyses for this experiment. Consider results of statistical tests performed earlier. Let $\alpha = 0.05$.

23. Refer to Problem 19 above and perform the most appropriate post-ANOVA analyses for this experiment. Consider results of statistical tests performed earlier. Let $\alpha = 0.05$.

24. Refer to Problem 20 above and perform the most appropriate post-ANOVA analyses for this experiment. Consider results of statistical tests performed earlier. Let $\alpha = 0.05$.

25. Suppose the data given below resulted from a 2x3 factorial experiment in a CR design. Table entries are means, each based on observations from $r = 3$ EU.

| | Factor B | | |
Factor A	12%	19%	26%
4 sec.	18	25	6
8 sec.	20	28	11

(a) Calculate numerical values for the trend components of the main effects and interaction.
(b) Calculate the sum of squares for each trend component.

26. Suppose the data given below resulted from a 2x4 factorial experiment in a CR design. Table entries are means, each based on observations from $r = 5$ EU.

| | Factor B | | | |
Factor A	2 oz	5 oz	8 oz	11 oz
Early	10	16	17	21
Late	18	21	20	25

(a) Calculate numerical values for the trend components of the main effects and interaction.
(b) Calculate the sum of squares for each trend component.

27. In a human engineering experiment, worker's efficiency was measured for 4 levels of sound intensity each played at 3 intermitent times (for example, played for 2 minutes, off for 2 minutes, and so on.) Entries in the table below are mean efficiency ratings, each mean based on the performance and output of r = 4 workers.

Intensity (Factor A)

Factor B		30	40	50	60
Time	2	14	17	18	15
(Minutes)	4	11	12	21	21
	6	10	13	20	24

(a) Calculate numerical values for the trend components of the main effects and interaction.
(b) Calculate the sum of squares for each trend component.

28. Refer to Problem 27 and suppose that Exp. Error sum of squares was 168.6. Make tests of the trend components.

29. In the application of hardening agents to ferrous materials (drill bits, bolts, camera parts, etc.), Wahl and Etchells [6] discussed two types of quenching: water bath and "salt" cool bath. Specimens made under each type of quenching were subjected to various lengths of testing with the following average weight losses (in mg.) reported:

	Quenching	
Testing time (hours)	Water	Salt
3	7.0	6.5
5	10.8	7.8
8	20.5	11.6
11	30.0	18.5
15	48.8	24.7

Assume that each mean is based on the results of r = 3 specimens and that Exp. Error sum of squares is 1442.

(a) Construct the ANOVA table under a fixed effects model.
(b) Make appropriate tests of treatment components.
(c) For significant treatment components, give tables of mean differences and standard deviations.

30. Refer to Problem 29 above and perform the most appropriate post-ANOVA analyses for this experiment. Consider results of statistical tests performed earlier. Let $\alpha = 0.05$.

31. The utilization of 3 energy diets by each of four breeds of domestic rabbits was investigated by Grobner, et. al. [12]. The diets (representing low, medium, high energy levels) contained 2000, 2500, and 3000 Kilo-calories of digestible energy per kilogram of diet. Six mature, non-pregnant does of each breed were used (two assigned at random to each energy level.) The following average daily feed intakes (in grams) were reported:

Breed	Energy Level		
	L	M	H
1	60	56	52
2	158	146	145
3	224	210	175
4	235	231	196

   (a) Set up the ANOVA table under a fixed effects model. Take the Exp. Error sum of squares to be 14,246.
   (b) Make tests of treatment components. Interpret.

32. Refer to the diet experiment of the previous problem and perform post-ANOVA analyses given that the four breeds are:

1	Dwarf size
2,3	Intermediate size
4	Giant size

   Consider results of statistical tests performed earlier and let $\alpha = 0.05$.

33. A 2x3x5 factorial was conducted in a CR design with $r = 3$ EU per treatment. Give the partial ANOVA (sources, df, and EMS) if factors A and C are random and factor B is fixed.

34. A 3x4x3 factorial was conducted in a CR design with $r = 2$ EU per treatment. Give the partial ANOVA (sources, df, and EMS) if factor A is fixed and factors B and C are random.

35. In a pallet research study, wood pallets were made for each combination of four factors:

   A (6 Types of Wood): Eastern Oak, yellow poplar, silver maple, yellow pine, Douglas fir, white spruce

B (6 Types of Fasteners):

    0.120 helically threaded stiffstock nail
    0.112   "   "   "   "
    0.105   "   "   nail
    0.120   "   "   hardened steel nail
    0.120 annularly   "   "   "   "
    15 guage, 7/16 inch crown, coated staple

C (3 Deckboard Thicknesses): 3/8, 5/8, 7/8 inches

D (3 Numbers of Fasteners per Joint): one, two, three

(Courtesy of Dr. M. White, Forest Products Dept., Va. Tech.)

(a) How many treatments are there in this experiment?
(b) What are the EU in this experiment?
(c) Give the linear model for this experiment. The pallets are made and tested in a completely randomized fashion but there were 8 replicates of each treatment.
(d) Give the partial ANOVA, sources, df and EMS, under the fixed effects model.

36. Refer to the pallet study of Problem 35 and suppose the ANOVA revealed significant differences only for the main effects (that is, no interactions were significant.) What post ANOVA procedures if any, would be most appropriate?

37. In an experiment to study the effect of urethane finish on cotton fabric, Wolfgang [7] reported the following tear strengths for the 7 treatments indicated:

Control	138.3
Hard Urethane	
1%	111.0
5%	105.2
10%	91.8
Soft Urethane	
1%	106.4
5%	103.4
10%	105.8

Assume each mean is the result of tests made on 5 pieces of fabric and take the Exp. Error sum of squares to be 2229.0.

(a) Construct the ANOVA table under a fixed effects model.
(b) Make appropriate tests of treatment components.

38. Refer to the previous problem and perform appropriate post-ANOVA analyses for this experiment. Consider results of statistical tests performed earlier.

39. Impact toughness tests were made on three types of steel when water cooled after tempering at various temperatures. Lei, et. al. [3] reported the following mean impact toughness values for the 11 treatments indicated:

Steel	Tempering Temperature (°C)				
	450	500	550	600	650
40Cr	7.3	7.3	13.3	14.4	14.7
40CrNi			13.4		
40CrNiMo	5.6	6.0	7.8	11.1	10.5

Suppose each mean is based on the results of r = 3 specimens and that Exp. Error sum of squares is 351.7

(a) Complete the ANOVA table under a fixed effects model.
(b) Make tests of treatment components.

40. Refer to the previous problem and perform appropriate post-ANOVA analyses for this experiment. Consider results of statistical tests performed earlier. Let α = 0.05.

41. A 2x4x2x3 factorial set of treatments were used in a CR design. Factors A and D were fixed while factors B and C were random. Each treatment appeared on 2 EU. Give the EMS for the sources of variation indicated.

Source	EMS
A	
B	
AB	
AC	
BC	
ABC	
AD	

42. Repeat Problem 41 if only factor B is considered fixed.

43. Oliveira, et. al., [5] conducted a 2^4 factorial experiment to investigate austenite grain size of high speed steel. The factors were:

A, pre-austenitization temperature, 890°C, 940°C
B, temper-annealing temperature, 750°C, 800°C
C, time, 4 hours, 24 hours
D, hardening temperature, 1150°C, 1200°C

Two specimens were subjected to each combination of these factors, giving austenite grain sizes (in μm):

			d_1	d_2
a_1	b_1	c_1	7.3, 7.5	9.9, 13.5
		c_2	8.4, 11.3	12.1, 12.9
	b_2	c_1	9.5, 9.5	10.4, 12.6
		c_2	9.5, 12.9	10.0, 13.3
a_2	b_1	c_1	7.9, 9.6	13.7, 13.9
		c_2	9.3, 9.7	12.2, 12.3
	b_2	c_1	8.7, 10.6	10.4, 11.7
		c_2	10.0, 10.3	11.3, 12.8

(a) Complete the ANOVA table under a fixed effects model.
(b) Make tests of treatment components.

44. Refer to the previous problem and perform appropriate post-ANOVA analyses for this experiment. Consider results of statistical tests performed earlier. Let $\alpha = 0.05$.

REFERENCES

[1] Graybill, F., *An Introduction to Linear Statistical Models, Vol. 1*, McGraw-Hill, New York, 1961.
[2] Grobner, M.A., K.L. Robinson, P.R. Cheeke, and N.M. Patton, Utilization of low and high energy diets by dwarf(Netherland Dwarf), intermediate (Mini Lop, New Zealand White) and giant (Fleming Giant) breeds of rabbits., *Jour. App. Rabbit Res.*, Vol. 8, 1985.
[3] Lei, T. C., C. H. Tang, and M. Su, New mechanism of high temperature (reversible) temper brittleness of alloy steel., *Heat Treatment Shanghai*, 1983.
[4] Lentner, M., Listing expected mean square components., *Biometrics*, Vol. 33, 1965.

[5]   Oliveira, M. M., I. M. Martins, and H. Carvalhinhos, Temper-
      annealing of high-speed steel AISI M41, *Heat Treatment*,
      1980.
[6]   Wahl, G., and I. V. Etchells, Salt bath ferritic nitrocar-
      burizing: an economic and versatile heat treatment process
      which solves a multitude of design problems., *Heat Treatment*
      1981.
[7]   Wolfgang, W.G., Urethanes as textile finishes., *Jour. Coated
      Fibrous Materials*, Vol. 1, 1972.

## Chapter 6

RANDOMIZED COMPLETE BLOCK DESIGNS

6.1 Introduction

Our discussions in previous chapters dealt with various aspects of treatment and response design components (defined in Section 1.6.) Mean separation procedures, contrasts, and factorials are some of the treatment design aspects while subsamplings belong to the response design category. Even though previous discussions were limited to completely randomized (CR) designs, the material of Chapters 3, 4 and 5 is applicable to most other experimental designs. As we apply this material to designs other than CR, we will indicate necessary modifications.

   In this and following chapters we will be discussing the experimental design category of research. Linear models must now contain experimental design components. Including these design components in an experimental model requires certain information about the experimental units (EU), information unavailable in the CR designs. One particular arrangement of the EU leads to the randomized complete block (RCB) designs, the subject of this chapter.

6.2 Some Design Considerations

In the CR designs, we assume that EU are homogeneous with respect to their potential effect on our response variable. For example, if plots of land are used as EU in a yield study of different cabbage varieties, a CR design is appropriate only if all plots have the same degree of fertility. Or if we are planning a greenhouse experiment to investigate the effects of different types of light on the growth of pine seedlings, we would obtain seedlings (the EU) of the same age, the same height, the same varietal seed stock, and so on.

Homogeneous effects of EU in the CR designs were assumed because we had no information of any consequence about the EU which could have been incorporated in the experimental plan. But what if we were told that a fertility gradient is present in the land that we wish to use for our cabbage experiment? In this case we expect certain plots of land (the EU) to have a different impact on yield and growth responses. Adjacent plots of land should have about the same influence on these responses whereas plots some distance apart likely have a different influence. And in the pine seedling experiment mentioned earlier, if uniform trees are not available, the researcher might expect different effects depending upon the initial height or age of the trees.

Similar concerns exist for many other experiments. An educator might expect different test scores from different school systems. An engineer might expect different strengths from batches of steel made on different days. An animal scientist might expect litter mates to show more similar reactions to drug injections or stress situations than those exhibited by animals from different litters.

Certain information (soil test data, different batches, litter mates, and so on) about the EU can be used to identify groups of EU which will provide homogeneous influences on our response variable. The available information dictates meaningful groupings or arrangements of the EU. In exchange for the time and effort put forth in grouping or arranging the EU, the researcher expects an improvement in the quality of inferences about the treatments. We now define these basic groups of EU.

Definition 6.1

A *block* is group of EU which provide homogeneous effects on a response variable. A *complete block* is a homogeneous group of EU upon which the t treatments appear equally often (usually only once.)

The notion of blocking refers to specific groupings(arrangements) of the EU in which subsets of homogeneous units are identified. Quite often the EU are naturally grouped by some criteria (age, location, initial height, and so on.) At other times the grouping is done by the researcher on the basis of available information.

Example 6.1

Four strips of land are available for use in an experiment. Suppose the strips are separated but otherwise similar (same location, terrain, and so on.) Soil tests have provided the following information on soil pH.

pH = 6.2

pH = 6.9

pH = 8.1

pH = 8.8

Whenever pH has an effect on the variable of interest, the researcher should consider the four strips of land as blocks. If each strip of land can be subdivided into t plots (EU), each strip could be used as a complete block. ▓

While most blocks contain only t EU, there is no reason why a block cannot contain 2t EU (in which case each treatment would appear twice) or 3t EU, and so on. Unless specified otherwise, we shall assume that blocks contain only t EU and will refer to these as basic RCB designs. A later section will address block sizes of mt, where m > 1.

In a number of instances, homogeneous blocks of t EU cannot be obtained. When a block contains less than t EU, and therefore cannot contain all treatments, we have an *incomplete block*. Designs having incomplete blocks are covered later.

An RCB design is appropriate if we can arrange the EU into homogeneous blocks according to their effect on our response variable. When properly constructed, each block should consist of homogeneous EU, or at least EU that are as homogeneous as possible. In practice, perfect homogeneity is rarely, if ever, attained. Furthermore, blocks should be constructed so that EU of two different blocks are as heterogeneous as possible. As we achieve these two criteria to a greater degree, we increase the precision of the experiment. On the other hand, if we completely ignored the heterogeneity among the EU and used a CR design (instead of a more appropriate RCB design,) the residuals would include the ignored block differences. Stated in another way, blocking removes an identified source of variability from the Exp. Error variation of a CR design. Thus, we say that *blocking is a form of error control*.

At this point, one might ask, "Why not always use an RCB design, even if we must randomly form blocks?" If we use r blocks, we will lose r-1 degrees of freedom from Exp. Error (had a CR design been used instead).   Because of this, the EU need to be arranged in blocks so that the heterogeneity will cause a reduction in the Exp. Error SS large enough to compensate for the loss in degrees of freedom.   So, if we block when all EU are homogeneous, we would inflate the Exp. Error MS and obtain less precise results than from the corresponding CR design.   This is due to the reduction in degrees of freedom without a comparable reduction in the Exp. Error SS.

Sometimes a researcher suspects heterogeneity among the EU but does not know specific sizes and locations of differences. The most effective blocking requires this knowledge.   The ideal way of obtaining this information is from *uniformity trials*, a preliminary experiment in which all EU are subjected to uniform conditions (a single treatment, the same management practices, environmental conditions, and so on.)   The effect of each EU can be determined and used in the formation of blocks.   Note that the use of a common treatment removes any possibility of treatments playing a role in the blocking process.   Blocking is a feature of the EU and not the treatments.   A given set of blocks can be used with many different treatments.

6.3   The Experimental Plan

We assume an experiment is planned to investigate t treatments which may or may not be structured.   Blocks containing t EU each are assumed to be available.   The t treatments are randomly assigned to the EU within each block, with the randomization done independently for each block.   This represents a restriction on the randomization process in that there is not complete freedom in allocating the treatments--all treatments are forced to occur equally often (usually once) within each block.

Example 6.2

An experiment is conducted with t = 5 treatments, say A, B, C, D, E, and r = 4 blocks.   One possible randomized allocation of the treatments is the following:

Block 1	B	D	C	A	E
2	B	A	E	D	C
3	C	D	E	B	A
4	D	B	A	E	C

Note that each treatment appears once in each block and the treatments are randomized within each block. ▓

The random allocation of treatments to EU in an RCB design is most easily accomplished by using random permutations. Tables A.2 and A.3 contain random permutations of 7 and 12 integers, respectively. From a randomly chosen starting point in the appropriate table, a random permutation is obtained for each block. If permutations of a table contain more than t integers, simply omit those greater than t.

Example 6.3

Consider the experiment of the previous example: t = 5 treatments in r = 4 blocks. We need 4 permutations "of 5 integers." Table A.2 gives random permutations "of 7 integers" so we may use these if 6's and 7's are omitted. From the (arbitrary) starting point, row 29 and column 11, using columns 11, 12, 13, and 14, we obtain the following permutations:

$$
\begin{array}{l}
7\ 4\ 3\ 1\ 2\ 5\ 6 \rightarrow\ 4\ 3\ 1\ 2\ 5 \\
1\ 7\ 5\ 4\ 3\ 2\ 6 \rightarrow\ 1\ 5\ 4\ 3\ 2 \\
1\ 7\ 6\ 2\ 4\ 3\ 5 \rightarrow\ 1\ 2\ 4\ 3\ 5 \\
6\ 2\ 1\ 5\ 3\ 7\ 4 \rightarrow\ 2\ 1\ 5\ 3\ 4 \quad ▓
\end{array}
$$

Because all t treatments appear in each complete block of homogeneous units, any contrast among treatments should reflect only treatment and extraneous components. The contribution due to any block would be eliminated under the assumption of homogeneous EU within blocks. Referring to Example 6.2, let us consider the contrast $\mu_A - 2\mu_C + \mu_D$ among the treatment means. To estimate the value of this contrast we would add the means of treatments A and D and subtract twice the mean of treatment C. In so doing, we note that Block 1 effect enters twice (once with A, once with D) and is removed twice (with C). Thus, the effect of Block 1 will not appear in the above contrast. By the same reasoning we see that none of the other block effects will appear either. With homogeneous EU within blocks, each block effect contributes equally to each treatment. We give a definition of this idea.

Definition 6.2

When one factor contributes equally to all levels of a second factor, the first factor is said to be *balanced out* of the second factor.

In an RCB design, therefore, the block factor is balanced out of the treatment factor. If one or more observations cannot be obtained in a block design, some of the balance feature is lost. With no missing values, the RCB is a *balanced design*.

6.4   Advantages and Disadvantages of RCB Designs

RCB designs have found widespread acceptance in many branches of scientific research since their discussion and analysis by Fisher [3] in 1926. The continued popularity of RCB designs is due to their advantages, the more important ones being:

(i) Straightforward analysis. Even with missing observations in some of the blocks, a meaningful analysis may be possible.
(ii) More accurate results. When a significant blocking can be done, differences due to EU are eliminated from treatment contrasts.
(iii) Increased sensitivity. Variability due to heterogeneous groups of EU is removed from Exp. Error.
(iv) Flexibility. Subject to conditions for a balanced design and available resources, there is no limitation on the number of treatments and/or blocks.

Major disadvantages of the RCB designs are:

(i) If t is large, homogeneous blocks may be difficult to set up. The more EU per block, the greater is the chance of their being heterogeneous.
(ii) If block and treatment effects interact (that is, they are not additive), the RCB analysis is not appropriate. See Section 6.13 for one way to address this problem.

6.5   Linear Model and Assumptions

The experimental observations from a basic RCB design are assumed to be explained by the linear model

(6.1) $y_{ij} = \mu + \tau_i + \rho_j + \epsilon_{ij}; \quad i = 1, 2, ---, t$
$\quad\quad\quad\quad\quad\quad\quad\quad\quad\quad\quad\quad\quad\quad\quad j = 1, 2, ---, r$

where

μ = overall mean, a constant

(6.2)

$\tau_i = \mu_{i.} - \mu$ = effect due to the i-th treatment

$\rho_j = \mu_{.j} - \mu$ = effect due to the j-th block

ϵ_{ij} = random component explaining all extraneous variation

Note that the linear model (6.1) does not include block-treatment interaction components (see the second disadvantage stated in the previous section.) We emphasize strongly that "no block-treatment interaction" is a necessary assumption of all RCB designs. For additional discussion on this, refer to Sections 6.8 and 6.13.

For an RCB design, each block-treatment combination defines a population of possible observations. For valid inferences under any RCB model, the residual components are assumed to be independently and identically distributed normal variables:

(6.3) ϵ's are i.i.d. $N(0,\sigma_\epsilon^2)$ for all i,j

Other assumptions depend upon the type of block and treatment factors. For fixed effects models, the additional assumptions are

(6.4) $\sum_i \tau_i = 0$ and $\sum_j \rho_j = 0$

For a random effects model, the additional assumptions are

τ's are i. i. d. $N(0,\sigma_\tau^2)$

(6.5) ρ's are i. i. d. $N(0,\sigma_\rho^2)$

The τ's, ρ's, and ϵ's are independently distributed.

There are two mixed models: one having fixed treatment and random block effects, the other having fixed block and random treatment effects. The appropriate segments of (6.4) and (6.5) make up the assumptions for these two models.

6.6   Data Tabulation and Least Squares Estimates

As experimental data are collected in an RCB design, they may be recorded conveniently in a tabulation as shown below.

Table 6.1   Data Tabulation for RCB Designs

Block	1	2	---	j	---	r	Trt. totals
Treatment 1	y_{11}	y_{12}	---	y_{1j}	---	y_{1r}	$y_{1.}$
2	y_{21}	y_{22}	---	y_{2j}	---	y_{2r}	$y_{2.}$
.							
.							
i	y_{i1}	y_{i2}	---	y_{ij}	---	y_{ir}	$y_{i.}$
.							
.							
t	y_{t1}	y_{t2}	---	y_{tj}	---	y_{tr}	$y_{t.}$
Bk.totals	$y_{.1}$	$y_{.2}$	---	$y_{.j}$	---	$y_{.r}$	$y_{..}$

The border totals are useful in forming estimates and calculating sums of squares.   First of all, these totals can be converted to means:

$$\bar{y}_{i.} = y_{i.}/r = \text{i-th treatment mean}$$

(6.6) $$\bar{y}_{.j} = y_{.j}/t = \text{j-th block mean}$$

$$\bar{y}.. = y../rt = \text{overall mean}$$

For least squares estimation in an RCB design, the system of normal equations is

$$rt\hat{\mu} + r \sum_i \hat{\tau}_i + t \sum_j \hat{\rho}_j = y..$$

(6.7) $$r\hat{\mu} + r\hat{\tau}_i + \sum_j \hat{\rho}_j = y_{i.} \quad \text{for } i = 1, 2, \ldots, t$$

$$t\hat{\mu} + \sum_i \hat{\tau}_i + t\hat{\rho}_j = y_{.j} \quad \text{for } j = 1, 2, \ldots, r$$

Two constraints must be imposed on this system of equations to obtain a solution.   Using the constraints

(6.8) $$\sum_i \hat{\tau}_i = 0 \quad \text{and} \quad \sum_j \hat{\rho}_j = 0$$

gives the least squares estimates

(6.9) $\hat{\tau}_i = \bar{y}_{i.} - \bar{y}_{..}$

and

(6.10) $\hat{\rho}_j = \bar{y}_{.j} - \bar{y}_{..}$

The estimates of the treatment parameters measure the deviation of the observed treatment mean from the overall mean. Then, from (6.2), we see that the difference of two treatment means, say $\mu_{i.} - \mu_{h.}$, may be estimated in either of two ways:

(6.11) $\hat{\mu}_{i.} - \hat{\mu}_{h.} = \bar{y}_{i.} - \bar{y}_{h.} = \hat{\tau}_i - \hat{\tau}_h$

Care must be exercised in considering and interpreting the observed block effects. In many experiments, the researcher knows there are differences among blocks so the only interest is in eliminating their effects from the analysis. Fertility gradients in agricultural (land) experiments often are of this nature. Batch effects in industrial experiments are another example.

There is another problem with block effects. This is due to the restriction on randomization imposed by an RCB design. The treatments are randomized within each block but blocks are not randomized. Consequently, outside influence may have a spurious correlation with observations within a block; that is, observed block effects may be due to outside influences in addition to differences in experimental material. Therefore, a "block effect" may represent a conglomerate of effects.

Unrestricted randomization (as in a CR design) would have diminished the possibility of an outside factor affecting an entire block of EU. But in most RCB experiments, there is little or no interest in making inferences about blocks anyway so the problem is not crucial. When properly constructed, blocks are expected to show different effects. Blocks are included in an experiment to account for variation in experimental material, and included in such a way that their effect is balanced out of treatments.

Occasionally though, an experimenter might wish to calculate estimates of the block effects to access the effectiveness of setting up blocks, particularly if any outside influences are believed to affect all blocks about equally. Subject to the above remarks, estimates of the block effects would be calculated as indicated in (6.10) above.

Example 6.4

In the early stages of processing, natural fibers (such as cotton and wool) require cleaning. A textile specialist investigated 4 cleaning processes for wool. Because different batches of wool are received (from different ranchers, suppliers, and so on), batches were taken to be blocks. Wool from 5 different batches was obtained. After removing foreign debris, the wool from each batch was thoroughly mixed, and a equal amount was assigned to each process. The losses in weight (in mg.) after cleaning and drying were:

Batch	1	2	3	4	5	$y_{i.}$
Process 1	21	36	25	18	22	122
2	26	38	27	17	26	134
3	16	25	22	18	21	102
4	28	35	27	20	24	134
$y_{.j}$	91	134	101	73	93	492 = y..

We assume that the weight loss, y_{ij}, can be represented as the sum of 4 components: a common value, an effect due to the process used, an effect due to the particular batch, and a residual quantity. In model form, this representation is

$$y_{ij} = \mu + \tau_i + \rho_j + \epsilon_{ij}; \quad i = 1, 2, 3, 4 \\ j = 1, 2, 3, 4, 5$$

The researcher was interested only in these 4 cleaning processes so fixed process (treatment) effects are assumed. Block effects probably would be considered random.

From the observed data we calculate the following estimates:

(i) the overall mean,

$$\bar{y}.. = 492/20 = 24.6$$

(ii) the treatment means,

$$\bar{y}_{1.} = 122/5 = 24.4 \qquad \bar{y}_{2.} = 26.8$$
$$\bar{y}_{3.} = 20.4 \qquad \bar{y}_{4.} = 26.8$$

(iii) the block means,

$$\overline{y}_{.1} = 91/4 = 22.75 \qquad \overline{y}_{.2} = 33.5 \qquad \overline{y}_{.3} = 25.55$$

$$\overline{y}_{.4} = 18.25 \qquad\qquad \overline{y}_{.5} = 23.25$$

(iv) the treatment effects,

$$\hat{\tau}_i = \overline{y}_{i.} - \overline{y}.. \rightarrow -0.2, \ 2.2, \ -4.2, \ 2.2$$

(v) the block effects,

$$\hat{\rho}_j = \overline{y}_{.j} - \overline{y}.. \rightarrow -1.85, \ 8.9, \ 0.65, \ -6.35, \ 1.35$$

Note that the estimated treatment and block effects each sum to zero; that is,

$$\sum_i \hat{\tau}_i = 0 \quad \text{and} \quad \sum_j \hat{\rho}_j = 0$$

which are the least squares constraints. ※

6.7.1   Sum of Squares Calculations: Basic RCB Design

The RCB model (6.1) may be considered a simple extension of the basic CR model (2.1) by including an additional component for blocking.   Consequently,   the ANOVA for an RCB experiment must reflect an additional source of variation due to the blocking.

In terms of calculations, only one additional sum of squares is required. We now need sums of squares for sources of variation under the broad headings of blocks, treatments, Exp. Error and total.   The formulas for these SS are the same for all four RCB models discussed in Section 6.5.   The general definitional and calculating formulas are

(6.12) $C = \text{correction term} = y^2../rt$

(6.13) $SSY = \text{total SS}$

$$= \sum_{ij}\sum(\overline{y}_{ij} - \overline{y}..)^2 = \sum_{ij}\sum y_{ij}^2 - C$$

(6.14)   SSBL = block SS

$$= \Sigma\Sigma(\bar{y}_{.j} - \bar{y}..)^2 = \frac{1}{t} \Sigma_j y^2_{.j} - C$$

(6.15)   SST = treatment SS

$$= \Sigma\Sigma(\bar{y}_{i.} - \bar{y}..)^2 = \frac{1}{r} \Sigma_i y^2_{i.} - C$$

(6.16)   SSE = Exp. Error SS

$$= \Sigma\Sigma(y_{ij} - \bar{y}_{i.} - \bar{y}_{.j} + \bar{y}..)^2 = SSY - SSBL - SST$$

Example 6.5

The SS calculations for the RCB experiment dealing with the four
cleaning processes are given.   The raw data, block and treatment
totals appeared in Example 6.4

Correction term: $C = (492)^2/20 = 12,103.2$

Total: $SSY = (21)^2 + (36)^2 + \ldots + (24)^2 - C = 724.8$

Blocks: $SSBL = [(191)^2 + (134)^2 + (101)^2 + (73)^2 + (93)^2]/4 - C$

$$= 500.8$$

Treatments: $SST = [(122)^2 + (134)^2 + (102)^2 + (134)^2]/5 - C$

$$= 136.8$$

Exp. Error: $SSE = 724.8 - 500.8 - 136.8 = 87.2$ ✹

The analysis of variance table for an RCB design provides a
convenient summarization of the above calculations.   Table   6.2
illustrates this summarization for the basic RCB.   The EMS   are
given for both fixed and random treatment effects.   In view   of
remarks made in the previous section, we will not give EMS   for
blocks.

Table 6.2   ANOVA for a Basic RCB Design

Source	df	SS	MS	EMS Fixed	EMS Random
Blocks	r-1	SSBL	MSBL	---	---
Trt.	t-1	SST	MST	$\sigma^2_\epsilon + r\kappa^2_\tau$	$\sigma^2_\epsilon + r\sigma^2_\tau$
Exp.Error	(r-1)(t-1)	SSE	MSE	σ^2_ϵ	σ^2_ϵ
Total	rt-1	SSY			

Example 6.6

The ANOVA table for the wool cleaning experiment can be formed
from the calculations in the previous example. Recall that fixed
processing (treatment) effects were assumed earlier.

ANOVA of Cleaning Processes for Wool

Source	df	SS	MS	EMS
Batches	4	500.8	125.2	---
Processes	3	136.8	45.6	$\sigma^2_\epsilon + 5\kappa^2_\tau$
Exp.Error	12	87.2	7.3	σ^2_ϵ
Total	19	724.8		※

6.7.2   Inferences Under the Fixed Effects Model

We suppose all assumptions in (6.3) and (6.4) hold.   Inferences
about treatments are of primary interest.   Tests, confidence
intervals, orthogonal and multiple comparisons are the common
inferences.   All procedures and techniques presented in Chapters
3 and 4 are applicable for inferences about treatment means in an
RCB experiment.   The presence of the block factor has no direct
effect on the inference process, unless there are missing values
or some of the assumptions do not hold.

The test of equal treatment means, equivalently no treatment
effects, is a test of

(6.17) $H_0: \mu_{1.} = \mu_{2.} = \cdots = \mu_{t.}$

or stated in terms of treatment effects

(6.18) H_0: All $\tau_i = 0$

The statistic for testing the above hypothesis is

(6.19) $F = MST/MSE$

which has the F distribution with $(t-1)$ and $(r-1)(t-1)$ degrees of freedom. The critical region for this test corresponds to the upper area of this F distribution.

Example 6.7

A test of equal mean weight losses for the cleaning processes would be:

$$H_0: \tau_1 = \tau_2 = \tau_3 = \tau_4 = 0$$

H_1: At least two τ_i are non-zero.

Test statistic: $F = MST/MSE$ with 3 and 12 df

$F = 45.6/7.3 = 6.25$

The observed significance level is $P < 0.05$. ✴

Data presentation might include a table of differences of pairs of treatment means and standard deviations. In a basic RCB experiment, the difference of two observed treatment means is $\bar{y}_{i.} - \bar{y}_{h.}$ with an estimated standard deviation of

(6.20) $s_{\bar{y}_{i.} - \bar{y}_{h.}} = \sqrt{2MSE/r}$

which is the same for all pairs of treatment means.
    A $100(1-\alpha)\%$ CI for a specific mean difference, say $\mu_{i.} - \mu_{h.}$, may be constructed from the limits

(6.21) $\bar{y}_{i.} - \bar{y}_{h.} \pm t_{\alpha/2} \, s_{\bar{y}_{i.} - \bar{y}_{h.}}$

where $t_{\alpha/2}$ is the upper $\alpha/2$ percentage point of the Student t distribution with (r-1)(t-1) degrees of freedom, those of MSE.

Example 6.8

A table of mean differences for the wool cleaning experiment would be

i	h	$\bar{y}_{i.} - \bar{y}_{h.}$
1	2	-2.4
1	3	4.0
1	4	-2.4
2	3	6.4
2	4	0
3	4	-6.4

The standard deviation of the difference of any two treatment means is estimated by

$$s_{\bar{y}_{i.} - \bar{y}_{h.}} = \sqrt{2(7.3)/5} = 1.71$$

A 99% CI for the difference of the last two treatment means, $\mu_{4.} - \mu_{3.}$, is

$$\bar{y}_{4.} - \bar{y}_{3.} \pm t_{0.005}(1.71) = 6.4 \pm 3.05(1.71)$$

$$= 1.2;\ 11.6.$$

The experimenter can be 99% sure that the difference in the mean weight loss for the third and fourth processes is between 1.2 and 11.6 mg. ▓

6.7.3 Inferences Under the Random Effects Models

As we have indicated before, the random effects models do not occur very often in practice. Should an experiment be conducted assuming random treatment effects, inference about the variance component σ_τ^2 would be of interest. A test of the hypothesis

$$H_0: \sigma_\tau^2 = 0 \qquad (6.22)$$

would be made by using the same test statistic as in the fixed effects case; namely,

(6.23) $F = MST/MSE$

where F has $t-1$ and $(t-1)(r-1)$ degrees of freedom. The critical region corresponds to the upper area of this F distribution.

Following the procedure discussed in Section 2.10, with the appropriate modification of degrees of freedom, an approximate CI can be constructed for σ_τ^2. And as in the CR design, an exact CI can be constructed for σ_ϵ^2, should one be desired.

6.8 Nonadditivity of Blocks and Treatments

For simplicity, nonadditivity will be discussed for the basic RCB design. With appropriate modification of notation, the ideas are applicable to RCB designs other than the basic setup.

An important assumption required for validity of the RCB analysis is "no block-treatment interaction." This is equivalent to saying that block and treatment effects are additive and that the experimental data can be represented by a linear model such as the one given in (6.1). When additivity does not hold, we should consider a linear model of the following form

(6.24) $y_{ij} = \mu + \tau_i + \rho_j + (\tau\rho)_{ij} + \epsilon_{ij};\ \begin{array}{l} i = 1, 2, \text{---}, t \\ j = 1, 2, \text{---}, r \end{array}$

where $(\tau\rho)_{ij}$ is the interaction component for treatments and blocks. Without getting unduly complicated at this time, let us simply say that the interaction and residual components are "inseparable". (See Chapter 9 for further discussion.) As a result, these two components are represented collectively by a single source of variation whose expected mean square will be a function of both interaction and residual components. Depending upon the fixed-random nature of components in (6.24), tests of treatment components may not be possible.

How can we investigate additivity of blocks and treatments, or the lack of it? A simple plot of the treatment observations per block, on a single graph, can be done as a first step. No statistical significance can be attached to this procedure but it might provide some insight into the matter.

Example 6.9

For the RCB design of Example 6.4 dealing with cleaning processes (treatments) and batches (blocks), we present a plot of the data.

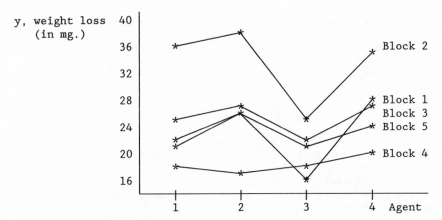

y, weight loss (in mg.)

Exact parallelism is not present,   nor should it be expected, because each plotted point represents a single observation from a different population. From the above graph, it is difficult to say whether the departures from parallelism are within the   realm of sampling variation or whether nonadditivity of block and treatment effects is significant. ▓

A statistical test for additivity is available. For the basic RCB model (6.1), Tukey [4] has presented a test procedure based on a subdivision of the residual sum of squares under the basic additive model (6.1).   Unfortunately though, Tukey's test is designed to investigate one specific pattern of interaction; namely, $(\tau\rho)_{ij}$ in model (6.24) given by the product $\phi\tau_i\rho_j$, where ϕ is some constant. Should other patterns be present, this test would not detect them. Secondly, this is not a very robust test; it is quite sensitive to departures from the RCB assumptions. Nevertheless, it may be used when nonadditivity is suspected, particularly if MSE appears to be large under the model (6.1).

As proposed by Tukey to measure the nonadditivity of treatment and block effects, a sum of squares with one degree of freedom is extracted from SSE, where SSE is calculated under the basic additive model (6.1). The sum of squares for nonadditivity is

$$(6.25) \qquad SSNA = rt[\underset{ij}{\Sigma\Sigma}\ y_{ij}(\bar{y}_{i.} - \bar{y}..)(\bar{y}_{.j} - \bar{y}..)]^2/(SST)(SSBL)$$

where SSBL and SST are, respectively, block and treatment sum of squares. Then, the "remainder" sum of squares is

$$(6.26) \qquad SSRem = SSE - SSNA$$

which has $q = (r-1)(t-1)-1$ degrees of freedom. For the hypothesis

H_0: Block and Treatment effects are additive

Tukey's test statistic is

$$(6.27) \qquad F_{1,q} = \frac{SSNA}{SSRem/q}$$

The critical region corresponds to the upper area of the $F_{1,q}$ distribution.

The (unsquared) numerator of SSNA can be calculated more simply from an equivalent expression involving totals rather than mean deviations. The sum is

$$(6.28) \quad \sum_{ij} y_{ij}(\bar{y}_{i.} - \bar{y}..)(\bar{y}_{.j} - \bar{y}..) = [\sum_{ij} y_{ij} y_{i.} y_{.j} - y.. (SSBL+SST+C)]/rt$$

where C is the general correction term.

Example 6.10

Tukey's test of additivity will be applied to the data of the RCB experiment dealing with cleaning processes. From Examples 6.4 and 6.5, we have

C = 12,103.2	SSBL = 500.8
SST = 136.8	SSE = 87.2

The table of data and totals is

Batches	1	2	3	4	5	$y_{i.}$
Processes 1	21	36	25	18	22	122
2	26	38	27	17	26	134
3	16	25	22	18	21	102
4	28	35	27	20	24	134
$y_{.j}$	91	134	101	73	93	

$$\sum_{ij} y_{ij} y_{i.} y_{.j} = 21(122)(91) + 36(122)(134) + \ldots + 24(134)(93)$$

$$= 6,274,864$$

Then

$$\text{SSNA} = \frac{[6,274,864 - 492(500.8 + 136.8 + 12,103.2)]^2}{20(500.8)(136.8)}$$

$$= (6390.4)^2/1,370,188.8$$
$$= 29.80$$

and

$$\text{SSRem} = 87.2 - 29.8 = 57.4$$

The F statistic for testing additivity is

$$F_{1,11} = (29.8)/(57.4/11) = 5.71$$

which gives an observed significance of $0.01 < P < 0.05$. ※

What should the researcher do if evidence of nonadditivity exists? Surely the researcher will want to investigate the plot of treatment versus blocks and try to come up with a reason for the inconsistent behavior of treatments in the different blocks. This may be valuable information in that one subset of treatments might be better for some conditions (blocks) while a different subset of treatments might be preferred in other conditions.

When nonadditivity is present in an RCB design, the basic analysis is at best only approximate and could provide quite erroneous inferences about treatments. For certain types of nonadditivity, a transformation of the data might be considered (see Section 2.14). Before doing a transformation of the data, it would be advisable to plot the data and investigate the patterns as discussed in the previous paragraph.

In some instances, the researcher may opt to re-design the experiment to investigate the nature of the "block-treatment" interaction. This could be done with a generalized RCB design, see Section 6.13.

6.9 Missing Values in an RCB Design

The analysis presented earlier for the basic RCB assumed that every treatment appeared once in every block. Occasionally, one or more observations are missing from the final data set. This may occur because (i) an animal dies before the experiment is completed, (ii) a plot is flooded, (iii) a wrong treatment is applied to an EU, or other reasons.

As indicated in the discussion following Definition 6.2, each missing value destroys the balance feature of an RCB design. Some treatment contrasts will be contaminated with block effects. Steps should be taken then to minimize the effect of missing data, especially as it affects inferences about treatments.

An acceptable procedure for analyses of experiments with a "few" missing values is to insert estimates which minimize the Exp. Error sum of squares. When more than a few missing values are present, the experimenter might consider the elimination of treatments or blocks in an effort to perform a "better" analysis.

For one missing value in a basic RCB design, the procedure has the following general format:

(i) insert an arbitrary symbol, say M, for the missing observation,

(ii) determine the value of M which will minimize the Exp. Error sum of squares,

(iii) insert the calculated value of M for the missing value and calculate the "usual" sum of squares, and

(iv) reduce the degrees of freedom for total and Exp. Error each by one.

While this procedure minimizes SSE, it does (upwardly) bias the treatment sum of squares. Fortunately, a simple subtraction will correct the bias introduced by M; this correction is indicated below.

Suppose the missing observation is y_{gh}, for treatment g and block h. Then we define the following totals

$$T_g = \text{total of all non-missing observations for treatment g}$$

(6.29) B_h = total of all non-missing observations for block h

$$G = \text{grand total of all non-missing observations}$$

Then, the calculated values of M which minimizes SSE is

(6.30) $$M = [rB_h + tT_g - G]/(r - 1)(t - 1)$$

The usual treatment sum of squares, SST, calculated after M has been inserted, needs to be corrected for the bias by subtracting

(6.31) $$\text{Bias} = [B_h - (t-1)M]^2/t(t-1)$$

The corrected treatment mean square has expected value as given in Table 6.2. An exact F test can then be made for treatments.

Example 6.11

The data for the RCB experiment dealing with cleaning processes is reproduced below with a value missing.

Batch	1	2	3	4	5	$y_{i.}$
Process 1	21	M	25	18	22	$86=T_1$
2	26	38	27	17	26	134
3	16	25	22	18	21	102
4	28	35	27	20	24	134
$y_{.j}$	91	98	101	73	93	$456=G$
		$(=B_2)$				

From (6.30), we calculate: $M = [5(98) + 4(86) - 456]/12 = 31.5$
Inserting this value for M in the above table, we calculate the SS by the usual formulas obtaining:

$$C = (487.5)^2/20 = 11882.81$$

$$SSY = 12,524.25 - 11882.81 = 641.44$$

$$SSBL = 12,307.56 - 11882.81 = 424.75$$

$$SST = 12,024.45 - 11882.81 = 141.64$$

$$SSE = 641.44 - 424.75 - 141.64 - 75.05$$

$$\text{Correction for Bias} = [98 - 3(31.5)]^2/4(3) = 1.02$$

$$\text{Corrected Treatment SS} = 141.64 - 1.02 = 140.62$$

$$MSE = 75.05/11 = 6.82$$

$$MST = 140.62/3 = 46.87$$

$$F = 46.85/6.82 = 6.87 \text{ with 3 and 11 df.}$$

The observed significance level is $P < 0.01$. ✳

We shall present a general method of dealing with (any number of) missing values when we discuss the analysis of covariance, see Chapter 9.

6.10   Efficiency of the Basic RCB to a CR Design

A natural question to ask is whether an RCB design is better than a CR design. Were the additional time required in setting up blocks and the extra calculations worthwhile? This becomes a more important issue when a researcher anticipates repeating a similar experiment at one or more future times. Such is the case when doing long-term research in psychological testing, engineering production, agronomic crops, and so on.

Remember that when blocking is done effectively, we expect more precise inferences from an RCB analysis than from a CR analysis. The best way to compare these two designs then would be to compare the inferences obtained when each design is used with the same treatments and experimental material. Needless to say, this would be an inefficient use of resources. A researcher usually conducts one experiment---most likely, an RCB if blocking can be justified. We therefore need some way to assess whether the RCB experiment has afforded us an appreciable gain in the quality of our inferences.

One way to compare the precision of two experimental designs is to compare the variances of treatment means for the 2 designs. The number of observations per treatment mean must be the same for both designs, so equivalently we may compare the Exp. Error mean squares of the ANOVAS. The only problem is that we have an ANOVA only for the completed RCB experiment. Consequently, we need to estimate an Exp. Error mean square for a CR design that could have been conducted with the same experimental material.

Treatments play no role in the formation of blocks (see the remarks at the end of Section 6.2.) Consequently, they should play no role in evaluating the precision of RCB and CR designs either. We therefore assume that both designs would have been completed with zero treatment effects. With treatment variation nonexistent, treatment SS differ from Exp. Error SS only because of sampling variation. Thus, SST should be combined with SSE. And had the experiment been conducted as a CR, the block SS would automatically be a part of the Exp. Error SS. Thus, an Exp. Error MS for a CR design estimated from a completed RCB design would be

$$(6.32) \qquad MSE* = \frac{(r-1)MSBL + (t-1)MSE + (r-1)(t-1)MSE}{(r-1) + (t-1) + (r-1)(t-1)}$$

$$= \frac{(r-1)MSBL + r(t-1)MSE}{rt - 1}$$

And with no treatment effects, the Exp. Error MS for the RCB design would be

$$(6.33) \qquad MSE^{**} = \frac{(t-1)MSE + (r-1)(t-1)MSE}{(t-1) + (r-1)(t-1)}$$

$$= MSE.$$

We now define the efficiency of an RCB design relative to a CR design as

$$(6.34) \qquad RE(RCB \text{ to } CR) = MSE^*/MSE^{**}$$

$$= \frac{(r-1)MSBL + r(t-1)MSE}{(rt-1)MSE}$$

The RE often is expressed as a percentage. Any value greater than 1 (or 100%) indicates that the RCB is more efficient than a CR design. The value of RE may be interpreted as the ratio of treatment replications needed for the two designs to give equal variances of treatment means. For example, if the value of RE in (6.34) was 2, the CR design would require twice as many replications to achieve the same variance of a treatment mean as the completed RCB did.

Example 6.12
For the wool cleaning experiment described in Example 6.4, an RCB design was used. From the ANOVA table in Example 6.6, we have MSE = 7.3, MSBL = 125.2 with 4 df, and treatments with 3 df. Thus, by (6.34), we obtain a relative efficiency measure of

$$RE(RCB \text{ to } CR) = \frac{4(125.2) + 15(7.3)}{19(7.3)} = 4.4$$

A value of 4.4 or 440% indicates that the RCB is considerably more efficient than a CR design. One would need 4.4 times as many replications of treatments with a CR design to achieve the same variance of a treatment mean as with the given RCB design. This is not surprising because we saw earlier that the blocks (batches) accounted for a major portion of the total variation. ✳

6.11 Factorial Treatments in an RCB Design

Factorial treatments in an RCB design present no new concepts or ideas. We measure block variation as in the basic RCB but the "overall treatment variation" is partitioned into components to measure main effects and interactions. An extensive discussion of factorial treatments occurred in Chapter 5; all of that material is applicable here.

To briefly illustrate some of the ideas, consider a set of t = ab factorial treatments to be investigated in an RCB design. The linear model for such an experiment would be

(6.35) $y_{ijk} = \mu + \alpha_i + \beta_j + (\alpha\beta)_{ij} + \rho_k + \epsilon_{ijk}$

where α_i, β_j and $(\alpha\beta)_{ij}$ are, respectively, components for the main effect and interaction of treatment factors A and B, ρ_k is a block component, and ϵ_{ijk} is a residual component.   A partial ANOVA (sources, df, SS and EMS) for this experiment follows.

Table 6.3   ANOVA for a Two Factor Factorial in an RCB Design

Source	df	SS	Fixed	Random
Blocks	r-1	SSBL	---	---
A	a-1	SSA	$\sigma_\epsilon^2 + rb\kappa_A^2$	$\sigma_\epsilon^2 + r\sigma_{AB}^2 + rb\sigma_A^2$
B	b-1	SSB	$\sigma_\epsilon^2 + ra\kappa_B^2$	$\sigma_\epsilon^2 + r\sigma_{AB}^2 + ra\sigma_B^2$
AB	(a-1)(b-1)	SSAB	$\sigma_\epsilon^2 + r\kappa_{AB}^2$	$\sigma_\epsilon^2 + r\sigma_{AB}^2$
Exp.Error	(r-1)(ab-1)	SSE	σ_ϵ^2	σ_ϵ^2
Total	rab-1	SSY		

The EMS given in Table 6.3 are for fixed or random treatment components. For the two mixed models, the EMS would follow rules given in the previous chapter.
    The sums of squares for A, for B, and for the AB interaction represent an orthogonal partitioning of the "overall treatment" or " AB subclass" sum of squares. Thus,

(6.36)   SST = SAB = SSA + SSB + SSAB

where SSA,   SSB, and SAB are calculated from the border and cell totals of the AB subclass.

6.12   Subsampling in RCB Designs

The researcher may find it necessary or advantageous to obtain
two or more observations per EU of an RCB design.   Subsampling
may be done in any of the RCB experiments: the basic designs,
ones having factorial treatments, and so on.

There   may   be several stages of   subsampling, as   indicated
earlier for the CR designs, although a single stage occurs with
the greatest frequency. Of course, the extent of subsampling may
need to be curtailed in an experiment having a large number of
factorial treatments.

Subsampling refers only to the response design portion of the
experimental model and analysis.   The ideas of Section 2.10 are
applicable here with slight modifications to account for the
presence of the blocking factor. Due to the balance requirements
of RCB designs, only the equal numbers case will be considered.

We will briefly discuss the basic RCB design when one   stage
of subsampling is included.   The linear model is

$$(6.37) \qquad y_{ijk} = \mu + \tau_i + \rho_j + \epsilon_{ij} + \delta_{ijk}; \quad \begin{array}{l} i = 1, 2, \cdots, t \\ j = 1, 2, \cdots, r \\ k = 1, 2, \cdots, s \end{array}$$

where we are assuming s subsamples per EU. The first 3 components
on the right-hand-side of (6.37) have the same definitions as in
(6.2).   The ϵ's measure variation among mean observations of the
EU and represent Exp. Error variation.   The δ's measure variation
within the EU (among the subsamples) and represent sampling error
variation.

In addition to the various assumptions about the block and
treatment parameters, the usual assumptions about the two "error
components" are

ϵ's are i.i.d. $N(0, \sigma_\epsilon^2)$

(6.38) δ's are i.i.d. $N(0, \sigma_\delta^2)$

The ϵ's and δ's are distributed
independently of each other.

We need not discuss the sums of squares calculations for this
design as they are analogous to what was done for single stage
subsampling in CR designs, Section 2.11.1. The major difference
is that now block variation is being extracted from Exp. Error
variation.   We give the ANOVA table to illustrate the general
idea.

Table 6.4 ANOVA for RCB with Subsampling: Fixed Treatment Effects

Source	df	SS	MS	EMS
Blocks	r-1	SSBL	MSBL	- - -
Treatments	t-1	SST	MST	$\sigma_\delta^2 + s\sigma_\epsilon^2 + rs\kappa_\tau^2$
Exp.Error	(r-1)(t-1)	SSE	MSE	$\sigma_\delta^2 + s\sigma_\epsilon^2$
Subsampling	rt(s-1)	SSS	MSS	σ_δ^2
Total	rts-1	SSY		

6.13   The Generalized RCB Design

An extension of the basic RCB design occurs with the use of blocks of size mt, where m is a positive integer greater than 1. To preserve balance, the researcher would randomly assign each treatment to m EU within each block. Randomization is done anew in each block. Such a setup is a generalized RCB design. Wilk [5] and Addelman [1] have discussed this design.

While this seems like a simple modification of the basic RCB design, the use of the generalized RCB design is very minimal. As might be anticipated, the researcher usually has difficulty in constructing or obtaining homogeneous blocks of mt EU each, even for m = 2.

A generalized RCB design permits the investigation of block-treatment interaction. Recall: each block-treatment combination defines a population of observations. Thus, when each treatment occurs m times within a block, the Exp. Error variability can be estimated from certain contrasts while other contrasts provide an estimate of block-treatment interaction---and these estimates will be orthogonal. This removes the problem of "inseparable" components indicated in our discussion of (6.24).

It is very important to realize the difference between the generalized RCB design and an RCB design with subsampling. The former design has mt EU in each block. Each treatment is randomly assigned to m EU within each block. A *single observation* is obtained from each EU. The latter design has t EU in each block. Each treatment is randomly assigned to a *single EU* within each block. But from each EU, s = m observations are obtained. In each design, a total of rmt observations is obtained, but in different manners as indicated. Even so, the subsampling design does not allow the investigation of interaction---the subsamples of each EU are not independent observations from the corresponding block-treatment population.

The linear model for a generalized RCB design is

$$(6.39) \qquad y_{ijk} = \mu + \tau_i + \rho_j + (\tau\rho)_{ij} + \epsilon_{ijk}; \quad \begin{aligned} i &= 1, 2, ---, t \\ j &= 1, 2, ---, r \\ k &= 1, 2, ---, s \end{aligned}$$

which is seen to be an extension of (6.24) to account for the additional EU per block. In addition to assumptions about the treatment and block factors, the usual assumption about the residual components is

$$(6.40) \qquad \epsilon\text{'s are i.i.d. } N(0, \sigma_\epsilon^2)$$

One notes that model (6.39) has a direct correspondence to a model for a "two factor factorial in a CR design." Consequently, sums of squares for a generalized RCB design may be calculated as in a factorial experiment. We give the ANOVA table under the fixed model.

Table 6.5 ANOVA for Generalized RCB Design

Source	df	SS	MS	EMS
Blocks	r-1	SSBL	MSBL	---
Treatments	t-1	SST	MST	$\sigma_\epsilon^2 + rm\kappa_\tau^2$
B x T	(r-1)(t-1)	SSBxT	MSBxT	$\sigma_\epsilon^2 + m\kappa_{\tau\rho}^2$
Exp.Error	rt(m-1)	SSE	MSE	σ_ϵ^2
Total	rtm-1	SSY		

There are three other possibilities depending upon the fixed or random nature of blocks and treatments. The ANOVA table as shown in Table 6.5 remains unchanged except for the EMS. The rules for finding EMS in factorials are not applicable here because of the nonrandomized nature of blocks. The interested reader may refer to Addelman [2]. We merely give a summary of the test procedures. In all four cases, the interaction would be tested by an F statistic having MSE in the denominator. A test of no treatment effects is made by an F statistic having MSE in the denominator if blocks have fixed effects but having MSBxT in the denominator if blocks have random effects.

PROBLEMS

1. An RCB design was conducted with 6 treatments and 8 blocks. Both treatments and blocks were fixed factors.
    (a) Give the linear model for this experiment. Explain terms used in the model and indicate ranges of subscripts.
    (b) Give assumptions required for valid tests and CI from this experiment.

2. Repeat Problem 1 for both factors random.

3. Repeat Problem 1 when treatments have a 3x2 factorial structure (factors A and B, respectively). All treatment and block factors have fixed effects.

4. Repeat Problem 3 for random treatment factors.

5. The partial ANOVA table given below was obtained from an RCB design.

Source	df	SS	MS	EMS
Blocks		200		
Treatments	4			
Exp. Error			10	
Total	29	550		

    (a) Complete the ANOVA table under the fixed effects model.
    (b) Test the hypothesis of zero treatment effects.
    (c) Give the estimated standard deviation of the difference of two observed treatment means.

6. Refer to the partial ANOVA table given in Problem 5.
    (a) Complete the ANOVA table under the random effects model.
    (b) Test the hypothesis of zero variance for the population of treatment effects.
    (c) Estimate the variance components.

7. The partial ANOVA table given below was obtained from an RCB design.

Source	df	SS	MS	EMS
Blocks	5			
Treatments		308		
Exp. Error			16	
Total	23	978		

    (a) Complete the ANOVA table under the random effects model.
    (b) Test the hypothesis of zero variance for the population
        of treatment effects.
    (c) Estimate the variance components.

8. Refer to the partial ANOVA table given in Problem 7.
    (a) Complete the ANOVA table under the fixed effects model.
    (b) Test the hypothesis of zero treatment effects.
    (c) Give the estimated standard deviation of the difference
        of two observed treatment means.

9. An engineer is evaluating four types of filters for their
   ability to remove impurities from engine oil. In addition to
   tests made at the engineer's research lab, three independent
   testing laboratories have been asked to make the same tests.
   The same brand of engine oil will be used for all tests.   An
   identical amount of impurities will be added to each quart of
   the oil.   The amount of impurities removed by a given filter
   will be determined by a standard process. The following data
   were obtained:

	Filters			
	A	B	C	D
Lab 1	28.8	31.0	30.4	27.3
Lab 2	28.4	29.4	30.5	27.1
Lab 3	29.6	30.8	31.7	28.0
Lab 4	28.9	29.9	31.0	27.4

   Consider that labs(blocks) have random effects and filters
   have fixed effects.
    (a) Give the linear model for this experiment. Explain terms
        used in the model and give ranges on subscripts.
    (b) What assumptions are required for valid tests and CI from
        an ANOVA of this data?
    (c) Set up the ANOVA under the above model and assumptions.
    (d) Test equality of mean impurities removed by the filters.
    (e) Give the estimated standard deviation of the difference
        of two observed treatment means.

10. Refer to the filtering experiment of the previous problem.
    Assuming no structure among the treatments
    (a) perform a Duncan's mean separation with $\alpha = 0.05$.
    (b) use Hsu's method to specify a subset of treatments con-
        taining the population having the greatest mean filtering
        ability.   Also give the SCI of the other treatment means
        with the greatest. Use a coverage assurance of 0.95.

11. An agronomist studied the yield of soybeans under various weed defoliage schemes. The experiment was conducted in an RCB design with four locations taken as blocks. The yields per plot (in pounds) were:

	Defoliage Schemes								
Location	1	2	3	4	5	6	7	8	9
1	12.1	13.0	11.5	12.6	14.3	12.7	10.0	10.7	3.0
2	12.9	12.0	11.3	10.0	13.7	12.6	9.3	10.0	7.3
3	12.0	13.5	8.4	8.7	12.6	12.0	12.7	10.5	5.5
4	14.1	9.6	13.5	13.0	14.0	9.6	8.8	8.9	4.9

(a) Give the linear model for this experiment and explain in practical terms what each model component represents.
(b) Construct the ANOVA table under the fixed effects model.
(c) Test the equality of mean yields for the 9 defoliage treatments.
(d) Give the estimated standard deviation of the difference of two observed treatment means.

12. Refer to the results of the previous problem and perform the following post-ANOVA analyses.
(a) Suppose the defoliage treatments have no structure. Test differences of pairs of treatment means by Duncan's procedure, with $\alpha = 0.05$.
(b) Suppose that treatment #9 is a control (no defoliage applied) but no other treatment structure is present. Perform a Dunnett's analysis with $\alpha = 0.05$.

13. Four drivers volunteer to drive their car in a study of mileage obtained under four different brands of gasoline. Each driver uses each brand in 100 miles of driving (over a fixed course, at fixed speeds, and so on.) The mileage per gallon was calculated for each run, giving the following:

	Brand			
Driver	A	B	C	D
1	12.6	10.9	11.3	13.2
2	16.4	16.1	17.6	17.9
3	13.0	10.6	11.8	12.6
4	19.2	18.0	18.9	19.9

(a) Give the linear model for this experiment. Explain terms used in your model.
(b) Set up a complete ANOVA under the fixed model.

(c) Test equality of treatment means. Interpret in practical terms.

(d) Give the estimated standard deviation of the   difference of two observed treatment means.

(e) As you begin to do the ANOVA, one of your co-workers questions whether a more extensive model should be used, one taking into account possibly different car effects. Comment on the co-worker's concern.

14. Refer to the mileage experiment of the previous problem.

(a) Suppose that the treatments have no structure. Use Hsu's method to specify a subset of treatments containing the population with the greatest mean mileage per gallon. Also give the SCI of the other treatment means with the greatest. Use a coverage assurance of 0.95.

(b) Suppose the 4 brands of gasoline are defined as follows:

A	Ewing with chemical JR
B	Ewing without chemical JR
C	Barnes without chemical CB
D	Barnes with chemical CB

Give a set of meaningful orthogonal contrasts and test each for significance.

15. A horticulturist performed a field experiment to study the effect of fungicide treatments applied to plots upon which azaleas were to be grown. The fungicides were applied to plots before inoculation in 4 complete block arrangements. Uniform plants were inoculated, planted, and after several weeks they were dug up and root weights determined. The results (in grams) were:

			Treatment		
Block	1	2	3	4	5
1	14	21	19	22	24
2	13	18	14	21	18
3	11	23	18	22	17
4	10	21	15	18	17

(a) Construct the ANOVA table under the mixed model with fixed treatments and random blocks.

(b) Test equality of mean root weights under the fungicide treatments.

(c) Give the estimated standard deviation of the difference of two observed treatment means.

16. Refer to the fungicide experiment of the previous problem.
   (a) If blocks were locations, different branch stations of a university's experimental farm system, which of the RCB assumption(s) might be of concern to the horticulturist?
   (b) Suppose the fungicide treatments were:

      1   Control, no fungicide applied
      2   Fungicide T
      3   Fungicide N, 1 oz.
      4   Fungicide N, 2 oz.
      5   Fungicide N, 3 oz.

   Investigate the treatment effects by the most meaningful set of orthogonal contrasts. Make tests of the contrasts.

17. The data recorded below was from a 2x3 factorial conducted in an RCB design with 2 blocks.

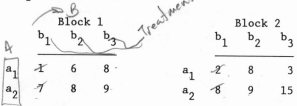

	Block 1					Block 2		
	b_1	b_2	b_3			b_1	b_2	b_3
a_1	1	6	8		a_1	2	8	3
a_2	7	8	9		a_2	8	9	15

   (a) Set up a complete ANOVA for this experiment, under the fixed effects model.
   (b) Test hypotheses of treatment components.   Interpret the results.
   (c) Plot the mean profiles for the two levels of factor A. Does the information obtained from this plot appear to be consistent with the findings in part b)?

   Give the estimated standard deviation of the difference of

   (d) two observed B means.
   (e) two observed AB means at a fixed level of A.

18. Refer to the results of the previous problem. Suppose post-ANOVA procedures will be done only for significant treatment components.

   (a) If the levels of factor B have no structure, how would a mean separation procedure (LSD, Tukey's, Duncan's, S-N-K) be performed?
   (b) If the levels of factor B have a gradient structure (equally spaced) how would a trend analysis be performed.

19. Suppose the four treatments (filters) in Problem 9 have a factorial structure defined by

$$A = a_1b_1 \qquad B = a_1b_2 \qquad C = a_2b_1 \qquad D = a_2b_2$$

where Factor 1 represents manufacturer and Factor 2 the amount of filtering material.
(a) Re-do the ANOVA presented earlier to account for this factorial treatment structure.
(b) Test hypotheses of treatment components.

20. Suppose the 4 treatments (brands of gasoline) in Problem 13 have a factorial structure with

$$A = a_1b_1 \qquad B = a_1b_2 \qquad C = a_2b_1 \qquad D = a_2b_2$$

where Factor 1 represents the company while Factor 2 the octane rating (b_1 = 85, b_2 = 89.) Note the difference between this treatment structure and that given in Problem 14(b).

(a) Re-do the ANOVA presented earlier to account for the factorial treatment structure.
(b) Test hypotheses of treatment components.

21. For the agronomist studying defoliage schemes, Problem 11, calculate the efficiency of the conducted RCB design relative to a CR design.

22. For the horticulturist studying the fungicides, Problem 15, calculate the efficiency of the conducted RCB design relative to a CR design.

23. For the engineer evaluating oil filters, Problem 9, calculate the efficiency of the conducted RCB design relative to a CR design.

24. For the experiment investigating gasoline mileage, Problem 13, calculate the efficiency of the conducted RCB design relative to a CR design.

25. A horticulturist is studying the effects of four growth regulators on chrysanthemums. There are 43 fairly uniform chrysanthemum plants available for the experiment and sufficient bench space in the greenhouse to place the 43 plants. There are several problems, however: (i) the bench space is disconnected, occuring at 12 places throughout the greenhouse (ii) there is room for at least 6 plants at each place, (iii) a ventilating fan is located at one end of the greenhouse,

and   (iv) the door to the greenhouse is located at the   other end.

Design what you believe would be the best experiment for   the horticulturist to use in this situation.   Give enough detail so   the horticulturist can perform the experiment.

26. A plant scientist is studying the effect of 5 inoculants on a commonly used legume.   He has 37 greenhouse pots upon which there is a fairly uniform stand of the legume.   He has   been given some space in the departmental greenhouse (among other people's experiments). He has listed the following potential problems:
   (i) the   available   space is disconnected,   it is in six   6 different locations of the greenhouse,
   (ii) at each of   these locations, there is room for 7 of the pots,
   (iii) the heater is located in the center of the greenhouse,
   (iv) some juveniles have thrown   rocks and cracked four   of the   glass   panes of the greenhouse (the glass   is   not missing, only cracked),
   (v) the entrance is   through a   storage shed (no   glass   or windows) attached to one end of the greenhouse.

Design an experimental layout for   this scientist,   including a   scheme showing the allocation of treatments and a skeleton ANOVA.

27. An engineer is planning an experiment to study the   strength of concrete "stepping blocks" made by 6 formulations: 2 stone sizes each with 3 cement mixtures.   With a small mixer,   the engineer   can   mix one batch of each formulation on   any   one day.   The smallest batch contains enough material to make   5 blocks.   The engineer plans to repeat the process on a number of "successive days not exceeding 10 days".

Design   an   experiment for the engineer.   Include   in   your report (i) the linear model,   (ii) the partial ANOVA,   (iii) treatment randomizations, and (iv) other issues or concerns.

28. A   textile chemist is   planning an experiment to study   color retention   of   a   new fabric when subjected to 2   different washing temperatures, 2 different rinsing temperatures, and 2 different   amounts   of   bleach (giving   8   treatments.)   The research   lab has 10 identical washing machines reserved   for the   chemist's   use.   A large supply of the   new   fabric   is available.   Two replicates of the treatments can be completed

in the morning and another two replicates can be completed in
the afternoon hours of any day. Each replicate uses different
pieces of the material. Color retention will be measured
after a given batch has been washed 20 times (that is, no
measurements are made after any of the first 19 washings of
any set of material.)

Design an experiment for the chemist. Include in your report
(i) the linear model, (ii) the partial ANOVA, (iii) treatment
randomizations, and (iv) any other issues or concerns.

29. An agronomist brings you an experimental plan that was
    developed for a greenhouse study. Specifically, 8 treatments
    were planned in an RCB design with r = 4 blocks. The random-
    ized allocation of treatments within ech block is shown.
    After talking with the agronomist, you discover that (i) all
    32 EU are located in one area of the work bench, (ii) the
    greenhouse is uniformly controlled with respect to humidity,
    temperature, lighting, and so on, and (iii) the greenhouse is
    used only by the agronomist for his research projects.

    Comment on the design of this experiment and make suggestions
    for improvement, if any are necessary.

30. A plant pathologist brings you a growth chamber experimental
    plan to investigate two levels of ozone at each of two humid-
    ities. The plan is to use 60 uniform plants of one species,
    15 plants to be placed in each of 4 growth chambers. One
    treatment (level of ozone, percentage humidity) will be ran-
    domly assigned to one growth chamber. After 4 weeks, a
    measurement will be made on each plant within each growth
    chamber. The pathologist wonders if location within a growth
    chamber should be used as a blocking factor.

    Comment on the design of this experiment and make suggestions
    for improvement, if any are necessary.

31. A group of scientists brings you an experimental plan for a
    large scale project to study the establishment of ground
    cover on reclaimed mine land. They plan to use a 4x3x3
    factorial set of treatments in an RCB design with r = 3
    blocks. The factors are

    A   4 soil compositions of soil base
    B   3 soil compositions of topsoil
    C   3 densities of initial cover

Each plot within a block is to be the size of an "average building lot" (that is, about 1/5 acre.) After 6 months, the percentage ground cover will be determined for each plot.

Comment on the design of this experiment and make suggestions for improvements, if any are necessary.

32. A marketing firm brings you an experimental plan to study the effects of 3 advertising campaigns for savings institutions. Their plan was to use an RCB design with 6 blocks defined as follows:

> Small Savings and Loan Companies
> Large Savings and Loan Companies
> Small State Banks
> Large State Banks
> Small National Banks
> Large National Banks

The terms small and large refer to the dollar amounts of savings on hand prior to the start of the advertising campaigns. Each advertising compaign is to be randomly assigned to one institution selected at random within each block category. After two weeks of advertising, the dollar volume of new savings will be determined for each participating institution.

Comment on the design of this experiment and make suggestions for improvement, if any are necessary.

33. A home economist is studying the moistness of cakes baked by 4 variations of a basic recipe. Four identical ovens are available for use in the experiment. For each recipe, a cake is mixed. These cakes are placed one to an oven, baked, and cooled. Each cake is then divided in thirds and a moisture reading is obtained on each part. This set-up was repeated on each of 6 days (taken as blocks.)

(a) Give the linear model for this experiment. Explain terms used in the model and specify ranges on the indices.
(b) Give the partial ANOVA, sources, df and EMS. Assume these are the only recipes of interest.

34. Engineers for a state highway department conducted an experiment to investigate the visibility of 4 different paints under consideration for marking lanes on highways. Eight different highways were chosen for testing; on one section of

each,   the   4 paints were randomly applied to a strip   across
the highway.   After two months, six visibility readings were
made on each strip, at randomly chosen sites.

(a) Give the linear model for this experiment.   Explain terms
    used in the model and specify ranges on the subscripts.
(b) What assumptions are necessary for valid tests and CI in
    an ANOVA.
(c) Give the   partial ANOVA,   sources,   df and   EMS.   Assume
    these are the only paints of interest.

35. For the cake moistness experiment conducted as described in
    Problem 33, the following partial ANOVA was obtained:

ANOVA for Cake Moistness

Source	df	SS
Days		
Recipes		
Exp. Error		
Samplings		
Total		

Means for the 4 Recipes
Means for the 6 Days

(a) Complete the ANOVA table assuming fixed recipes.
(b) Test the equality of mean moistness for the 4 recipes.
(c) Give the estimated   standard deviation for the difference
    of observed mean moistness for two different recipes.

36. For the paint testing experiment conducted as   described   in
    Problem 34, the partial ANOVA was obtained:

ANOVA for Paint Visibility

Source	df	SS
Highways		186.44
Paints		77.28
Exp. Error		
Samplings		
Total		

Means for the 4 Paints: 8.11   8.63   ·10.52   9.15
Means for the 8 Highways: 8.86   9.11   8.67   9.20   9.63
                          8.54 10.02   9.54

(a) Complete the ANOVA table assuming fixed paints.
(b) Test equality of mean visibilities for the 4 paints.
(c) Give the estimated standard deviation for the difference of observed mean visibilities for two different paints.

37. An engineer studied four electroplating processes. Wire rods, 1/4 inch in diameter, were electroplated by each process. Raw materials (the rods) were available from 5 different suppliers and it was decided to use this as a blocking factor. Uniform rods were randomly selected from each supplier's material and one of these was electroplated by each process. The thickness of the electroplating was determined for three randomly selected locations of each rod, giving (in 100 mm.):

| Process | Supplier | | | | |
	1	2	3	4	5
1	18.6	19.7	19.2	18.3	19.1
	19.2	19.2	19.4	18.0	19.6
	19.0	18.9	19.4	18.4	19.7
2	21.4	21.8	19.9	22.4	20.6
	22.0	21.8	20.1	22.1	21.2
	21.3	21.7	19.6	22.3	21.3
3	17.4	18.3	18.7	20.1	19.4
	17.6	18.3	18.7	20.2	19.3
	17.9	18.3	18.9	20.4	19.6
4	19.3	20.6	20.0	19.6	20.3
	19.7	20.1	20.4	19.5	20.3
	19.8	19.7	20.5	19.9	20.8

(a) Give the linear model for this experiment.   Define each term used.
(b) Give the ANOVA table, including EMS, under your model.
(c) Test that the mean thickness of the plating is the same for all processes.   Interpret your results.
(d) Give the estimated standard deviation of the difference between any two process mean thicknesses.

38. A plant scientist is studying the effect of irrigation and fertilization on the water retention of leaves on tobacco plants. He uses two levels of irrigation (low and medium) and three amounts of fertilizer (low, medium, and high). These six treatment combinations were randomized in three complete blocks and after ten weeks, two samples of leaves were randomly selected from each plot for analysis. The water retention values are presented below. (These have been coded by subtracting a constant value from each observation. This will not affect most results, specifically F-ratios, differences of means, their variances, and so on.)

	Block					
Treatment	1		2		3	
a_1b_1	2	3	3	2	3	4
a_1b_2	3	2	4	4	5	4
a_1b_3	4	6	5	4	5	5
a_2b_1	3	2	2	3	3	3
a_2b_2	4	3	3	5	3	4
a_2b_3	6	4	5	6	6	5

(a) Give the linear model for this experiment. Explain in words what each term in your model refers to.
(b) Set up a complete ANOVA assuming fixed treatment effects.
(c) Test treatment components at the 5% level. Interpret the results in practical terms.

REFERENCES

[1] Addelman, S., "Analyze 'em the way you randomize 'em," SUNY Buffalo, Research Report No. 29, 1969.
[2] Addelman, S., The generalized randomized block design, *The American Statistician*, Vol. 23, 1969.
[3] Fisher, R. A., The arrangement of field experiments, *Jour. Ministry Agr.*, Vol. 33, 1926.

[4] Tukey, J.W., One degree of freedom for additivity, *Biometrics* Vol. 18, 1949.
[5] Wilk, M. B., The randomization analysis of a generalized randomized block design, *Biometrika*, Vol. 42, 1955.

repeated measure Chp 7 Wiener

NESTED FACTOR EXPERIMENTS

7.1 Introduction

The factorials represent a particular class of experiments having treatments defined by multiple factors. Recall two prominent, identifying features of factorials: (i) The crossing of factors --- every level of each factor occurs with all levels of every other factor, and (ii) The interaction of two or more factors could be investigated.

There is a second class of experiments with multiple factors of interest, but the factors are not crossed. We begin with a definition for two factors.

Definition 7.1

If each level of factor A contains *different* levels of factor B, we say that *factor B is nested within factor A.*

Experiments with two or more factors satisfying this definition are called nested factor experiments or nested factor designs. Other than the common element of having several factors, the nested experiments differ from the factorials with respect to both features mentioned above. The levels of nested factors do not appear in all combinations; in fact, the levels of most factors change for every combination of levels of the other factors. Without this crossing of factors (as occurs in the factorials), there can be no interactions, of course.

A familiar example of nested factors is that afforded by states of the U.S. and their counties. There are different counties within different states. They would be different even if they have the same names, as for example, Montgomery counties in Virginia and Maryland. So in an experiment where both counties and states are factors of interest, we have "counties nested within states." The nesting process may continue; we might include cities nested within counties, and additionally, streets nested within cities. The number of nested factors ultimately used in an experiment is governed primarily by cost, size, and time considerations.

The factors in a nested experiment can be viewed as forming a hierarchy. From each level of a "first stage" or "major" factor, the experimenter selects levels of a "second stage" factor. Then from each of these selected levels, the experimenter selects levels of a "third stage" factor, and so on. For this reason, nested experiments also have been called "hierarchal designs" and "subsampling experiments".

Example 7.1

An industrial company is planning a new product that requires aluminum wire stock (15 millimeters in diameter). The company has identified three manufacturers and has asked each to supply six rolls of wire randomly selected from its inventory. Then from each roll, random sections of wire will be selected and analyzed for tensile strength.

This is a 3 factor nested experiment with major factor being manufacturer, rolls nested within manufacturers, and sections nested within rolls. The purchaser most likely would consider manufacturers a fixed factor and the 2 nested factors random.

Variability among rolls of a given manufacturer as well as variability among sections(within rolls) are measures of interest as they would indicate consistency of the wire stock. ※

Example 7.2

A soil scientist (see Edmonds, et. al. [1]) is studying the mineral content of three soil formations (major factor) in a particular county. Three sites are randomly selected within each formation. Then, two small areas are randomly selected at each site. A soil sample is obtained at each area for mineral analysis.

The second stage factor, sites, and the third stage factor, areas, provide the scientist with information to investigate variability among sites as well as the variability within sites (that is, among areas).

The soil scientist probably would consider soil formations a fixed factor and the two nested factors random. ※

As we have indicated in the two preceding examples, the first stage or major factor often is assumed fixed and the nested factors random. Occasionally several of the upper stage factors are assumed fixed with the remaining lower stage factors random. It seems inconceivable, in practical situations, to have a fixed nested factor at any stage below a random nested factor.

Example 7.3

The following illustrate additional situations of nested factor experiments:

(1) A sociologist studying cultural practices might use A = locations (in a single country), B = tribes within locations, and C = people within tribes.
(2) An educator might use A = schools (in a single school district), B = classes within each school, and C = students within classes.
(3) In a business efficiency study, an accountant might use A = banks, B = branches within banks, and C = employees within branches.
(4) An animal scientist might use A = breeds of cattle, B = animals within each breed, and C = randomly selected skin locations on each animal. ▓

Subsampling in [E.U.]

Nesting occurs also when subsampling is present in the basic experimental designs. The sampling units are nested within the EU. But such subsamplings do not represent factors of interest; rather, they are aspects of the response design. And as the earlier models reflected, components were included to account for variation among the subsamples.

7.2 Linear Model and Assumptions: Three Factors

First we consider a linear model for an experiment having three factors. Suppose factor A has a levels, factor B has b levels nested within each level of A, and factor C has c levels nested within each level of B. Then, the model is

$$(7.1) \qquad y_{ijk} = \mu + \alpha_i + \beta_{ij} + \epsilon_{ijk}; \quad \begin{array}{l} i = 1, 2, ---, a \\ j = 1, 2, ---, b \\ k = 1, 2, ---, c \end{array}$$

where

μ = overall mean, a constant

(7.2)

α_i = effect due to the i-th level of A

β_{ij} = effect due to the j-th level of B within the i-th level of A

ϵ_{ijk} = residual component for the (ijk)-th observation

These components may be written in terms of means as follows:

(7.3)

$\alpha_i = \mu_{i..} - \mu$ → means of factor B's in each A

$\beta_{ij} = \mu_{ij.} - \mu_{i..}$ → mean of factor in each level of B

where $\mu_{ij.}$ represents the mean for the j-th level of B within the i-th level of A, and $\mu_{i..}$ represents the mean for the i-th level of A.

At this time we should point out that care must be exercised when making inferences about factor C; if indeed, any should be made at all. The reason is that any extraneous variation is combined with the factor C effects. We note that model (7.1) does not contain a component that specifically identifies factor C effects. Rather, an overall residual component, an ϵ, is used to represent the composite of C effects and any other unexplained variation.

With fixed A effects, the assumptions accompanying model (7.1) are

(7.4)

(i) $\sum\limits_{i} \alpha_i = 0$

(ii) the β's are i.i.d. $N(0, \sigma_\beta^2)$ for each i

(iii) the ϵ's are i.i.d. $N(0, \sigma_\epsilon^2)$ for each i and j

(iv) the β's and the ϵ's are distributed independently of each other.

Example 7.4

The soil scientist of Example 7.2 would consider a linear model

$$y_{ijk} = \mu + \alpha_i + \beta_{ij} + \epsilon_{ijk}; \quad \begin{array}{l} i = 1, 2, 3 \\ j = 1, 2, 3 \\ k = 1, 2 \end{array}$$

where μ = overall mean, a constant

α_i = fixed effect due to the i-th soil formation

β_{ij} = random effect due to the j-th selected site within the i-th formation

ϵ_{ijk} = random component explaining all extraneous variation of the (i,j,k)-th observation

For inference purposes, the assumptions would be

$\sum\limits_i \alpha_i = 0$

The β's are i.i.d. $N(0,\sigma_\beta^2)$

The ϵ's are i.i.d. $N(0,\sigma_\epsilon^2)$

where the β's and ϵ's are distributed independently of each other. ※

7.3 Analysis of Three Factor Nested Experiments

A three factor nested experiment is one for which the model (7.1) is appropriate. The recorded data and pertinent totals for a nested experiment having a = 2, b = 4, and c = 3 appear in the table below.

Table 7.1 Observations and Totals: Three Factor Nested Experiment

	A_1				A_2			
	B_{11}	B_{12}	B_{13}	B_{14}	B_{21}	B_{22}	B_{23}	B_{24}
C_{ij1}	y_{111}	y_{121}	y_{131}	y_{141}	y_{211}	y_{221}	y_{231}	y_{241}
C_{ij2}	y_{112}	y_{122}	y_{132}	y_{142}	y_{212}	y_{222}	y_{232}	y_{242}
C_{ij3}	y_{113}	y_{123}	y_{133}	y_{143}	y_{213}	y_{223}	y_{233}	y_{243}
$y_{ij.}$	$y_{11.}$	$y_{12.}$	$y_{13.}$	$y_{14.}$	$y_{21.}$	$y_{22.}$	$y_{23.}$	$y_{24.}$
$y_{i..}$	$y_{1..}$				$y_{2..}$			

$y_{...}$

Only the totals shown in Table 7.1 are needed to complete the analysis. These are totals for each level of A, totals for each level of B within A, and the overall total.

7.3.1   Sums of Squares and the ANOVA Table

From the data and totals displayed in Table 7.1, the sums of squares are calculated as follows:

(7.5) $C = y^2_{...}/abc$ = general correction term

$$\left[\frac{(GT)^2}{abc} \right]$$

(7.6) $SSY = \underset{ijk}{\Sigma\Sigma\Sigma}\, y^2_{ijk} - C$ = total sum of squares

(7.7) $SSA = \frac{1}{bc} \underset{i}{\Sigma}\, y^2_{i..} - C$ = sum of squares for A

(7.8) $SSB(A) = \frac{1}{c} \underset{ij}{\Sigma\Sigma}\, y^2_{ij.} - C - SSA$ = sum of squares for B within A

(7.9) $SSE = SSY - SSA - SSB(A)$ = residual sum of squares

In view of the remarks given earlier, the sum of squares for the "third stage" is labeled SSE rather than SSC(B)(A). Upon observation, one notes that the sums of squares calculations exhibit a telescoping pattern: At the second stage, the sum of squares for the first stage is subtracted. At the third stage, the sums of squares for the two upper stages are subtracted. With these subtractions, all effects due to factors above a particular stage are eliminated, as they should be. This telescoping pattern extends to any number of nested factors.

Under the assumptions in (7.4), the ANOVA table is as follows.

Table 7.2   ANOVA for a Three Factor Nested Experiment

Source	df	SS	MS	EMS
A	a-1	SSA	MSA	$\sigma^2_\epsilon + c\sigma^2_\beta + bc\kappa^2_A$
B(A)	a(b-1)	SSB(A)	MSB(A)	$\sigma^2_\epsilon + c\sigma^2_\beta$
Residual	ab(c-1)	SSE	MSE	σ^2_ϵ
Total	abc-1	SSY		

Example 7.5

One of the minerals measured by the soil scientist, Examples 7.2
and 7.4, was calcium. Recorded as milligrams/gram (mg/g.), the
data were:

Site	Formation 1			Formation 2			Formation 3		
	1	2	3	1	2	3	1	2	3
Area 1	.283	.317	.577	.041	.196	.025	.237	.228	.028
2	.336	.336	.621	.051	.219	.046	.221	.258	.046
$y_{ij.}$	.619	.653	1.198	.092	.415	.071	.458	.486	.074
$y_{i..}$		2.470			0.578			1.018	
$y_{...}$					4.066				

Sums of squares calculations are

$$C = (4.066)^2/18 = 0.9185$$

Total: $SSY = (0.283)^2 + (0.317)^2 + \cdots + (0.046)^2 - C$

$$= 1.4448 - 0.9185 = 0.5263$$

Formations: $SSA = [(2.47)^2 + (0.578)^2 + (1.018)^2]/6 - C$

$$= 1.2452 - 0.9185 = 0.3267$$

Sites/Formations: $SSB(A) = [(0.619)^2 + \cdots + (0.074)^2]/2 - C - SSA$

$$= 1.4410 - 0.9185 - 0.3267$$

$$= 0.1958$$

Residual: $SSE = 0.5263 - 0.3267 - 0.1958$

$$= 0.0038$$

The following ANOVA table can be set up.

ANOVA of Calcium in Soil Formations

Source	df	SS	MS	EMS
Formations	2	0.3267	0.1634	$\sigma_\epsilon^2 + 2\sigma_\beta^2 + 6\kappa_\alpha^2$
Sites(F)	6	0.1958	0.0326	$\sigma_\epsilon^2 + 2\sigma_\beta^2$
Residual	9	0.0038	0.0004	σ_ϵ^2
Total	17	0.5263		※

7.3.2 Tests of Hypotheses

Hypotheses for a nested factor experiment are quite similar to those for a factorial experiment---stating equality of means, or equivalently zero effects, for fixed factors and stating zero variances for random factors.

We consider tests of hypotheses for an experiment having model (7.1) and assumptions (7.4). That is, we are assuming only fixed effects for the major stage factor. Then a test of primary interest is that of equal mean responses for the levels of factor A. In symbolic form, one way to state the hypotheses would be

(7.10) $H_0: \alpha_i = 0$ for all i   vs. $H_1: \alpha_i \neq 0$ for some i

From the EMS given in Table 7.2, we see that the appropriate test statistic is

(7.11) $F = MSA/MSB(A)$

which has the F distribution with (a-1) and a(b-1) degrees of freedom. The critical region is the upper tail area of this F distribution.

For a random second stage factor, the appropriate hypotheses would be

(7.12) $H_0: \sigma_\beta^2 = 0$ vs. $H_1: \sigma_\beta^2 > 0$

Again from the EMS column of Table 7.2, we can specify the test statistic; namely,

(7.13) $F = MSB(A)/MSE$

which has the F distribution with $a(b-1)$ and $ab(c-1)$ degrees of freedom. The critical region is the upper tail area of this F distribution. The null hypothesis in (7.12) is equivalent to saying there is no variation among the levels of factor B at each level of factor A. That is, the F statistic (7.13) simultaneously tests no variation in B across all levels of A.

Example 7.6

The soil scientist would test the equality of mean calcium for the three formations. For this test, we have

$$H_0: \alpha_i = 0 \text{ for all } i \quad \text{vs.} \quad H_1: \alpha_i \neq 0 \text{ for some } i$$

$$F_{2,6} = MSA/MSB(A) = 0.1634/0.0326 = 5.01$$

This gives an observed significance level of $0.05 < P < 0.10$.
     A test of no variation among calcium content for areas within formations would be

$$H_0: \sigma_\beta^2 = 0 \quad \text{vs.} \quad H_1: \sigma_\beta^2 > 0$$

$$F_{6,9} = MSB(A)/MSE = 0.0326/0.0004 = 81.5$$

This calculated value indicates an observed significance level of $P < 0.005$. There is reason to believe that calcium content differs among areas within at least one formation. ✱

7.3.3 Estimation of Effects and Variance Components

Estimates of means for the various subdivisions can be obtained from the totals shown in Table 7.1. We have

$$\bar{y}... = y.../abc = \text{overall sample mean}$$

(7.14) $\bar{y}_{i..} = y_{i..}/bc = \text{sample mean for the i-th level of A}$

$\bar{y}_{ij.} = y_{ij.}/c = \text{sample mean for the j-th level of B}$ at the i-th level of A

Estimates for fixed effect parameters can be obtained from these means, see (7.3). Thus, if factor A is fixed, estimates of the A effects are

$$(7.15) \qquad \hat{\alpha}_i = \bar{y}_{i..} - \bar{y}_{...}$$

The difference of two A level means, say $\theta_1 = \mu_{i..} - \mu_{h..}$, would be estimated by

$$(7.16) \qquad \hat{\theta}_1 = \bar{y}_{i..} - \bar{y}_{h..} = \hat{\alpha}_i - \hat{\alpha}_h$$

with an estimated standard deviation of

$$(7.17) \qquad s_{\bar{y}_{i..} - \bar{y}_{h..}} = \sqrt{2MSB(A)/bc}$$

Then, under assumptions (7.4), a CI for θ_1 can be constructed from the t distribution with $a(b-1)$ degrees of freedom.

In those experiments where the second stage factor is fixed, estimates of the β's would be given by

$$(7.18) \qquad \hat{\beta}_{ij} = \bar{y}_{ij.} - \bar{y}_{i..}$$

The difference of two B level means at the same level of A, say $\theta_2 = \beta_{ij} - \beta_{ih}$ would be estimated by

$$(7.19) \qquad \hat{\theta}_2 = \bar{y}_{ij.} - \bar{y}_{ih.} = \hat{\beta}_{ij} - \hat{\beta}_{ih}$$

with an estimated standard deviation of

$$(7.20) \qquad s_{\bar{y}_{ij.} - \bar{y}_{ih.}} = \sqrt{2MSE/c}$$

Then, a CI for θ_2 can be constructed from the t distribution with $ab(c-1)$ degrees of freedom.

Example 7.7

The soil scientist, Example 7.2, assumed fixed formations (factor A). Hence, estimates of formation effects and CI for differences of these effects may be of interest.

From the totals given in Example 7.5, we can calculate the formation effects

$$\hat{\alpha}_1 = (2.470/6) - (4.066/18)$$

$$= 0.412 - 0.226 = 0.186$$

$$\hat{\alpha}_2 = (0.578/6) - 0.226$$

$$= 0.096 - 0.226 = -0.130$$

$$\hat{\alpha}_3 = (1.018/6) - 0.226$$

$$= 0.170 - 0.226 = -0.056$$

These estimates indicate that calcium content is above average for formation 1 and below average for formations 2 and 3.

The difference in mean calcium content for formations 1 and 2 would be $\theta = \mu_1. - \mu_2.$ which is estimated by

$$\hat{\theta} = \hat{\alpha}_1 - \hat{\alpha}_2 = 0.186 - (-0.130) = 0.316$$

with an estimated standard deviation of $s_{\hat{\theta}} = \sqrt{2(0.0326)/6} = 0.104$

Using the t distribution with 6 degrees of freedom, a 95% CI for θ is obtained as follows:

$$\hat{\theta} \pm s_{\hat{\theta}} t_{0.975} = 0.316 \pm 0.104(2.45)$$

$$= 0.316 \pm 0.255$$

$$= (0.061 \; ; \; 0.571)$$

The scientist can be 95% confident that the difference in mean calcium content for formations 1 and 2 ranges from 0.061 to 0.571 mg/g. ※

We now consider estimation of variance components associated with random factors. These estimates are calculated from the differences of mean squares as indicated by the EMS column of the ANOVA table.

The EMS of Table 7.2 are for the most common situation: fixed major factor A and random nested factors B and C. For this type of experiment, estimation of σ_ϵ^2 and σ_β^2 might be of interest. Equating mean square entries with estimates of the EMS components gives

(7.21)
$$\hat{\sigma}_\epsilon^2 = MSE$$
$$\hat{\sigma}_\beta^2 = [MSB(A) - MSE]/c$$

The variance component σ_β^2 refers to the variation among the β's within a given level of A. And we have assumed this variation to be homogeneous for all levels of A. The quantity $\hat{\sigma}_\beta^2$ given in (7.21) is the estimate of this common variance.

We have indicated earlier that σ_ϵ^2 represents the extraneous variation, including variation due to factor C. The quantity $\hat{\sigma}_\epsilon^2$ is the estimate of this variance component.

Example 7.8

From the ANOVA table given in Example 7.5, the soil scientist would estimate the two variance components by

$$\hat{\sigma}_\epsilon^2 = 0.0004$$
$$\hat{\sigma}_\beta^2 = [0.0326 - 0.0004]/2 = 0.0161$$

These two values indicate a very small amount of variation among sites within a given area (measured by 0.0004) with considerably more variation among areas within a given formation (measured by 0.0161). ▓

For a three factor nested experiment having all factors random, the general ANOVA given in Table 7.2 is appropriate if κ_α^2 in the EMS column is replaced by σ_α^2. In such an experiment, the estimator of the major stage variance component would be

(7.22) $$\hat{\sigma}_\alpha^2 = [MSA - MSB(A)]/bc$$

7.3.4   Contrasts and Mean Separation Procedures

For the fixed factors of a nested experiment the researcher may wish to consider specific contrasts or comparisons of pairs of means. These inferences are accomplished by the procedures of Chapters 3 and 4. Because most nested experiments have fixed effects factors only for one or two "upper stages," the need for applying these techniques is somewhat limited. And as usual, contrasts would be more appropriate when structure is present among the (fixed) factor levels while multiple comparison and range procedures would be considered when structure is lacking.

A basic quantity required for contrasts and mean separation procedures is the standard deviation of the difference of two observed means. Thus, for inferences about the A level means in a three factor nested experiment, the standard deviation is given in (7.17).

7.3.5   Unequal Numbers

We first make the following observation in the equal numbers case: A three factor nested design corresponds to a CR design with subsampling.   Compare the model and assumptions given by (7.1) thru (7.4) with those given by (2.21) thru (2.23). The major factor A of the nested design corresponds to treatments in the CR design.   The second stage factor B corresponds to EU in the CR design.   Finally, factor C corresponds to the subsamples. Clearly then, the only difference in the two designs is whether the "second and third stages" represent factors of interest or response design components. This translates into differences in the definition and interpretation of components under the models for the two designs.   All other aspects are equivalent: sums of squares calculations, ANOVA table, test statistics, and so on.

Likewise, when unequal numbers are present in a three factor nested design, a correspondence can be made to an unequal numbers CR design with subsampling.   Consequently, all calculations of Section 2.10.2 are appropriate for an unequal numbers nested design.   With the appropriate change of "factor labels" and "model components", Table 2.4 gives the ANOVA for the unequal numbers, three factor nested design. Finally, calculation of EMS coefficients would be done as in (2.34) and (2.35).

7.4   Nested Experiments Having Four or More Factors

All material presented in the previous section for three factor nested experiments can be extended in a straightforward manner to handle nested experiments having more than three factors.   We merely list some of the major aspects to illustrate what is involved.

For a four factor nested experiment, the linear model would be

$$(7.23) \qquad y_{ijkm} = \mu + \alpha_i + \beta_{ij} + \gamma_{ijk} + \epsilon_{ijkm}; \quad \begin{array}{l} i = 1, 2, \text{---}, a \\ j = 1, 2, \text{---}, b \\ k = 1, 2, \text{---}, c \\ m = 1, 2, \text{---}, d \end{array}$$

where μ, α_i, and β_{ij} have the same definitions as in (8.2) and

$$\gamma_{ijk} = \text{effect due to the k-th level of factor C within the j-th level of B within the i-th level of A}$$

(7.24)

$$\epsilon_{ijkm} = \text{residual component for (ijkm)-th observation}$$

Now ϵ_{ijkm} represents a composite of factor D effects and any other extraneous variation. See the comments following (7.2).

For the experiments having major factor A fixed and nested factors B,C, and D random, the assumptions are

(i) $\sum\limits_{i} \alpha_i = 0$

(ii) the β's are i.i.d. $N(0,\sigma_\beta^2)$ for each i

(7.25)   (iii) the γ's are i.i.d. $N(0,\sigma_\gamma^2)$ for each i and j

(iv) the ϵ's are i.i.d. $N(0,\sigma_\epsilon^2)$ for each i, j, and k

(v) the β's, γ's, and ϵ's are distributed independently of each other.

As one sees from model (7.23), four subscripts are required for identification of the observations (three were required in Table 7.1). Then to complete the sums of squares calculations, we need the overall total as well as totals for

(i) each level of A
(ii) each level of B within A
(iii) each level of C within B within A.

For completeness, we give the sums of squares formulas:

(7.26) $C = y^2.../abcd$ = general correction term

(7.27)   Total: $SSY = \sum\sum\sum\sum_{ijkm} y^2_{ijkm} - C$

(7.28)   Factor A: $SSA = \dfrac{1}{bcd} \sum_{i} y^2_{i...} - C$

(7.29)   Factor B within A: $SSB(A) = \dfrac{1}{cd} \sum\sum_{ij} y^2_{ij..} - C - SSA$

(7.30)   Factor C within B within A:

$$SSC(B)(A) = \dfrac{1}{d} \sum\sum\sum_{ijk} y^2_{ijk.} - C - SSB(A) - SSA$$

(7.31)   Residual: $SSE = SSY - SSA - SSB(A) - SSC(B)(A)$

Notice that the telescoping pattern alluded to after (7.9) is present in the above sums of squares calculations. With these calculations, and assumptions (7.25), we present the ANOVA table.

Table 7.3   ANOVA for a Four Factor Nested Experiment

Source	df	SS	MS	EMS
A	$a-1$	SSA	MSA	$\sigma^2_\epsilon + d\sigma^2_\gamma + cd\sigma^2_\beta + bcd\kappa^2_\alpha$
B(A)	$a(b-1)$	SSB(A)	MSB(A)	$\sigma^2_\epsilon + d\sigma^2_\gamma + cd\sigma^2_\beta$
C(B)(A)	$ab(c-1)$	SSC(B)(A)	MSC(B)(A)	$\sigma^2_\epsilon + d\sigma^2_\gamma$
Residual	$abc(d-1)$	SSE	MSE	σ^2_ϵ
Total	$abcd-1$	SSY		

Several features of Table 7.3 are worth noting. Inspection of the EMS reveals that a *telescoping effect* occurs, with one component dropped on each EMS from the top. But, the degrees of freedom "telescope" in the opposite direction: from top to bottom, there is an increase in the symbols entering the products.

PROBLEMS

1. An engineer investigated four liquid coolants for gasoline
    engines. Twenty gasoline engines (of a single manufacturer,
    same model, size, and so on) were obtained and each coolant
    was put in 5 randomly selected engines. Each engine was run
    at a fixed speed for two hours; during the second hour, the
    coolant temperature was recorded at three randomly chosen
    times.
    (a) Give the linear model for this experiment. Explain terms
        used and specify ranges of subscripts.
    (b) Give the partial ANOVA, sources, df, and EMS. Assume
        these are the only 4 coolants of interest to the engineer.
    (c) What assumptions are required for valid tests and CI
        based on the ANOVA?

2. A manufacturing firm is negotiating with a new supplier of
    wire stock. It is very important that the wire be uniform in
    tensile strength. The supplier agrees to send a random
    sample of 6 rolls of wire stock from its inventory for test-
    ing by the firm. From each roll, 10 two-foot sections of
    wire are randomly selected for testing. On each of these
    sections, the tensile strength is determined for 3 randomly
    selected locations.
    (a) Give the linear model for this experiment. Explain terms
        used and specify ranges of subscripts.
    (b) Give the partial ANOVA, sources, df, and EMS, under the
        random model.
    (c) What assumptions are required for valid ANOVA tests?

3. In a greenhouse experiment, a scientist investigated the
    absorption of a chemical used in herbicides. A dozen pin oak
    seedlings were potted and placed in the greenhouse. A recom-
    mended amount of herbicide was applied to each pot. After
    one week, three branches were randomly selected from each
    seedling and on each of these branches, three leaves were
    randomly selected for analysis. The amount of chemical was
    determined in each leaf.
    (a) Give the linear model for this experiment. Explain terms
        used and specify ranges on subscripts.
    (b) Give the partial ANOVA, sources, df, and EMS, under the
        random model.
    (c) What assumptions are required for valid ANOVA tests.

4. A plant scientist investigated plant uptake of heavy metals
    (for example, nickel) when 4 rates of sludge applications
    were used. For a single variety of sweet corn, a total of 40

plants were established. The plants were individually potted
and 10 were randomly chosen to receive each rate of sludge.
After a designated period of time following the sludge application, three leaves were randomly selected from each plant
and analyzed for presence of the heavy metal. In addition to
differences in uptake due to the different rates of sludge,
there was interest in variation among plants treated alike as
well as variation among leaves of the same plant.

(a) Give the linear model for this experiment. Explain terms
    used and specify ranges on subscripts.
(b) Give the partial ANOVA, sources, df, and EMS, under the
    model specifying fixed rates of sludge.
(c) What assumptions are required for valid ANOVA tests.

5. A textile specialist is investigating the variability in
    length of wool fibers obtained from 4 breeds of sheep.   For
    each breed, five ranchers were randomly selected. At each of
    these ranches, a random sample of 5 sheep was selected.   On
    each selected sheep, the wool length was determined at 4
    randomly selected sites.

    (a) Give the linear model for this experiment. Explain terms
        used in the model and specify ranges of subscripts.
    (b) Give the partial ANOVA, sources, df, and EMS. Assume
        these 4 breeds are the only ones of interest.
    (c) What assumptions are required for valid tests and CI
        based on the ANOVA.

6. The department of education in a particular state is invest-
    igating the role of educational TV habits of fourth graders
    in 5 school districts.   Three schools were randomly selected
    within each district and 3 fourth grade classes were randomly
    selected within each of these schools.   From each selected
    class, a random sample of 12 students was selected and each
    was asked how many hours were spent each week watching educa-
    tional TV.

    (a) Give the linear model for this experiment. Explain terms
        used in the model and specify ranges of subscripts.
    (b) What assumptions are necessary for valid tests and CI
        based on an ANOVA?
    (c) Give the partial ANOVA, sources, df, and EMS. Assume the
        5 school districts are the only ones of interest.

7. For the coolant experiment, Problem 1, the engineer obtained
    the following partial ANOVA:

ANOVA for Engine Coolants

Source	df	SS
Coolants		178.02
Engines(C)		125.42
Observations(E)		148.80
Total		

(a) Complete the ANOVA under the model and assumptions given in Problem 1.
(b) Test equality of mean temperatures for the 4 coolants.
(c) Estimate the variance components so one can compare variation among temperatures within a given engine as well as for the engines within a coolant.

8. Refer to the ANOVA given in the previous problem.

(a) Assuming no structure among the four coolants, perform a Duncan's mean separation analysis with $\alpha = 0.05$.
(b) Suppose the coolants have the following structure

    I   alcohol base
    II   glycol base, Company A
    III glycol base, Company A (different brand name)
    IV   glycol base, Company B

Give a set of meaningful, orthogonal contrasts and test each for significance.

9. For the greenhouse experiment conducted in Problem 3, the scientist constructed the following partial ANOVA:

ANOVA for Absorption of Herbicide

Source	df	SS	MS	EMS
Seedlings		2.416		
Branches(S)		1.807		
Leaves(B)		2.383		

(a) Complete the ANOVA table under the model and assumptions given in Problem 3.
(b) Estimate the variance components in an effort to pinpoint variation in the amounts of absorptions.
(c) Test the hypotheses of zero variance components for absorptions of seedlings, and for branches on the same seedlings.

10. After conducting the experiment described in Problem 2,   the firm constructed the following partial ANOVA:

ANOVA for Tensile Strength of Wire Stock

Source	df	SS	MS	EMS
Rolls		410.7		
Sections(R)		1689.7		
Locations(S)		1026.0		

(a) Complete the ANOVA under the model and assumptions given in Problem 2.
(b) Estimate the variance components in an effort to pinpoint non-uniformity of tensile strengths.
(c) Test the hypotheses of zero variance components for rolls, and for sections on the same roll.

11. Three coffee makers (comparable models of different manufacturers) were tested for their ability to maintain temperature after brewing stopped.   Six coffee makers of each brand were used.   An eight cup pot of coffee was brewed in each and the temperature was determined 30 minutes after brewing stopped. To assess consistency of individual coffee makers,   each was tested 3 times, giving the following

Brand	Unit	Temperatures (°F)
1	1	168,169,172
	2	172,171,174
	3	175,172,172
	4	174,171,173
	5	173,175,177
	6	169,170,169
2	1	180,182,178
	2	179,178,181
	3	182,178,183
	4	177,175,176
	5	175,180,178
	6	178,179,182
3	1	182,184,181
	2	184,182,182
	3	183,185,179
	4	187,181,182
	5	186,184,183
	6	180,182,185

(a) Set up the ANOVA for this experiment, under fixed brands.

(b) Test equality of mean temperatures for the 3 brands.

(c) Estimate variance components for temperatures among units of the same brand and for temperatures within units.

12. A food processing plant has contracted with a number of growers for purchase of their broccoli. The processor asked each grower to harvest when clusters of three heads would weigh between one and one-third pounds (341 grams) and one and one-half pounds (384 grams.) Each grower was supplied with a set of boxes for use at harvest time; each box was to be packed with 18 clusters. Different varieties used by the growers, delays in harvesting, the inability of growers to judge the head weights, and other factors lead to variation of cluster weights. The processor decided to obtain information on the sources and magnitude of the variation in the delivered broccoli. As growers brought their harvest to the processing plant, certain growers were selected at random for evaluation of cluster weights. From each selected grower, four boxes were selected at random, and from each of these boxes, three clusters were randomly selected for weighing. During one week, the following was recorded (in grams):

	Grower 1	Grower 2	Grower 3
Box 1	352, 369, 383	339, 367, 328	376, 359, 388
2	365, 372, 329	358, 349, 377	337, 361, 354
3	348, 340, 362	350, 366, 387	326, 374, 361
4	359, 371, 351	338, 373, 345	378, 362, 340

(Courtesy of Dr. C. Coale, Ag. Econ. Dept., Va. Tech.)

The processor is interested in measuring variability among all growers supplying broccoli to the plant as well as the variability among the clusters of each grower. Therefore, a random effects model is appropriate.

(a) Give the linear model that the processor should use. Explain terms used.

(b) Construct the ANOVA table for this data.

(c) Estimate appropriate variance components.

REFERENCE

[1]   Edmonds, W. J., S. Iyengar, L. Zelazny, M. Lentner, and C. Peacock., Variability in Family Differentia of Soils in a Second Order Survey Mapping Unit., Soil Sci. Soc. of Amer. Jour., Vol. 46, 1982.

## Chapter 8

LATIN SQUARE DESIGNS

8.1   Introduction

Information about differences in experimental material should be incorporated into the experimental plan if possible. Whether such information leads to better quality inferences may depend upon the nature of the information, rearrangement of EU, and so on. As we saw in Chapter 6, blocking in RCB designs illustrates a broad class of situations where the heterogeneity of EU is taken into account by the researcher. Recall that blocking is a device for error control, removing identified variation which otherwise would be a part of Exp. Error variation.

Effective blocking in an RCB design is possible when we have information about an aspect(factor) of the experimental material, information which identifies a source of variation among the EU. But what if we have information about two or more aspects of the experimental material? Might we be able to effectively utilize all of this information?

Let us consider some examples. If we were considering a field experiment with plots of land and were told that the plots were situated on different terrains as well as having different pH values, we might set up an RCB using either one of the two aspects: pH or terrain. Or, we might set up a different type of experiment with a "double blocking" where both aspects are used simultaneously.

In another experiment, suppose a company is planning an extensive advertising campaign over a period of several weeks. Later weeks of the campaign are expected to have a different impact on sales than earlier weeks. Additionally it is known that shopping habits generally differ considerably among the early part of each week, mid-week and weekend. Here we have information about two time aspects, "weeks" and "portions of weeks", both of which probably have an effect on sales.

We first consider experiments having a set of t unstructured treatments and t^2 EU which can be grouped by 2 blocking factors. These are the *basic* Latin Square (LS) designs discussed in the next five sections. Later sections will deal with modifications of the basic LS design.

8.2   The Experimental Plan

Because double blocking is an important aspect of LS designs, we begin by giving some examples of EU arrangements having this feature. As originally proposed, LS designs consisted of square arrays of EU with Latin letters representing treatments, hence the name.

Example 8.1

LS designs were introduced for more efficient analysis of certain agricultural experiments. A (square) plot of land having an East-West fertility gradient as well as a North-South fertility gradient could be subdivided into EU for which two blockings can be identified.

```
              Gradient 1 →
                                     Row blockings
Gradient 2   ┌─────────────────────┐
     ↓       │                     │      1
             ├─────────────────────┤
             │                     │      2
             ├─────────────────────┤
             │                     │      3
             ├─────────────────────┤
             │                     │      4
             └─────────────────────┘
               1   2   3   4
             Column blockings           ▓
```

LS designs have been applied to many diverse problems since their initial uses in agriculture. In particular, industrial, biological, educational, business and behavioral research have utilized the LS designs as well as extensions and variations. An LS design does not require a square array of EU. Indeed, the only requirement is that the two blockings be "balanced out" with respect to each other (Definition 6.2). This is to say, any contrast among either blocking must be free of the other blocking effects. Such blockings are "orthogonal". The following example illustrates a "non-square" LS design.

Example 8.2

An LS design would be considered for a marketing research experiment to study the effect of an extensive three week advertising campaign. Sales are expected to increase over the 3 weeks as the

campaign progresses.   In addition,   sales might be different for
the first part of each week (Mon-Tues),   the middle part of each
week (Wed-Thur), and the last part of each week (Fri-Sat).   Thus,
we have two blockings in a continuous line of EU as illustrated:

Week 1			Week 2			Week 3			Blocking 1
F	M	L	F	M	L	F	M	L	Blocking 2

Note that any contrast among Weeks (Blocking 1),   such as
"Week 3-Week 2" would be free of F,M,L effects.   Likewise, a
constrast among F,M,L (Blocking 2),   such as "F-2M+L" would be
free of Week effects. ▓

We now suppose there are t treatments under investigation. In
order for all block effects to be balanced out of treatment
contrasts,   each block must contain t EU so all treatments can
appear in each block.   For an LS design then, we need t EU within
each level of both blocking factors. As a result, we must have t^2
EU.   In summary then, an experimenter would use an LS design when
there are (i) t treatments to be investigated and (ii) t^2 EU
available which can be blocked in two orthogonal ways.   The two
blockings often are labeled Row blocks and Column blocks,   even
when the EU do not form a square array.

The arrangement given earlier in Example 8.1 is appropriate
for investigating 4 treatments,   such as varieties,   types of
irrigation, and so on.   And the 9 EU appearing in Example 8.2 are
appropriate for 3 types of advertising or 3 other treatments.

For convenience,   it is helpful to consider a square array of
EU (which can always be set up artificially,   if necessary).   The
t treatments are randomly assigned to the EU subject to the
restrictions that every treatment must appear equally often in
every row and in every column.   Under this type of treatment
allocation, row and column effects will be balanced out of treatment comparisons.

When Latin letters, A, B, C, and so on, are used to represent
treatments,   a basic Latin square can be defined as an array where
each letter occurs once in every row and once in every column.
Thus,   randomization of treatments to the EU in an LS experiment
is equivalent to randomly selecting a txt Latin square.   But all
Latin squares of a given size can be generated from a special
subset: those having letters in the first row and first column in
alphabetical order.   These are called Latin squares in standard
order. For sizes 2x2 and 3x3,   there is a single Latin square in

standard order. For size 4x4, there are 4 Latin squares of standard order. These 6 Latin squares are given in Table A.4.

When all txt Latin squares of standard order are available, the randomization proceeds as follows:

   (i) randomly select one of the squares of standard order,
  (ii) randomize all columns of the selected square,
 (iii) randomize all rows of the selected square, except the first, and
  (iv) randomly assign the treatments to the letters.

All Latin squares can be generated without randomizing the first row. Specifically, each txt square of standard order will generate $t!(t-1)!$ different Latin squares. Random permutations from Tables A.2 and A.3 are useful in carrying out steps (ii), (iii), and (iv).

For Latin squares of sizes greater than 4x4, the number of squares of standard order increases dramatically: for 5x5, there are 56; for 6x6, there are 9308, and so on. Table A.4 contains one Latin square of standard order for sizes 5x5 thru 10x10. Additionally, it is easy to generate one square of standard order as follows. First write down the letters of the first row, in alphabetical order. The second row is obtained from the first by a "one letter shift to the left" with the first letter placed in the extreme right column. Then, the third row is obtained from the second in the same manner, and so on. The 5x5 Latin square of standard order obtained in this manner is:

A	B	C	D	E
B	C	D	E	A
C	D	E	A	B
D	E	A	B	C
E	A	B	C	D

If only a limited number of Latin squares of standard order is available for selection at step (i), in the randomization process discussed above, all rows are randomized in step (iii). This does not generate all possible Latin squares but it does increase the number considerably.

8.3   Advantages and Disadvantages of LS Designs

The important advantages of LS designs are:

(i) When 2 sources of heterogeneity can be identified among the EU, more accurate treatment comparisons will result if two effective blockings can be set up.

(ii) Greater sensitivity because row and column variation is removed from Exp. Error. This implies a loss in degrees of freedom which must be compensated for by a reduction in the Exp. Error SS when variation among the rows and columns is removed.

(iii) Fairly easy to analyze---compared to the RCB design, only one additional sum of squares is required.

(iv) Several LS designs of the same size can be combined. This is important for small LS designs where the Exp. Error degrees of freedom are small.

Even though these advantages would appear to tip the scales in favor of LS designs, there are several disadvantages which limit the use of these designs. The major disadvantages are:

(i) The number of rows, columns, and treatments must be equal. Often, it is difficult to obtain homogeneous blocks in "two dimensions", especially if the number of treatments is large.

(ii) Small designs provide only a few degrees of freedom for estimation of Exp. Error variation. As the designs get larger, we may have heterogeneity within either blocking.

(iii) The LS design is not appropriate if row, column, and treatment effects are non-additive; that is, if they interact in any combination.

8.4 Linear Model and Assumptions: Basic LS Designs

The linear model used to explain observations from a basic LS experiment is

$$(8.1) \qquad y_{ijk} = \mu + \rho_i + \gamma_j + \tau_k + \epsilon_{ijk}; \quad \begin{array}{l} i = 1, 2, ---, t \\ j = 1, 2, ---, t \\ k = 1, 2, ---, t \end{array}$$

where μ = overall mean, a constant

ρ_i = effect due to the i-th row

γ_j = effect due to the j-th column

τ_k = effect due to the k-th treatment

ϵ_{ijk} = residual component for the (i,j,k)-th observation.

Assumptions necessary for valid inferences in an LS design are the same as for an RCB design with the requisite modification to account for two blocking factors. For example, a mixed model with fixed rows and treatments and random columns would require the following assumptions:

The ϵ's are i.i.d. $N(0,\sigma_\epsilon^2)$

$$\sum_i \rho_i = 0 \quad \text{and} \quad \sum_j \beta_j = 0$$

(8.2)   The γ's are i.i.d. $N(0,\sigma_\gamma^2)$

The ϵ's and γ's are distributed independently of each other.

There are no interactions among rows, columns, and treatments.

We have listed one of eight possibilities; each of rows, columns, and treatments could be fixed or random.

8.5   Estimation and Calculations: Basic LS Designs

Only two subscripts are necessary to specify any observation because only one treatment occurs for a given row and column. Once we fix values of the subscripts i and j (rows and columns), the third subscript k is useful only to indicate which treatment has been randomly assigned to (i,j)th cell. Therefore, the three subscripts have a dependency which implies we must exercise care in forming the totals and means. We cannot sum over all three subscripts; summing over any two subscripts gives the overall total. The various totals used in forming estimates and in calculation of sums of squares for an LS design are:

$$y... = \sum_{ij} y_{ijk} = \text{overall total}$$

$$y_{i..} = \sum_j y_{ijk} = \text{i-th row total}$$

(8.3)

$$y_{.j.} = \sum_i y_{ijk} = \text{j-th column total}$$

$$y_{..k} = \text{k-th treatment total}$$

Due to the restricted randomization of treatments within rows and columns, no simple expression is possible for the treatment totals. This will be apparent when an example is presented.

The earlier discussion about estimation of block effects is appropriate here (see Section 6.5). Thus, in LS designs there would be little, if any, interest in estimation of either row or column effects.

The treatment effects, the τ's, would be estimated from the observed treatment and overall means. For the k-th treatment, the estimate is

$$(8.4) \qquad \hat{\tau}_k = \bar{y}_{..k} - \bar{y}...$$

where $\bar{y}_{..k} = y_{..k}/t$ and $\bar{y}... = y.../t^2$

Example 8.3

A manufacturing firm is investigating the quality of raw material from 4 different suppliers. Four machine operators are selected to manufacture a particular component from raw materials of each supplier. Each operator produced only one component on a given day. To account for possible day and operator effects, a Latin square design was used. For suppliers A, B, C and D, the breaking strengths, y, of the components were:

		Days			Operator Totals, $y_{i..}$
	1	2	3	4	
Operator	B	C	A	D	
1	810	1080	700	910	3,500
	C	D	B	A	
2	1100	880	780	600	3,360
	D	A	C	B	
3	840	540	1055	830	3,265
	A	B	D	C	
4	650	740	1025	900	3,315
Day Totals, $y_{.j.}$ →	3400	3240	3560	3240	13,440 = y...

Supplier (treatment) Totals, $y_{..k}$

A	2490	B	3160
C	4135	D	3655

The letters in the table tell which raw material was used by an operator on a particular day. The location of these letters is a function of the particular randomization. Note the calculation of the treatment totals, one can not calculate them from any fixed pattern of either rows or columns.

Estimates of the 4 supplier effects are calculated by the mean difference as indicated in (8.4):

$$\hat{\tau}_1 = 2490/4 - 13440/16 = 622.5 - 840 = -217.5$$

$$\hat{\tau}_2 = 3160/4 - 840 = 790 - 840 = -50$$

$$\hat{\tau}_3 = 4135/4 - 840 = 1033.75 - 840 = 193.75$$

$$\hat{\tau}_4 = 3655/4 - 840 = 913.75 - 840 = 73.75$$

One can see from these estimates that there is considerable variation in the effect due to these four suppliers of the raw materials. ✷

From the raw data and the totals in (8.3) we can readily calculate all necessary sums of squares. These calculations are

(8.5) $C = y^2_{...}/t^2$ = general correction term

(8.6)   Rows: $SSR = \underset{ij}{\Sigma\Sigma}(\bar{y}_{i..} - \bar{y}...)^2 = \frac{1}{t} \underset{i}{\Sigma} y^2_{i..} - C$

(8.7)   Columns: $SSC = \underset{ij}{\Sigma\Sigma}(\bar{y}_{.j.} - \bar{y}...)^2 = \frac{1}{t} \underset{j}{\Sigma} y^2_{.j.} - C$

(8.8)   Treatments: $SST = \underset{ik}{\Sigma\Sigma}(\bar{y}_{..k} - \bar{y}...)^2 = \frac{1}{t} \underset{k}{\Sigma} y^2_{..k} - C$

(8.9)   Total: $SSY = \underset{ij}{\Sigma\Sigma}(y_{ijk} - \bar{y}...)^2 = \underset{ij}{\Sigma\Sigma} y^2_{ijk} - C$

(8.10)   Exp.Error: $SSE = SSY - SSR - SSC - SST$

The above expressions reflect the need of only two subscripts for calculation purposes. These calculation are summarized in the following analysis of variance table.

Table 8.1 Analysis of Variance for a Basic LS Design

Source	df	SS	MS	Fixed	Random
				EMS	
Rows	t-1	SSR	MSR	- - -	- - -
Columns	t-1	SSC	MSC	- - -	- - -
Treatments	t-1	SST	MST	$\sigma_\epsilon^2 + t\kappa_\tau^2$	$\sigma_\epsilon^2 + t\sigma_\tau^2$
Exp.Error	(t-1)(t-2)	SSE	MSE	σ_ϵ^2	σ_ϵ^2
Total	t^2-1	SSY			

Example 8.4

The sums of squares calculations for the breaking strength data of the previous example are now given.

$$C = (13,440)^2/16 = 11,289,600$$

$$\text{Day SS} = [(3400)^2 + (3240)^2 + (3560)^2 + (3240)^2]/4 - C$$

$$= 11,307,200 - 11,289,600 = 17,600$$

$$\text{Operator SS} = [(3500)^2 + (3360)^2 + (3265)^2 + (3315)^2]/4 - C$$

$$= 11,297,262.5 - 11,289,600 = 7,662.5$$

$$\text{Supplier SS} = [(2490)^2 + (3160)^2 + (4135)^2 + (3655)^2]/4 - C$$

$$= 11,660,737.5 - 11,289,600 = 371,137.5$$

$$\text{Total SS} = (810)^2 + (1080)^2 + \text{- - -} + (900)^2 - C$$

$$= 11,723,250 - 11,289,600 = 433,650$$

$$\text{Exp. Error SS} = 433,650 - 371,137.5 - 7,662.5 - 17,600$$

$$= 37,250$$

Under fixed treatment effects, the analysis of variance table is:

ANOVA for Breaking Strengths

Source	df	SS	MS	EMS
Days	3	17,600	5,867	---
Operators	3	7,662	2,554	---
Suppliers	3	371,138	123,712	$\sigma_\epsilon^2 + 4\kappa_\tau^2$
Exp. Error	6	37,250	6,208	σ_ϵ^2
Total	15	433,650		※

8.6   Inference Procedures: Basic LS Design

Ordinarily one does not perform tests for the row and column blockings for a number of reasons. Validity of such tests is an issue because rows and columns are not randomized in LS designs. This is the same concern expressed in RCB designs.

Inferences about treatments is of primary interest. As in the earlier designs, the inference is different for the random and fixed effects cases.

In the fixed effects case, the hypothesis of interest is

$$(8.11) \qquad H_0: \tau_k = 0 \quad \text{for all } k$$

The test statistic for this hypothesis is

$$(8.12) \qquad F = MST/MSE$$

which has $(t-1)$ and $(t-1)(t-2)$ degrees of freedom. And when appropriate, all post-anova procedures of Chapters 3 and 4 are applicable to treatments used in LS designs. Particular contrasts of interest are differences of pairs of treatment means; for example, $\mu_{..k} - \mu_{..h} = \tau_k - \tau_h$, having estimate of

$$(8.13) \qquad \hat{\tau}_k - \hat{\tau}_h = \bar{y}_{..k} - \bar{y}_{..h}$$

which has an estimated standard deviation of

$$(8.14) \qquad s_{\bar{y}_{..k}-\bar{y}_{..h}} = \sqrt{2MSE/t}$$

Example 8.5

We use the breaking strength data of Example 7.3 to illustrate the ideas discussed above. To test equality of supplier means, we have:

H_0: The mean breaking strengths are the same for the four suppliers.

H_1: The mean breaking strengths are different for the four suppliers.

Test statistic: $F_{3,6}$ = Supplier MS/MSE

$F_{3,6}$ = 123,712/6,208 = 19.93

The observed significance level is P < 0.005.

The four supplier means are:

A	22.5	B	790
C	1033.75	D	913.75

The standard deviation of the difference of two supplier means is, by (8.14):

$$s_{\bar{y}_{..k}-\bar{y}_{..h}} = \sqrt{2(6,208)/4} = 55.71 \text{ ※}$$

In the random effects case, the hypothesis of interest is

$$(8.15) \qquad H_0: \sigma_\tau^2 = 0$$

The test statistic for testing this hypothesis is the same one as given in (8.12). Mean separation procedures and contrasts usually are not applied to random effects. Rather, estimation of variance components σ_ϵ^2 and σ_τ^2 are of interest, and are accomplished by the procedures of Section 2.10.

8.7 Missing Values in Basic LS Designs

The principle set forth in Section 6.8 for the RCB designs will be followed here: estimation of missing values so that the Exp. Error sum of squares is minimized. For a single missing value in a basic LS design, the estimate to achieve such a minimization is

268   Latin Square Designs

(8.16) $M = [t(R_g + C_h + T_u) - 2G]/(t-1)(t-2)$

where

R_g = total for the row where the missing value occurs

C_h = total for the column where the missing value occurs

T_u = total for the treatment where the missing value occurs

G = grand total of all non-missing observations

Using M in the analysis results in a treatment sum of squares which is upwardly biased by the amount

(8.17)   Bias = $[G - R_g - C_h - (t-1)T_u]^2/(t-1)(t-2)$

After the treatment sum of squares is corrected for the bias, division by t-1 gives an "unbiased treatment mean square", say MST*. Then, remembering that degrees of freedom for Exp. Error and Total are each reduced by one for estimating the missing value, an unbiased (exact) test of treatments can be made using the F statistic

(8.18) $F = MST*/MSE$

which has degrees of freedom of (t-1) and (t-1)(t-2)-1.

8.8   Efficiency of Basic LS Designs

The concept of an efficiency measure was introduced in Section 6.10 for comparing RCB designs relative to CR designs. Now we wish to consider efficiency measures for LS designs.

Instead of using an LS design, we could have used an RCB design or a CR design had one or both of the blocking factors been ignored. Thus, we have three possibilities: efficiency of the "completed LS design" relative to

   (i) an RCB design using only the column blocks (ignoring row blocks)
  (ii) an RCB design using only the row blocks (ignoring column blocks)
 (iii) a CR design (that is, had neither blocking been used.)

We assume dummy treatment effects for each pair of designs being compared. Then, the efficiency of a completed LS design relative to "an RCB design that would result had row blocks not been used" is estimated by

$$(8.19) \qquad RE(LS \text{ to } RCB, \text{no rows}) = \frac{MSR + (t-1)MSE}{tMSE}$$

Had column blocks not been used, the relative efficiency can be obtained from (8.19) by replacing MSR by MSC to give

$$(8.20) \qquad RE(LS \text{ to } RCB, \text{no columns}) = \frac{MSC + (t-1)MSE}{tMSE}$$

Finally, the efficiency of a completed LS design relative to a CR design would be estimated by

$$(8.21) \qquad RE(LS \text{ to } CR) = \frac{MSR + MSC + (t-1)MSE}{(t+1)MSE}$$

Example 8.6

Referring to the ANOVA given in Example 8.4, for the breaking strength analysis conducted under an LS design, we obtain the following efficiencies relative to "reduced RCB designs":

$$RE(LS \text{ to } RCB \text{ with Operator Blocks}) = [5867 + 3(6208)]/4(6208)$$
$$= 0.986$$

$$RE(LS \text{ to } RCB \text{ with Day Blocks}) = [2554 + 3(6208)]/4(6208)$$
$$= 0.853$$

Thus, the completed LS design is almost as efficient as an RCB design using operator blocks (but ignoring day blocks.) But the completed LS design is about 15% less efficient than an RCB design using day blocks (but ignoring operator blocks.) ▓

8.9   Subsampling in LS Designs

Several observations are made on each EU. It is assumed that the subunits giving these observations are selected at random on each EU. The reasons for and implications of obtaining multiple observations on each EU are the same as those discussed for CR and RCB designs (Sections 2.11 and 6.12.) The analysis of an LS design with subsampling is almost identical to that for an RCB design with subsampling. The only differences are due to the inclusion of a second blocking factor in the LS design. Therefore, we will present only the model and ANOVA table.

Variation among observations on a given unit is accounted for by a "sampling error" component in the model. Thus, for t treatments and s subsamples per unit, the linear model is

$$(8.22) \quad y_{ijkm} = \mu + \rho_i + \gamma_j + \tau_k + \epsilon_{ijk} + \delta_{ijkm}; \quad \begin{aligned} i &= 1, 2, ---, t \\ j &= 1, 2, ---, t \\ k &= 1, 2, ---, t \\ m &= 1, 2, ---, s \end{aligned}$$

where δ_{ijkm} measures variation among observations on the same unit, and all other components have the same meaning as in model (8.1).

For inference purposes, the assumptions of a basic LS are required. Specific assumptions depend upon the random or fixed nature of the effects. Additionally, it is assumed that the sampling components, the δ's, are independent and identically distributed normal variables with mean zero and variance σ_δ^2, and distributed independently of the ϵ's. Assuming fixed treatment effects, the ANOVA table would be the following.

Table 8.2   ANOVA for LS Design with Subsampling

Source	df	SS	MS	EMS
Rows	t-1	SSR	MSR	---
Columns	t-1	SSC	MSC	---
Treatments	t-1	SST	MST	$\sigma_\delta^2 + s\sigma_\epsilon^2 + st\kappa_\tau^2$
Exp. Error	(t-1)(t-2)	SSE	MSE	$\sigma_\delta^2 + s\sigma_\epsilon^2$
Sampling	$t^2(s-1)$	SSS	MSS	σ_δ^2
Total	st^2-1	SSY		

8.10 Factorial Treatments in LS Designs

The use of factorial treatments in LS designs is restricted, for the most part, to "small factorials" because of the difficulty of attaining homogeneity within the double blockings. In view of the limited occurrence, and the fact that this type of experiment differs from an "RCB design with factorial treatments" only in having a second blocking factor, there seems little need in presenting a detailed discussion. The ideas of Section 6.11 are easily modified to handle an experiment in this category.

8.11 Latin Rectangle Designs

A major disadvantage of LS designs is the requirement of an equal number of rows, columns, and treatments. Because of this, small LS designs have few degrees of freedom for Exp. Error, and then low precision, generally. If the number of rows or columns can be increased, the Exp.Error degrees of freedom increase accordingly, and so should the precision. We assume that only one of the blocking factors is increased, but to maintain the necessary balance features, the increased number of blocks must be some multiple of the number of treatments. For definiteness, suppose the column blocks are increased to give a rectangular array of EU of size t x mt. With the double blocking of EU, the t treatments are randomly allocated to EU such that every treatment appears once in every column and m times in every row. A design having this structure is called a *Latin rectangle*.

Example 8.7

A 3x6 Latin rectangle design with t = 3 treatments might appear as follows, after randomization:

A	C	C	B	B	A
C	B	A	C	A	B
B	A	B	A	C	C

Such a set-up would allow for removal of variation among the EU due to the three row blocks and the six column blocks. ✖

Example 8.8

In Example 8.3, a manufacturing firm used a 4x4 LS design to investigate breaking strengths of components made from different sources of raw materials (treatments). Blocking factors were Days and Operators.

Suppose the firm had decided to conduct the experiment with the same 4 operators but could extend it over 8 days.   A 4x8 Latin rectangle design could be used and would allow for removal of variation due to 4 operators and 8 days. ✖

The linear model for a t x mt Latin rectangle design is

(8.23) $y_{ijk} = \mu + \rho_i + \gamma_j + \tau_k + \epsilon_{ijk}$; $\begin{array}{l} i = 1,\ 2,\ ---,\ t \\ j = 1,\ 2,\ ---,\ mt \\ k = 1,\ 2,\ ---,\ t \end{array}$

One notes that this model is identical to the model (8.1) for the basic LS design, except for the range of the column index.

Assumptions required for valid inferences in a Latin rectangle design are identical to those for LS designs.

Sums of squares calculations are analogous to those for the basic LS design.   Equations (8.5) thru (8.10) are appropriate if a factor of m is included in the divisors of (8.5), (8.6) and (8.8).   The ANOVA table for a Latin rectangle design appears below.

Table 8.3   ANOVA for a Latin Rectangle Design

Source	df	SS	MS	EMS Fixed	EMS Random
Rows	t-1	SSR	MSR	- - -	- - -
Columns	mt-1	SSC	MSC	- - -	- - -
Treatments	t-1	SST	MST	$\sigma_\epsilon^2 + mt\kappa_\tau^2$	$\sigma_\epsilon^2 + mt\sigma_\tau^2$
Exp. Error	(mt-2)(t-1)	SSE	MSE	σ_ϵ^2	σ_ϵ^2
Total	$mt^2 - 1$	SSY			

With the appropriate modification of degrees of freedom, tests and inferences are otherwise the same as for the basic LS design. The Latin rectangle has (m-1)t(t-1) additional degrees of freedom for Exp. Error compared to a basic LS design with the same number of treatments.

Example 8.9

Suppose the firm, Example 8.3, conducted its experiment for 8 days but kept the same 4 operators and 4 suppliers of raw materials

				Days					
	1	2	3	4	5	6	7	8	Totals
Operator	B	C	A	D	D	C	A	B	
1	810	1080	700	840	840	1050	775	805	6970
	C	D	B	A	A	D	B	C	
2	1100	880	780	600	670	930	720	1035	6715
	D	A	C	B	C	B	D	A	
3	840	540	1055	830	980	700	810	610	6365
	A	B	D	C	B	A	C	D	
4	650	740	1025	900	860	730	970	900	6775
Totals	3400	3240	3560	3240	3350	3410	3275	3350	26825

Supplier (treatment) Totals, $y_{..k}$

A	5275	B	6245
C	8170	D	7135

Estimates of the 4 supplier effects are:

$$\hat{\tau}_1 = 5275/8 - 26825/32 = 659.4 - 838.3 = -178.9$$

$$\hat{\tau}_2 = 6245/8 \quad - 838.3 \quad = 780.6 - 838.3 = -57.7$$

$$\hat{\tau}_3 = 8170/8 \quad - 838.3 \quad = 1021.2 - 838.3 = 183.0$$

$$\hat{\tau}_4 = 7135/8 \quad - 838.3 \quad = 891.9 - 838.3 = 53.6$$

The sums of squares calculations are:

$$C = (26825)^2/32 = 22,486,895$$

$$\text{Day SS} = [(3400)^2 + \cdots + (3370)^2]/4 - C$$
$$= 22,506,881 - 22,486,895 = 19,986$$

$$\text{Operator SS} = [(16970)^2 + \cdots + (6775)^2]/8 - C$$
$$= 22,510,746 - 22,486,895 = 23,851$$

Supplier SS = $[(5275)^2 + \text{---} + (7135)^2]/8$ - C

$\qquad$ = 23,060,346 - 22,486,895 = 573,451

$\quad$ Total SS = $(10)^2 + \text{---} + (900)^2$ - C

$\qquad$ = 23,185,425 - 22,486,895 = 698,530

Exp. Error SS = 698,530 - 19,986 - 23,851 - 573,451 = 81,242

Under fixed treatment effects, the analysis of variance table is:

ANOVA for Breaking Strengths

Source	df	SS	MS	EMS
Days	7	19,986	2,855.1	---
Operators	3	23,851	7,950.3	---
Suppliers	3	573,451	191,150.3	$\sigma^2_\epsilon + 8\kappa^2_\tau$
Exp. Error	18	81,242	4,513.4	σ^2_ϵ
Total	31	698,530		

For a test of equal mean breaking strengths for the 4 suppliers, we have

$$H_0: \tau_k = 0 \quad \text{for all k}$$

$$H_1: \tau_k \neq 0 \quad \text{for some k}$$

Test statistic: $F_{3,18}$ = Supplier MS/Exp. Error MS

The calculated value of this statistic is F = 191,150.3/4,513.4 = 42.4 which gives an observed significance level of $P < 0.005$. Thus, there is evidence of differences in the mean breaking strengths of the four suppliers.
$\quad$ The supplier means are

A	659.4	B	780.6
C	1021.2	D	891.9

The standard deviation of the difference of two supplier means is

$$s_{\bar{y}_{..k} - \bar{y}_{..h}} = \sqrt{2(4,513.4)/8} = 33.6 (\text{psi.})$$

8.12   The Simple Cross-over Design

An experimental setup that closely resembles the Latin   rectangle is the simple cross-over design.   Only mt experimental units are required with t measurements obtained on each.   The t treatments are   randomly assigned to each experimental unit over a series of t   time   periods.   Other names have been used for this   type   of design, including change-over, switch-back, and reversal.

Cross-over designs are used in a variety of   situations.   In animal science experiments, each animal frequently receives every treatment in dietary analyses, production studies, and so on.   In psychological   research,   human or animal subject often   receives every   stimuli being investigated.   In marketing   research,   all displays   (treatments)   are used in each   store.   In industrial research,   the   same   machine   might be used for   all   production processes (treatments).

Due to the nature of cross-over designs,   repeated use of the same experimental units in succeeding time periods,   one needs to be concerned about possible carry-over or residual effects.   That is,   a treatment applied in one time period may continue to exert some   of its effect during part of the next   time   period.   This would   contaminate treatment effects unless the residual   effects can be eliminated,   or accounted for, in the design and analysis. In   many   cases,   they can be eliminated by using a   rest   period between   each pair of adjacent time periods.   No treatment (or a standard treatment) would be applied to all units during all rest periods.   The   researcher must decide if residual   effects have been   eliminated,   based upon the nature of the   treatments,   the length   of the rest period,   and other information.   The   entire duration   of an experiment might be limited so that rest   periods may not be feasible.

The simple cross-over design assumes no carry-over or residual effects.   Then,   under certain treatment allocations, the simple cross-over   design is equivalent to a Latin rectangle (or LS if m = 1).   All   treatments are applied to   each   experimental   unit (animal, store, machine) so each unit can be regarded as a block. If we insist on each treatment occurring m times within each time period,   treatment   allocations of a Latin rectangle are present. The   two   blocking factors are "time periods"   and   "experimental units".

Further   discussion of the simple cross-over design will   not be necessary.   The model, analysis, and inferences are equivalent to those for the Latin rectangle.

The   cross-over design with residual effects present will not be   discussed in this book,   the interested reader may   refer   to Federer [2] or Cochran and Cox [1].

8.13 Repeated Latin Square Designs

Latin square and Latin rectangle designs control for two sources of nuisance variation. A Latin rectangle might be preferred because it provides more degrees of freedom for measuring Exp. Error variation. In some experimental situations, limitations of personnel, equipment, time or space might make it impossible to conduct an experiment larger than a basic LS design.

We may find it feasible to repeat a basic LS design, each repetition with all new experimental material. Notice how this differs from the structure of a Latin rectangle. The columns of a t x 2t Latin rectangle, for example, could always be rearranged to give the appearance of two t x t Latin squares, however, the rows would be common to both squares: The rows and columns must all be different for an experiment to be classified as a repeated Latin square.

Example 8.10

If the manufacturing firm repeated the experiment described in Example 8.3 a second time with 4 different operators, we would have a repeated LS experiment. A total of 8 operators as well as 8 days would be used in this experimental setup--only 4 operators were used in the Latin rectangle described in Example 8.8. ✳

In a repeated LS experiment, the researcher actually has conducted a series of small (basic LS) experiments. Must these be analyzed individually or can their information be combined into a single analysis? The basic LS experiments may have been conducted at different locations, under different environments, and so on. The researcher must decide whether or not important differences exist among the small experiments. If so, will a combined analysis take these differences into account, or would they be addressed more fully by a series of individual analyses? Indeed, a researcher might profit from doing both individual and combined analyses. Individual analyses follow the procedures for basic LS designs discussed in the first part of this chapter. The remainder of this section is devoted to combined analyses.

We assume the basic LS design is repeated s times. The same t treatments are used in each of the s squares; thus, treatments are crossed with squares so we may need to consider if "treatment x square" interaction could exist.

In a repeated LS design, rows and columns are not crossed with squares because there are different rows and columns for each square. In other words, "rows" and "columns" are nested within squares of the repeated LS design. This is reflected in the linear model which we now give

(8.24) $y_{gijk} = \mu + S_g + R_{gi} + C_{gj} + \tau_k + (S\tau)_{gk} + \epsilon_{gijk}$

$$\text{for } g = 1, 2, ---, s$$
$$i = 1, 2, ---, t$$
$$j = 1, 2, ---, t$$
$$k = 1, 2, ---, t$$

where

μ = overall mean

S_g = effect of the g-th square

R_{gi} = effect of the i-th row within the g-th square

C_{gj} = effect of the j-th column within the g-th square

τ_k = effect of the k-th treatment

$(S\tau)_{gk}$ = interaction component for the g-th square and k-th treatment

ϵ_{gijk} = residual component of the (gijk)-th observation

For valid inferences of treatment parameters of model (8.24), we make the following assumption about the residual components:

(8.25)   The ϵ's are i.i.d. $N(0,\sigma_\epsilon^2)$

Additional assumptions concerning the fixed/random effects of squares, rows, columns and treatments are made as the particular situation dictates.

For the model (8.24), we now give the totals and related sums of squares calculations to complete the ANOVA. The totals are:

overall: $y.... = \underset{gij}{\Sigma\Sigma\Sigma} y_{gijk}$

g-th square: $y_{g...} = \underset{ij}{\Sigma\Sigma} y_{gijk}$

(8.26a)

i-th row of g-th square: $y_{gi..} = \underset{j}{\Sigma} y_{gijk}$

j-th column of g-th square: $y_{g.j.} = \underset{i}{\Sigma} y_{gijk}$

k-th treatment: $y_{\ldots k} = \underset{gi}{\Sigma\Sigma}\, y_{gijk}$

(8.26b)

(g,k)-th cell, square-treatment subclass: $y_{g..k} = \underset{i}{\Sigma}\, y_{gijk}$

Then, the sums of squares are:

(8.27) $C = y^2_{\ldots}/st^2$

(8.28) Squares: $SSSq = \dfrac{1}{t^2}\underset{g}{\Sigma}\, y^2_{g\ldots} - C$

(8.29) Rows(Sq): $SSR(Sq) = \dfrac{1}{t}\underset{gi}{\Sigma\Sigma}\, y^2_{gi..} - \dfrac{1}{t^2}\underset{g}{\Sigma}\, y^2_{g\ldots}$

(8.30) Columns(Sq): $SSC(Sq) = \dfrac{1}{t}\underset{gj}{\Sigma\Sigma}\, y^2_{g.j.} - \dfrac{1}{t^2}\underset{g}{\Sigma}\, y^2_{g\ldots}$

(8.31) Treatments: $SST = \dfrac{1}{st}\underset{k}{\Sigma}\, y^2_{..k} - C$

(8.32) Sq x Trt: $SSSqxT = \dfrac{1}{t}\underset{gk}{\Sigma\Sigma}\, y^2_{g..k} - C - SSSq - SST$

(8.33) Total: $SSY = \underset{gij}{\Sigma\Sigma\Sigma}\, y^2_{gijk} - C$

(8.34) Exp. Error: $SSE = SSY - SSSq - SSR(Sq) - SSC(Sq) -$

$SST - SSSqxT$

The ANOVA summary is given is Table 8.4 for model (8.24) when squares and treatments are assumed to have fixed effects.

In certain experimental situations, a researcher may replace model (8.24) with one omitting the interaction component $(S\tau)_{gk}$. This would be appropriate when treatments are expected to respond consistently from square to square. Which model is used depends on the information known by the researcher. When the interaction component is omitted from the model, the corresponding df and SS are included in Exp. Error.

Table 8.4   **ANOVA** for Repeated Latin Squares

Source	df	SS	MS	EMS
Squares	s-1	SSSq	MSS	- - -
Rows(Sq.)	s(t-1)	SSR(Sq)	MSR	- - -
Columns(Sq.)	s(t-1)	SSC(Sq)	MSC	- - -
Treatments	t-1	SST	MST	$\sigma_\epsilon^2 + st\kappa_\tau^2$
Sq x T	(s-1)(t-1)	SSSqxT	MSST	$\sigma_\epsilon^2 + t\kappa_{S\tau}^2$
Exp.Error	s(t-1)(t-2)	SSE	MSE	σ_ϵ^2
Total	$st^2 - 1$	SSY		

8.14 Extended LS Designs

A basic LS design may be defined as a t x t square into whose cells we place t letters of one language. We will use the first t letters of the Latin language and insert them in such a way that each occurs once in every row and once in every column. The letters then indicate which treatment would be assigned to that row-column cell.

If $t \geq 3$, letters of 2 or more languages can be placed in the cells of a t x t square such that: (i) the letters of each language form an LS design and (ii) the letters of every pair of languages are orthogonal; that is, each letter of any language appears in precisely one cell with each letter of every other language. For any value of t, it has been shown that conditions (i) and (ii) hold only when the letters of t-1 or fewer languages are placed in the cells of a t x t square.

Any square containing two or more languages of orthogonal letters will be called an *extended Latin square*. Another name for these arrays is *orthogonalized Latin squares*. Those with only two languages, usually Greek and Latin, form a special set called Graeco-Latin squares (GLS). A GLS of size 6x6 does not exist, Fisher and Yates [4].

Example 8.11

A 3x3 Graeco-Latin square is

$$A\alpha \quad B\gamma \quad C\beta$$

$$B\beta \quad C\alpha \quad A\gamma$$

$$C\gamma \quad A\beta \quad B\alpha$$

The Latin letters form a 3x3 Latin square. So do the Greek letters. Furthermore, letter A occurs in one cell with each Greek letter. So does letter B, and letter C.
 A 4x4 GLS is

$$C\alpha \quad D\beta \quad B\gamma \quad A\delta$$

$$D\delta \quad C\gamma \quad A\beta \quad B\alpha$$

$$B\beta \quad A\alpha \quad C\delta \quad D\gamma$$

$$A\gamma \quad B\delta \quad D\alpha \quad C\beta$$

The letters of a third language could be placed (orthogonally) in the cells of this 4x4 array but not in the cells of a 3x3 array.※

Example 8.12

The following array represents a 5x5 extended LS with 3 languages (Greek, Latin and "numbers").

$$A1\alpha \quad B4\gamma \quad C2\epsilon \quad D5\beta \quad E3\delta$$

$$B2\beta \quad C5\delta \quad D3\alpha \quad E1\gamma \quad A4\epsilon$$

$$C3\gamma \quad D1\epsilon \quad E4\beta \quad A2\delta \quad B5\alpha$$

$$D4\delta \quad E2\alpha \quad A5\gamma \quad B3\epsilon \quad C1\beta$$

$$E5\epsilon \quad A3\beta \quad B1\delta \quad C4\alpha \quad D2\gamma \text{ ※}$$

Extended LS designs are not used very often in experimental research. This is because of two primary requirements: (i) the multi-way blockings must be orthogonal, and (ii) there can be no interactions among the blocking factors and treatments. When these requirements are met, there is a substantial reduction in the number of EU needed. Any LS or extended LS requires only t^2

EU.   Using all combinations of rows, columns, and treatments of
an LS would require t^3 EU. For a GLS, using all combinations of
rows, columns, Greek and Latin letters would require t^4 EU. In
general, the required number of EU is t^{L+2}, where L is the number
of languages appearing in the array.

The analysis of any extended LS design is straightforward. The
only change from an analysis of a basic LS design is the presence
of additional blocking factors. An additional source of variation
and sum of squares is required for each blocking factor of an
extended LS design.

PROBLEMS

1. An experiment to investigate the effects of 6 treatments was
    conducted in a 6x6 Latin Square design. The blocking factors
    were days and operators.

    (a) Give the linear model for this experiment. Explain terms
        used in the model and specify ranges on the subscripts.
    (b) Give the partial ANOVA, sources, df, and EMS. Assume
        fixed treatment effects.
    (c) What assumptions are necessary for valid tests and CI
        based on the ANOVA.

2. An experiment to investigate the effects of 8 treatments was
    conducted in an 8x8 Latin Square design. The blocking factors
    were days and working hours of a day.

    (a) Give the linear model for this experiment. Explain terms
        used and specify ranges on the subscripts.
    (b) Give the partial ANOVA, sources, df, and EMS. Assume
        fixed treatment effects.
    (c) What assumptions are necessary for valid tests and CI
        based on the ANOVA.

3. Repeat Problem 1 for an experiment with treatments having a
    3x2 factorial structure.

4. Repeat Problem 2 for an experiment with treatments having a
    2x2x2 factorial structure.

5. Repeat Problem 1 if the experimenter records 3 observations
    from randomly selected portions of each EU.

6. Repeat Problem 2 if the experimenter records 2 observations
    from randomly selected portions of each EU.

7. A greeting card company is the studying the effect of four
    different types of displays on the sales of its cards. They
    decide to test each display for one week at each of four

locations within a store in a Latin square arrangement. The following partial ANOVA resulted(for dollar volume of sales):

ANOVA for Sales of Greeting Cards

Source	df	SS	MS	EMS
Weeks		37.2		
Locations		29.7		
Displays		53.1		
Exp. Error		22.8		

(a) Complete the above ANOVA under a fixed effects model.
(b) Make a test of equal mean sales for the 4 displays.
(c) Give the estimated standard deviation for the difference of observed mean sales for two different displays.

8. For the greeting card experiment of the previous problem, calculate the efficiency of the conducted LS design relative to an RCB design using only blocks of

(a) Weeks   (b) Locations.

9. For the greeting card experiment of Problem 7, calculate the efficiency of the conducted LS design relative to a CR design (that is, using neither blocking factor.)

10. A manufacturer studied the effect of six music intensities on the output of a certain product.   The design was a 6x6 Latin square with blocking factors of operators and time of day. The number finished components was measured for a 20 minute period. The following partial ANOVA resulted:

ANOVA for Production under Music Intensities

Source	df	SS	MS	EMS
Operators		8.35		
Times		17.44		
Music Int.		12.09		
Exp. Error		16.34		

(a) Complete the ANOVA under a fixed effects model.
(b) Test for equality of mean outputs under the six music intensities.
(c) Give the estimated standard deviation for the difference of observed mean output for any two music intensities.

11. A poultry science professor used diets with low, medium and high concentrations of protein to see if there were effects on the amount of food intake in leghorn chickens. Space limitations were such that the cages had to be stacked on top of one another and in front of each other (but with some spacing.) This arrangement introduced two blockings: (i) the height of a cage was important because of a temperature diff- erential (temperature affects food intake), and (ii) the depth of a cage was important because there were windows only on the front side of the cages (the amount of light also affects food intake.) Therefore, cages were stacked 3 high and 3 deep (with spacers) and 10 chickens were randomly assigned to each cage. Treatments were then assigned to the cages according to the LS design below and after one week, total food intake of each cage was measured with the results given in the table (in ounces):

| | | Height | |
Depth	Bottom Row	Middle Row	Top Row
Front Row	M(96)	H(81)	L(106)
Middle Row	H(94)	L(116)	M(114)
Back Row	L(100)	M(91)	H(89)

(Courtesy of Dr. L. Potter, Poultry Science Dept., Va. Tech)

(a) Complete the ANOVA under the fixed effects model.
(b) Test equality of mean food intakes for the concentrations of protein.

12. A plant breeder conducted an experiment to compare the yields of 3 new varieties and a standard variety of peanuts. The varieties were assigned to the plots in an LS arrangement because the land had slight sloping from East to West and differences in available nitrogen from North to South. The variety and its yield per plot (in pounds) are given in the table below:

| | | Slopings | | |
Nitrogen	1	2	3	4
1	C 26.7	A 19.7	B 28.0	D 29.8
2	A 23.1	B 20.7	D 24.9	C 29.0
3	B 28.3	D 20.1	C 29.0	A 27.3
4	D 25.1	C 17.4	A 28.7	B 34.1

(a) Set up the ANOVA table for this experiment (under the fixed model).

(b) Test for equality of mean yields for the varieties.
(c) Assume that variety A is a standard variety and use Dunnett's procedure to assess the value of the other varieties. Let $\alpha = 0.05$.

13. For the poultry science experiment of Problem 11, calculate the efficiency of the conducted LS design relative to a CR design.

14. For the plant breeder's experiment of Problem 12, calculate the efficiency of the conducted LS design relative to

(a) an RCB design using only blocks of East-West slopings.
(b) a CR design.

15. An experiment was conducted to investigate 4 treatments in a 4x12 Latin rectangle design. The blocking factors were 4 operators and 12 days.

(a) Give the linear model for this experiment. Explain terms used in the model and specify ranges on the subscripts.
(b) Give the partial ANOVA, sources, df, and EMS. Assume treatments are fixed.

16. A human nutrition researcher conducted an experiment to determine acceptability of cakes baked with sucrose substitutes as the sweetening agent. Specifically there were 6 recipes formed by combinations of 3 sweeteners:

> 100% sucrose
> 75% high fructose corn syrup, 25% sucrose
> 75% crystalline fructose, 25% sucrose

and 2 leavening agents:

> baking soda
> baking soda plus "additional acid"

A panel of 6 taste testers were used to evaluate various characteristics of the cakes. On each of six days, cakes were baked for each of the six recipes. A Latin square arrangement was used with days and testers as the blocking factors. Let the variable of interest be y , a taste score.

(Courtesy of Dr. Janet Johnson, Human Nutrition and Foods Dept., Va. Tech.)

(a) What are the EU in this experiment?
(b) Give the linear model for this experiment. Explain terms used and give ranges on the subscripts.
(c) Give the partial ANOVA, sources, df and EMS, assuming fixed treatment effects.

17. Repeat Problem 15 for treatments having a 2x2 factorial structure.

18. The greeting card company, Problem 7, conducted a second, more extensive experiment in a larger city using the same 4 types of displays, the same 4 locations within a store, but extended over an 8 week period. The following partial ANOVA resulted (for dollar volume of sales):

ANOVA for Sales of Greeting Cards

Source	df	SS	MS	EMS
Weeks		60.9		
Locations		32.7		
Displays		48.6		
Exp. Error		54.9		

(a) Complete the above ANOVA under a fixed effects model.
(b) Test equality of mean sales for the 4 displays.
(c) Give the estimated standard deviation for the difference of observed mean sales for two different displays.

19. After the poultry science professor completed the experiment in Problem 11, additional space became available so a second, larger experiment was conducted. The same three height layers were necessary, but instead of the depth layers, the cages could be placed side by side. There still was a light grad-ient as the cages moved away from the windows when placed side by side. There was ample room to place 6 cages side by side. Thus a Latin Rectangle design was used. All other conditions were the same as in the first experiment (10 chickens per cage, the same 3 treatments.) The data were:

Height	First	Second	Third	Fourth	Fifth	Sixth
		Side by Side Positions				
Top	M 99	L101	L104	H 87	M 96	H 84
Middle	L110	H 88	H 91	M100	L102	M 94
Bottom	H 90	M104	M 98	L100	H 86	L 97

(Courtesy of Dr. L. Potter, Poultry Science Dept., Va. Tech)

(a) Set up the ANOVA table assuming fixed treatments.
(b) Test equality of mean food intakes for the 3 protein concentrations.
(c) Give the estimated standard deviations of the difference of observed mean intakes for two protein concentrations.

20. Suppose the greeting card company ran an experiment with the same 4 displays in a second (and different type) store during the four week period following the experiment in the first store.   Four locations were used within the second store but because this was a different type of store, these four locations were different from the locations of the first store.

   (a) Give the linear model for this experiment. Explain terms used in the model and specify ranges on the subscripts.
   (b) Give the partial ANOVA, sources, df, and EMS. Assume a fixed effects model.
   (c) What assumptions are necessary for valid tests and CI based on the ANOVA.

REFERENCES

[1]   Cochran, W.G. , and G.M. Cox, *Experimental Designs, 2nd. ed.*, Wiley, New York, 1957.
[2]   Federer, W.T. , *Experimental Design*, Macmillan, New York, 1955.
[3]   Fisher, R.A. , The arrangement of field experiments, *Jour. Ministry Agr.*, Vol. 33, 1926.
[4]   Fisher, R.A. , and F. Yates, The 6x6 latin squares, *Cambridge Phil. Soc. Proc.*, Vol. 30, 1934.

## Chapter 9

ANALYSIS OF COVARIANCE

9.1 Introduction

Control of "nuisance variation" is achieved by a number of design considerations. If ignored, such variations contaminate treatment effects and inflate Exp. Error variation, both of which lead to poor quality inferences.

As we have seen in Chapters 6 and 8, blocking is one way of accounting for nuisance variation known to exist among EU. Blocking is a form of error control (local control) that requires arrangement of EU prior to allocation of treatments. Due to incomplete knowledge about experimental material, an insufficient amount of material, or other reasons, it may be impossible to construct blocks. And blocking cannot be done in another class of experiments - those where factors exert their influence only after the experiment is underway.

Let us consider a situation where a blocking problem might exist. Suppose a researcher is studying a variable y, the yield of some crop. Surely yield is affected by a number of factors, in particular by

(9.1)   x = number of plants per plot.

We are saying that the value of y is some function of x so ideally we would strive to have an equal number of plants per plot. But inequality in x often occurs because not all planted seeds germinate. To overcome this, one might consider planting extra seeds and thinning plants after emergence. Even starting with equal numbers of plants is no guarantee all will survive and contribute to the yield y. Maybe the soil on some plots is of poor quality, affecting germination and the ability to support a certain number of plants. Blocking the plots according to soil quality cannot be done if such information is unavailable. Even if the necessary information is available, we may have trouble setting up homogeneous or complete blocks.

Similar problems of controlling factors are common in many other experimental situations. Researchers have had to cope with this predicament when dealing with such factors as

x = amount of rainfall

x = weight of volunteers in an exercise study

x = pre-scores for a group of children to be used in an educational testing program

x = initial height of seedling plants

x = initial weight of an experimental animal

x = value of back-orders at the end of the fiscal year

(9.2)

and so on. We are not saying that a researcher will always have a problem controlling the variables listed above. For the most part, it depends upon the nature and availability of experimental material. In some experiments, factors such as initial height and initial weight could be used to set up blocks; in others, they could not.

In summary then, there are experimental situations where the following issue must be addressed: certain nuisance variables (influential factors) cannot be controlled during the experiment but their effect must be taken into account to achieve quality inferences. Subject to certain conditions, it is often possible to utilize information obtained during the experiment at the analysis and inference stage. The remainder of this chapter will focus on one procedure of utilizing secondary information of the scope and nature indicated above.

9.2   The Role of Auxiliary Variables

An experiment can have both controlled and uncontrolled factors. Those factors not controlled during an experiment, but included in the analysis, are called *auxiliary variables* or *covariates*. Both terms are used widely and interchangeably.   The latter term is reflected in the name of the technique.

The analysis of covariance (ANACOVA) is a procedure that permits us to make adjustments for the effects of auxiliary variables.   It is the combination (or merging) of two widely used statistical techniques: regression and analysis of variance. ANACOVA is concerned with experiments where treatment effects are to be analyzed---this is the analysis of variance portion of the procedure. And in addition to the treatment effects, the response variable is influenced by one or more auxiliary variables---this functional relationship represents the regression portion of the procedure. As we will see later, it is mandatory that treatments have no influence on the auxiliary variables.

Before we look at a detailed example let us make one further observation. The effect of covariates enters the analysis through appropriate terms in the experimental model. Covariate terms permit the adjustments alluded to earlier.   We will discuss this concept again when we consider models and the analysis.

Example 9.1

To better illustrate certain ideas and concepts, we will use a modified version of an actual experiment.   A forest biometrician wished to investigate the effects of three types of fertilizer on the growth of oak seedlings.   Fifteen seedlings of the same age were obtained.   The forester noticed that the initial heights were not the same but no attempt was made to set up blocks according to this factor.   Five seedlings were randomly assigned to each fertilizer treatment and after six months, the growth (final height minus initial height) was determined for each tree giving, in mm:

Fertilizer

A	B	C
39	38	19
32	40	21
41	33	15
27	34	23
41	39	19

These growths are plotted below in Figure 9.1.

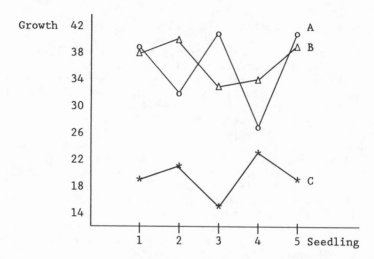

Figure 9.1   Seedling Growths from Three Fertilizers

Ignoring initial heights and concentrating only on the graph of plotted growths, three statements can be made:

(1) There is considerable variability of growth measurements within each of the three groups.

(2) The within group variability is largest for fertilizer A and about equal for the other two fertilizers.

(3) Fertilizers A and B give larger average growths and they appear to be equal.

For purposes of discussion, let us now suppose that the growth, say y, of this type of oak seedling is related to the initial height, say x, linearly and specified by $y = 50 - x$ for each of the three types of fertilizer. We also assume fertilizer effects of $\tau_1 = 8$, $\tau_2 = 0$ and $\tau_3 = -8$. The initial heights and growths are given below in the format (x,y):

Fertilizer

A	B	C
(20,39)	(13,38)	(25,19)
(25,32)	(12,40)	(22,21)
(17,41)	(15,33)	(27,15)
(30,27)	(17,34)	(21,23)
(14,41)	(14,39)	(25,19)

Examination of the above data reveals that initial heights are
heterogeneous for each fertilizer, especially for A. Fertilizer B
has most of the smallest heights. For the three fertilizer
treatments, the average initial heights are $\bar{x}_{1.} = 21.2$, $\bar{x}_{2.} = 14.2$
and $\bar{x}_{3.} = 24.0$. Past experience with biological experiments tell
us that seedlings with a smaller initial height tend to have a
larger growth (that is, shorter seedlings tend to exhibit a
compensatory gain), so we see that the apparent good performance
under fertilizer B is partly due to the shorter seedlings which
were allocated to this fertilizer. And while it is not obvious at
this time, a large portion of the Exp. Error variation which
would occur in a CR analysis of this data is attributed to the
great variation in initial heights. The CR analysis has no way to
utilize this information about initial heights and simply ignores
it.
    The plot of growth as a function of initial height, x, is now
given.

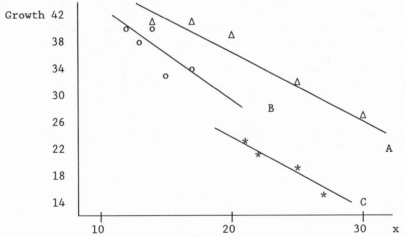

Figure 9.2   Plot of Seedling Growths versus Initial Height

    From this graph it is clear that most of the variation in the
growth measurements is explained by the linear relationship. In
other words, if we ran a linear regression of growth on initial
height for each fertilizer, the regression function would account
for a large percentage of the total variation; or, the residual
variation about each regression line would be small. When we
analyze the above data by ANACOVA, using initial height as the
covariate, we will be able to show that fertilizer A has the
greatest effect. Recall the values of the r's given earlier. ▓

The analysis of covariance makes an adjustment for differences in values of the covariate. For the example just discussed, an adjustment would be made for initial height. In our earlier discussion, the values of the treatment effects and the exact relationship between the response and the covariates were assumed known. Usually, these are estimated from the data. The next section treats the adjustments in more detail.

9.3   Adjustments to Sums of Squares in Regression Models

Upon first encounter, the formulas used in calculating various sums of squares in ANACOVA appear to be cumbersome as well as lacking in intuitive meaning. To aid in understanding covariance adjustments, we will review the general adjustment process in regression models. Recall simple linear regression for a moment. The means and total variation of the variable of interest, y, are each adjusted by subtracting a quantity based on the auxiliary variable, x. We will review these adjustments shortly.

In any regression experiment, we can think of a sequence of models in which each successive model adds one more parameter to the preceding model. For such a sequence of models, one can test the significance of including each additional parameter after adjusting for the influence of the other parameters in the model. We will see that this influence is measured from two successive models in the sequence.

As a simple example, suppose we look at a regression study with two independent variables, x_1 and x_2. One sequence of possible models is

$$y = \beta_0 + \epsilon_1$$

(9.3)
$$y = \beta_0 + \beta_1 x_1 + \epsilon_2$$

$$y = \beta_0 + \beta_1 x_1 + \beta_2 x_2 + \epsilon_3$$

To indicate the difference of the residual components under the three models, we have added a subscript on ϵ.

We can always subdivide the total observed variation in y into two components:

(9.4)

    (1) a reduction due to the model; that due to fitting all model components except ϵ, and

    (2) the Exp. Error (or residual) variation, only that corresponding to ϵ.

This is true for models in ANOVA, regression, or ANACOVA.

Under the first model listed in (9.3), the reduction due to the model is that due to fitting the single parameter β_0. We will denote this reduction by $R(\beta_0)$. The component for the Exp. Error variation will be denoted by SSE1. These two components represent a subdivision of the total observed variation; namely,

$$(9.5) \qquad \sum_i y_i^2 = R(\beta_0) + \text{SSE1}$$

where $\sum_i y_i^2$ is the total (uncorrected) sum of squares of y.

For the second model listed in (9.3), the reduction due to the model will be denoted by $R(\beta_0, \beta_1)$ and measures the influence of both parameters in explaining the response y. The Exp. Error for the second model will be SSE2, and again these components represent a subdivision of the total observed variation:

$$(9.6) \qquad \sum_i y_i^2 = R(\beta_0, \beta_1) + \text{SSE2}$$

Notice now that we have quantities measuring the joint impact of β_0 and β_1 as well as the impact of β_0 only. The difference of these two quantities, $R(\beta_0, \beta_1) - R(\beta_0)$, measures the impact of β_1 over and above that provided by β_0 only. This difference is called the reduction "due to β_1 adjusted for β_0" and is denoted by $R(\beta_1 | \beta_0)$. From (9.5) and (9.6), we see that

$$(9.7) \qquad R(\beta_1 | \beta_0) = R(\beta_0, \beta_1) - R(\beta_0) = \text{SSE1} - \text{SSE2}$$

Similarly for the third model, the reduction due to the model would be $R(\beta_0, \beta_1, \beta_2)$ and the Exp. Error would be SSE3. Then, the reduction "due to β_2 adjusted for β_0 and β_1" would be

$$(9.8) \qquad R(\beta_2 | \beta_0, \beta_1) = R(\beta_0, \beta_1, \beta_2) - R(\beta_0, \beta_1) = \text{SSE2} - \text{SSE3}$$

Now let us interpret the models in (9.3) and the above reductions in the context of covariance analysis. Suppose that variable x_2 represents a "treatment factor" while x_1 represents a "second factor" which influences our response, y. Assume that no other factors can be identified as influencing our response; that is, all other variation is unexplained (extraneous.)

We would like to measure the treatment effects free of all other model components. Thus, we would like to ascertain the effect of x_2 (the treatment component) after eliminating the effects of the other two model components. What we really want, then, is the reduction specified in (9.8). We see that this reduction is equivalent to two differences: one involving total reductions under two of the models in (9.3), the other involving "residual" sums of squares for the same two models. Of interest here is the implication of the latter difference. The ϵ's of the third model in (9.3) represent only extraneous variation because all other identified variation is accounted for by components in the model. Thus, we may interpret SSE3 as an "error sum of squares." But the ϵ's of the second model in (9.3) represent "treatment effects" as well as the extraneous components (but not those represented by the model component $\beta_1 x_1$). Thus, we may interpret SSE2 as a "treatment plus error sum of squares." Therefore, we may express the reduction in (9.8) in the general format

(9.9)   "treatment plus error SS " minus "error SS".

Let us introduce a notation S[H] to denote a sum of squares for a factor H adjusted for "some auxiliary variable." In the present context, we wish to find S[T], "a sum of squares for treatments adjusted for the variate x_1." Then, the expression (9.9) can be written as

(9.10)   S[T] = S[T + E] - S[E].

Expression (9.10) gives the general format of calculations for obtaining "adjusted treatment SS". The right hand side of (9.10) provides a simple algorithm for calculating "adjusted treatment SS" in the analysis of covariance. It eliminates the need for having to fit two different models and finding the difference of their residual sums of squares.

We can obtain a general expression for the covariance

adjustments, which are actually linear regression adjustments, by examining the sums of squares from simple linear regression. The first two models in (9.3) represent the sequence of models we would need to fit. Dropping the subscript on x_1 for simplicity, we have

(9.11) $$SSE1 = \sum_i (y_i - \beta_0)^2 = \sum_i (y_i - \bar{y})^2 = SSY$$

(9.12) $$SSE2 = \sum_i (y_i - \hat{\beta}_0 - \hat{\beta}_1 x)^2 = SSY - \hat{\beta}_1^2 SSX$$

where

(9.13)
$$\hat{\beta}_0 = \bar{y} - \hat{\beta}_1 \bar{x}$$

$$\hat{\beta}_1 = SPXY/SSX$$

with $$SPXY = \sum_i (x_i - \bar{x})(y_i - \bar{y}) = \text{sum of cross-products for x and y}$$

Replacing the estimate of $\hat{\beta}_1$ in (9.12), we obtain

(9.14) $$SSE2 = SSY - (SPXY)^2/SSX$$

from which we get a sum of squares for "fitting β_1 adjusted for all other parameters in the model"

(9.15) $$S[T] = S[T+E] - S[E] = SSY - [SSY - (SPXY)^2/SSX]$$
$$= (SPXY)^2/SSX.$$

In experiments where there is a single linear covariate, the last expression in (9.15) represents an algorithm for making covariate adjustments.

We may interpret the expression in (9.14) as the "observed variation in y adjusted for the auxiliary variable." As we see, the adjustment is equal to "the squared sum of cross products divided by the sum of squares of the auxiliary variable." This provides the general algorithm for making covariate adjustments in experiments where there is a single linear covariate.

9.4   Models and Assumptions for Analysis of Covariance

The linear models given below are for the analysis of covariance when a single linear covariate is present.   Each model will be discussed in a following section.   The subscripts and their ranges are the same as for the corresponding ANOVA model.

CR Design: $y_{ij} = \mu + \tau_i + \beta(x_{ij} - \bar{x}..) + \epsilon_{ij}$

RCB Design: $y_{ij} = \mu + \tau_i + \rho_j + \beta(x_{ij} - \bar{x}..) + \epsilon_{ij}$

LS Design: $y_{ijk} = \mu + \tau_i + \rho_j + \gamma_k + \beta(x_{ijk} - \bar{x}...) + \epsilon_{ijk}$

Two-Factor Factorial in an RCB Design:

$$y_{ijk} = \mu + A_i + B_j + (AB)_{ij} + \rho_k + \beta(x_{ijk} - \bar{x}...) + \epsilon_{ijk}$$

The assumptions for each model are a combination of those required in analysis of variance and in regression,   plus some additional assumptions.   Basically,   we assume that the residual components, the ϵ's, are i.i.d. $N(0,\sigma_\epsilon^2)$ and that the x values are fixed(that is, measured without distributional errors). These are the minimum assumptions from ANOVA and regression.   Usual fixed/random assumptions would be made about the treatment and block components appearing in a model.   Additionally in ANACOVA, we assume that (i) treatment factors do not affect the values of the covariate and (ii) all populations have a common slope, β, when variable y is regressed on the covariate. If either of these latter assumptions do not hold,   the "covariance adjustments" would not be made.   Instead,   the y data would be analyzed by a regular ANOVA.   A later section deals with preliminary tests of these assumptions.

9.5.1   ANACOVA in a CR Design

We now give the basic calculations needed to set up the analysis of covariance table for a CR design.   Required are calculations for y, the variable of interest, a second set of calculations for x,   the covariate, and a third set of calculations for the cross-products of x and y.
    The total,   treatment,   and Exp.   Error sums of squares for variable x are, respectively:

$$SSX = \sum_{ij}\sum x_{ij}^2 - x_{..}^2/n \quad \text{where } n = rt$$

(9.16)
$$TXX = \frac{1}{r}\sum_i x_{i.}^2 - x_{..}^2/n$$

$$EXX = SSX - TXX$$

We perform similar calculations for y:

$$SSY = \sum_{ij}\sum y_{ij}^2 - y_{..}^2/n$$

(9.17)
$$TYY = \frac{1}{r}\sum_i y_{i.}^2 - y_{..}^2/n$$

$$EYY = SSY - TYY$$

Finally, we compute the sum of cross-products for x and y:

$$SPXY = \sum_{ij}\sum x_{ij}y_{ij} - x_{..}y_{..}/n$$

(9.18)
$$TXY = \frac{1}{r}\sum_i x_{i.}y_{i.} - x_{..}y_{..}/n$$

$$EXY = SPXY - TXY$$

To summarize these calculations and to illustrate the covariance adjustments, an analysis of covariance table, denoted by Part A, is presented. Remaining calculations, given in a table denoted by Part B, are made after determining that covariance adjustments are appropriate.

Table 9.1 Analysis of Covariance for CR Design--Part A

		SS and SP			Adjust.	From Regression	
Source	df	XX	XY	YY	$(XY)^2/XX$	Dev.	MS[adj]
Treatments	t-1	TXX	TXY	TYY			
Error	n-t	EXX	EXY	EYY	$(EXY)^2/EXX$	S[E]	MS[E]
Treatments + Error	n-1	SXX	SXY	SYY	$(SXY)^2/SXX$	S[T+E]	

where

$$SXX = TXX + EXX \qquad SXY = TXY + EXY \qquad SYY = TYY + EYY$$

(9.19) $$S[E] = EYY - (EXY)^2/EXX \qquad S[T+E] = SYY - (SXY)^2/SXX$$

$$MS[E] = S[E]/(n-t-1)$$

In practice, the last segment of Part A (labeled Treatment + Error) would be completed only after one is satisfied that all covariance assumptions hold. This issue is addressed more fully in the next subsection.

Examination of the last three columns of Table 9.1, Part A, and the expressions given in (9.19) reveal that these entries are calculated from the following general expressions:

$$Adjust. = (XY)^2/XX$$

(9.20) $$Dev. = YY - Adjust.$$

$$MS = Dev./(df-1)$$

when entries of the XX, XY, YY, and df columns of the given row are inserted. These general expressions hold for all covariance analyses. See also the discussion following (9.15).

Example 9.2

Let us set up the ANACOVA table for the data presented in Example 9.1. Sums of squares for x and y should present no difficulty. The total, treatment, and Exp. Error SS for each variable would be calculated as in Chapter 2, now using equations (9.16) and (9.17). For the cross-products, we have

$$x_{1.} = 106 \qquad x_{2.} = 71 \qquad x_{3.} = 120$$

$$y_{1.} = 180 \qquad y_{2.} = 184 \qquad y_{3.} = 97$$

$$TXY = (1/5)[(106)(180) + (71)(184) + (120)(97)] - (297)(461)/15$$

$$= 8756.8 - 9127.8 = -371.0$$

$$SPXY = 8554 - 9127.8 = -573.8$$

$$EXY = -573.8 - (-371.0) = -202.8$$

ANACOVA on Growth--Part A

Source	df	XX	XY	YY	Adj.	Dev.	MS
Treatments	2	254.8	-371.0	964.9			
Error	12	201.6	-202.8	230.0	204.0	26.0	2.36
T+E	14	456.4	-573.8	1194.9	721.4	473.5	

The entries in the last three columns are calculated as follows:

$$204 = (-202.8)^2/201.6 \quad \text{and} \quad 721.4 = (-573.8)^2/456.4$$

$$26 = 230 - 204 \quad \text{and} \quad 473.5 = 1194.9 - 721.4$$

$$2.36 = 26.0/11 \ \text{❋}$$

9.5.2 Some Preliminary Tests

The last segment of Part A of the ANACOVA table, as well as the
entire Part B, contain the entries needed to test equality of
adjusted y treatment means. Before completing these entries, we
need to discuss some preliminary tests useful in checking the
validity of ANACOVA assumptions.   All rules and comments made
in Section 1.13 about preliminary tests are appropriate here.
     Preliminary analyses might include tests of (i) equal Exp.
Error variances for all populations (treatment groups in CR
designs),   (ii) equal slopes for all populations,   (iii) common
slope of zero, (iv) equality of unadjusted x-treatment means, and
(v) equality of unadjusted y-treatment means.   These tests are
not   made in the order listed;   in fact,   some are required to be
made in a specific sequence, and some may not be made in a given
experiment. We will discuss each test and indicate where it fits
in the overall scheme.
     A reference to Figure 9.2 will be helpful in discussing the
test listed in (i).   Recall that ANACOVA was described as fitting
a linear regression to the data of each population (treatment
group.) The Exp. Error variation for each treatment group would
be the residual variation about that group's regression line.
These residual variances could be used in a test for homogeneity
of variances discussed in Section 2.13.   For other than the CR
designs,   there usually is insufficient data to make this test.
For example, in an RCB design, one would need 2 or more observations in each cell of the block-treatment array (these cells

define the populations.) The issue of equality of these variances should be resolved before proceeding to the tests in (ii) and (iii) which require this equality.

Another look at Figure 9.2 reveals that equality of slopes for the various populations (treatment groups) is equivalent to "no interaction between treatments and the covariate". Therefore, the concept of adjustments in sequential models, as discussed in Section 9.3, can be applied here to develop a procedure for testing equality of slopes. Because the test procedure is fairly complicated and because most statistical computing packages will perform the test, we will present only the highlights here. Basically, we would fit two linear models: one containing terms for the interaction of treatments with the covariate, the other not containing these terms. The second model results from a null hypothesis

(9.21) H_0: No interaction of x and treatments.

which, as we have indicated, is equivalent to

(9.22) H_0: Equal slopes for the treatment groups.

Then, if we modify (9.9) to read

(9.23)   "interaction plus error SS" minus "error SS"

and calculate these two sum of squares from the models described above, we could form an F statistic to test the null hypothesis in (9.22). This is the procedure discussed in Section 9.3. If there is concern about unequal slopes, this preliminary test should precede the test of "zero common slope" to be discussed next.

The first two rows of Table 9.1, Part A, contain all necessary quantities to perform the last three tests in the list given earlier. We first consider the test of zero common slope. The null hypothesis is

(9.24) H_0: $\beta = 0$

The test statistic is an F variable formed from entries on the Exp. Error line of Table 9.1. Specifically

(9.25) $F = (EXY)^2/(EXX)MS[E]$

which has degrees of freedom of 1 and n-t-1. If we fail to reject the null hypothesis in (9.24), we have no evidence of a non-zero slope. When we arrive at such a conclusion, we would not make the covariance adjustments because adjustments to a line with "zero slope" are just random adjustments. Accordingly, the researcher would make a test of unadjusted y treatment means.

A test about the equality of unadjusted x-treatment means is considered if there is some question of treatments affecting the covariate. Here one uses the x-data, the XX column of Table 9.1, Part A, and makes a "regular ANOVA test" on treatments. The null hypothesis is

(9.26) H_0: All x-treatment means are equal.

The test statistic is

$$(9.27) \qquad F = \frac{TXX/(t-1)}{EXX/(n-t)}$$

which has degrees of freedom of (t-1) and (n-t). If we should reject H_0 and conclude that the x-treatment means differ---implying that treatments have caused the x-values to change---we would not make the covariance adjustments. When treatments are affecting the covariate x, the effects of treatments and x are interrelated; the covariance adjustment to remove the effect of x is tantamount to removing some of the treatment effect. A test of equality of unadjusted y-means would be made when there is evidence that treatments are affecting the covariate. Note that the hypothesis in (9.26) would not be tested if treatments cannot possibly affect the value of the covariate, when for example, the covariate is a pre-experiment score.

The test of unadjusted y-treatment means would be made in the two situations indicated in the preceding paragraphs. Sometimes, this test is made to gain insight into the treatment effects before and after adjustment for the covariate. This is a "regular ANOVA test" on the y-data, using the YY column of Table 9.1

Example 9.3

Two of the preliminary tests described above will be performed for the growth data of Example 9.1. The test of equality of unadjusted x-treatment means will not be made because treatments (fertilizers) cannot affect the covariate (initial height of the seedlings).

To test for zero slope, we have

$$H_0: \beta = 0 \quad \text{vs.} \quad H_1: \beta \neq 0$$

Test Statistic: F = (Adjustment for Error)/MS[E]

$$F = 204/(2.36) = 86.3$$

which gives an observed significance level of P < 0.005.   At the 0.5% level, we have evidence of a nonzero slope.   (From this test, covariance adjustment of y-treatments is appropriate.)
A test of unadjusted y-treatment means would be

H_0: Unadjusted y-treatment means are equal.

H_1: Unadjusted y-treatment means are unequal.

Test Statistic: $F = \dfrac{TYY/(t-1)}{EYY/(n-t)} = MST/MSE$

$$F = \frac{964.9/2}{230.0/12} = 25.2$$

which gives an observed significance level of P < 0.005.   At the 0.5% level, there is evidence of difference among the unadjusted y-treatment means. ❊

9.5.3   Adjusted Treatment Means

We now consider Part B of the analysis of covariance table, that portion containing entries for testing the equality of adjusted y-treatment means, adjusted for the covariate.   Part B can be set up easily from Part A.   Recall the basic equation for obtaining adjusted y-treatment sums of squares:

(9.28)   S[T] = S[T+E] - S[E]

Table 9.2 Analysis of Covariance for a CR Design--Part B

Source	df	SS[adj]	MS[adj]
Treatments	t-1	S[T]	MS[T]
Error	n-t-1	S[E]	MS[E]

The null hypothesis of interest in an ANACOVA experiment is

(9.29) H_0: Adjusted y-treatment means are equal.
                All $\tau_i = 0$. (equivalently)

The test statistic is

(9.30) $F = MS[T]/MS[E]$

which, under H_0 and the assumptions stated earlier, has the F distribution with (t-1) and (n-t-1) degrees of freedom. The critical region is the upper tail of this F distribution.

At this time we point out an equivalent interpretation for the test of (9.29), the test of equal adjusted y-treatment means. Refer to Figure 9.2. When all regression lines have equal slopes, the test of equal y-treatment means is equivalent to a test that all regressions have a common intercept. If this holds, all regression lines would coincide.

Example 9.4

For the growth data of Example 9.1, Part B of the analysis of covariance is:

ANACOVA of Growth Data--Part B

Source	df	SS[adj]	MS[adj]
Fertilizers	2	447.5	223.75
Error	11	26.0	2.36

A test of equality of adjusted mean growths for the three fertilizers is:

      H_0: Adjusted mean growths are equal for the three
            fertilizers.

      H_1: Adjusted mean growths differ for the three
            fertilizers.

Test Statistic: $F = MS[T]/MS[E]$,   with 2 and 11 df

$$F = 223.75/2.36 = 94.67$$

which gives an observed significance of $P < 0.005$. There is sufficient evidence of unequal adjusted growth means for the three fertilizers. ✱

When we reject the null hypothesis (9.29), there is interest in the adjusted y-treatment means. Estimates of these means are provided by

(9.31)   Adj. $\bar{y}_{i.} = \bar{y}_{i.} - \hat{\beta}(\bar{x}_{i.} - \bar{x}_{..})$

where

(9.32) $\hat{\beta} = EXY/EXX$

is the estimator of the common slope. Also of interest are the adjusted treatment effects, the τ's, whose estimates are

(9.33) $\hat{\tau}_i = (Adj. \bar{y}_{i.}) - \bar{y}_{..}$

Finally, estimates of differences of adjusted means are given by

(9.34)   Adj. $\bar{y}_{i.} - Adj. \bar{y}_{h.} = (\bar{y}_{i.} - \bar{y}_{h.}) - \hat{\beta}(\bar{x}_{i.} - \bar{x}_{h.})$.

$$= \hat{\tau}_i - \hat{\tau}_h$$

The estimated variance of the difference of two adjusted means is

(9.35) $V(Adj. \bar{y}_{i.} - Adj. \bar{y}_{h.}) = MS[E]\{\frac{2}{r} + (\bar{x}_{i.} - \bar{x}_{h.})^2/EXX\}$

We note that this estimated variance changes for most pairs of adjusted means because of the last term in the brackets of (9.35). Because of these unequal variances, multiple comparison procedures must be modified as with unequal replications, (see Chapter 3.)

Example 9.5

Here we calculate the adjusted means, the differences of adjusted
means and the estimated variances of adjusted growth means. From
Example 9.2, we obtain

$$\bar{x}_{1.} = 21.2 \qquad \bar{x}_{2.} = 14.2 \qquad \bar{x}_{3.} = 24 \qquad \bar{x}.. = 19.8$$

$$\bar{y}_{1.} = 36 \qquad \bar{y}_{2.} = 36.8 \qquad \bar{y}_{3.} = 19.4 \qquad \bar{y}.. = 30.73$$

$$\hat{\beta} = 202.8/201.6 = -1.006$$

$$MS[E] = 26/11 = 2.36 \quad \text{and} \quad EXX = 201.6.$$

Then, the adjusted growth means are

$$\text{Adj. } \bar{y}_{1.} = 36 + 1.006(21.2 - 19.8) = 37.41$$

$$\text{Adj. } \bar{y}_{2.} = 36.8 + 1.006(14.2 - 19.8) = 31.17$$

$$\text{Adj. } \bar{y}_{3.} = 19.4 + 1.006(24 - 19.8) = 23.63.$$

The adjusted treatment effects are

$$\hat{\tau}_1 = 37.41 - 30.73 = 6.68$$

$$\hat{\tau}_2 = 31.17 - 30.73 = 0.44$$

$$\hat{\tau}_3 = 23.63 - 30.73 = -7.10$$

The estimated variances of the differences of adjusted growth
means are, from (9.35):

$$V(\text{Adj. } \bar{y}_{1.} - \text{Adj. } \bar{y}_{2.}) = 2.36[2/5 + (21.2 - 14.2)^2/201.6] = 1.52$$

$$V(\text{Adj. } \bar{y}_{1.} - \text{Adj. } \bar{y}_{3.}) = 2.36[2/5 + (21.2 - 24.0)^2/201.6] = 1.04$$

$$V(\text{Adj. } \bar{y}_{2.} - \text{Adj. } \bar{y}_{3.}) = 2.36[2/5 + (14.2 - 24.0)^2/201.6] = 2.07 \text{❋}$$

9.6.1 ANACOVA in an RCB Design

When a covariate is measured in an RCB design, the analysis
proceeds as in a CR design with additional calculations for
blocks. The model for analysis of covariance in an RCB design
was given in the first part of Section 9.4.

For completeness, we list all sums of squares calculations

needed to construct Part A of the ANACOVA table.   The sums of squares for the covariate x are:

Correction Term: $C_x = x_{..}^2/rt$

Total: $\quad SSX = \sum\limits_{ij} x_{ij}^2 - C_x$

(9.36) Treatment: $\quad TXX = \dfrac{1}{r} \sum\limits_{i} x_{i.}^2 - C_x$

Blocks: $\quad RXX = \dfrac{1}{t} \sum\limits_{j} x_{.j}^2 - C_x$

Exp. Error: $\quad EXX = SSX - TXX - RXX$

Similar calculations for the response variable, y, are:

$$C_y = y_{..}^2/rt$$

$$SSY = \sum\limits_{ij} y_{ij}^2 - C_y$$

(9.37) $\quad TYY = \dfrac{1}{r} \sum\limits_{i} y_{i.}^2 - C_y$

$$RYY = \dfrac{1}{t} \sum\limits_{j} y_{.j}^2 - C_y$$

$$EYY = SSY - TYY - RYY$$

For the sum of cross-products, the calculations are:

$$C_{xy} = x_{..}y_{..}/rt$$

$$SPXY = \sum\limits_{ij} x_{ij}y_{ij} - C_{xy}$$

(9.38) $\quad TXY = \dfrac{1}{r} \sum\limits_{i} x_{i.}y_{i.} - C_{xy}$

$$RXY = \dfrac{1}{t} \sum\limits_{j} x_{.j}y_{.j} - C_{xy}$$

$$EXY = SPXY - TXY - RXY.$$

The above calculations are summarized below in Table 9.3.

Table 9.3   Analysis of Covariance for an RCB Design--Part A

Source	df	XX	XY	YY	Adjust.	SS[Adj]	MS[adj]
Blocks	r-1	RXX	RXY	RYY			
Treatments	t-1	TXX	TXY	TYY			
Error	(r-1)(t-1)	EXX	EXY	EYY	$(EXY)^2/EXX$	S[E]	MS[E]
T + E	r(t-1)	SXX	SXY	SYY	$(SXY)^2/SXX$	S[T+E]	

where

$$SXX = TXX + EXX \qquad SXY = TXY + EXY \qquad SYY = TYY + EYY$$

(9.39) $S[E] = EYY - (EXY)^2/EXX \qquad S[T+E] = SYY - (SXY)^2/SXX$

$$MS[E] = S[E]/\{(r-1)(t-1)-1\}$$

9.6.2 Preliminary Tests

All test procedures and discussion given in Section 9.5.2 for the CR design are applicable to RCB designs.   Unfortunately though, tests of equal residual variances and equal slopes cannot be made unless several observations are available from each population (block-treatment combination).   Preliminary tests of unadjusted x-treatment means (when appropriate) and of zero common slope do not require multiple observations.

When applying the test procedures from Section 9.5.2 to RCB designs, minor changes in notation must be made, notably the degrees of freedom.

9.6.3   Adjusted Treatment Means

After being satisfied that the ANACOVA assumptions are met, one proceeds with the covariance adjustments so that Part B can be completed.   With the basic equation for calculating the adjusted y-treatment sum of squares, S[T] = S[T+E] - S[E], Part B of the analysis of covariance table may be constructed easily once Part A is available. The adjusted sums of squares are presented as Part B in Table 9.4 below.

Table 9.4 Analysis of Covariance Table--Part B--RCB Design

Source	df	SS[adj.]	MS[adj.]
Treatments	t-1	S[T]	MS[T]
Error	(r-1)(t-1)-1	S[E]	MS[E]

From Table 9.4 we can test the hypothesis of most interest, the equality of adjusted treatment means. The null hypothesis is

$$H_0: \tau_1 = \tau_2 = \cdots = \tau_t = 0$$

(9.40)   (or equivalently)

$$H_0: \text{All adjusted y-treatment means are equal.}$$

The test statistic is

(9.41) $F = MS[T]/MS[E]$

with (t-1) and [(r-1)(t-1)-1] degrees of freedom.

Example 9.6

A corn breeder tested four newly developed varieties in an RCB experiment. Blocks represented different fields. In addition to recording the yield per plot, y, the number of plants harvested per plot, x, was recorded. The data were:

				Variety						
	I		II		III		IV		Totals	
Block	x	y	x	y	x	y	x	y	x	y
1	40	320	37	282	32	290	41	273	150	1165
2	32	300	34	278	32	283	42	271	140	1132
3	38	325	41	290	39	310	40	283	158	1208
4	42	341	30	270	33	265	36	266	141	1142
5	35	316	45	293	37	296	37	280	154	1185
Totals	187	1602	187	1413	173	1444	196	1373	743	5832

Sum of squares calculations for yield, y, are:

$SSY = (320)^2 + (300)^2 + \text{---} + (280)^2 - (5832)^2/20$

$\quad = 1,709,164 - 1,700,611.2 = 8,552.8$

$TYY = [(1602)^2 + (1413)^2 + (1444)^2 + (1373)^2]/5 - 1,700,611.2$

$\quad = 1,706,647.6 - 1,700,611.2 = 6,036.4$

$RYY = [(1165)^2 + (1132)^2 + \text{---} + (1185)^2]/4 - 1,700,611.2$

$\quad = 1,701,575.5 - 1,700,611.2 = 964.3$

$EYY = 855.8 - 6036.4 - 964.3 = 1552.1$

Sum of squares calculations for the number of plants, x, are:

$SSX = (40)^2 + (32)^2 + \text{---} + (37)^2 - (743)^2/20$

$\quad = 27,925 - 27,602.4 = 322.6$

$TXX = [(187)^2 + (187)^2 + (173)^2 + (196)^2]/5 - 27,602.4$

$\quad = 27,656.6 - 27,602.4 = 54.2$

$RXX = [(150)^2 + (140)^2 + \text{---} + (154)^2]/4 - 27,602.4$

$\quad = 27,665.2 - 27,602.4 = 62.8$

$EXX = 322.6 - 54.2 - 62.8 = 205.6$

Sum of product calculations are:

$SPXY = 40(320) + (32)(300) + \text{---} + (37)(280) - (743)(5832)/20$

$\quad = 217,147 - 216,658.8 = 488.2$

$TXY = [(187)(1602) + \text{---} + (196)(1373)]/5 - 216,658.8$

$\quad = 216,545 - 216,658.8 = -113.8$

$RXY = [(150)(1165) + \text{---} +(154)(1185)]/4 - 216,658.8$

$\quad = 216,901.5 - 216,658.8 = 242.7$

$EXY = 488.2 - (-113.8) - 242.7 = 359.3$

Analysis of Covariance for Corn Varieties--Part A

Source	df	XX	XY	YY	Adj.	SS[Adj]	MS[Adj]
Blocks	4	62.8	242.7	964.3			
Varieties	3	54.2	-113.8	6036.4			
Error	12	205.6	359.3	1552.1	627.9	924.2	84.0
Var. + E	15	259.8	245.5	7588.5	232.0	7359.5	

A test of zero slope is:

$$H_0: \beta = 0 \quad \text{vs.} \quad H_1: \beta \neq 0$$

$$F = \text{Error Adjustment}/\text{MS[E]} \quad \text{with} \quad 1 \text{ and } 11 \text{ df.}$$

$$F = 627.9/(924.2/11) = 627.9/84 = 7.47$$

which gives an observed significance level of $0.01 < P < 0.025$. There is evidence of a significant slope, so we proceed with the adjustment.

Analysis of Covariance for Corn Varieties--Part B

Source	df	SS[adj]	MS[adj]
Varieties	3	6432.3	2144.1
Error	11	924.2	84.0

A test of equal adjusted variety means is:

$$H_0: \tau_i = 0 \quad \text{for all } i$$
$$H_1: \tau_i \neq 0 \quad \text{for some } i$$

$$F = (\text{Adj. Variety MS})/\text{MS[E]}, \quad \text{with } 3 \text{ and } 11 \text{ df}$$

$$F = 2144.1/84 = 25.5$$

which gives an observed significance level of $P < 0.005$.

There is evidence of differences among the mean yield of the four new varieties (when adjusted for the existing stand). ✻

9.7   ANACOVA for a Latin Square Design

We will not give a presentation of ANACOVA in a Latin Square because it is so similar to the presentation for the RCB design. Other than two sources of variation for row and column blockings, and resulting changes in certain degrees of freedom, there is little else that differs from the entire Section 9.6. Actually, if we gave the formulas, the analysis would appear much more complicated than it really is.

9.8   ANACOVA for a Two-Factor Factorial in an RCB Design

The analysis of covariance for factorial experiments is slightly more involved than for a comparable experiment with unstructured treatments. We must calculate additional sums of squares to account for the factorial structure of treatments. We consider only the case of two factors in an RCB design. The extension to more than two factors or to other designs is straightforward.

The linear model for a two-factor ANACOVA in an RCB design appeared in Section 9.4. As in any covariance analysis, sums of squares are required for the covariate x and response variable y as well as sum of cross-products for x and y. Because it is necessary to identify the variable and the factorial structure, the notation appears cumbersome but a careful analysis will reveal its necessity and usefulness.

For the covariate x, the sums of squares are calculated according to the following expressions:

$$\text{Correction Term:} \quad C_x = x^2_{\cdot\cdot\cdot}/rab$$

$$\text{Total:} \quad SSX = \sum_{ijk} x^2_{ijk} - C_x$$

$$\text{Block:} \quad RXX = \frac{1}{ab} \sum_k x^2_{\cdot\cdot k} - C_x$$

(9.42)

$$\text{AB Subclass:} \quad SABX = \frac{1}{r} \sum_{ij} x^2_{ij\cdot} - C_x$$

$$\text{Main Effect A:} \quad AXX = \frac{1}{rb} \sum_i x^2_{i\cdot\cdot} - C_x$$

$$\text{Main Effect B:} \quad BXX = \frac{1}{ra} \sum_j x^2_{\cdot j\cdot} - C_x$$

$$\text{AB Interaction:} \quad ABXX = SABX - AXX - BXX$$

$$\text{Exp. Error:} \quad EXX = SSX - RXX - SABX$$

The comparable SS for y, the variable of interest, are:

$$C_y = y^2../rab$$

$$SSY = \underset{ijk}{\Sigma\Sigma\Sigma}\ y^2_{ijk} - C_y$$

$$RYY = \frac{1}{ab}\ \underset{k}{\Sigma}\ y^2_{..k} - C_y$$

$$SABY = \frac{1}{r}\ \underset{ij}{\Sigma\Sigma}\ y^2_{ij.} - C_y$$

(9.43)

$$AYY = \frac{1}{rb}\ \underset{i}{\Sigma}\ y^2_{i..} - C_y$$

$$BYY = \frac{1}{ra}\ \underset{j}{\Sigma}\ y^2_{.j.} - C_y$$

$$ABYY = SABY - AYY - BYY$$

$$EYY = SSY - RYY - SABY$$

Finally, the sums of cross-product calculations are:

$$C_{xy} = x...y.../rab$$

$$SPXY = \underset{ijk}{\Sigma\Sigma\Sigma}\ x_{ijk}y_{ijk} - C_{xy}$$

$$RXY = \frac{1}{ab}\ \underset{k}{\Sigma}\ x_{..k}y_{..k} - C_{xy}$$

$$SABXY = \frac{1}{r}\ \underset{ij}{\Sigma\Sigma}\ x_{ij.}y_{ij.} - C_{xy}$$

(9.44)

$$AXY = \frac{1}{rb}\ \underset{i}{\Sigma}\ x_{i..}y_{i..} - C_{xy}$$

$$BXY = \frac{1}{ra}\ \underset{j}{\Sigma}\ x_{.j.}y_{.j.} - C_{xy}$$

$$ABXY = SABXY - AXY - BXY$$

$$EXY = SPXY - RXY - SABXY$$

The above sums of squares and cross-products are summarized in Part A of the analysis of covariance table appearing below.

Table 9.5 ANACOVA for Two-Factorial in an RCB Design--Part A

Source	df	XX	XY	YY	Adjust.	SS[Adj]	MS[Adj]
Blocks	r-1	RXX	RXY	RYY			
A	a-1	AXX	AXY	AYY			
B	b-1	BXX	BXY	BYY			
AB	(a-1)(b-1)	ABXX	ABXY	ABYY			
Error	(r-1)(ab-1)	EXX	EXY	EYY	$(EXY)^2/EXX$	S[E]	MS[E]
A+E		SAx	SAxy	SAy	$(SAxy)^2/SAx$	S[A+E]	
B+E		SBx	SBxy	SBy	$(SBxy)^2/SBx$	S[B+E]	
AB+E		SABx	SAXxy	SABy	$(SABxy)^2/SABx$	S[AB+E]	

where

$$SAx = AXX + EXX \qquad SAxy = AXY + EXY \qquad SAy = AYY + EYY$$

$$SBx = BXX + EXX \qquad SBxy = BXY + EXY \qquad SBy = BYY + EYY$$

$$SABx = ABXX + EXX \qquad SABxy = ABXY + EXY \qquad SABy = ABYY + EYY$$

$$S[A+E] = SAy - (SAxy)^2/SAx \qquad S[B+E] = SBy - (SBxy)^2/SBx$$

$$S[AB+E] = SABy - (SABxy)^2/SABx$$

$$MS[E] = S[E]/\{(r-1)(ab-1)-1\}$$

Then, the factorial effects and interaction adjusted for the effect of the covariate are obtained as follows:

$$S[A] = S[A+E] - S[E]$$
(9.45)
$$S[B] = S[B+E] - S[E]$$
$$S[AB] = S[AB+E] - S[E]$$

Part B of the analysis of covariance now has the form given below.

Table 9.6   ANACOVA for Two-Factor Factorial---RCB Design--Part B

Source	df	SS[adj]	MS[adj]
A	a-1	S[A]	MS[A]
B	b-1	S[B]	MS[B]
AB	(a-1)(b-1)	S[AB]	MS[AB]
Error	(r-1)(ab-1)-1	S[E]	MS[E]

9.9 Missing Value Analyses by ANACOVA

The analysis of covariance offers a unified approach to missing
value analyses. This has been discussed by Coons [1]. The procedure is quite simple; just introduce a dummy covariate for each
missing value. Specifically, for a single missing value in any
design:

   (i) Assign a value of y = 0 where the variable of interest
      has a missing value.

  (ii) Introduce a dummy covariate, x, and assign it values of

$$x = -1 \text{ where the missing y occurs}$$

$$= \ 0 \text{ for all other values of y}$$

 (iii) Do a regular ANACOVA for the given design, including
      adjustments of Exp. Error and treatments.

The estimate of the missing value, say M, which would have
been obtained by general missing value procedure leading to a
minimized Exp. Error sum of squares (see Section 6.9) is the
estimate of the "common slope"; that is,

(9.46) $M = \hat{\beta}$

And as discussed in Section 6.9, the use of a missing value
estimate in an ANOVA results in a biased treatment sum of
squares. The covariance adjustment of the treatment sum of
squares makes the necessary correction for this bias. In other
words, the test of adjusted treatments by ANACOVA provides an
unbiased test of treatment means.

Because this is a specialized use of ANACOVA, many of the
assumptions of ANACOVA are irrelevant. In particular, assumptions
related to the regressor x need not be of concern when ANACOVA is
used to accomplish a missing value analysis. Only the usual ANOVA
assumptions must be satisfied.

PROBLEMS

1. An agronomist conducted an experiment to compare yields, y,
    of six varieties of cotton. The design was an RCB with 6
    blocks. The number of mature plants per plot, x, was used as
    a covariate.

    (a) Give the linear model for this experiment. Explain terms
        used in the model and specify ranges on the subscripts.
    (b) What assumptions are necessary for valid inferences under
        an ANACOVA? Assume fixed treatment effects.

2. Repeat Problem 1 if treatments have a 2x3 factorial structure.

3. An educator conducted an experiment to evaluate four training
    procedures. The experiment was conducted in a large city
    where 8 elementary schools, each having at least 4 sixth
    grade classes, were available. The 4 training procedures
    were randomly assigned to 4 classes at each school; there-
    fore, schools serve as blocks. The measurement of interest
    was y, the mean score for the class on an examination at the
    end of the training procedure. To account for possibly
    different backgrounds of the classes, all classes were given
    a pre-training examination. The mean score that a class
    received on this examination was taken as a covariate, x.

    (a) Give the linear model for this experiment. Explain terms
        used in the model and specify ranges on the subscripts.
    (c) What assumptions are necessary for valid inferences under
        an ANACOVA? Assume fixed treatment effects.

4. Repeat Problem 3 if treatments have a 2x2 factorial structure.

5. The ANACOVA table given below was calculated for the variety
    experiment of Problem 1.

ANACOVA For Cotton Variety Yields

Source	df	XX	XY	YY
Blocks	5	2008	3610	5113
Varieties	5	713	1624	4628
Exp. Error	25	2200	3778	7464

Unless stated otherwise, required assumptions are satisfied.

(a) Estimate the common slope.

(b) Test the hypothesis of zero common slope.

(c) Is there evidence that treatments are affecting the covariate? Justify your answer.

(d) Is there evidence of different mean yields of cotton for the 6 varieties? Justify your answer.

6. Refer to information given in the previous problem for the cotton variety experiment. Additional information is given below for the treatment means:

Variety (i)	$\bar{x}_{i.}$	$\bar{y}_{i.}$
1	44	158
2	38	140
3	46	170
4	41	155
5	49	174
6	51	167

(a) Give the adjusted variety means for cotton yields.

(b) Give a table of differences of adjusted variety mean yields and their estimated standard deviations.

7. The ANACOVA table given below was calculated for the educational training experiment described in Problem 3.

ANACOVA For Training Procedures

Source	df	XX	XY	YY
Schools	7	984	1145	2036
Trainings	3	328	628	1382
Exp. Error	21	1343	1971	3359

Unless stated otherwise, required assumptions are satisfied.

(a) Estimate the common slope.

(b) Test the hypothesis of zero common slope.

(c) Is there evidence that treatments are affecting the covariate? Justify your answer.

(d) Is there evidence of different mean scores for the training procedures? Justify your answer.

8. Refer to information given in the previous problem for the educational training experiment. Additional information is given below for the treatment means:

Training	$\bar{x}_{i.}$	$\bar{y}_{i.}$
1	68	74
2	73	89
3	77	91
4	72	85

(a) Give adjusted mean scores for the training procedures.
(b) Give a table of differences of adjusted mean scores and their estimated standard deviations.

9. A testing agency was funded by the federal government to investigate different applicators for spraying insecticides by aircraft. There were four different applicators but some had several speeds giving 11 "applicator treatments." The applicators were mounted on an airplane and used throughout the experiment. On a "calm day" one spraying was made for each of the 11 treatments. The variable measured was the amount of insecticide, y, landing on a unit area of special collection paper placed on the ground below the flight path of the airplane. Even though calm days were used, variable wind and slight gusts were expected; this was accounted for by a covariate, x, the wind speed on the flight path. The following table resulted:

ANACOVA for Insecticide Applications

Source	df	XX	XY	YY
Days	4	110	137	391
Treatments	10	221	469	1215
Exp. Error	40	617	892	1439

(a) Estimate the common slope.
(b) Test the hypothesis of zero common slope.
(c) Should the agency be concerned about the assumption that treatments do not affect the covariate? Explain.
(d) Is there evidence that the applicators have different mean amounts of insecticide landing on the collection paper? Justify your answer.

10. A merchandising company having a large number of outlet stores undertook an extensive advertising campaign. Four types of advertising were used:

(1) Newspaper
(2) Mail
(3) Radio
(4) Television

A total of 32 comparable outlet stores (with respect to floor space, inventory, and so on) were selected and each type of advertising was randomly assigned to 8 outlet stores. The outlets were separated so that the advertising of one outlet did not overlap that of another outlet. The average weekly sales (in dollars) for one month prior to the campaign was taken as a covariate. The average weekly sales for the month of the campaign was the variable of interest.

The following partial ANACOVA resulted:

ANACOVA for Weekly Sales

Source	df	XX	XY	YY
Advertisings	3	16321	23815	35887
Exp. Error	28	100217	108612	136157

(a) Estimate the common slope.
(b) Test the hypothesis of zero common slope.
(c) Is there evidence that treatments are affecting the covariate? Justify your answer.
(d) Is there evidence that average weekly sales during the campaign, when adjusted for pre-campaign sales, differ for the four types of advertising? Justify your answer.

REFERENCE

[1] Coons, I., The analysis of covariance as a missing plot technique., *Biometrics*, Vol. 13, 1957.

## Chapter 10

CONFOUNDING SYSTEMS

10.1   Introduction

As we have noted on several ocassions, the control of nuisance variation (equivalently, error control) is a major objective of experimental design. We have considered two design techniques that are used extensively for control of nuisance variation: the use of blocking factors and the use of covariates.

In the randomized block and Latin square designs, we insisted on using complete blocks; that is, where all treatments occur with every level of any blocking factor. This was done to insure a balanced design, one having the important advantage: "all treatment contrasts are free of block effects." But for various reasons, a researcher may have difficulties obtaining complete blocks. We shall discuss some of these reasons in the next section. We begin with a definition.

Definition 10.1

An *incomplete block* is a set of k homogeneous EU, where k < t, the number of treatments.

The use of incomplete blocks in an experiment has several implications. First, some treatments are missing in every block. Which treatments should we place in a given block? Are some block compositions better than others? Second, if we insist on having r replications of each treatment, we may need a large number of blocks. For example, with t = 10 treatments in blocks of size 2, five blocks are required just for one replication of the treatments. Is a sufficient number of blocks available for several replicates of the treatments? And generally, more blocks would require more work. Are the resources (time, personnel, and so on) sufficient to handle such a number of blocks? Third, some treatment contrasts will contain "incomplete block effects." Which ones? What, if anything, can be done to minimize these "unwanted" effects? These questions should be kept in mind as one continues through this chapter. Answers to some are given; others need to be addressed by the researcher.

10.2 Some Design Considerations

Even when variation among experimental material is identified, there may be reasons why complete blocks cannot be formed. The number of treatments may be so large that homogeneous blocks of such a size are impractical or impossible. Plant breeders, in particular, are faced with such a problem for they frequently investigate a dozen new varieties and all possible combinations of parental crosses of these varieties. Maybe only blocks of 15 homogeneous plots (EU) of land can be delineated whereas t could be as large as 144. Likewise, an engineer dealing with 5 factors in the study of polymer production has at least 32 treatments. But in most industrial experiments, blocks of this size usually are difficult to manage and require extensive space and equipment. Any experimenter dealing with a "large factorial" would face a similar problem.

Another reason for using incomplete blocks might be the inability to effectively manage complete blocks. If treatment application is tedious or time consuming, it might be advisable to apply only some of the treatments at one time. This is true especially if a time effect is introduced or if experimenter fatigue might introduce a lack of uniformity when dealing with a complete set of treatments.

In other experimental situations, incomplete blocks occur naturally. The use of automobiles to investigate brands of tires would imply blocks of size 4 (the wheels) unless position was also important. Thus, more than 4 brands of tires would call for the use of incomplete blocks. The use of human subjects often provides incomplete blocks of size 2. Each person has 2 eyes, 2 ears, 2 arms, 2 legs, and so on, which might be the experimental units. Similarly, animal subjects provide incomplete blocks of size 2 or 4 --- 2 eyes, 4 legs, and so on. Also in animal experiments, litters are taken to be blocks quite often because litter mates are expected to be somewhat homogeneous due to their similar genetic background. In engineering experiments, a batch of material might serve as a block; but a batch might not be large enough to accomodate all treatments. In an educational experiment, classes within each school might serve as blocks; but the number of classes might be smaller than the number of treatments.

Finally, we point out that an experimenter might end up with
an incomplete block design even if it was not originally designed
as such. During the course of the experiment, parts of complete
blocks may be lost --- this might occur if

(i) flooding washes out some of our plots
(ii) some of our experimental animals die
(iii) observations or identification of some units may be lost,
and so on.

Even though an incomplete block could occur due to an accident,
as indicated above, we will be interested primarily in designing
experiments having incomplete blocks of specific sizes and
compositions. In the remainder of this chapter we consider incomplete block designs and study some special systems for factorial
treatments.   Later chapters cover additional incomplete block
designs, including general treatments.
    Whether incomplete blocks occur by accident or by design,
some of the balance feature is lacking. The next section
addresses the impact which incomplete blocks have on the balance
features and how this affects the inference process. A number of
questions come to mind in this regard. Do different block sizes
produce different degrees of imbalance? Do different patterns of
"imbalance" have different impacts on our inferences?   Can we
identify the specific imbalance?   If so, it seems we should be
able to do something about it.   Can we make design modifications
to minimize the effects of imbalance?

10.3 Confounded Effects

Before we formally define this concept, let us discuss a simple
example to gain some understanding of the potential problem.
Suppose flooding washed out several plots in an established RCB
experiment leaving us with the following incomplete block design:

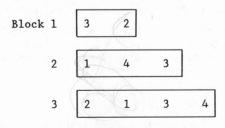

Block 1 | 3   2 |

      2 | 1   4   3 |

      3 | 2   1   3   4 |

where the numbers represent the treatments originally assigned to
the EU. Some contrasts among the treatment means will be free of

block effects while others will not.   For the contrast of treat-
ment means, $\theta = \mu_{1.} - \mu_{4.}$, we would use an estimator $C_1 = \bar{y}_{1.} - \bar{y}_{4.}$
based on the available data.   But we note that C_1 is free of all
block effects because

  (i) from block 1, no data are used in calculating either
      $\bar{y}_{1.}$ or $\bar{y}_{4.}$ so the effect of the first block cannot
      enter C_1

  (ii) from block 2, one observation is used in calculating
       each of $\bar{y}_{1.}$ and $\bar{y}_{4.}$, so the effect of the second block
       enters and leaves C_1 when these 2 means are subtracted,

 (iii) from block 3, one observation is used in calculating
       each of $\bar{y}_{1.}$ and $\bar{y}_{4.}$, so the effect of the third block
       enters and leaves C_1 when these 2 means are subtracted.

But for the contrast $\theta_2 = \mu_{1.} - \mu_{2.}$, we would use the estimator C_2
$= \bar{y}_{1.} - \bar{y}_{2.}$ which contains effects of blocks 1 and 2 because:

  (i) from block 1, one observation is used in calculating $\bar{y}_{2.}$
      but no observations are used in calculating $\bar{y}_{1.}$, so the
      effect of the first block is present in C_2 when these
      two means are subtracted,

  (ii) from block 2, one observation is used in calculating $\bar{y}_{1.}$
       but no observations are used in calculating $\bar{y}_{2.}$, so the
       effect of the second block is present in C_2 when these
       two means are subtracted, and

 (iii) from block 3, one observation is used in calculating
       each of $\bar{y}_{1.}$ and $\bar{y}_{2.}$, so the effect of the third block
       enters and leaves C_2 when these 2 means are subtracted.

We   therefore don't know what portion of the observed value of C_2
is due to treatment effects and what portion is due to these two
block effects.   This treatment contrast is contaminated with
block effects. So are others. We now define this concept of
contaminated effects.

Definition 10.2

Two factors are said to be _confounded_ if their effects
are inter-related in such a way that the amount attributable
to each factor are inseparable.

definition of confounding

Therefore, we can say that an incomplete block experiment, whether occurring by accident or by design, has *some* treatment components confounded with incomplete block effects. A designed experiment, unlike the one at the beginning of this section, would have some systematic features: (i) probably blocks of equal sizes, (ii) certain block compositions, (iii) equal replications of treatments, and so on.

10.4 Incomplete Block Designs for Factorial Experiments

For the RCB experiment given in the previous section, confounding was caused by an unforeseen event---the flooding of plots. And we noted specific information about treatments which was lost because of the confounding. If different plots had been washed out, then different information about treatments might have been lost. As one might surmise, the specific information lost depends upon the specific treatment patterns appearing in the completed experiment.

Suppose then at the designing stage of an experiment we know that complete blocks cannot be constructed. By allowing certain treatments to appear together in the incomplete blocks we can govern what treatment information is confounded and what is not. We therefore should have the confounded effects represent the treatment contrasts of least interest to us.

In factorial experiments we generally confound higher order interactions. This is done in accordance with the following basic principle: As the number of factors increase, their dependency tends to decrease (that is, higher order interactions tend to be less significant than lower order interactions). Of course, the researcher's knowledge and objectives would take precedence over this general confounding proposal.

10.5 Incomplete Block Designs for 2^n Factorial Experiments: Complete Confounding

In this section we shall discuss confounding plans for factorial experiments where each treatment factor has only 2 levels, called low and high levels, in general. With these special factorials, the confounding plans are simple yet complete enough to exhibit the basic ideas of confounding.

We can use a very simplified notation to represent the two levels of any factor: the absence/presence of the corresponding lower-case letter. For example, the presence of the letter "a" would immply the high level of Factor A while the absence of "a" would imply the low level of A. The treatment having the low level of all factors will be represented by (1) rather than "no letter."

For a 2x2 factorial, the four treatments are given below in the earlier notation and in the simplified format:

(10.1)

$$a_1b_1 \rightarrow (1) \qquad\qquad a_1b_2 \rightarrow b$$

$$a_2b_1 \rightarrow a \qquad\qquad a_2b_2 \rightarrow ab$$

10.5.1   IB Designs for 2^2 Factorials

This is, of course, the simplest factorial we can have --- four treatments having a 2 x 2 factorial structure. One might think it unnecessary to even consider confounding with such a small number of treatments. After all, we only need blocks of 4 EU each. On the contrary, <u>blocks of 2 EU may be more homogeneous than blocks of size 4 EU.</u> <u>This is not an uncommon situation when we are dealing with such experimental material as (1) identical twins, (2) halves of potatoes (or other similar items), (3) the right and left arms of people, and so on.</u>
With four treatments and blocks of size two, we will need two blocks for each complete replication of the treatments. We now must decide which treatments are to appear in each block. the four treatments were given earlier in (10.1). Suppose for the time being that we decide to use 1 replication of the treatments with the following allocation to incomplete blocks:

(10.2)

	Block 1	(1)	a
	Block 2	b	ab

Let us now consider the three treatment contrasts which measure the main effects A and B, and the interaction of A and B. These contrasts were given in Chapter 5, equation (5.10) in terms of subclass means. Incorporating the notation change indicated in (10.1), we may represent the estimates in the simple format

(10.3)

$$-(1) - b + a + ab = C_1 \qquad \text{Main effect of A}$$

$$-(1) + b - a + ab = C_2 \qquad \text{Main effect of B}$$

$$(1) - b - a + ab = C_3 \qquad \text{Interaction of A and B}$$

where we are letting the symbols (1), a, b, and ab represent both treatment designations and their respective means. [Here each mean is based on a single observation because we have only one replicate.]

By an analysis similar to that used in Section 10.3, we note that the first and third contrasts are free of block effects but the second one is not. In fact, if we let B_1 and B_2 denote the totals of the blocks in (10.2), the second contrast in (10.3) is

because subtract one element from another in side the block

(10.4) $C_2 = B_1 - B_2$

As expressed in (10.3), C_2 represents a contrast of the 2 levels of factor B; as expressed in (10.4), C_2 represents a contrast of the 2 incomplete blocks. By Definition 10.2, we can say that the main effect for B is confounded with these two incomplete blocks.

We also may establish a link between the degrees of freedom and sums of squares for confounded effects and incomplete blocks. In the above example, a 2 x 2 factorial, each main effect and interaction has one degree of freedom. Recall from our earlier work on contrasts that these sums of squares can be calculated as

$$SS_A = SS[C_1] = C_1^2/[1 + 1 + 1 + 1]$$

(10.5) $$SS_B = SS[C_2] = C_2^2/[1 + 1 + 1 + 1]$$

$$SS_{AB} = SS[C_3] = C_3^2/[1 + 1 + 1 + 1]$$

where C_1, C_2, and C_3 are given in (10.3).

There is a single degree of freedom among the two incomplete blocks in (10.2). Consequently, the sum of squares for incomplete blocks may be calculated from the contrast of block totals in (10.4). We indicated above that this contrast also represents the main effect for B; thus, the sum of squares for incomplete blocks is identical to the sum of squares for the confounded effect. In general when the sums of squares and degrees of freedom for incomplete blocks and certain effects are identical, we say that these effects are *completely confounded with incomplete blocks*. For the confounded scheme using the 2 blocks described above, the treatment portion of the ANOVA is

Source	df
(10.6) Blocks (= Main Effect B)	1
A	1
AxB	1

Instead of the confounding scheme used in (10.2), we could have used either of two other schemes; namely,

	Scheme 2	Scheme 3
Block 1	(1) b	(1) ab
Block 2	a ab	a b

(10.7)

Using Scheme 2 results in main effect A completely confounded with blocks whereas the interaction of A and B is completely confounded with blocks under Scheme 3.

Which of the 3 confounding schemes to be used in a particular experiment will depend upon which 2 treatment components are most important, or on which component one is willing to relinquish information.

10.5.2   IB Designs for 2^3 Factorials

With three factors each at two levels, there are eight treatment combinations. Blocks of size 2, 4 or 8 are possibilities; using blocks of size 8 result in a complete block experiment.

Using blocks of size 4 would require 2 blocks for a replicate of the treatments. Because there is 1 degree of freedom among the 2 blocks, one main effect or interaction must be confounded with incomplete blocks of each complete replicate of the treatments. Suppose we decide to confound the ABC interaction with the incomplete blocks. The contrast to measure the ABC interaction is

$$(10.8) \quad C_{ABC} = -(1) + a + b + c - ab - ac - bc + abc$$

The treatments having the same sign in the contrast for ABC must appear together in an incomplete block. Therefore to have the ABC interaction confounded with incomplete blocks, one replicate (unrandomized) would be

	Block 1	Block 2
	a	(1)
(10.9)	b	ab
	c	ac
	abc	bc

We note that the contrast for measuring the ABC interaction, as given in (10.8), can be expressed also as $C_{ABC} = B_1 - B_2$, where B_1 and B_2 are the totals for blocks 1 and 2 in (10.9). The sum of squares for the confounded effect, ABC in this case, would be identical to the block sum of squares (both obtainable from the contrast $B_1 - B_2$).

We now turn our attention to confounding of a 2^3 factorial in blocks of size 2. Four incomplete blocks would be required for each replicate of the treatments. Consequently, 3 treatment degrees of freedom will be confounded in each replicate. While it is not obvious, only 2 confounded effects need to be specified for setting up the incomplete block compositions. The third confounded effect will be automatically specified; this will be discussed later. Suppose then that we agree to confound the interactions AB and AC. The four block compositions are the treatments of these two contrasts that enter

(10.10)
(1) positively in both AB and AC
(2) positively in AB, negatively in AC
(3) negatively in AB, positively in AC
(4) negatively in both AB and AC.

The contrasts for AB and AC are

(10.11)
$$C_{AB} = (1) - a - b + c + ab - ac - bc + abc$$
$$C_{AC} = (1) - a + b - ab - c + ac - bc + abc$$

Examination of the contrasts in (10.11) reveals that the four block compositions should be

Block 1	Block 2	Block 3	Block 4
(1) abc	a bc	c ab	b ac

Now, if we let B_1, B_2, B_3, and B_4 denote these block totals, the two contrasts in (10.11) can be written as

(10.12)
$$C_{AB} = B_1 - B_2 + B_3 - B_4$$
$$C_{AC} = B_1 - B_2 - B_3 + B_4$$

A third comparison of the block totals orthogonal to the ones in (10.12) would be

$$(10.13) \qquad C_3 = B_1 + B_2 - B_3 - B_4$$

If we check the contrasts for the remaining treatment components in the 2^3 factorial, we would discover that C_3 is the contrast for measuring the BC interaction.   Now we discuss a simpler way of determining the third confounded effect when two confounded effects are specified. We begin with a definition.

Definition 10.3

In the 2^n factorials, if two effects are confounded, so is their product.   Any exponent in such a product is replaced by the remainder upon division by 2. The resulting product is called the *generalized interaction* of the two confounded effects.

Replacing an exponent by its remainder upon division by 2 is due to properties of "mathematical group theory." We will mention and utilize certain properties of this group theory but an in-depth knowledge is not necessary. This group theory is really the basis for constructing confounding schemes, block compositions and their generalized interactions (and not just for the 2^n factorials, as we will see later.) For the above example with AB and AC confounded, the generalized interaction is $ABAC = A^2BC = BC$ where we have replaced A^2 by $A^0 = 1$, because the exponent of A has a zero remainder upon division by 2.

In the above confounding system we chose to confound AB and AC giving a generalized interaction of BC.   Had we initially chosen to confound AB and BC, we see that the generalized inter-action is $ABBC = AB^2C = AC$.   The same three treatment components would be confounded no matter which pair of AB, AC and BC were initially chosen to construct the confounding system.   This will be the case always because a chosen set of treatment components plus their generalized interactions form a "closed set" under the rule stated in Definition 10.3.

Example 10.1

In a 2^3 factorial experiment, suppose it is necessary to use blocks of size 2.   If we decide to invoke the basic principle mentioned in the beginning of Section 10.3, we would confound the

ABC interaction and one of the two-factor interactions, say AC. But then the generalized interaction is $ABCAC = A^2BC^2 = B$. There is no way to avoid the confounding of a main effect in this experiment when we insist on confounding the ABC interaction. The researcher must decide which treatment components are most important; these should not be confounded with incomplete blocks. Using only treatment components of lesser interest, there usually are various confounding schemes which might be considered.✱

10.5.3 IB Design for a General 2^n Factorial

In a general 2^n factorial, n factors each at 2 levels, we might consider experiments with blocks of size 2, 4, 8, ---, 2^{n-1}. One complete replicate of the treatments requires 2^n EU. So if blocks of size 2^p are used, we need $2^n/2^p$ or $2^{n-p} = 2^w$ blocks. When 2^w blocks are needed for a complete replicate of the treatments, w confounded effects must be chosen --- those chosen cannot be generalized interaction of each other. For the 2^3 factorial of the previous section, $2^2 = 4$ blocks were required; w = 2 effects were chosen for confounding.

If the experimenter suspects that certain factorial components are unimportant or negligible, they are candidates for being confounded effects. Otherwise, an experimenter probably would want to confound higher order interactions under the general principle stated earlier. One must be careful though and not choose all high order interactions because their generalized interactions will be main effects or low order interactions.

And as indicated before, the w chosen effects together with their generalized interactions form a closed set in the sense that any subset of w effects (which are not generalized interactions of each other) from this set would generate the same closed set of confounded effects.

Example 10.2

In a 2^5 factorial to be conducted in $2^3 = 8$ incomplete blocks of size 4 EU each, we need to choose 3 effects to be confounded. If we choose ABCD, ACDE and BCDE, the 4 generalized interactions also confounded are:

$$ABCDACDE = BE \qquad\qquad ACDEBCDE = AB$$

$$ABCDBCDE = AE \qquad\qquad ABCDACDEBCDE = CD$$

These 7 effects form a closed system in the sense that the same 7

confounded effects would result from any three effects, none of
the three being generalized interactions of the other two chosen
effects.   Thus, AE, BE and ABCE would generate the same set of
confounded effects. So would AE, BE and CD, and so on. ⬛

10.6 Partial Confounding for the 2^n Factorials

In Section 10.5, we discussed completely confounded systems for
the 2-level factorials.   This means that the same set of effects
is confounded with incomplete blocks in every complete replicate
of the factorial treatments.   A better way to proceed might be
the following: Confound different effects in different complete
replicates of the treatments. All treatment components can be
estimated, free of block effects, if sufficient replicates are
used.
   The analysis of any partially confounded system is very
straightforward.   Simply   estimate any effect (and calculate its
sum of squares) only from replicates where it is not confounded.
The next example gives the basic ideas of partial confounding in
a 2^2 factorial experiment where all effects can be estimated free
of block effects.   Example 10.4 illustrates the analysis for such
an experiment.

Example 10.3

Suppose an experiment is being conducted with 2x2 factorial
treatments and blocks of size 2.   Each complete replicate of the
treatments   requires   2 blocks.   A decision is made to use   the
following three replicates:

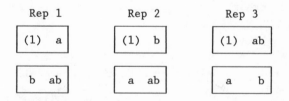

Then, main effect B is confounded in Rep 1, main effect A in
Rep 2, and the AB interaction in Rep 3.   This implies we can
estimate (free of any block effects):   main effect B from Reps 2
and 3;   main effect A from Reps 1 and 3; the AB interaction from
Reps 1 and 2.   The ANOVA table has the following form:

ANOVA for Partially Confounded 2^2 Factorial

Source	df
Reps	2
Blocks(Reps)	3
A	1*
B	1*
AB	1*
Exp. Error	3
Total	11

The first two sources, labeled "Days" and "Blocks within Days" represent a subdivision of the variation among the 6 incomplete blocks. From (7.7) and (7.8), the sums of squares for these two sources are seen to be related by

$$\text{SS Blocks(Day)} = \text{SBD} - \text{SS Days}$$

where SBD is the Block-Day subclass sum of squares. The asterisks are used to indicate confounded effects whose sums of squares are calculated from only some of the replicates. Note that the use of additional replicates would change the degrees of freedom for all non-treatment components but not those of treatments. Also changed would be the replicates from which treatment estimates and their sums of squares would be calculated.✺

Example 10.4

A chemist conducted an experiment to determine the yield of a chemical produced under two catalysts (factor A), each at two percentages (factor B.) Due to available equipment and the nature of the analyses required, only two of the treatment combinations could be handled at one time. Any two could be completed in 3-4 hours so one complete replicate could be done on a single day. Because environmental conditions might have a slight effect on the yields, it was decided to use "mornings" and "afternoons" as blocks. The experiment was conducted on 3 different days using the confounded scheme of Example (10.3). Yields (in grams) were:

	Day 1		Day 2		Day 3	
AM	ab 34.5	b 29.6	b 31.1	(1) 30.4	(1) 27.7	ab 35.2
PM	(1) 38.1	a 32.2	ab 34.8	a 32.9	b 31.8	a 30.5

Let us first calculate the sums of squares for Days, Blocks within Days, and Total.

$$C = (34.5 + 29.6 + \ldots + 30.5)^2/12 = (378.8)^2/12 = 11957.45$$

Total SS: $(34.5)^2 + \ldots + (30.5)^2 - C = 12025.5 - C = 68.05$

The subclass for Days-Blocks is:

	Day 1	Day 2	Day 3
AM	64.1	61.5	62.9
PM	60.3	67.7	62.3
Day Totals	124.4	129.2	125.2

Day SS: $[(124.4)^2 + (129.2)^2 + (125.2)^2]/4 - C = 11960.76 - C$
$$= 3.31$$

Block(Day) SS: $[(64.1)^2 + \ldots + (62.3)^2]/2 - 11957.45 - 3.31$
$$= 11974.07 - 11960.76 = 13.31$$

To estimate the treatment components, refer to the contrasts in (10.3). For main effect A, we use Days 1 and 3:

$$C_A = -(1) + a - b + ab$$
$$= -27.9 + 31.35 - 30.7 + 34.85 = 7.6$$

For main effect B, we use Days 2 and 3:

$$C_B = -(1) - a + b + ab$$
$$= -29.05 - 31.7 + 31.45 + 35.0 = 5.7$$

For the AB interaction, we use Days 1 and 2:

$$C_{AB} = (1) - a - b + ab$$
$$= 29.25 - 32.55 - 30.35 + 34.65 = 1.0$$

The sums of squares for the treatment components are:

$$SSA = C_A^2/[1/2 + 1/2 + 1/2 + 1/2] = (7.6)^2/2 = 28.88$$

where the divisors of 2 result from $r = 2$ observations per mean.

$$SSB = C_B^2/2 = (5.7)^2/2 = 16.24$$

$$SSAB = C_{AB}^2/2 = (1.0)^2/2 = 0.50.$$

By subtraction, the Exp. Error sum of squares is

$$SSE = 68.05 - 3.31 - 13.31 - 28.88 - 16.24 - 0.5 = 5.81$$

ANOVA for Chemical Yields

Source	df	SS	MS	F
Days	2	3.31	- - -	- - -
Blocks(Days)	3	13.31	- - -	- - -
Catalysts	1	28.88	28.88	14.89
Amounts	1	16.24	16.24	8.37
CxA	1	0.50	0.50	< 1
Exp. Error	3	5.81	1.94	
Total	11	68.05		

The observed significance levels are, respectively, $0.025 < P < 0.05$; $0.05 < P < 0.10$; and $P > 0.10$.❆

For the partially confounded experiments discussed in the last two examples, we saw that each treatment effect could be calculated from 2 of the 3 replicates. Thus, every treatment effect is based upon the same amount (2/3) of the total information available in the experiment. This seems to indicate some degree of "balance" among the treatments. We will return to this idea and these examples in later chapters.

Example 10.5

We now consider a 2^3 factorial to be conducted in blocks of size 4. A complete replicate of the treatments will require 2 blocks. One main effect or interaction needs to be confounded in each replicate so let us confound the interactions AB, AC, and ABC in replicates 1,2, and 3, respectively. The block compositions can be obtained from (10.7) and (10.10). Each of these 3 interactions can be estimated from two replicates where they are unconfounded while the remaining four treatment effects can be estimated from all three replicates. The partial ANOVA for this experiment would be:

ANOVA for Partially Confounded 2^3 Factorial

Source	df
Reps	2
Blocks(Reps)	3
A	1
B	1
AB	1*
C	1
AC	1*
BC	1
ABC	1*
Exp.Error	11

The   interactions AB,   AC and ABC are based on 2/3 of   the   total experimental   information   while   all other treatment effects   are based on full information. ▓

10.7   General Method of Constructing Incomplete Blocks : 2^n Confounding Systems

For   any 2^n factorials experiment,   the blocks compositions for confounded systems could be obtained from the plus/minus signs of the contrasts for those effects that we elect to confound.   This is   the procedure that we emphasized in the earlier part of   this chapter.   Even from the simple cases that we considered,   it   is clear that the method will become more tedious and time-consuming for   larger   factorials.   Consider the effort required to   write down contrasts,   such as those in (10.10),   when there are   seven factors.   Each contrast contains 128 treatments.

   More   critical   than the time element,   though,   is that   the method using plus/minus signs does not extend to other   factorial systems,   such as the 3^n series,   the 5^n series,   and so on.   We therefore   need to consider a more general method of constructing incomplete blocks which will extend to other systems.   To develop this   general method,   we will need to introduce some new   ideas, including another representation of factor levels.   We begin with a definition.

Definition 10.4

For a positive integer p,   the *residues modulo p (mod p)* are 0, 1, 2, ..., p-1. For any integer m, the residue (mod p) is equal to the remainder when m is divided by   the integer p. If m has r for its residue (mod p), we write m=r(mod p).

Example 10.6

The residues (mod 2) are 0 and 1. Any even integer has a residue (mod 2) equal to 0, while any odd integer has a residue (mod 2) equal to 1.   Thus, 3=1(mod 2), 10=0(mod 2), 15=1(mod 2), 2=0(mod 2), and so on.

The residues (mod 3) are 0, 1, and 2. Every integer that is a multiple of 3 has residue (mod 3) equal to 0. All other integers have residues (mod 3) equal to 1 or 2. Thus, 5=2(mod 3), 10=1(mod 3), 15=0(mod 3), 439=1(mod 3), and so on. ✷

The residues (mod p) provide a very convenient representation of the p levels for a factor. Thus, in the 2^n series, the levels of each factor can be represented by 0 and 1, the residues (mod 2). For the i-th factor, we may think of the residues as being values of a variable x_i. A typical treatment will be denoted by $x_1 x_2 \cdots x_n$.

Example 10.7

The four treatments of a 2^2 factorial may be represented by

$x_1 x_2$	Earlier Designation
00	(1)
10	a
01	b
11	ab

And, for a 2^3 factorial, the 8 treatments may be represented by

$x_1 x_2 x_3$	Earlier Designation
000	(1)
100	a
010	b
110	ab
001	c
101	ac
011	bc
111	abc ✷

We now will show that the contrast for each main effect and interaction in a 2^n factorial can be obtained from a specific equation in x_1, x_2, ..., x_n. An examination of the treatments listed in Example 10.7 reveals that the contrast for main effect A (the first factor) is the difference between "the treatments with a 1 in the *first* position and those with a 0 in the *first* position." This is true not only for the two factorials listed in this example but for any 2^n factorial. Likewise, for any 2^n factorial, the contrast for main effect B (the second factor) is the difference between "those treatments with a 1 in the *second* position and those with a 0 in the *second* position." Therefore

(10.14)
$$C_A = [\text{trts. with } x_1 = 1 (\text{mod } 2)] - [\text{trts. with } x_1 = 0 (\text{mod } 2)]$$
$$C_B = [\text{trts. with } x_2 = 1 (\text{mod } 2)] - [\text{trts. with } x_2 = 0 (\text{mod } 2)]$$

Similar results hold for any main effect---we merely use the x_i corresponding to the factor involved.

But how can we represent the contrasts for interactions? Consider the AB interaction. The treatments shown in Example 10.6 reveal that the contrast for AB can be expressed as

(10.15)
$$C_{AB} = [\text{trts. with } x_1 + x_2 = 0 (\text{mod } 2)] -$$
$$[\text{trts. with } x_1 + x_2 = 1 (\text{mod } 2)]$$

Similar results hold for the other two factor interactions--- merely use the two x's corresponding to the factors involved.

The expressions in (10.14) and (10.15) can be extended to any number of factors by including the x's corresponding to the factors involved. We give a general rule for these expressions.

Rule 10.1

In any factorial, each treatment component can be related uniquely to a linear modular equation in x_1, x_2, ..., x_n whose coefficients are the respective exponents of the factor letters. In a 2^n factorial, each x_i has a coefficient of 0 or 1 because each factor letter appears in every treatment component with an exponent of either 0 or 1.

From the expressions in (10.13) and (10.14), one sees that the contrasts for treatment components in 2^n factorials also can be described as contrasts among the residues of the corresponding modular equation. This is true in general. The residues of 0 and 1 correspond to the plus/minus signs considered earlier. Which residue corresponds to plus signs depends upon the number of x's with non-zero coefficients in the modular equation. When this number of x's is odd, the residues of 1 correspond to plus; when this number of x's is even, the zero residues correspond to plus.

Example 10.8

In a 2^6 factorial, the ABD interaction is related to a modular equation "$x_1 + x_2 + x_4 \pmod 2$." Thus

$$C_{ABD} = [\text{trts. with } x_1 + x_2 + x_4 = 1 (\text{mod } 2)] -$$

$$[\text{trts. with } x_1 + x_2 + x_4 = 0 (\text{mod } 2)]$$

The BCDF interaction is related to a modular equation of the form "$x_2 + x_3 + x_4 + x_6 \pmod 2$"; namely,

$$C_{BCDF} = [\text{trts. with } x_2 + x_3 + x_4 + x_6 = 0 (\text{mod } 2)] -$$

$$[\text{trts. with } x_2 + x_3 + x_4 + x_6 = 1 (\text{mod } 2)]$$

Notice how the residues enter positively and negatively in these two contrasts. See the statement preceding this example.▨

We now describe a general method of constructing incomplete blocks in the 2^n series. The following rule governs the block compositions.

Rule 10.2

Using modular equations related to confounded treatment components, those treatments satisfying the equations for each "combination of residues" are placed together in an incomplete block. No equation of a generalized interaction may be used.

If a single treatment component is confounded with incomplete blocks in a 2^n factorial, there are only two combinations of

residues: the appropriate modular equation may be set equal to
either 0 or 1, the residues(mod 2). When two treatment components
must be selected for confounding, the corresponding two modular
equations each may be set equal to 0 or 1 giving 4 "combinations"
of residues (and consequently, 4 incomplete blocks.) Generalized
interactions must be ignored, as they were in the earlier methods
of construction, because their modular equations are directly
related to the equations for the selected components.   One may
wish to calculate the generalized interactions, though, to see if
some of their modular equations might be easier to work with.
Generally, those with the fewest x's are easiest to use.

Example 10.9

Suppose a 2^3 factorial is to be conducted in blocks of size 4
with ABC confounded.   Then, the block compositions are given by
the treatments satisfying the following two modular equations:

$$x_1 + x_2 + x_3 = 0(\text{mod } 2) \quad \text{and} \quad x_1 + x_2 + x_3 = 1(\text{mod } 2).$$

For one replicate of the treatments, the two blocks are:

000	100
110	010
101	001
011	111

If the same factorial is to be conducted in blocks of size 2,
we might elect to confound the AB and AC interactions.   For one
replicate of the treatments, the block compositions are given by
the following four pairs of modular equations:

(i) $x_1 + x_2 = 0$ (ii) $x_1 + x_2 = 0$

       $x_1 + x_3 = 0$ $x_1 + x_3 = 1$

(iii) $x_1 + x_2 = 1$ (iv) $x_1 + x_2 = 1$

       $x_1 + x_3 = 0$ $x_1 + x_3 = 1$

The four blocks are

(1) | 000 | (2) | 001 | (3) | 010 | (4) | 100 |
|-----|
| 111 | | 110 | | 101 | | 011 |

Example 10.10

Suppose a 2^5 factorial is to be conducted in 2^3 blocks. We need to choose 3 treatment effects for confounding, none of which are generalized interactions of each other. If the ABD, the BCE, and the ACE interactions are chosen for confounding, the 8 blocks would be determined by 3 modular equations each containing 3 x's. The four generalized interactions of these effects are: ACDE, BCDE, AB, and D. Therefore, the block compositions could be found equivalently, and more easily, from the 3 modular equations for AB, D, and BCE, as well as from other sets of 3 equations. ✳

Solving the various sets of modular equations, although not too difficult in the 2^n series, can become time consuming. We therefore consider one further simplification in the construction of incomplete blocks. The block which contains the low level of all factors is the one for which the modular equations of all confounded effects are set equal to zero. This is a very special block from which all other blocks can be obtained quite easily. We first define this special block.

Definition 10.5

When all modular equations related to confounded treatment effects are set equal to zero, the resulting incomplete block is called the *intrablock subgroup*. For simplicity, we shall call it *the intrablock*.

All incomplete blocks for one replicate of the treatments can be generated from the intrablock by *position-wise modular addition*; that is, reducing each position to residues (mod 2) without any carryover to other positions. The steps are:

  (i) Determining any treatment not appearing in the intrablock.
 (ii) Add the modular representation of this treatment position-wise to each treatment appearing in the intrablock. This will generate one of the other blocks.
(iii) Repeat these 2 steps until all treatments are represented.

Consider the intrablock of size 4 given in the first part of Example 10.9. We note that treatment "100" does not appear in the intrablock. Upon "adding 100" position-wise to each of the four treatments in the intrablock, we obtain

$$000 + 100 \rightarrow 100 \qquad 101 + 100 \rightarrow 001$$
$$110 + 100 \rightarrow 010 \qquad 011 + 100 \rightarrow 111$$

We leave it to the reader to verify that the same four treatments would have been obtained had any of the other 3 treatments (010 or 001 or 111) been used as the addend.

Example 10.11

A 2^5 factorial is to be conducted in 8 blocks of size 4 each. Suppose we elect to confound the ABD, the BCE and the AC interactions (which are not generalized interactions of each other). The intrablock subgroup is generated by the modular equations

$$x_1 + x_3 = 0 \ (\text{mod } 2)$$

$$x_1 + x_2 + x_4 = 0 \ (\text{mod } 2)$$

$$x_2 + x_3 + x_5 = 0 \ (\text{mod } 2)$$

and is found to be

$$
\begin{array}{c}
00000 \\
01011 \\
11100 \\
10111
\end{array}
$$

A second block can be generated by modular addition of "10000" to each intrablock treatments giving

$$
\begin{array}{c}
10000 \\
11011 \\
01100 \\
00111
\end{array}
$$

A third block can be generated by modular addition of "01000" to each intrablock treatment giving

$$
\begin{array}{c}
01000 \\
00011 \\
10100 \\
11111
\end{array}
$$

And so on. ▓

10.8   Incomplete Block Designs for General Factorials

We will devote very little time to incomplete block designs other than the 2^n series covered in the previous sections. In those discussions we introduced some basic ideas of algebraic group theory, notably modular equations and a modular arithmetic. But, additional concepts would be required for an efficient coverage of the general case. Therefore we shall limit our coverage to a few general remarks, and illustrate with some examples for 3^n factorials. The reader is referred to John [2], Kempthorne [3], and Mann [4] for in-depth coverage of the general case.

Consider a general p^n series of factorial experiments, n factors each at p levels, when p is a prime number. The levels of each factor can be represented by the residues (mod p). Each main effect has (p-1) degrees of freedom; each two factor inter- action has $(p-1)^2$ degrees of freedom, and so on. For a unique representation by modular equations (as in Rule 10.1), every interaction must be partitioned into components each having (p-1) degrees of freedom. In other words, every "treatment component" has (p-1) degrees of freedom. For uniqueness, the first letter of any component must have an exponent of one. If the first letter has an exponent greater than one, repeated squaring of the component and reducing each exponent (mod p) will eventually give an equivalent component whose first letter has exponent of one.

Example 10.12

Consider a 3^3 factorial experiment, factors A, B, and C each at 3 levels. We represent the levels of each factor by 0, 1, and 2, the residues (mod 3). Each main effect has 2 degrees of freedom and a related modular equation in a single x. Each two-factor interaction has 4 degrees of freedom which will be represented by two components, each with 2 degrees of freedom. For the AxB interaction, for example, the two components are AB and AB^2, under the unique representation scheme where the first letter has exponent of one. Each component of a two-factor interaction has a corresponding modular equation in two x's. For the AB and AB^2 components, these are

$$x_1 + x_2 (\text{mod } 3) \quad \text{and} \quad x_1 + 2x_2 (\text{mod } 3)$$

Notice that the coefficient of x_2 is the exponent of the corresponding factor letter (see Rule 10.1).

Finally, the three-factor interaction, AxBxC, has 8 degrees of freedom which will be represented by four components each with 2 degrees of freedom. These are

ABC	AB^2C
ABC^2	AB^2C^2

with corresponding modular equations

$$x_1 + x_2 + x_3 \,(\text{mod } 3) \qquad\qquad x_1 + 2x_2 + x_3 \,(\text{mod } 3)$$

$$x_1 + x_2 + 2x_3 \,(\text{mod } 3) \qquad\qquad x_1 + 2x_2 + 2x_3 \,(\text{mod } 3) \; *$$

Incomplete block designs for p^n factorials would utilize modular equations involving residues (mod p). Incomplete blocks of size p, p^2, ..., p^n might be used. When p^w blocks are required, the researcher must select w treatment components for confounding. The remaining components confounded with incomplete blocks are generalized interactions of the selected components. In the p^n series, there are p-1 generalized interactions of any 2 "treatment components," say X and Y, and are given by:

$$(10.16) \qquad\qquad XY,\ XY^2,\ \ldots,\ XY^{p-1}$$

The above statements all result from the mathematical group theory alluded to earlier. Additionally, Rule 10.2 can be used to construct incomplete blocks. It is valid for any p^n factorial. And, Definition 10.5 is valid for any p^n factorial if "confounded effects" are taken to be "treatment components" under the unique representation where each has (p-1) degrees of freedom. The following examples will illustrate these ideas.

Example 10.13

Let us consider a simple example of a 3^2 factorial on factors A and B. There are t = 9 treatments and suppose blocks of size 9 units cannot be constructed but blocks of size 3 can be.

Each of the two main effects has 2 degrees of freedom while their interaction has 4 degrees of freedom. A complete replicate of the treatment requires 3 blocks whereby 2 treatment degrees of freedom will be confounded. If we choose to confound a main effect, the block compositions would be easily formed by placing one level of the confounded main effect in each block. For example, confounding main effect A gives the 3 blocks

00	10	20
01	11	21
02	12	22

which correspond to the modular equations

$$x_1 = 0(\text{mod } 3), \qquad x_1 = 1(\text{mod } 3), \qquad \text{and} \quad x_1 = 2(\text{mod } 3).$$

Two orthogonal contrasts among the blocks would equivalently represent the 2 degrees of freedom for A.

If we decide that neither main effect should be confounded, then we need to confound 2 of the interaction degrees of freedom. But which 2?   As we have indicated earlier, the AxB interaction can be represented by the "components" AB and AB^2, each with 2 degrees of freedom and related modular equations

$$x_1 + x_2(\text{mod } 3) \qquad \text{and} \quad x_1 + 2x_2(\text{mod } 3)$$

One may confound either of these components with incomplete blocks, as it is customary to assume that both components play an equal role in estimating the AxB interaction. Therefore, we may estimate the AxB interaction from the unconfounded component, with 2 degrees of freedom. If we choose to confound AB^2, the blocks would be obtained from

$$x_1 + 2x_2 = 0, \ 1, \ 2(\text{mod } 3)$$

and are

00	01	02
11	12	10
22	20	21 ✳

Example 10.14

Suppose a 3^4 factorial is to be conducted with blocks of size 9. Then, $3^2 = 9$ blocks are needed for a complete replicate of the treatments.   The exponent of 2 (on the number of blocks) implies that two treatment components must be specified to generate the block compositions. Suppose we decide to confound the two interaction components AB^2C and ACD^2? Letting $X = AB^2C$ and $Y = ACD^2$? we can determine the generalized interactions with the aid of (10.16):

$$XY \rightarrow (AB^2C)(ACD^2) = A^2B^2C^2D^2 = ABCD$$

$$XY^2 \rightarrow (AB^2C)(ACD^2)^2 = A^3B^2C^3D^4 = BD^2.$$

where all exponents are converted to residues (mod 3). For unique representation, the first letter should have an exponent equal to one(see the remarks preceding Example 10.12.) Thus, $A^2B^2C^2D^2$ and B^2D are each squared to give the final form of the generalized interaction.✳

Partial confounding is an option available to an experimenter dealing with factorials other than those of the 2^n series. Thus, in the 3^2 experiment discussed in Example 10.13, component A could be confounded in one replicate, B confounded in a second, AB confounded in a third, AB^2 confounded in a fourth, and so on.

In the mixed or asymmetrical factorials, where the factors have different levels, the choice of incomplete blocks is more limited. Additionally, the construction of incomplete blocks and the analysis generally is more complicated. Factorials is this category include a 2x3, a 3x5, a 2x3x5, and so on. Information about some of these factorials may be found in Federer [1] and Kempthorne [3].

PROBLEMS

1. A 2^3 factorial experiment is planned but complete blocks of 8 EU cannot be obtained. Numerous blocks of 4 homogeneous EU are available. Using the method in Section 10.5, give the block compositions for one replicate of the treatments if

   (a) the ABC interaction is confounded with incomplete blocks
   (b) the AB interaction is confounded with incomplete blocks.

2. A 2^4 factorial experiment is planned but complete blocks of 16 EU cannot be obtained. Numerous blocks of 8 homogeneous EU are available. Using the method in Section 10.5, give the block compositions for one replicate of the treatments if

   (a) the ABCD interaction is confounded with incomplete blocks.
   (b) the ABC interaction is confounded with incomplete blocks.

3. A 2^3 factorial experiment is planned but complete blocks of 8 EU cannot be obtained. Only blocks of 2 homogeneous EU are available. Using the method in Section 10.5, give the block compositions for one replicate of the treatments if

   (a) the ABC and AB interactions are confounded with incomplete blocks.
   (b) the AB and AC interactions are confounded with incomplete blocks.

4. A 2^4 factorial experiment is planned but complete blocks of 16 EU cannot be obtained. Only blocks of 4 homogeneous EU are available. Using the method in Section 10.5, give the block compositions for one replicate of the treatments if

   (a) the ABCD and AB interactions are confounded with incomplete blocks.
   (b) the ABC and BCD interactions are confounded with incomplete blocks.

5. A researcher needs to conduct a 2^3 factorial experiment but cannot obtain blocks of 8 EU. Blocks of 4 homogeneous EU are obtained and the researcher arbitrarily assigns treatments to the blocks. What effect is confounded with incomplete blocks if one replicate of the treatments has block compositions:

(a)   Block 1 :   (1), b, ac, abc   Block 2 : a, c, ab, bc

(b)   Block 1 :   b, ab, bc, abc   Block 2 : (1), a, c, ac

6. A researcher needs to conduct a 2^4 factorial experiment but cannot obtain blocks of 16 EU. Blocks of 8 homogeneous EU are obtained and the researcher arbitrarily assigns treatments to the blocks. What effect is confounded with incomplete blocks if one replicate of the treatments has block compositions:

(a)   Block 1 : a, ab, ac,abc, bd,bcd,cd,d

Block 2 : (1), ad,abd, acd,abcd,b,c,bc

(b)   Block 1 : (1), a, abc,abd,acd,bc,bd,cd

Block 2 : ab,ac,ad,abcd,b,c,bcd,d

7. Give the generalized interactions when a 2^5 factorial is conducted and one complete replicate of the treatments occurs

(a) in 4 incomplete blocks with ACE and BDE confounded.
(b) in 8 incomplete blocks with ABC, CDE, and AD confounded.

8. Give the generalized interactions when a 2^6 factorial is conducted and one complete replicate of the treatments occurs

(a) in 8 incomplete blocks with ABF, CDE, and BD confounded.
(b) in 16 incomplete blocks with ABC, BCD, CDE and DEF confounded.

9. A 2^4 factorial was conducted in an incomplete block design. Four replicates were used, each with 2 incomplete blocks of 8 EU. The effects confounded were the BC interaction in Rep I, the ABC interaction in Rep II, the BCD interaction in Rep III, and the ABCD interaction in Rep IV.

(a) Give the partial ANOVA (sources, df) for this experiment.
(b) How much information do we have on each three factor interaction?

10. Repeat Problem 9 if five replicates were used with the ABCD interaction and each three factor interaction confounded in one replicate.

11. A 2^5 factorial experiment was conducted in an incomplete block design. Three replicates were used, each with 4 incomplete blocks of 8 EU. The effects chosen for confounding were

> Rep I   ABE, CDE
> Rep II   ABC, ADE
> Rep II   ACD, BDE

(a) Give the partial ANOVA (sources, df) for this experiment.
(b) How much information do we have on each three factor interaction?

12. Repeat Problem 11 if the chosen confoundings were

> Rep I   ABD, BCE
> Rep II   ABE, CDE
> Rep III   ACE, BCD
> Rep IV   ABC, CDE

13. An engineer used six furnaces to "bake" ceramic components. Components were produced under 2x2x2 factorial treatments: 2 percentages of silicon (A), 2 percentages of calcium (B), and 2 cooling processes (C). The variable of interest was a hardness index determined 24 hours after baking.

Only 4 components could be baked in a furnace at one time so the furnaces(which were identical) were arbitrarily paired for each replicate of the treatments. The furnaces, treatments, and observations were:

		Furnace			
1	2	3	4	5	6
(1) 17	ab 35	bc 29	abc 29	b 22	ac 28
bc 26	c 22	a 28	b 19	a 31	bc 24
abc 34	ac 32	ab 37	(1) 18	abc 39	ab 41
a 27	b 25	c 24	ac 30	c 21	(1) 14

(a) What treatment component is confounded in Furnaces 1 and 2? in Furnaces 3 and 4? in Furnaces 5 and 6?
(b) Set up the ANOVA for this experiment.
(c) Test treatment components. Interpret.

14. A nutrition researcher studied the effects of stress and diets on weight gain of guinea pigs. The 2x2x2 factorial treatments were: two levels of stress (length of disturbing the animal), two percentages of protein, and 2 levels of vitamin supplementation. Blocks were formed from litters, upon weaning, but only 4 uniform guinea pigs were selected per litter. Each guinea pig was placed in an individual cage

and assigned one of the treatments. Two litters received a replicate of the treatments. The litters, treatments, and weight gains, in grams, were:

			Litter				
1	2	3	4	5	6	7	8
abc 98	b 94	a 82	c 95	a 81	bc 107	ac 83	c 99
bc 110	ac 81	b 101	(1) 87	abc 94	ac 88	(1) 89	ab 96
(1) 96	c 102	bc 98	ab 96	b 91	ab 90	b 100	a 73
a 76	ab 91	ac 87	abc 92	c 100	(1) 94	abc 102	bc 101

(a) What treatment component is confounded in each pair of litters?
(b) Give the ANOVA for this experiment.
(c) Test treatment components. Interpret.

15. A 2^4 factorial experiment is conducted in incomplete blocks of 4 EU. Give the block compositions for one replicate of the treatments for each of the following intrablocks:

    (a) 0000   (b) 0000
        0011   1010
        1101   0111
        1110   1101

16. A 2^5 factorial experiment is conducted in incomplete blocks of 8 EU. Give the block compositions for one replicate of the treatments for each of the following intrablocks:

    (a) 00000   10110   (b) 00000   11100
        01101   11100   00011   10011
        00110   10011   01001   10110
        01001   11011   01100   11111

17. A 2^4 factorial experiment is conducted in incomplete blocks of 4 EU. Give the intrablock if one chooses to confound

(a) ABC and BCD with incomplete blocks.
(b) BD and ABC with incomplete blocks.

18. A 2^5 factorial experiment is conducted in incomplete blocks of 8 EU. Give the intrablock if one chooses to confound

(a) ABD and BCE with incomplete blocks.
(b) ABC and BCDE with incomplete blocks.

19. A 3^3 factorial experiment is conducted in incomplete blocks of 3 EU. Give the generalized interactions if all replicates have

(a) AB and BC confounded with incomplete blocks.
(b) ABC and BC confounded with incomplete blocks.

20. A 3^4 factorial experiment is conducted in incomplete blocks of 9 EU. Give the generalized interactions if all replicates have

(a) ABC and BCD confounded with incomplete blocks.
(b) BCD and AB^2CD confounded with incomplete blocks.

21. Give the intrablock for each experiment of Problem 19.

22. Give the intrablock for each experiment of Problem 20.

23. Give the block compositions for one complete replicate if a 3^3 factorial experiment conducted in incomplete blocks of 3 EU has intrablock

(a)	000	(b)	000
	111		121
	222		212

24. Give the block compositions for one complete replicate if a 3^4 factorial experiment conducted in incomplete blocks of 9 EU has intrablock

(a)	0000	1011	2022	(b)	0000	1011	2022
	0112	1120	2101		0110	1121	2102
	0221	1202	2210		0210	1201	2212

REFERENCES

[1]   Federer, W.T., *Experimental Design*, Macmillan, New York, 1955.

[2]   John, P.W.M., *Design and Analysis of Experiments*, Wiley, New York, 1971.

[3]   Kempthorne, O., *The Design and Analysis of Experiments*, Wiley, New York, 1952.

[4]   Mann, H. B., *Analysis and Design of Experiments*, Dover, New York, 1949.

## Chapter 11

SPLIT PLOT DESIGNS AND VARIATIONS

11.1 Introduction

General confounding schemes for factorial experiments were given in Chapter 10. There we indicated that the researcher should use a confounding system which will allow the estimation of the most important treatment components. Generally, high order interaction components would be confounded with incomplete blocks unless knowledge or information suggests otherwise. In some cases, the researcher may need to estimate specific high order interactions. In other cases, the nature of experimental material (particularly equipment) might dictate the confounding of main effects. The researcher really may have no choice because of limitations of material and equipment.

We now consider a broad class of factorial experiments where at least one main effect is confounded with incomplete blocks. This represents a special pattern of confounding. Split plot is a term applied to such experiments. The basic split plot design is presented in the next section. There are many variations of this basic design; we will discuss the major types in later sections of this chapter.

11.2.1 The Simple Split Plot Design

We first characterize the experimental settings that lead to a simple split plot design. There are 2 treatment factors: A having "a" levels and B having "b" levels. Blocks of b experimental units (EU) are available. For one reason or another, suppose that each level of factor A can be applied only to blocks of b units. As illustrations of this, consider the following:

(a) In agricultural experiments, irrigation systems must apply water to large areas of land which may be subdivided into smaller plots for applying varieties, fertilizers, or some other factor.

examples for split plot designs

(b)   In industry, a manufacturing process may be efficient only when applied to a large batch of raw material. For example furnaces are usually set at one temperature and a large batch of a metal alloy is melted at one time. The batch can be subdivided into small portions for application of a second factor (hardening agent, curing method, and so on.)

(c)   In educational research, teaching techniques generally need to be applied to entire classes of students. A class may be divided into subsections for additional training, or individual students may be used as units for a second factor.

In practice, the two factors of a split-plot experiment often are applied at different times. Thus, a simple (or basic) split-plot can be viewed as two "experiments" superimposed on each other. One "experiment" concerns the application of factor A to the "larger units of material" while the second "experiment" concerns the application of factor B to the "smaller units of material." In situation (a) above, the levels of factor A (irrigation systems) would be randomly assigned to the incomplete blocks (areas of land.) All levels of factor B (varieties, here) occur within each incomplete block; a separate randomization of the B levels would be performed within each incomplete block. In this split-plot experiment, the varieties would be applied first whereas the irrigation would be applied later, during the growing season.

It is important to note that a simple split-plot experiment is characterized by two randomization schemes, one for each treatment factor. In the agricultural experiment just discussed, we observed different types of randomizations for varieties and irrigation systems. They deal with different segments of the experimental material. Factor A levels are randomized to sets of contiguous "subplots"; factor B levels are randomized to these subplots. The presence of two types of randomizations in an experiment is often a clue that one is dealing with a split-plot design, or one of its variants.

In the simple split-plot, a total of ra incomplete blocks, each of size b, is required if each of the t = ab treatment combinations is to occur r times. We further assume that the incomplete blocks are arranged in r "replicates". The blocks of each replicate, while expected to exhibit some differences, are fairly homogeneous relative to blocks of other replicates. Quite often, replicates of incomplete blocks are formed on the basis of location, time, and so on.

The incomplete blocks may not be arranged in replicates; that is, into "homogeneous" sets that will accomodate the treatments. This variation of the split-plot is covered in Section 11.4.

Example 11.1

We give a simple split plot design with factor A having a = 3 levels and factor B having b = 4 levels. Two "replicates" of incomplete blocks are to be used (r=2). One experimental plan, in randomized form, for this simple split plot would be the following:

Each incomplete block contains 4 EU and therefore can contain only 4 of the 12 treatments. Furthermore, all EU within an incomplete block contain the same level of factor A. All levels of factor B appear in each block, and are randomized anew within each block.

The use of the 2 "replicates" would imply a certain degree of homogeneity within the 2 sets of IB (1, 2, 3 and 4, 5, 6) and heterogeneity between these two sets. ✳

Incomplete blocks usually are called *whole plots*. Units within each block are called *split plots*. Factor A is termed the whole plot factor and Factor B the split plot factor.

Applying a single level of factor A to the EU within a whole plot imposes restrictions on the randomization of treatments, the factorial combinations. Within each replicate, main effect A totals are also whole plot totals. This implies that contrasts among A effects will be contaminated by whole plot effects. In other words, factor A is confounded with whole plots.

Because each level of factor B occurs within every whole plot, it is clear that contrasts among B effects are free of whole plot effects. It is not obvious but it can be shown that contrasts among interaction components (of factors A and B) are also free of whole plot effects.

In the simple split plot experiment, the whole plots are really "experimental units" for the levels of factor A. We randomly allocate the levels of factor A to whole plots. So *prior* to assigning the levels of factor B, the experiment can be viewed as an RCB design with r replicates of the "a" treatments.

As mentioned earlier, the split plots are the experimental units for levels of factor B. In some experiments, the levels of factor B are applied first, and at a later time, the levels of factor A are applied, see illustration (a) above. If for some reason factor A is not applied, the experiment would become a *modified* RCB design with r replicates each containing the "b" treatments "a" times. (Because the randomization of treatments is too restrictive, this would not be a generalized RCB design.)

Let us further consider the "two RCB designs" described above. The one dealing with factor B affords considerably more precision than does the one dealing with factor A. This is due to the difference in the number of "experimental units" as well as the randomizations involved. The number of degrees of freedom for Exp. Error is greater for the design dealing with factor B. The simple split plot design can be viewed as a superimposing of these two RCB designs. Consequently, we encounter an experimental design where the two main effects have different precisions, have different error terms. We will discuss this later.

11.2.2 Simple Split Plot Model and Assumptions

The linear model for the simple split plot design is

(11.1)
$$y_{ijk} = \mu + \rho_i + \alpha_j + \delta_{ij} + \beta_k + (\alpha\beta)_{jk} + \epsilon_{ijk}$$

$$\text{for } i = 1, 2, \cdots, r$$
$$j = 1, 2, \cdots, a$$
$$k = 1, 2, \cdots, b$$

where

μ = overall mean, a constant

ρ_i = effect due to the i-th replicate

α_j = effect of the j-th level of A

δ_{ij} = whole plot error component

β_k = effect of the k-th level of B

$(\alpha\beta)_{jk}$ = interaction effect of the j-th level of A and the k-th level of B

$$\epsilon_{ijk} = \text{split-plot error component}$$

Basic assumptions for inferences under this design are:

(11.2)

(i) the δ's are i.i.d. $N(0,\sigma_\delta^2)$

(ii) the ϵ's are i.i.d. $N(0,\sigma_\epsilon^2)$

(iii) the δ's and ϵ's are distributed independently of each other.

Assumptions about the replicate and treatment components depend upon the fixed or random nature of these components.

11.2.3 The ANOVA for the Simple Split Plot Design

The sum of squares calculations for a simple split plot design are almost identical to those for a two factor factorial in an RCB design. Only one additional sum of squares is required. Specifically, we have

(11.3) $C = y_{...}^2/rab = \text{general correction term}$

(11.4) $SSY = \underset{ijk}{\Sigma\Sigma\Sigma}\ y_{ijk}^2 - C = \text{Total SS}$

(11.5) $SSR = \dfrac{1}{ab}\ \underset{i}{\Sigma}\ y_{i..}^2 - C = \text{Replicate SS}$

(11.6) $SSA = \dfrac{1}{rb}\ \underset{j}{\Sigma}\ y_{.j.}^2 - C = \text{SS for A}$

In order to compute the sum of squares for the whole plot error, we first compute the sum of squares for the whole plots which may be interpreted also as the "replicate-factor A" subclass SS

(11.7) $SRA = \dfrac{1}{b}\ \underset{ij}{\Sigma\Sigma}\ y_{ij.}^2 - C = \text{Whole Plot SS} = SSWP$

This is the additional sum of squares referred to above. Then, we easily calculate

(11.8) $SSE1 = SSWP - SSA - SSR = \text{Whole Plot error SS}$

The sums of squares for the split plot factor, B, and the AxB interaction are calculated as in any factorial

(11.9) $SSB = \dfrac{1}{ra} \sum_{k} y^2_{..k} - C = SS$ for B

(11.10) $SAB = \dfrac{1}{r} \sum_{jk} y^2_{.jk} - C = AB$ Subclass SS

(11.11) $SSAB = SAB - SSA - SSB = AB$ Interaction SS

Finally, the second error sum of squares, also called the split plot error, is obtained by subtraction

(11.12) $SSE2 = SSY - SSR - SSA - SSE1 - SSB - SSAB =$ Split plot
                                                    error SS

[Note: Some authors use "Error (a)" for Whole Plot Error and "Error (b)" for Split Plot Error. We will not use these designations as there may be confusion when we consider some of the variations in later sections.]

The sums of squares calculations are summarized in the following table.

Table 11.1 Partial ANOVA for Simple Split Plot Design

Source	df	SS	EMS (Fixed Treatment Effects)
Reps	r-1	SSR	- - - - -
A	a-1	SSA	$\sigma^2_\epsilon + b\sigma^2_\delta + rb\kappa^2_A$
WP Error	(r-1)(a-1)	SSE1	$\sigma^2_\epsilon + b\sigma^2_\delta$
B	b-1	SSB	$\sigma^2_\epsilon + ra\kappa^2_B$
AxB	(a-1)(b-1)	SSAB	$\sigma^2_\epsilon + r\kappa^2_{AB}$
SP Error	(r-1)a(b-1)	SSE2	σ^2_ϵ
Total	rab-1	SSY	

Example 11.2

The superintendent of the city schools in a large midwestern city decided to investigate the effect of three teaching methods, (1) classical instruction, (2) T.V. instruction, and (3) T.V. plus classical instruction on achievement in mathematics concepts. However, he also was concerned about the effect of the use of calculators on achievement in mathematics concepts. Because the teaching methods must naturally be applied to a block of students the experiment was set up in the following way. Three schools within the system were chosen at random and 3 classes of fourth graders were chosen at random within each school. (This could be done because each elementary school had at least 4 sections of the fourth grade.) One of the 3 teaching methods was randomly assigned to the 3 classes and one-half of each class was randomly selected to use calculators, the other half could not. At the end of the instruction each student was given a test and then the average test score of all students with a given treatment combination was recorded. (Note: we should assume equal numbers of children in each treatment combination.)

This is a split-plot design with whole plot factor A = teaching method, split plot factor B = type of calculation and replicates = schools. The following represents the unrandomized design setup:

	Classical	T.V.	T.V. + Classical
Rep 1	calc / no calc	calc / no calc	calc / no calc
Rep 2	calc / no calc	calc / no calc	calc / no calc
Rep 3	calc / no calc	calc / no calc	calc / no calc

Model: $y_{ijk} = \mu + \rho_i + \alpha_j + \delta_{ij} + \beta_k + (\alpha\beta)_{jk} + \epsilon_{ijk}$; $\begin{array}{l} i = 1,2,3 \\ j = 1,2,3 \\ k = 1,2 \end{array}$

Data

		Classical	T.V.	T.V. + Cl.	Rep Totals
Rep 1	C	72	61	65	
	NC	78	63	67	406
Rep 2	C	80	64	71	
	NC	85	63	73	436
Rep 3	C	75	58	64	
	NC	79	59	67	402
Method, A Totals		469	368	407	1244 = y...

The sums of squares calculations are:

$$C = (1244)^2/18 = 85,974.22$$

$$\begin{aligned} SSR &= [(406)^2 + (436)^2 + (402)^2]/6 - C \\ &= 86,089.3 - 85,974.22 = 115.11 \end{aligned}$$

$$\begin{aligned} SSA &= [(469)^2 + (368)^2 + (407)^2]/6 - C \\ &= 86,839 - 85,974.22 = 864.78 \end{aligned}$$

$$\begin{aligned} SRA &= [(150)^2 + (124)^2 + \text{---} + (131)^2]/6 - C \\ &= 86,978 - 85,974.22 = 1,003.78 \end{aligned}$$

$$SSE1 = 1,003.78 - 115.11 - 864.78 = 23.89$$

$$\begin{aligned} SSB &= [(610)^2 + (634)^2]/9 - C \\ &= 86,006.22 - 85,974.22 = 32.00 \end{aligned}$$

$$\begin{aligned} SAB &= [(227)^2 + \ldots + (207)^2]/3 - C \\ &= 86,885.33 - 85,974.22 = 911.11 \end{aligned}$$

$$SSAB = 911.11 - 864.78 - 32.00 = 14.33$$

$$\begin{aligned} SSY &= (72)^2 + (61)^2 + \text{---} + (67)^2 - C \\ &= 87,028 - 85,974.22 = 1,053.78 \end{aligned}$$

ANOVA for Education Study

Source	df	SS	MS	EMS
Reps.	2	115.11	57.55	---
T. Methods	2	864.78	432.39	$\sigma_\epsilon^2 + 2\sigma_\delta^2 + 6\kappa_T^2$
WP Error	4	23.89	5.97	$\sigma_\epsilon^2 + 2\sigma_\delta^2$
Calculation	1	32.0	32.00	$\sigma_\epsilon^2 + 9\kappa_C^2$
T x C	2	14.33	7.17	$\sigma_\epsilon^2 + 3\kappa_{TC}^2$
SP Error	6	3.67	0.61	σ_ϵ^2
Total	17	1053.78		※

11.2.4   Inferences for the Simple Split Plot Design

Under the assumptions stated in (11.2), F tests of treatment components can be conducted. From the EMS in Table 11.1, we note that the F statistic for testing main effect A utilizes the whole plot error while tests of the main effect B and AB interaction utilize the split plot error. The hypotheses are

(11.13)
$$H_0:\text{All } (\alpha\beta)_{ij} = 0 \qquad H_0:\text{All } \alpha_j = 0 \qquad H_0:\text{All } \beta_k = 0$$
$$H_1:\text{Some } (\alpha\beta)_{jk} \neq 0 \qquad H_1:\text{Some } \alpha_j \neq 0 \qquad H_1:\text{Some } \beta_k \neq 0$$

The respective test statistics are

(11.14) $F = MSAB/MSE2$ $F = MSA/MSE1$ $F = MSB/MSE2$

with respective degrees of freedom

(11.15)
$$(a-1)(b-1); \ (r-1)a(b-1) \qquad (a-1); \ (r-1)(a-1)$$
$$(b-1); \ (r-1)a(b-1)$$

All of these tests have critical regions corresponding to upper tail areas of the appropriate F distribution.

Example 11.3

Let us return to the education study of the previous example. To make tests of no interaction of teaching methods and calculations, equal mean scores for the 3 teaching methods, and equal mean scores for calculations, the F statistics are, respectively:

$$F = \frac{7.17}{0.61} = 11.73 \qquad F = \frac{432.39}{5.97} = 72.40 \qquad F = \frac{32.00}{0.61} = 52.36$$

which give observed significance levels of $P < 0.01$, $P < 0.005$, and $P < 0.005$, respectively. There is strong evidence to indicate that main effects for both T and C exist, and lesser evidence that interaction may be present. A plot of the means follows.

Mean Scores

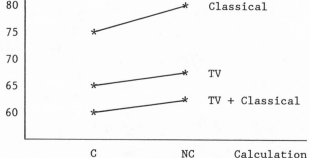

Combining the ANOVA results and the implications of the above plotting it is apparent that the TxC interaction is the least significant treatment component. In fact, based on these findings one likely would be willing to assert that the classical teaching method is substantially better than the other two and students who do not use calculators generally understand mathematical concepts better than those who do use calculators. The inference might be made here that not using calculators forces the student to grasp basic concepts in order to do the arithmetic properly. A word of caution should also be inserted. The experimenter did not make clear whether the same teacher ran all the classes or if each class had a different teacher. If the latter is true then the effect of teaching and type of instruction are confounded and serious questions arise as to the validity of our conclusions about types of instruction. ⋇

Inferences about pairs of means are frequently of interest.
We will need estimated standard deviations for the differences of
various pairs of means. Because whole plot and split plot treatment comparisons have different precisions (different error
terms), it follows that several variances are involved. For the
simple split plot design, the estimated standard deviations are
(i) for the difference of any pair of A means, say $\bar{y}_{.j.} - \bar{y}_{.h.}$

(11.16) $\sqrt{2MSE1/rb}$

(ii) for the difference of any pair of B means, say $\bar{y}_{..k} - \bar{y}_{..h}$

(11.17) $\sqrt{2MSE2/ra}$

(iii) for the difference of a pair of AB means at a fixed level
   of factor A, say $\bar{y}_{.jk} - \bar{y}_{.jh}$

(11.18) $\sqrt{2MSE2/r}$

(iv)   for the differences of a pair of AB means, at the same or
   different levels of B, say $\bar{y}_{.jk} - \bar{y}_{.hk}$ or $\bar{y}_{.jk} - \bar{y}_{.hg}$

(11.19) $\sqrt{2[MSE1+(b-1)MSE2]/rb}$

For the first three cases, inferences can be made by Student
t distributions, under the assumptions in (11.2). The degrees of
freedom of these t variable are equal to those of the mean square
appearing in the denominator of the t expression.
For the last case, a few words are in order. The indicated
differences of means can be shown to include A effects as well as
AB interaction components. It is for precisely this reason that
the standard deviation given in (11.19) is based on a weighted
average of whole plot and split plot error mean squares. Because
of this, exact inferences are not possible for these particular
mean differences. Approximate inferences may be made using the t
distribution with (approximate) degrees of freedom calculated as
in Section 2.12; that is, by Satterthwaite's procedure.

Example 11.4

We first calculate the standard deviations for the education study of Example 11.2. These are:

for difference of mean scores for	Standard Deviation
two teaching methods	$\sqrt{2(5.97)/6} = 1.41$
two types of calculation	$\sqrt{2(0.61)/9} = 0.37$
two types of calculation for a specific teaching method	$\sqrt{2(0.61)/3} = 0.64$
two teaching methods for the same or different types of calculation	$\sqrt{2[0.61+(2)5.97]/6} = 2.04$

For inferences in the first three cases, the t variables would have 4, 6, and 6 degrees of freedom, respectively.

A 95% confidence interval for the difference in mean scores for the classical and TV teaching methods is

$$(58.6 - 48) \pm 1.41(2.78) = 10.6 \pm 3.9$$

We can be 95% confident that the difference in mean scores under the classical and TV teaching methods lies between 6.7 and 14.5.※

11.3   Split Plot Design With Three or More Treatment Factors

This modification of the simple split plot design occurs in the treatment design, not in the experimental design. We now consider experiments where either whole plot treatments or split plot treatments (or both) have a factorial structure.

No new concepts are introduced with these designs. We merely incorporate factorial partitioning of whole plot treatments or split plot treatments wherever appropriate. Only a few additional sums of squares would need to be calculated; those for the extra factorial components.

The following example illustrates the types of experiments belonging to this category. The plot compositions are shown for each of the three possibilities.

Example 11.5

For definiteness, suppose factors A, B, C, and D have levels 3, 2, 3, and 2, respectively. (Not all factors will be used in every experiment discussed below.)

First, if whole plot treatments are factorial combinations of A and B while split plot treatments are the levels of factor C, then one replicate would have the following appearance.

That is, the 6 combinations $(a_j b_k)$ are randomly assigned to whole plots and the 3 levels c_m are randomly assigned within each whole plot.

If the whole plot treatments are levels of factor A while split plot treatments are factorial combinations of factors B and C, then one replicate would have the following appearance:

Here, the levels of factor A are randomized to the whole plots while all "factorial combinations" of factors B and C are random- ized anew within each whole plot.

Finally, if whole plot treatments are factorial combinations of factors A and B while split plot treatments are factorial combinations of factors C and D, then one replicate would have the following appearance:

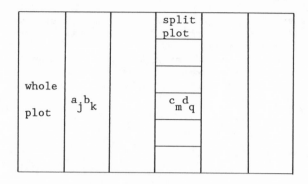

Now the factorial combinations of factors A and B are randomized
to the whole plots while all factorial combinations of factors C
and D are randomized anew within each whole plot. ▓

    The linear model for any experiment in this general category
is easily obtained from model (10.1) by incorporating the appropriate factorial structure.   For the first plan in Example 11.5,
with only whole plot treatments having a factorial structure, the
linear model would be

$$(11.20) \quad y_{ijkm} = \mu + \rho_i + \alpha_j + \beta_k + (\alpha\beta)_{jk} + \delta_{ijk} +$$

$$\gamma_m + (\alpha\gamma)_{jm} + (\beta\gamma)_{km} + (\alpha\beta\gamma)_{jkm} + \epsilon_{ijkm}$$

where the whole plot error components, the δ's, and the split
plot error components, the ϵ's, satisfy assumptions analogous to
those stated in (11.2). The ANOVA for such an experiment is given
below in Table 11.2.
    Note that the whole plot treatment components, main effects A
and B and the AB interaction, are tested against the whole plot
error mean square while the remaining treatment components (split
plot factor C and its interaction with whole plot components) are
tested against the split plot error mean square.   This is true
only for fixed treatment effects. If any treatment factors have
random effects, some of the κ^2's in Table 11.2 would be σ^2's components, and additional σ^2 components would be present by the
rules of Section 5.8.1. Then, for certain factorial components,
exact F tests do not exist; one would have to use Satterthwaite's
procedure, (see Section 2.12.)

Table 11.2 ANOVA for Split Plot Design: Factorial WP Treatments

Source	df	MS	EMS (Fixed Treatments)
Replicates	r-1	MSR	- - - - -
A	a-1	MSA	$\sigma_\epsilon^2 + c\sigma_\delta^2 + rbc\kappa_A^2$
B	b-1	MSB	$\sigma_\epsilon^2 + c\sigma_\delta^2 + rac\kappa_B^2$
AB	(a-1)(b-1)	MSAB	$\sigma_\epsilon^2 + c\sigma_\delta^2 + rc\kappa_{AB}^2$
WP Error	(r-1)(ab-1)	MSE1	$\sigma_\epsilon^2 + c\sigma_\delta^2$
C	c-1	MSC	$\sigma_\epsilon^2 + rab\kappa_C^2$
AC	(a-1)(c-1)	MSAC	$\sigma_\epsilon^2 + rb\kappa_{AC}^2$
BC	(b-1)(c-1)	MSBC	$\sigma_\epsilon^2 + ra\kappa_{BC}^2$
ABC	(a-1)(b-1)(c-1)	MSABC	$\sigma_\epsilon^2 + r\kappa_{ABC}^2$
SP Error	(r-1)ab(c-1)	MSE2	σ_ϵ^2
Total	rabc-1		

11.4 Split Plot Design Without Replicates

We assume that the whole plots possess no identifiable variation which can be used to form replicates. In Example 11.2, we would have no replicates if the three schools were assumed to have classes with the same degree of homogeneity/heterogeneity. If we felt that schools within this system had similar characteristics, this would not be an unreasonable assumption.

Without replicates, the randomization of whole plot treatments is less restrictive. Each level of factor A is assigned to r whole plots completely at random. All levels of the split plot factor B are randomized within each whole plot, as in the simple split plot design. It should be clear that the confounding of treatment effects is identical to what we had in the simple split plot design.

The linear model for this variation of the basic split plot design is

(11.21) $y_{ijk} = \mu + \alpha_j + \delta_{ij} + \beta_k + (\alpha\beta)_{jk} + \epsilon_{ijk}$; $\begin{aligned} i &= 1,2,---,r \\ j &= 1,2,---,a \\ k &= 1,2,---,b \end{aligned}$

where all components have the same definitions as in (11.1). We see that model (11.21) can be obtained from (11.1) by simply omitting the replicate component, ρ. Basically, this component is now a part of the whole plot error variation. Realizing this fact aids in the sums of squares calculations. All sums of squares are obtained as in the simple split plot with the following exceptions: (i) there are no Replicate SS, as in (11.5), and (ii) the whole plot error sum of squares is given by

(11.22)   SSE1 = SSWP - SSA

Different whole plots occur for each level of factor A implying that "whole plots are nested within A", see Section 7.1. This nesting feature is reflected in both the sum of squares and degrees of freedom for whole plot error. See (11.22) and the following table.

Table 11.3 Partial ANOVA for Split Plot Design Without Rep.

Source	df	SS	EMS(Fixed Treatments)
A	a-1	SSA	$\sigma^2_\epsilon + b\sigma^2_\delta + rb\kappa^2_A$
WP Error	a(r-1)	SSE1	$\sigma^2_\epsilon + b\sigma^2_\delta$
B	b-1	SSB	$\sigma^2_\epsilon + ra\kappa^2_B$
A x B	(a-1)(b-1)	SSAB	$\sigma^2_\epsilon + r\kappa^2_{AB}$
SP Error	(r-1)a(b-1)	SSE2	σ^2_ϵ
Total	rab-1	SSY	

Example 11.6

For comparative purposes, let us modify the education study of Example 11.2. Suppose the 6 classes all come from a single school and are homogeneous enough that replicates cannot be identified. Then, the replicate sum of squares would not be calculated; the

numerical   value obtained earlier (Rep SS = 115.11) would be part
of   the whole plot error.   All other sums of squares would   have
numerical values as stated in   Example 11.2.   Now the ANOVA table
would be as follows:

ANOVA for Education Study

Source	df	SS	MS	EMS
T. Methods	2	864.78	432.39	$\sigma_\epsilon^2 + 2\sigma_\delta^2 + 6\kappa_T^2$
WP Error	6	139.00	23.33	$\sigma_\epsilon^2 + 2\sigma_\delta^2$
Calculations	1	32.00	32.00	$\sigma_\epsilon^2 + 9\kappa_C^2$
T x C	2	14.33	7.17	$\sigma_\epsilon^2 + 9\kappa_{TC}^2$
SP Error	6	3.67	0.61	σ_ϵ^2
Total	17	1053.78		

The   same   conclusions would result as stated   in   Example   11.2.
Here,   the   difference of two teaching method means would have   a
different standard deviation; namely,

$$\sqrt{2MSE1/6} = \sqrt{2(23.33)/6} = 2.79. \text{※}$$

11.5   The Split Block Design

This   modification   of   the simple split plot design   goes   by   a
variety of names: "split plot in strips", "two-way whole plots",
"subunits   in strips",   and so on.   The split block designs   are
used   primarily in agricultural experiments.   In the most   basic
setting, there are two factors, say A and B.   Factor A is applied
to   whole   plots as in the simple split plot   design.   But   then
factor   B   is   applied to "whole plots" (or "strips")   which   are
orthogonal to the whole plots of factor A.   The following figure
illustrates   the   factor level allocations for this   basic   split
block design.

Figure 11.1   Basic Split Block Design: Factor Level Allocations

Factor A might be irrigation systems; each usually require a large amount of land. Factor B might be herbicide spraying which ordinarily would be applied with equipment to large areas of land to avoid excessive turning, crop damage, and so on. Thus, the levels of these two factors would need to be applied as indicated in Figure 11.1 to keep from confounding A and B effects. The levels of factor A are randomized to the vertical strips (whole plots) while levels of factor B are randomized to the horizontal strips (whole plots orthogonal to vertical whole plots.) Neither factor is randomized within the whole plots of the other factor (which was true for factor B in the simple split plot design.)

Example 11.7

A split block design having a factor A with 3 levels and a factor B with 4 levels could have the following randomized treatment allocations for one replicate of the experiment:

a_2b_3	a_3b_3	a_1b_3
a_2b_1	a_3b_1	a_1b_1
a_2b_4	a_3b_4	a_1b_4
a_2b_2	a_3b_2	a_1b_2

We see from Figure 11.1 or from Example 11.7 that levels of factor A are confounded with the vertical whole plots while levels of factor B are confounded with horizontal whole plots. But it can be shown that the AB interaction is not confounded with either set of whole plots. Thus, main effect A, main effect B, and the AB interaction are measured with different precisions.

The linear model for the basic split block design is

$$(11.23) \quad y_{ijk} = \mu + \rho_i + \alpha_j + \delta_{ij} + \beta_k + \eta_{ik} + (\alpha\beta)_{jk} + \epsilon_{ijk}$$

$$\text{for } i = 1, 2, \text{---}, r; \ j = 1, 2, \text{---}, a; \ k = 1, 2, \text{---}, b$$

where μ = overall mean

ρ_i = i-th replicate effect

α_j = effect of the j-th level of factor A

δ_{ij} = error component for factor A

β_k = effect of the k-th level of factor B

η_{ik} = error component for factor B

$(\alpha\beta)_{jk}$ = interaction component for j-th level of A and k-th level of B

ϵ_{ijk} = residual component, the error component for the AB interaction.

For inference purposes, the following assumptions are made

(11.24)

    (i)   the δ's are i.i.d. $N(0,\sigma_\delta^2)$

    (ii) the η's are i.i.d. $N(0,\sigma_\eta^2)$

    (iii) the ϵ's are i.i.d. $N(0,\sigma_\epsilon^2)$

    (iv) the δ's, η's and ϵ's are distributed independently of each other.

Additional assumptions are made about the treatment components, depending upon their fixed or random nature.

The ANOVA table for the basic split block design appears below. Except for an additional error term, this ANOVA is the same as for the simple split plot design.

Table 11.4 ANOVA for Split Block Design

Source	df	SS	EMS(Fixed Treat.)
Reps	r-1	SSR	---
A	a-1	SSA	$\sigma_\epsilon^2 + b\sigma_\delta^2 + rab\kappa_A^2$
WP Error 1	(r-1)(a-1)	SSE1	$\sigma_\epsilon^2 + b\sigma_\delta^2$
B	b-1	SSB	$\sigma_\epsilon^2 + a\sigma_\eta^2 + ra\kappa_B^2$
WP Error 2	(r-1)(b-1)	SSE2	$\sigma_\epsilon^2 + a\sigma_\eta^2$
AB	(a-1)(b-1)	SSAB	$\sigma_\epsilon^2 + r\kappa_{AB}^2$
Error 3	(r-1)(a-1)(b-1)	SSE3	σ_ϵ^2
Total	rab-1	SSY	

All sums of squares in Table 11.4 are calculated as for the
simple split plot design except for the 3 error sums of squares.
We now have "two whole plot SS", one for vertical whole plots and
one for horizontal whole plots.   A slight change of notation in
equation (11.7) gives for vertical whole plots

(11.25) $SSWP1 = \frac{1}{b} \sum_{ij} y_{ij.}^2 - C =$ "Whole Plot 1" SS

Then, the "whole plot error 1" sum of squares is

(11.26)   SSE1 = SSWP1 - SSR - SSA

Similarly for the horizontal whole plots, we have

(11.27) $SSWP2 = \frac{1}{a} \sum_{ik} y_{i.k}^2 - C =$ "Whole Plot 2" SS

from which we obtain "whole plot error 2" sum of squares

(11.28)   SSE2 = SSWP2 - SSR - SSB

Lastly, the Error 3 sum of squares may be obtained by subtraction

(11.29)   SSE3 = SSY - SSR - SSA - SSE1 - SSB - SSE2 - SSAB

Example 11.8

A turf specialist is studying the durability of six varieties of
turf grass in combination with three levels of compacting (none,
slight and moderate).   Sufficient land was available at three
locations (replicates) for use in the study.   In each replicate,
the turf varieties were established on six plots as indicated
below (unrandomized):

V1	V2	V3	V4	V5	V6

The compacting machine could not be maneuvered easily within the
whole plots established for varieties, so it was necessary to run
the compacting machine in strips indicated below (unrandomized)

None
Slight
Moderate

Note: The compacting was not applied until some time after the varieties were well established.

One of the variables of interest was the amount of dry matter from a sample taken on each subunit.   The results were (in grams):

		Variety					
		1	2	3	4	5	6
Rep	Compacting						
	N	10.3	9.7	11.2	10.8	10.5	9.9
1	S	9.8	10.1	11.0	10.4	10.6	9.5
	M	9.0	9.6	10.8	10.1	9.8	11.0
	N	11.8	10.3	12.1	12.3	11.8	10.6
2	S	10.7	11.6	11.9	11.8	11.7	10.1
	M	10.1	10.9	12.1	11.0	10.3	9.2
	N	10.2	10.1	11.6	11.2	10.6	10.3
3	S	9.5	10.7	10.8	9.9	10.5	9.4
	M	9.7	9.3	11.2	9.6	10.4	10.3

(Courtesy of Dr. John Hall, Agronomy Dept., Va. Tech.)

Variety (V) Totals: 91.1, 92.3, 102.7, 97.1, 96.2, 90.3
Compacting (C) Totals: 195.3, 190.0, 184.4
Rep Totals: 184.1, 200.3, 185.3

The sums of squares calculations are:

$$C = (569.7)^2/54 = 6010.34$$

$$SSR = [(184.1)^2 + (200.3)^2 + (185.3)^2]/18 - C = 9.05$$

$$SSV = [(91.1)^2 + \cdots + (90.3)^2]/9 - C = 12.19$$

$$SSE1 = [(29.1)^2 + \cdots + (30)^2]/3 - C - 9.05 - 12.19 = 2.81$$

$$SSC = [(195.3)^2 + (190)^2 + (184.4)^2]/18 - C = 3.30$$

$$SSE2 = [(62.4)^2 + \cdots + (60.5)^2]/6 - C - 9.05 - 3.30 = 0.93$$

$SVC = [(32.3)^2 + \text{---} + (30.5)^2]/3 - C = 19.94$

$SSVC = 19.94 - 12.19 - 3.3 = 4.45$

$SSY = 6046.51 - C = 36.17$

$SSE3 = 36.17 - 9.05 - 12.19 - 2.81 - 3.30 - 0.93 - 4.45$
$\quad = 3.44$

ANOVA for Dry Weight of Turf Grasses

Source	df	SS	MS	EMS
Rep	2	9.05	4.52	
Varieties	5	12.19	2.44	$\sigma_\epsilon^2 + 3\sigma_\delta^2 + 9\kappa_V^2$
Error 1	10	2.81	0.28	$\sigma_\epsilon^2 + 3\sigma_\delta^2$
Compacting	2	3.30	1.15	$\sigma_\epsilon^2 + 6\sigma_\eta^2 + 18\kappa_C^2$
Error 2	4	0.93	0.23	$\sigma_\epsilon^2 + 6\sigma_\eta^2$
V x C	10	4.45	0.44	$\sigma_\epsilon^2 + 3\kappa_{VC}^2$
Error 3	20	3.44	0.17	σ_ϵ^2
Total	53	36.17		

Tests of varieties, compacting, and their interaction give calculated F values of 8.71, 5.00, and 2.59, respectively, with observed significance levels $P < 0.005$, $0.05 < P < 0.10$, and $0.025 < P < 0.05$. The significance of the interaction can be interpreted by saying that the different amounts of compacting do not have the same effects on the different varieties. A plot of the variety-compacting means could be made to further analyze this interaction. ✸

For a split block design, the standard deviations are more complicated than for a simple split plot design. This is due to the restrictions on the randomizations, and the resulting three error terms. Using the mean squares in Table 11.4, the estimated standard deviations for differences of pairs of means are

(11.30) $\sqrt{2MSE1/rb}$; for two A means, say $\bar{y}_{.j.} - \bar{y}_{.h.}$

(11.31) $\sqrt{2MSE2/ra}$; for two B means, say $\bar{y}_{..k} - \bar{y}_{..h}$

(11.32) $\sqrt{2[MSE1+(b-1)MSE2]/rb}$; for two AB means at the same
               level of B, say $\bar{y}_{.jk} - \bar{y}_{.hk}$

(11.33) $\sqrt{2[MSE1+(a-1)MSE2]/ra}$; for two AB means at the same
               level of A, say $\bar{y}_{.jk} - \bar{y}_{.jh}$

(11.34) $\sqrt{2[aMSE1+bMSE2+(ab-a-b)MSE3]/rab}$; for two means
                                                         having different
                                                         levels of both
                                                         A and B, say
                                                         $\bar{y}_{.jk} - \bar{y}_{.hg}$

Only ratios formed with the differences and standard deviations
in (11.30) and (11.31) have exact t distributions. For the
standard deviations composed of combinations of mean squares, the
last 3 above, we cannot form exact t variables. Approximations
following the procedure of Section 2.12 may be utilized for these
three cases.

Example 11.9

For the turf grass study discussed in the preceding example, the
estimated standard deviations are:

difference of	standard deviation
two variety means	$\sqrt{2(0.28)/9} = 0.25$
two compacting means	$\sqrt{2(0.23)/18} = 0.16$
two varietal means at the same compaction	$\sqrt{2[0.28+2(0.17)]/9} = 0.37$
two compacting means at the same variety	$\sqrt{2[0.23+5(0.17)]/18} = 0.35$
two means, different varieties and different compaction	$\sqrt{2[6(0.28)+3(0.23)+9(0.17)]/54}$ $= 0.38$

11.6 The Split-Split Plot Design

This is an extension of the simple split plot design rather than
a modification of the treatment or experimental design structure.

We have whole plot units to which the levels of factor A are randomly applied. Each whole plot is subdivided into b split plots to which the levels of factor B are randomly applied. Then each split plot is subdivided into c split-split plots to which the levels of factor C are randomly applied. We emphasize that a separate randomization is performed within each whole plot, and within each split-plot.

To summarize, a split-split plot design consists of three treatment factors; one factor applied to whole plot units, a second factor applied to split plot units, and a third factor applied to split-split plot units. Figure 11.2 illustrates the allocation of these factors to the experimental material.

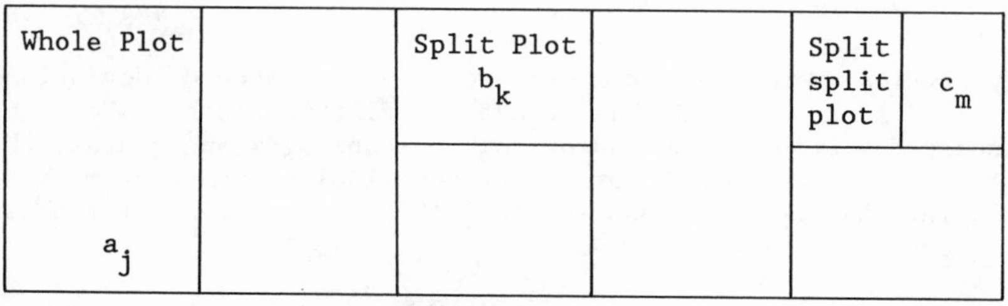

Whole Plot		Split Plot b_k		Split split plot	c_m
a_j					

Figure 11.2   Split-split Plot Design: Allocation of Factor Levels

Example 11.10

In an agricultural experiment, whole plot treatments might be irrigation systems, split plot treatments might be varieties of a crop, and split-split plot treatments might be fertilizers. If the fertilizer can be applied by hand or with a small applicator, small plots would suffice. The varieties and irrigations would be applied with larger equipment and ordinarily would require larger amounts of land.

In an industrial setting, whole plot treatments might be large batches of material prepared in furnaces at different temperatures. After a batch is removed from a furnace, it may be subdivided for application of different cooling temperatures. Each portion of cooled material may be further subdivided for application of anti-oxidizing agents. ※

Example 11.11

Let us give the randomized allocation of treatments for a single replicate of a split-split plot design for the industrial experiment described in the previous example. Suppose there are a = 3 furnace temperatures, b = 2 cooling temperatures, and c = 3 anti-oxidizing agents. We have

a_2		
b_2c_1	b_2c_2	b_2c_3
b_1c_3	b_1c_1	b_1c_2

a_3		
b_1c_2	b_1c_3	b_1c_1
b_2c_3	b_2c_1	b_2c_2

a_1		
b_2c_2	b_2c_1	b_2c_3
b_1c_1	b_1c_3	b_1c_2

A given batch, formed under furnace temperature a_j, was divided into two split plots which were subjected to one of the cooling temperature, b_k. The material of a given cooling temperature was divided into three split-split plots to which the anti-oxidizing agents were applied. ▓

The linear model for a split-split plot design is

$$(11.35) \qquad y_{ijkm} = \mu + \rho_i + \alpha_j + \delta_{ij} + \beta_k + (\alpha\beta)_{jk} + \eta_{ijk} + \gamma_m$$
$$+ (\alpha\gamma)_{jm} + (\beta\gamma)_{km} + (\alpha\beta\gamma)_{jkm} + \epsilon_{ijkm}$$

$$\text{for } i = 1, 2, \cdots, r; \; j = 1, 2, \cdots, a$$
$$k = 1, 2, \cdots, b; \; m = 1, 2, \cdots, c$$

For inferences, we make the following assumptions about the error components:

(i) the δ's are i.i.d. $N(0,\sigma_\delta^2)$

(ii) the η's are i.i.d. $N(0,\sigma_\eta^2)$

(11.36)

(iii) the ϵ's are i.i.d. $N(0,\sigma_\epsilon^2)$

(iv) the δ's, the η's, and the ϵ's are distributed independently of each other.

Additional assumptions would be made about treatment components, depending upon their fixed or random nature.

To complete the analysis, we need various sums of squares. Those for treatment components, main effects and interactions, should present no difficulties, they are calculated as explained in Chapter 5 for any 3 factor factorial. We therefore concentrate on the sums of squares for the three error components.   For the whole plot error sum of squares, we calculate

$$(11.37) \qquad SSE1 = \frac{1}{bc} \underset{ij}{\Sigma\Sigma} \, y^2_{ij..} \, - \, C \, - \, SSR \, - \, SSA$$

For the split plot error sum of squares, we calculate

$$(11.38) \qquad SSE2 = \frac{1}{c} \underset{ijk}{\Sigma\Sigma\Sigma} \, y^2_{ijk.} \, - \, C \, - \, SSR \, - \, SSA \, - \, SSE1 \, - \, SSB \, - \, SSAB$$

The split-split plot error sum of squares is, by subtraction

$$(11.39) \qquad \begin{aligned} SSE3 = & \, SSY \, - \, SSR \, - \, SSA \, - \, SSE1 \, - \, SSB \, - \, SSAB \\ & - \, SSE2 \, - \, SSC \, - \, SSAC \, - \, SSBC \, - \, SSABC \end{aligned}$$

Table 11.5 ANOVA Table for Split-Split Plot Design

Source	df	SS	EMS(Fixed Treatments)
Reps	$r-1$	SSR	$- - -$
A	$a-1$	SSA	$\sigma^2_\epsilon + c\sigma^2_\eta + bc\sigma^2_\delta + rbc\kappa^2_A$
WP Error	$(r-1)(a-1)$	SSE1	$\sigma^2_\epsilon + c\sigma^2_\eta + bc\sigma^2_\delta$
B	$b-1$	SSB	$\sigma^2_\epsilon + c\sigma^2_\eta + rac\kappa^2_B$
AxB	$(a-1)(b-1)$	SSAB	$\sigma^2_\epsilon + c\sigma^2_\eta + rc\kappa^2_{AB}$
SP Error	$(r-1)a(b-1)$	SSE2	$\sigma^2_\epsilon + c\sigma^2_\eta$
C	$c-1$	SSC	$\sigma^2_\epsilon + rab\kappa^2_C$
AxC	$(a-1)(c-1)$	SSAC	$\sigma^2_\epsilon + rb\kappa^2_{AC}$
BxC	$(b-1)(c-1)$	SSBC	$\sigma^2_\epsilon + ra\kappa^2_{BC}$
AxBxC	$(a-1)(b-1)(c-1)$	SSABC	$\sigma^2_\epsilon + r\kappa^2_{ABC}$
Error 3	$(r-1)ab(c-1)$	SSE3	σ^2_ϵ
Total	$rabc-1$	SSY	

Inferences about pairs of means may be of interest. With three factors, many different types of pairs may be considered. And due to the restrictions on randomizations, and the resulting three error terms, there are twelve different standard deviations. Using the mean squares given in Table 11.5, estimated standard deviations for differences of pairs of means are

(11.40) $\sqrt{2MSE1/rbc}$; for two A means: $\bar{y}_{.j..} - \bar{y}_{.h..}$

(11.41) $\sqrt{2MSE2/rac}$; for two B means: $\bar{y}_{..k.} - \bar{y}_{..h.}$

(11.42) $\sqrt{2MSE2/rc}$; for two AB means,
same level of A: $\bar{y}_{.jk.} - \bar{y}_{.jh.}$

(11.43) $\sqrt{2[MSE1+(b-1)MSE2]rbc}$; for two AB means, different
levels of A: $\bar{y}_{.jk.} - \bar{y}_{.hk.}$ or $\bar{y}_{.jk.} - \bar{y}_{.hg.}$

(11.44) $\sqrt{2MSE3/rab}$; for two C means: $\bar{y}_{...m} - \bar{y}_{...h}$

(11.45) $\sqrt{2MSE3/rb}$; for two AC means,
same level of A: $\bar{y}_{.j.m} - \bar{y}_{.j.h}$

(11.46) $\sqrt{2[MSE1+(c-1)MSE3]/rbc}$; for two AC means, different
levels of A: $\bar{y}_{.j.m} - \bar{y}_{.h.m}$ or $\bar{y}_{.j.m} - \bar{y}_{.h.g}$

(11.47) $\sqrt{2MSE3/ra}$; for two BC means,
same level of B: $\bar{y}_{..km} - \bar{y}_{..kh}$

(11.48) $\sqrt{2[MSE2+(c-1)MSE3]/rab}$; for two BC means, different
levels of B: $\bar{y}_{..km} - \bar{y}_{..hm}$ or $\bar{y}_{..km} - \bar{y}_{..hg}$

(11.49) $\sqrt{2MSE3/r}$; for two ABC means, same level of
both A and B: $\bar{y}_{.jkm} - \bar{y}_{.jkh}$

(11.50) $\sqrt{2[MSE2+(c-1)MSE3]/rc}$; for two ABC means, same
level of A only: $\bar{y}_{.jkm} - \bar{y}_{.jhg}$

(11.51) $\sqrt{2[MSE1+(b-1)MSE2+b(c-1)MSE3]/rbc}$; for any other pair
pair of ABC means.

The remarks following (11.19) are applicable here.

PROBLEMS

1. A split plot design was conducted with whole plot treatment factor A at 5 levels and split plot treatment factor B at 4 levels. There were 3 replicate groups for the experiment. All treatment factors were assumed fixed.

    (a) Give the linear model for this experiment and specify ranges on the subscripts.
    (b) Give the partial ANOVA (sources, df, and EMS) for this experiment.
    (c) What assumptions are necessary for valid tests and CI based on the ANOVA.

2. Repeat Problem 1 when the split plot treatments are 2x2 factorial, say factors B and C.

3. A split plot design was conducted with whole plot treatments being the combinations of a 2x3 factorial, say on factors A and B, and split plot treatments being levels of factor C at 4 levels. Two replicate groups were used in the experiment. All treatment factors were assumed fixed.

    (a) Give the linear model for this experiment. Specify ranges on the subscripts.
    (b) Give the partial ANOVA (sources, df and EMS) under the above model and usual split-plot analysis.

4. Re-do Problem 3 under all conditions stated there except that split plot treatments have a 2x2 factorial structure, say on factors C and D.

5. An engineer investigated the strength of steel made with all combinations of 3 percentages of carbon and 4 cooling techniques. Batches of steel could be made with a given percentage of carbon. After molds were filled from a batch, four were selected at random and one subjected to each of the cooling techniques. On a given day, three batches of steel could be made, one for each percentage of carbon. This was repeated on each of 4 days. The variable of interest was y, the strength of the molded steel product. Assume fixed treatment effects.

    (a) Give the linear model for this experiment. Explain terms used in the model and specify ranges on the subscripts.
    (b) What assumptions are necessary for valid tests and CI based on the ANOVA?
    (c) Give the partial ANOVA under the above linear model and assumptions.

6. A seafood specialist investigated bacterial growth, y, in oysters subjected to 5 lengths of cold storage each at 3 storage temperatures. Nine cold storage units were available; three were randomly selected and set at each temperature. Three batches of oysters were obtained at different times and from different sources. Each batch was considered to be a replicate so after obtaining an initial bacterial count, 15 packages were formed from the batch. Five of these packages were randomly assigned to one cold storage unit of each temperature. At the end of each designated length of storage, one package was randomly selected from each storage unit and a bacterial count taken. The difference between this count and the ititial count gave the bacterial growth, y. Assume fixed treatment effects.

(a) Give the linear model for this experiment. Explain terms used and specify ranges on the subscripts.
(b) What assumptions are necessary for valid tests and CI based on the ANOVA?
(c) Give the partial ANOVA under the above linear model and assumptions.

7. A food science researcher investigated the tenderness, y, of beef steaks from cattle that were fed for 45 days on one of the following finishing rations:

a_1 75% grain, 25% hay

a_2 50% grain, 50% hay

a_3 25% grain, 75% hay

Twelve ranchers agreed to participate in the experiment (four for each ration) and each supplied a carcass. Four sirloin steaks were obtained from each carcass, to these the four tenderizers were randomly assigned. All steaks were cooked uniformly and the tenderness determined. Finishing rations and tenderizers were assumed fixed. There are no replicate groups in this experiment. (Courtesy of Drs. K. Bovard and P. Graham, Animal and Food Science Depts., Va. Tech.)

(a) Give the linear model for this experiment. Specify ranges on the subscripts.
(b) Give the partial ANOVA under the above model and the usual assumptions.

8. Suppose the seafood specialist obtained the three batches of oysters at the same time. The technician working for the seafood specialist mixed all three batches together "to make the whole process simpler". An initial bacterial count was

made for the mixed batch and 45 packages were formed; five
were randomly assigned to each storage unit.   What impact
would the technician's modification have on the experimental
design and analysis? Give the linear model and partial ANOVA
for this experiment if different from the earlier ones.

9. A home economics graduate student was studying the effects of
    two different temperature factors on the moistness of cakes.
    One was a baking temperature with 5 levels; the second was
    the temperature of the milk (3 levels) when added to the cake
    mix.   Fifteen ovens were not available so it was decided to
    use 5 ovens--one for each baking temperature but containing
    three cakes, one for each milk temperature.   Only 15 cakes
    could be baked on a given day so the experiment extended
    over 12 days (replicates).   Moistness readings were obtained
    on each cake and provided the following:

ANOVA for Cake Moistness

Source	df	SS	MS	EMS
Days		416.7		
Baking T.		107.6		
Days x Baking T.		328.1		
Milk T.		89.2		
Baking T. x Milk T.		1.9		
Residual		641.2		
Total		1584.7		

Assume random effects for days and fixed effects for baking
and milk temperatures.

(a) Complete the ANOVA table.
(b) Make appropriate tests of treatment components.

Give the estimated standard deviation of the difference of
two observed:

(c) baking temperature means.
(d) baking temperature means for a fixed milk temperature.

10. Refer to the experiment of the previous problem.   The mean
    cake moistness for the 3 milk temperatures were:

$$\bar{y}_{..k} \rightarrow \quad 24.6 \qquad 26.3 \qquad 25.7$$

(a) Suppose that the first temperature (k=1) was room temp-
    erature while the other two were "warmer" temperatures.
    Do a Dunnett's analysis on these means with $\alpha = 0.05$.

(b) Suppose instead that milk temperatures were 70°F, 80°F, and 90°F. Do a trend analysis with $\alpha = 0.05$.

11. A plant scientist conducted a split plot experiment at 5 locations (replicates) with 6 "Management Systems" as a whole plot factor M and a split-plot factor C which was "amount of compacting the soil," there were 4 levels of C. The variable of interest was the total yield, y, per plot for a 6 week period. Assume M and C are fixed factors. The following partial ANOVA resulted:

Source	df	SS	MS	EMS
Reps		1186.2		
M		503.1		
Error 1		158.7		
C		378.3		
MxC		349.6		
Error 2		191.6		

(a) Complete the ANOVA table.
(b) Make appropriate tests of treatment components.

Give the estimated standard deviation of the difference of two observed:

(c) management system means.
(d) management system means for a given amount of compacting the soil.

12. Refer to the experiment of the previous problem. The mean yields for the 4 soil compactions were (in grams, dry weight)

$$\bar{y}_{..k} \rightarrow \quad 110.8 \qquad 108.1 \qquad 105.8 \qquad 107.9$$

(a) If the first soil compaction (k = 1) is "no compaction" while the other 3 are unspecified amounts of compaction, do a Dunnett's analysis on these means. Let $\alpha = 0.05$.
(b) If the soil compactions are specified as 0, 2, 4, 6 passes across a plot with the compactor, do a trend analysis on the 4 compaction means. Let $\alpha = 0.05$.

13. A cotton breeder conducted an experiment to study the effects of 3 spacings and 4 levels of fertilization on the lint yield of a new variety of cotton. Larger plots were required for planting the cotton according to the spacings than for application of the fertilizer. In each replicate, whole plots were set up for planting by each spacing. A hand applicator was used to apply the levels of fertilizer within each whole plot (randomly.) The lint weights, in pounds, per plot were:

S-F-R	Weight	S-F-R	Weight	S-F-R	Weight
111	61.5	112	65.3	113	48.7
121	66.3	122	58.4	123	57.4
131	50.3	132	62.3	133	50.9
141	55.1	142	60.2	143	49.8
211	49.9	212	72.3	213	47.6
221	58.8	222	68.3	223	62.4
231	77.4	232	66.4	233	71.0
241	62.0	242	67.7	243	64.5
311	72.3	312	65.1	313	64.2
321	55.2	322	78.4	323	66.8
331	49.8	332	73.5	333	65.8
341	63.3	342	73.1	343	71.2

(a) Set up the ANOVA for this experiment. Assume fixed effects of treatment components.

(b) Test appropriate hypotheses of treatment components.

Give estimated standard deviation for the difference of two observed means:

(c) for 2 spacings (d) for 2 fertilizers at the same spacing.

14. A biologist conducted an experiment to investigate changes in the plankton food chain of a pond when a photosynthesis-inhibiting herbicide was introduced. One-gallon glass jugs were randomly filled with well-mixed pond water, dosed with one of the rates of herbicide, sealed, and suspended in the pond just below the water surface. Eight bottles were dosed with each rate: 0, 0.1, 0.5, and 1.0 mg. 1 liter. Bottles were given labels so that two of each dose could be removed at the end of the first day and weekly intervals thereafter. Four dominant species of rotifers, a major element of the plankton food chain in this pond, were counted when each bottle was removed, giving the following numbers/liter:

Week	Dose	Bottle	Number	Week	Dose	Bottle	Number
0	0	1	6718	2	0	1	3815
		2	6392			2	3382
0	0.1	1	5832	2	0.1	1	4373
		2	5569			2	3333
0	0.5	1	5924	2	0.5	1	4912
		2	6715			2	4427
0	1.0	1	6821	2	1.0	1	4900
		2	5871			2	4067
1	0	1	5166	3	0	1	2340
		2	5039			2	2881

Week-Dose-Bottle			Number	Week-Dose-Bottle			Number
1	0.1	1	5233	3	0.1	1	2453
		2	4350			2	2924
1	0.5	1	3953	3	0.5	1	2575
		2	4295			2	4211
1	1.0	1	4849	3	1.0	1	4451
		2	4763			2	4552

(Courtesy of Mr. David Jenkins, Biology Dept., Va. Tech)

(a) Give the ANOVA for this experiment. Assume fixed effects of treatment components.

(b) Test appropriate hypotheses of treatment components.

Give estimated standard deviations for the differences of observed means:

(c) for two doses

(d) for two doses in the same week.

15. A split block design was conducted with 2x3 factorial treatments (factors A, B) applied to whole plots with a third factor C, at 4 levels, applied to the second type of whole plots. There were three replicates. All treatment factors were fixed.

    (a) Give the linear model for this experiment. Specify ranges on the subscripts.

    (b) Give the partial ANOVA (sources, df, and EMS) under the above model and usual split block assumptions.

16. Repeat Problem 15 using all conditions stated there except that the second set of whole plot treatments have a 2x2 factorial structure, say factors C and D.

17. A split-split plot experiment was conducted in which whole plot treatments were 3 levels of a factor A, split plot treatments were 2 levels of a factor B while split-split plot treatments were 6 levels of a factor C. Two replicates were used.

    (a) Give the linear model for this experiment. Specify ranges on the subscripts.

    (b) Give the partial ANOVA (sources, df, and EMS) under the above model. Assume fixed treatment factors.

18. Repeat Problem 17 when split-split plot treatments have a 3x2 factorial structure, say factors C and D.

19. The cotton breeder, Problem 13, conducted a second experiment to investigate the seed index for two varieties of cotton

each grown under two irrigations and 4 fertilizers. Three replicates were used with irrigations applied to the whole plots in each, varieties applied randomly to the split plots within each whole plot, and fertilizers applied randomly within each split plot. The seed index, y, for each plot was:

Irr.	Var.	Fert.	Replicate 1	2	3
1	1	1	13.1	11.0	12.7
		2	12.7	11.1	12.8
		3	12.3	11.3	12.2
		4	10.3	11.6	11.6
	2	1	11.6	12.2	11.4
		2	12.3	11.6	12.3
		3	11.4	9.5	11.7
		4	11.0	11.1	11.6
2	1	1	12.1	12.6	12.1
		2	13.6	11.6	12.4
		3	13.2	12.1	11.6
		4	13.0	12.7	12.3
	2	1	9.1	11.4	11.1
		2	11.6	12.1	11.2
		3	11.4	11.1	10.3
		4	11.3	11.7	10.1

(a) Give the ANOVA for this experiment. Assume fixed effects for the treatment factors.
(b) Test hypotheses about the treatment components.

Give estimated standard deviations for the difference of observed mean seed indices for
(c) the 2 irrigations   (d) the 2 varieties
(e) any 2 fertilizers.

20. The following additional information is given about the cotton experiment of the previous problem:

Irrigations: every 5 days; every 10 days (double amount)
Varieties: a new variety (Problem 13); an established variety
Fertilizers (applied to give the following per acre rates)

80 lbs N, 80 lbs P_2O_5 60 lbs N, 40 lbs P_2O_5
80 lbs N, 60 lbs P_2O_5 40 lbs N, 20 lbs P_2O_5

Perform the most appropriate post-ANOVA analysis.

## Chapter 12

FRACTIONAL FACTORIAL EXPERIMENTS

12.1 Introduction

Analyses of factorial experiments have been discussed in several general contexts. Chapter 5 contained an extensive discussion of factorial treatments in general. Chapter 10 dealt with general confoundings of factorial treatment components in incomplete blocks. And confounding of factorial main effects with incomplete blocks was considered in Chapter 11 where split plot designs were the topic. All of these experiments had one important feature in common: a complete set of factorial treatments was investigated, and each treatment was replicated several times.

Now we wish to discuss factorial experiments where only a fraction of all possible factor level combinations are present in the experiment. And, except on rare ocassions, the treatments of the selected fraction appear in the experiment only once (that is, without replication). Experiments having this description are known as *fractional factorials, or fractional replicates*. They were first proposed by Finney [3], with generalizations and additional results given by Finney [4], Plackett [6] and Plackett and Burman [7].

12.2 Some Design Considerations

The most extensive use of fractional factorials has occurred in industrial research and development: manufacturing, chemical research, and engineering. In such experiments, most factors are highly controlled and therefore Exp. Error variability can be measured very accurately from only a few observations. Because this control of factors is lacking somewhat in the agriculture and home economics disciplines, fractional factorials are used to a lesser extent in such experiments. The lack of control is even more severe in the biological, socio-economic, and behavioral disciplines where fractional factorials are used rarely.

There are various reasons why a complete factorial may not be desirable or even necessary---reasons why fractional factorial experiments might be considered. First of all, the size of a large, complete factorial might pose severe problems at several stages of the experimental process, notably treatment allocation, construction of homogeneous blocks, and making the measurements at the proper time. Incomplete block designs of Chapter 10 may offer some relief of these problems.

Perhaps more importantly is the cost of conducting a "large" experiment. This certainly would be true for a chemist who would like to conduct an experiment to study the effects of 10 factors on the production of a "complex" chemical process. Having just two levels of each factor would necessitate a total of 1024 experimental runs for just one complete replicate. No doubt the cost would be prohibitive.

The time required for conducting a complete factorial may be too great. The possibility of changes in experimental conditions may increase drastically with an increase in overall completion time of the experiment. This, of course, could inflate Exp. Error variation making inferences less precise.

A complete factorial may be unnecessary because higher order interactions tend to be insignificant. Consider a complete 2^7 factorial which has 127 treatment components: 7 main effects, 21 two-factor interactions, 35 three-factor interactions, 35 four-factor interactions, 21 five-factor interactions, 7 six-factor interactions, and one seven-factor interaction. It is not very likely that all of these treatment components are significant, so a fractional factorial would suffice, provided that we can make a prudent judgment of the nonsignificant components.

A large complete factorial might provide a much greater precision than necessary. For example, a complete 2^5 factorial experiment provides a 16-fold replication of each main effect. For preliminary investigations (pilot studies, screenings) where only a rough idea is needed about the size of main effects, such high precision would be a waste of resources.

12.3 Basic Ideas of Fractional Replication

Without replication of treatments, the estimation of Exp. Error variability cannot be accomplished in the usual manner. Some new assumptions will be required, and we will return to this issue later.

Even more severe restrictions occur with respect to estimation when only a fraction of all treatments are used in an experiment. Because the number of observations in the entire experiment is less than the number of treatments, we certainly cannot estimate all treatment effects. When a researcher uses a 1/2 fraction of a 2^4 factorial, there is a total of 8 experimental observations. These provide a total of 7 degrees of freedom for estimation purposes. Thus, at most 7 treatment effects could be estimated (leaving no degrees of freedom for Exp. Error.)

Which treatment effects can be estimated in a fractional replicate? And, while it may not be obvious at this time, we will see shortly that certain treatment effects are confounded with each other! This is a new concept of confounding. All previously discussed confounding systems had treatment components confounded only with incomplete blocks, not with each other. So as one might suspect, the particular treatments appearing in any fractional replicate completely determine the confoundings among the treatment components.

Example 12.1

Suppose a researcher determines that only a 1/2 fraction of a 2^3 factorial can be conducted. There are 3 factors of interest, say A, B, and C, each at 2 levels. Of the 8 factorial treatments, which 4 should we use in our experiment?

Let us begin by first considering a complete 2^2 factorial for factors A and B. The 4 factorial treatments, their observations, and contrasts to estimate the three factorial components are:

Treatment	Observation	A	B	AB
(1)	y_1	-	-	+
a	y_2	+	-	-
b	y_3	-	+	-
ab	y_4	+	+	+

where the signs in each column indicate the contrast of the observations that would be used to estimate the factorial effect listed at the heading of the column. Thus, the A effect would be estimated by

$$- y_1 + y_2 - y_3 + y_4$$

We also note that the signs in each row of only the A and B columns can be used to denote the level of the respective factor:

"plus" for the high level, "minus" for the low level. For example observation y_2 resulted from the high level of A and the low level of B, the factorial treatment designated by "a".

So far we have no indication of how the levels of factor C appear in the experiment. Let us agree to include the levels of C according to the signs of the AB interaction. Thus, the high level of C is to occur on the EU giving y_1 and y_4 while the low level of C is to occur on the EU giving y_2 and y_3. This implies that the main effect C will be confounded with the AB interaction because each is calculated by the same contrast. If we replace AB in the above display by C, we get the design setup for a 1/2 fraction of the 2^3 factorial; namely,

Treatment	Observation	A	B	C
c	y_1	-	-	+
a	y_2	+	-	-
b	y_3	-	+	-
abc	y_4	+	+	+

Again, the signs in each column indicate the contrast of the observations used to estimate the factorial component listed at the heading of the column. Now, the signs in each row of the 3 columns headed A, B, C are used to designate the factor levels used in the experiment. The resulting factorial combinations are those given in the first column.

The contrasts appearing in the columns of the last display need some clarification. In a *complete* 2^3 factorial, one would estimate the main effect for A, for example, by a contrast

$$C_A = -(1) + a - b + ab - c + ac - bc + abc$$

where each symbol represents the mean of all observations having treatment designation equal to that symbol. But the fractional replicate above consists of only 4 of the 8 possible treatments, and each occuring only once. So to estimate A, we use only the treatments available. Deleting the 4 treatments which are not used in this half-replicate gives an estimate for A of

$$C_A = a - c - b + abc \rightarrow y_2 - y_1 - y_3 + y_4$$

as indicated above. Likewise, the contrast for each of the other factorial components would be based on the four available observations. After deleting the appropriate treatment designations, the contrasts for all 7 treatment components using the above half-replicate of the 2^3 factorial are

A	B	C	AB	AC	BC	ABC	Observation
-	-	+	+	-	-	+	y_1
+	-	-	-	-	+	+	y_2
-	+	-	-	+	-	+	y_3
+	+	+	+	+	+	+	y_4

Several points can be made:

(i) In addition to C being confounded with AB, we also have B confounded with AC, and A confounded with BC.

(ii) The three-factor interaction cannot be estimated at all, we may say it is confounded with the overall total (or mean.) ✻

Points similar to the ones just made in the previous example can be made in any fractional replicate. The following example illustrates the confounding and estimation discussed in general terms above.

Example 12.2

Suppose a researcher conducted the half-replicate of the 2^3 factorial described in the previous example. The data obtained were:

$$c \quad 60 = y_1 \qquad\qquad b \quad 66 = y_3$$

$$a \quad 55 = y_2 \qquad\qquad abc \quad 73 = y_4$$

The estimates of the main effects are

$$C_A = - 60 + 55 - 66 + 73 = 2$$

$$C_B = - 60 - 55 + 66 + 73 = 24$$

$$C_C = + 60 - 55 - 66 + 73 = 12$$

As we indicated in the previous example, A is confounded with BC, B is confounded with AC, and C is confounded with AB. Therefore

$$C_{BC} = 2 \qquad C_{AC} = 24 \qquad C_{AB} = 12$$

And the ABC interaction cannot be estimated.

The total sum of squares for the experimental observations is

$$SSY = 60^2 + 66^2 + 55^2 + 73^2 - (254)^2/4$$

$$= 16310 - 16219 = 181$$

The sums of squares for A, B, and C are

$$SSA = C_A^2/4 = (2)^2/4 = 1$$

$$SSB = C_B^2/4 = (24)^2/4 = 144$$

$$SSC = C_C^2/4 = (12)^2/4 = 36$$

And we see that SSA + SSB + SSC = 181 = SSY, as it should. The contrasts for A, B, and C used here represent a complete orthogonal set. Thus, all variation among the four observations is explained by these three contrasts. There are no degrees of freedom left for estimation of Exp. Error. ※

The subject of which effects can be estimated in a fractional replicate and which effects are confounded needs to be examined more closely. The next 2 definitions are given for this purpose.

Definition 12.1

In a fractional replicate, treatment components confounded with each other are said to be *aliases*. An *alias set* consists of all treatment components which are estimated by the same contrast of observations.

For the fractional replicate discussed in Examples 12.1 and 12.2, there are 3 alias sets: (A,BC), (B,AC), and (C,AB). Thus, each main effect is aliased with a two-factor interaction. Another way to represent an alias set is to link the members by equal signs; for example, A = BC would represent the first alias set above.

Each alias set specifies the combination of treatment effects that can be estimated. One can estimate only the sum of treatment components in each alias set. An alias set defined by A=BC implies that one can estimate A+BC only. We now define a very special alias set.

Definition 12.2

The alias set containing the overall mean, and therefore containing the treatment components that cannot be estimated, is called the *defining contrast*. We will denote a defining contrast by

(12.1) $I = X = Y = Z = \text{---}$

where (X,Y,Z,---) are symbols used for treatment components, including generalized interactions, that cannot be estimated. The symbol I corresponds to the overall mean so that (12.1) indicates which treatment components are confounded with the overall mean. The defining contrast completely determines the treatments forming the fractional replicate.

Example 12.3

The fractional replicate described in Example 12.1 consisted of the 4 treatments which entered positively in the ABC contrast. We remarked that the ABC interaction could not be estimated because it was confounded with the overall mean. Then for this fractional replicate, the defining contrast is I = ABC. ✦

In earlier discussions of confounding systems, we noted that certain confounded effects were used for specifying block compositions and determining generalized interactions. In fractional replication, it is the defining contrast that gives us similar information. In addition to specifying the treatments to be used in a fractional replicate, the defining contrast is used to list the alias sets. We need only "multiply" the defining contrast by the treatment component whose aliases we desire. Any treatment components "multiplied" by I remain unchanged. [In the 2-level factorials, any letter with an even power may be deleted, and any odd power may be changed to unity.]

Example 12.4

For the 1/2 fraction of Example 12.1 having defining contrast I = ABC, the alias sets may be formed as follows:

$$A(I) = A(ABC) = A^2BC \rightarrow A = BC$$
$$B(I) = B(ABC) = AB^2C \rightarrow B = AC$$
$$C(I) = C(ABC) = ABC^2 \rightarrow C = AB \;✦$$

Example 12.5

If a researcher conducts a 1/4 fraction of a 2^4 factorial and uses a defining contrast of

$$I = AB = CD = ABCD$$

the alias sets are:

$$A = B = ACD = BCD$$
$$C = ABC = D = ABD$$
$$AC = BC = AD = BD$$

One might question whether this is a reasonable design because main effects are confounded with each other in two alias sets while one alias set consists entirely of two factor interactions. But, using only 4 of the 16 possible treatments, we can estimate only 3 "effects". These are, as the alias sets indicate:

$$A + B + ACD + BCD$$
$$C + D + ABC + ABD$$
$$AC + AD + BC + BD. \text{※}$$

Example 12.6

Earlier we used the contrast for the ABC interaction to form a 1/2 replicate of a 2^3 factorial. The plus/minus signs of this contrast partition the 8 treatments into 2 sets of 4 each. The treatments entering this contrast positively: a,b,c,abc form the 1/2 fractional replicate of Example 12.1. The defining contrast for this fractional replicate is $I = ABC$. And from the alias sets listed earlier, we noted that we can estimate the three "effects"

$$A + BC \qquad B + AC \qquad C + AB$$

But we could have used the other 1/2 fraction, those 4 treatments entering the ABC contrast negatively; namely, (1), ab, ac, bc. To indicate this negative aspect, we write the defining contrast as $I = -ABC$. This is consistent with the interpretation attached earlier to the defining contrast: the estimates of the overall mean and the ABC interaction are identical, except for sign now.

The alias sets for this 1/2 fraction can be obtained in several ways. From "multiplications" of the defining contrasts we obtain

$$A = - BC$$
$$B = - AC$$
$$C = - AB$$

The alias sets also could be obtained from an inspection of the contrasts for the main effects and two factor interactions. Being estimated from only the 4 experimental treatments: (1), ab, ac, and bc, the contrast for A and BC would be identical except for sign.   Similarly for B and AC as well as for C and AB.   Thus with this 1/2 fraction, the possible estimates are

$$A - BC$$
$$B - AC$$
$$C - AB. \text{ } *$$

In any fractional replicate, there is a certain number of orthogonal contrasts among the treatments which can be estimated. Not surprisingly, this number is related to the fraction of the treatments used in the experiments.   Specifically, the number of orthogonal treatment contrasts that can be estimated is exactly equal to the number of alias sets.   Earlier we considered a special set of orthogonal treatment contrasts, or "effects" as we called them: the sum of the treatment components in the alias sets.   The value of using these contrasts, instead of other possible ones that could be considered, is that we can attach a meaningful interpretation to them, provided that certain basic assumptions hold.   Let us consider the 1/2 fraction of Example 12.1 where the alias sets were

$$A = BC$$
$$B = AC$$
$$C = AB$$

For the 4 treatments of this fraction, the estimates for A, B, and C would be made, respectively, by considering the orthogonal contrasts

$$a - b - c + abc$$
$$- a + b - c + abc$$
$$- a - b + c + abc.$$

But as we have noted several times already, these contrasts are really estimates of

$$A + BC$$
$$B + AC$$
$$C + AB$$

respectively.   Thus we do not know whether the first contrast is estimating the effect of factor A, effect of the BC interaction, or some combination thereof.   This is *always* a risk involved in a fractional replicate. Only when an alias set contains a single significant treatment component will the corresponding contrast estimate a single treatment component cleanly. Otherwise, the corresponding contrast will represent confounded effects. For additional problems associated with fractional factorials, the article by Box and Hunter [1] is recommended.

This brings us to the basic assumption underlying fractional replication: higher order interactions(usually 3 or more factors) are assumed to be nonexistent. Of course, this assumption must withstand the usual challenges, and may need to be addressed jointly by the researcher and a statistician. When this basic assumption does hold, the ideal is to construct a fractional replicate so that main effects and two factor interactions are each aliased only with higher order interactions. Finally, alias sets which consist only of higher order interactions may be pooled to provide an estimate of Exp. Error variability, under the basic assumption.

We have not achieved the ideal confounding patterns in any of the fractional replicate examples considered up to this point. As the number of factors increases, it becomes much easier to construct fractional replicates in which all main effects and two factor interactions are aliased with only high-order interactions and therefore can be estimated cleanly (the basic assumption), provided that a suitably "large" fraction can be used. But the fraction of a large factorial needed to achieve the "ideal alias sets" might be so great that unreasonably large block sizes would be required. Thus we have opposing demands on the fraction of treatments to be included in the experiment.

Various design plans for fractional replicates in the 2^n factorials are given in Connor and Young [2] and the NBS tables [5]. For each plan there is an indication of what treatment components can be estimated under the basic assumption.

12.4 General Method of Constructing Fractional Replicates: 2^n Factorials

Because every fractional replicate corresponds to one block of a confounded system, the general method of constructing confounded systems can be used to advantage here. To construct a particular fractional replicate, we need only find the treatments belonging to the appropriate block of a confounded system. The appropriate block is the one for which the modular equations correspond to components of the defining contrast. (Equations of generalized interactions may be ignored.) In the 2^n factorials, all treatment components of the defining contrast have either plus or minus signs attached. The modular equation for each component would be set equal to one of the residues (mod 2). See the discussion of residues following Rule 10.1.

More specifically, if we wish to construct a $1/2^p$ fraction of a 2^n factorial, the defining contrast will contain treatment components chosen for confounding(together with their generalized interactions.) Then, solutions of p modular equations would give the desired fractional replicate, p being the exponent of the fraction size. Equations corresponding to any p components of the defining contrast may be used, provided none is a generalized interaction. Usually we would opt to use equations with the fewest x's. The next three examples will illustrate these ideas.

Example 12.7

A 1/2 fraction of a 2^3 factorial having defining contrast I = ABC would consist of the four treatments given by

$$x_1 + x_2 + x_3 = 1(\text{mod } 2)$$

where the residue of 1 is necessary to give the factorial combinations which enter positively in ABC. The treatments are

$$
\begin{array}{ll}
100 \rightarrow a & 001 \rightarrow c \\
010 \rightarrow b & 111 \rightarrow abc
\end{array}
$$

as we also determined in Example 12.1. ✳

Example 12.8

Suppose a researcher wishes to use a $1/4 = 1/2^2$ fraction of a 2^4 factorial and sets up a defining contrast

$$I = ABC = - BCD = - AD$$

The specific treatments corresponding to this contrast are given by solutions of two modular equations. The simplest pair would involve equations for "-AD" and either three-factor interaction listed in I. We will use ABC and so the fraction is given by

$$x_1 + x_4 = 1(\text{mod } 2)$$

$$x_1 + x_2 + x_3 = 1(\text{mod } 2)$$

where the residues of 1 are necessary to give those treatments which enter AD negatively and ABC positively. The treatments are:

$$
\begin{array}{ll}
1000 \rightarrow a & 0011 \rightarrow cd \\
0101 \rightarrow bd & 1110 \rightarrow abc
\end{array}
$$

Had we decided to use the equations corresponding to components "ABC" and "-BCD" of the defining contrast, we would have used

$$x_1 + x_2 + x_3 = 1 \pmod 2$$

$$x_2 + x_3 + x_4 = 0 \pmod 2$$

The four treatments given above also satisfy these equations, as they must. ※

Example 12.9

A 1/8 fraction of a 2^6 factorial has a defining contrast of

$$I = ACD = -BDEF = -ABCEF = ABF = BCDF = -ADE = -CE$$

The 8 treatments corresponding to this defining contrast are generated by a set of 3 modular equations. We will use those for ABF, -ADE, and -CE:

$$x_1 + x_2 + x_6 = 1 \pmod 2$$

$$x_1 + x_4 + x_5 = 0 \pmod 2$$

$$x_3 + x_5 = 1 \pmod 2$$

The treatments of this fraction are:

101100	011000
100010	010110
110011	001001
111101	000111 ※

12.5 Fractional Replicates in the 3^n Series

We will briefly indicate how the results discussed for the 2^n series can be extended to provide fractional replicates for the 3^n factorial experiments. A fraction of size $1/3^r$, for $r = 1, 2, \cdots, n-1$, can be used. The defining contrast will contain r treatment components chosen by the researcher (see, together with their generalized interaction(s). Recall that, in the 3^n series, generalized interactions of treatment components X and Y would be XY and XY^2.

Example 12.10

[It may be helpful to review Example 10.11]   Consider a 1/3 fractional replicate of a 3^3 factorial having defining contrast

$$I = ABC^2$$

The alias sets are

$$A = A(ABC^2) = A(ABC^2)^2 \rightarrow \quad A = AB^2C = BC^2$$

$$B = B(ABC^2) = B(ABC^2)^2 \rightarrow \quad B = AB^2C^2 = AC^2$$

$$C = C(ABC^2) = C(ABC^2)^2 \rightarrow \quad C = AB = ABC$$

$$AC = AC(ABC^2) = AC(ABC^2)^2 \rightarrow AC = AB^2 = BC$$

The 9 treatments for a 1/3 fraction with this defining contrast are solutions of the modular equation

$$x_1 + x_2 + 2x_3 = 0 (\mod 3)$$

and are

000	101	210
011	120	202
022	112	221

   This is not a very appealing experimental set-up because we see from the alias sets that each main effect is confounded with one of the two-factor interaction components.   Only when all two and three factor interactions are nonexistent can the researcher estimate the main effects.   Also then, the fourth alias set, being only two-factor interaction components, can be taken to represent 2 degrees of freedom for estimating Exp. Error. ✖

12.6 The Latin Square Design as a Fractional Replicate

We stated in Section 8.4 that the Latin Square designs could be considered a very special type of incomplete block design. Now we will show that a pxp LS design can be viewed as a 1/p fraction of a p^3 factorial. Suppose that we define the 3 factors as

A = Row Blocks
B = Column Blocks
C = Treatments.

One complete replicate of these three factors would have p^3 combinations. But an LS design uses only p^2 of these combinations --- only one treatment (level of factor C) appears with each combination of levels of factors A and B.   Suppose then that we consider the 1/p fraction of this p^3 factorial with defining contrast of

(12.2)   I = ABC

Clearly, for every value of p > 2, each main effect is aliased with two and three factor interaction components.   Consequently, when interactions exist among the row, column, and treatment factors, a Latin Square design would have treatments confounded with certain row and column effects.

Example 12.11

Consider a 3^3 factorial experiment where factors A, B, and C correspond to row blocking, column blocking, and treatments, respectively.   One complete replicate of these factor levels would give 27 combinations.   For a 1/3 fractional replicate, or 9 of these combinations, suppose we use a defining contrast of

I = ABC

One possible fraction corresponding to this defining contrast would be the factorial combinations satisfying

$$x_1 + x_2 + x_3 = 0(mod\ 3)$$

These combinations are

000	102	201
012	111	210
021	120	222

which we might restructure as

Columns (factor B)

Rows (factor A)	0	1	2
0	0	2	1
1	2	1	0
2	1	0	2

Indeed, the levels of factor C (the treatments) form a Latin Square arrangement within this square. The use of a different residue on the right-hand-side of the modular equation above would merely give a different randomization pattern for the treatments within the square.

Now let us look at the alias sets. From the discussion in Section 12.5, these are

$$A = A(ABC) = A(ABC)^2 \rightarrow A = AB^2C^2 = BC$$

$$B = B(ABC) = B(ABC)^2 \rightarrow B = AB^2C = AC$$

$$C = C(ABC) = C(ABC)^2 \rightarrow C = ABC^2 = AB$$

$$AB^2 = AB^2(ABC) = AB^2(ABC)^2 \rightarrow AB^2 = AC^2 = BC^2$$

Factor C, having two degrees of freedom, represents "treatment effects" but we note that this is aliased with components of the "row x column" and "row x column x treatment" interactions. Unless these interactions are non-existent (the usual assumption in a Latin Square design), we see that treatment effects are confounded with the interaction components indicated. ※

12.7 Analyses of Fractional Replicates: 2^n Series

As indicated earlier, estimates of main effects and low order interactions, as well as tests of these treatment components, will be possible only if the basic assumption of fractional replication holds. In the ideal case where all interactions are nonexistent, the main effects can be estimated cleanly (free of any interaction components.) And the alias sets involving only interaction components provide a "pooled" estimate of Exp. Error. One must remember though, that these tests generally have low power due to the small number of degrees of freedom on Exp. Error.

The estimation of effects and calculation of sums of squares are accomplished quite easily once we identify a correspondence between fractional replicates and certain complete factorials. For example, a 1/2 replicate of a 2^4 factorial can be made to correspond to a 2^3 factorial; a 1/4 replicate of a 2^4 factorial to a 2^2 factorial, and so on. The correspondence is established in the following manner:

(i) Each alias set is represented by one of its components.

(ii) For a $1/2^r$ fraction, r factor letters are "discarded" temporarily. Which r letters are discarded is purely arbitrary. Every component containing any discarded letter(s) is discarded and therefore is ineligible to represent an alias set. Under this scheme, there will be precisely one component not discarded in each alias set.

Example 12.12

A 1/2 fraction of a 2^4 factorial with I = ABCD has alias sets

A = BCD	AB = CD	B = ACD
AC = BD	C = ABD	AD = B
D = ABC		

Here, r = 1, so we need to "discard" one factor letter. Let us "discard" factor D. Then, the alias sets will be represented, respectively, by the components

A	AB	B
AC	C	BC
ABC		

which are the 7 treatment components of a complete factorial on factors A, B, and C. ✻

Example 12.13

A 1/4 = 1/2^2 fraction of a 2^4 factorial with I = AB = CD = ABCD has alias sets

$$A = B = ACD = BCD$$
$$C = ABC = D = ABD$$
$$AC = BC = AD = BD$$

Two factor letters need to be "discarded"; let us discard letters A and C. Then, the alias sets will be represented, respectively, by the components

$$B$$
$$D$$
$$BD$$

which are the 3 treatment components of a complete 2^2 factorial on factors B and D. ✻

Note that the representation of the alias sets which we have described to establish the correspondence between fractional replicates and complete factorials is merely a device that we will utilize shortly for estimation of effects and calculation of sums of squares. The original alias sets must be retained for they tell us which linear combinations of treatment effects can be estimated. And, this "new" representation of alias sets could be misleading without reference to the original alias sets. In

Example 12.12, the "new" component ABC really represents the main effect D (under the basic assumption.)

Let us now indicate how the corresponence can be utilized to calculate estimates and sums of squares. Observed data of the fractional replicate may be viewed as observed data from the corresponding complete factorial if the indices of the discarded factors are now discarded. Then, one merely calculates estimates and sums of squares for the complete factorial (by any one of the several methods discussed earlier.) From the correspondence, the calculated quantities can be related back to the original alias sets.

Example 12.14

In Example 12.12, the 8 observations would be denoted by y_{ijkm} where the indices represent factors A, B, C, and D, respectively. When factor D is discarded, the fourth index, m, is discarded. The 8 observations, now identified by y_{ijk}, are considered to be one replicate of a complete 2^3 factorial on factors A, B, and C. One merely calculates the estimates and sums of squares for this complete factorial. These calculated quantities are related back to the original alias sets by the established correspondence; that is, which component of the complete factorial represents the alias set. ✳

Example 12.15

A manufacturer of solar panels is trying to develop new panels with greater heat collecting capabilities. They conducted an experiment to study the effects of the following 4 factors, each at 2 levels:

A: Type of solar fluid
B: Type of heat absorbing material(covering the collector insulation and piping)
C: Thickness of absorbing material
D: Size of piping

Because of costs and complexities of making new panels, it was decided to make only 4 different panels. Each panel could be tested with both types of solar fluid as any panel can be flushed clean after a particular type of fluid is run through it. Thus, the 8 panel-fluid combinations would be a 1/2 replicate of a 2^4 factorial. A defining contrast of I = BCD was used to generate the fraction. The alias sets are:

$$A = ABCD \qquad AB = ACD \qquad B = CD$$
$$AC = ABD \qquad C = BD \qquad AD = ABC$$
$$D = BC$$

The treatments used in this fraction are generated by the modular equation

$$x_2 + x_3 + x_4 = 1(\bmod 2)$$

For all fluid runs, heat was applied at a constant temperature for one-half hour, starting with a fixed fluid temperature. The response was y_{ijkm}, the temperature (degrees F) of the fluid leaving the collector at the end of the half-hour. The treatments and results were:

ijkm		y_{ijkm}	ijkm		y_{ijkm}
0100	b	120	0001	d	132
1100	ab	126	1001	ad	134
0010	c	115	0111	bcd	128
1010	ac	113	1111	abcd	131

If we discard factor D and all components in the alias sets that contain D, we obtain the correspondence with the complete 2^3 factorial on factors A, B, and C, as in Example 12.12. Then, discarding the fourth index, m, the contrasts to estimate the effects and their sums of squares are

Contrast	SS
A: -120+126-115+113-132+134-128+131 = 9	81/8 = 10.125
B: 120+126-115-113-132-134+128+131 = 11	15.125
C: -120-126+115+113-132-134+128+131 = -25	78.125
AB: -120+126+115-113+132-134-128+131 = 9	10.125
AC: 120-126-115+113+132-134-128+131 = 7	6.125
BC: -120-126-115-113+132+134+128+131 = 51	325.125
ABC: 120-126+115-113-132+134-128+131 = 1	0.125

Under the basic assumption, but now assuming all interactions to be nonexistent, an ANOVA table would be

ANOVA for Heating Capabilities of New Solar Panels

Source	df	SS	MS	F	P
A:Fluid	1	10.125	10.125	1.85	> 0.1
B:Materials	1	15.125	15.125	2.77	> 0.1
C:Thickness	1	78.125	78.125	14.31	< 0.05
D:Pipe Size	1	325.125	325.125	59.56	< 0.005
Exp. Error	3	16.375	5.458		
Total	7	444.875			

Based on this analysis, there seem to be no differences due to the type of solar fluid or heat absorbing material. The thickness of the heat absorbing material and size of the piping appear to be the important factors which influence the heating capability of these new panels. ※

12.8 Analysis of Fractional Replicates: 3^n Series

The basic ideas of analysis of fractional replicates in the 2^n series discussed in the previous section are directly applicable to the analysis of fractional replicates in the 3^n series. To begin with, every fractional replicate in this series corresponds to a complete 3-level factorial. For a $1/3^r$ fraction of a 3^n factorial, r factor letters need to be discarded temporarily to obtain a single component representation of each alias set. The components of this representation form a complete 3^{n-r} factorial. Then, if the indices for these same r factors are discarded, the data of the fraction appear as that for the complete factorial. Any method of analysis for a complete 3-level factorial may be used to calculate estimates and sums of squares. After these quantities are calculated, they are related back to the original alias sets by the established correspondence.

We shall not pursue this topic any further. The following example will illustrate the above ideas.

Example 12.16

Suppose a 1/3 fraction of a 3^3 factorial is to be conducted. The experimental data would consist of 9 observations, say y_{ijk}. If a defining contrast of $I = AB^2C$ is used, the alias sets are

$$A = ABC^2 = BC \qquad\qquad C = AB^2C = AB^2$$
$$B = AC = ABC \qquad\qquad AB = AC^2 = BC$$

When letter C is discarded, and any component containing this letter, the alias sets are represented, respectively, by

$$A \qquad\qquad AB^2$$
$$B \qquad\qquad AB$$

which are the components for a complete 3^2 factorial on factors A and B. If the third index, k, is now discarded, the data of the 1/3 fraction may be analyzed as that of the complete factorial. The estimates and sums of squares for the 4 components of the complete factorial can be related back to the original 4 alias sets by the above correspondence. ⚹

PROBLEMS

1. A researcher is planning to conduct a 1/2 replicate of a 2^4 factorial with I = ABCD. Use the ideas of Section 12.3 to give

   (a) the treatments of this fraction   (b) the alias sets.

2. A researcher is planning to conduct a 1/2 replicate of a 2^5 factorial with I = ABCDE. Use the ideas of Section 12.3 to give

   (a) the treatments of this fraction   (b) the alias sets.

3. A researcher is planning to conduct a 1/4 replicate of a 2^5 factorial with I = ABCD = BDE. Use the ideas of Section 12.4 to give

   (a) the treatments of this fraction   (b) the alias sets.

4. A researcher is planning to conduct a 1/4 replicate of a 2^6 factorial with I = ABDE = BCEF. Use the method in Section 12.4 to give

   (a) the treatments of this fraction   (b) the alias sets.

5. For the fractional replicate in Problem 3, give the partial ANOVA (sources and df) under the basic assumption. Discuss tests of hypotheses.

6. For the fractional replicate in Problem 4, give the partial ANOVA (sources and df) under the basic assumption. Discuss tests of hypotheses.

7. A researcher is planning to conduct a 1/3 replicate of a 3^3 factorial with I = ABC. Give

   (a) the treatments of this fraction   (b) the alias sets.

8. A researcher is planning to conduct a 1/3 replicate of a 3^4 factorial with I = ABCD. Give

    (a) the treatments of this fraction   (b) the alias sets.

9. A researcher is planning to conduct a 1/9 replicate of a 3^4 factorial with I = ABC = BCD. Give

    (a) the treatments of this fraction   (b) the alias sets.

10. A researcher is planning to conduct a 1/9 replicate of a 3^5 factorial with I = ABCD = BCDE. Give

    (a) the treatments of this fraction   (b) the alias sets.

11. A researcher conducted a 1/2 replicate of a 2^5 factorial using I = ABCDE as the defining contrast. The experimental results were:

c	26	b	25	abcde	36	bde	46
cde	29	acd	35	abe	23	a	32
abd	40	bcd	28	d	35	abc	27
ace	22	e	21	bce	37	ade	39

    (a) Specify the alias sets.
    (b) Estimate the effects and interactions under the assumption of no interactions involving three or more factors.
    (c) Give the ANOVA table under the assumption stated in (b).
    (d) Discuss tests of treatment components.

12. An experiment was conducted to study the factors which influence the release of radionuclides during an accident involving a nuclear reactor. The study dealt with a specific accident sequence for a boiling water reactor (BWR); many factors could possibly have an effect on the response y, the iodine-bromine radionuclides released during the accident.

    After an initial screening, eight factors believed to have the greatest effect on the response y, were:

        A : Fission Deposition
        B : Fission Source During Meltdown
        C : Fission Source During Vaporization
        D : Fuel Melting Point
        E : Heat Transfer Coefficient
        F : Meltdown Model
        G : Core Fragmentation in Water
        H : Fission Deposition in the Annulus

Each factor occurred at 2 levels but a complete 2^8 factorial could not be conducted as each observation was obtained from an extremely expensive and long running computer simulation. Only factors A, B, C, and H were expected to interact (and only pairwise.) Thus, a 1/16 fractional replicate with a defining contrast

$$I = ABDH = ACEH = BCFH = ABCG$$

would assure that these two factor interactions (as well as all main effects) would be aliased only with nonsignificant components, under the basic assumption. The observations were:

abdh	0.0036	(1)	0.2320
abef	0.0508	bcde	0.1460
abcdefgh	0.0004	bcfh	0.0009
abcg	0.2938	bdfg	0.0394
acdf	0.0411	begh	0.0060
aceh	0.0062	cdgh	0.0030
adeg	0.2852	cefg	0.1692
afgh	0.0022	defh	0.0037

(a) Set up the ANOVA table for this experiment.
(b) Make appropriate tests of treatment components.

REFERENCES

[1] Box, G. E. P., and J. S. Hunter, Multifactor experimental designs, *Ann. Math. Stat.*, Vol. 28, 1957.
[2] Connor, W. S., and S. Young, *Fractional Factorial Designs for Experiments with Factors at Two and Three Levels*, Nat. Bur. of Standards, App. Math. Ser. 58, 1961.
[3] Finney, D. J., The fractional replication of factorial arrangements, *Ann. Eugenics*, Vol. 12, 1945.
[4] Finney, D. J., Recent developments in the design of field experiments. III. Fractional replication, *Jour. Agr. Sci.*, Vol. 36, 1946.
[5] National Bureau of Standards, *Fractional Factorial Experiment Designs for Factors at Two Levels*, NBS App. Math. Ser. 48, 1957.
[6] Plackett, R. L., Some generalizations in the multifactorial design, *Biometrika*, Vol. 33, 1946.
[7] Plackett, R. L., and J. P. Burman, The design of optimum multifactorial experiments, *Biometrika*, Vol. 33, 1946.

## Chapter 13

GENERAL INCOMPLETE BLOCK DESIGNS

13.1   Introduction

In Chapter 10 we discussed incomplete block designs for factorial treatment experiments. Block sizes equal to the number of levels of some factor or multiple thereof result in a certain balance and make the analysis of those designs straightforward. Also, certain block compositions preserved information on treatment contrasts of most interest while sacrificing information on the treatment contrasts of lesser interest. As always, a researcher's knowledge of treatments and experimental materials is invaluable at the design stage of determining block sizes and compositions.

Now we consider IB designs having general treatments and block compositions. In the remainder of this chapter we assume a set of t unstructured treatments in an experimental situation that demands the use of incomplete blocks each containing k < t EU. We will not discuss experiments having variable block sizes; such material is beyond the scope of this book. An applied discussion, including an example, may be found in the article by Pearce [2]. For a more mathematical treatment, Kempthorne [1] may be consulted.

In any IB design, block and treatment effects are confounded. That is, contrasts of block effects are not orthogonal to certain contrasts of treatment effects. With factorial treatments the confounded effects can be identified (see Chapter 10.) This cannot be done in general IB designs because meaningful contrasts rarely exist among unstructured treatments.

As we will see in Section 13.4, the calculations for a general IB design are fairly extensive. This is a direct result of the confounding and therefore the need to remove block effects from treatment contrasts. Specific (new) calculations are required to accomplish this. Some simplifications occur when well-defined block-treatment patterns are present. Two special patterns will be considered in the next 2 chapters.

13.2   Linear Model and Assumptions for General IB Designs

Many of the ideas and concepts are easier to comprehend if we view a general incomplete block design as a two-factor experiment with unequal numbers but without interaction of the block and treatment factors. The specific treatments appearing in a given incomplete block are a function of the randomization process. To account for the particular block-treatment combinations appearing in the conducted experiment, an extra index is required. With these facts in mind, we now give the linear model for a general incomplete block design:

$$(13.1) \qquad y_{ijg} = \mu + \tau_i + \rho_j + \epsilon_{ijg}; \begin{array}{l} i = 1, 2, \text{---}, t \\ j = 1, 2, \text{---}, b \\ g = n_{ij} \end{array}$$

where $n_{ij} = 0$ if the i-th treatment does not appear in the j-th block and $n_{ij} = 1$ if the i-th treatment does appear in the j-th block. The notation y_{ij0} indicates non-existent observations which may be ignored in the analysis.

The assumptions accompanying the model (13.1) are the usual ones:

   (i) the ϵ's are independent and identically distributed with mean of zero and variance of σ^2_ϵ. For inferences, normality of ϵ's also would be assumed.
  (ii) Usually the τ's would be fixed effects.
 (iii) The ρ's may be fixed or random components.

We will say more about the third assumption when we consider the analysis in the next section.

13.3   Analyses of General IB Designs

Two general methods of analyzing incomplete block designs have been developed. Either fixed or random block effects may be assumed for the first method. Random block effects are required for the second.

The original and simplest method is known as an *intrablock analysis*. As the name suggests, only comparisons within blocks are used in the analysis. This method was introduced in 1936 by Yates [4] and is merely the usual least-squares analysis. An intrablock analysis may be completed whether block effects are assumed fixed or random. (See also Section 13.5)

The second and more extensive analysis, also introduced by
Yates [5] and [6], is known as *the analysis with recovery of
interblock information*. Because of block-treatment confounding,
certain contrasts among blocks contain information about the
treatment effects. Yates pointed out that the intrablock analysis
ignores this "interblock information." The analysis with recovery
of interblock information is simply the combination of estimates
obtained from "within block" and "among block" contrasts.   This
combination requires random block effects and is done according
to the principles of combining estimates, see Theorem 1.4. We
will discuss this *combined analysis* only for the special designs
of the next two chapters.

13.4   The Intrablock Analysis

The linear model appears in (13.1). Then, the sum of squares to
be minimized for a least-squares analysis is

(13.2) $\underset{ijg}{\Sigma\Sigma\Sigma}\ e^2_{ijg} = \underset{ijg}{\Sigma\Sigma\Sigma}(y_{ijg} - \hat{\mu} - \hat{\tau}_i - \hat{\rho}_j)^2$

The following normal equations result from this minimization:

$$N_{..}\hat{\mu} + \underset{i}{\Sigma} N_{i.}\,\hat{\tau}_i + \underset{j}{\Sigma} N_{.j}\hat{\rho}_j = y_{...}$$

(13.3) $N_{i.}\,\hat{\mu} + N_{i.}\,\hat{\tau}_i + \underset{j}{\Sigma} n_{ij}\hat{\rho}_j = y_{i..}$ for i = 1, 2, ---, t

$$N_{.j}\hat{\mu} + \underset{i}{\Sigma} n_{ij}\hat{\tau}_i + N_{.j}\hat{\rho}_j = y_{.j.}$$ for j = 1, 2, ---, b

where

$$N_{i.} = \underset{j}{\Sigma} n_{ij}$$

(13.4) $N_{.j} = \underset{i}{\Sigma} n_{ij}$

$$N_{..} = \underset{i}{\Sigma} N_{i.} = \underset{j}{\Sigma} N_{.j}$$

With our restriction of equal block sizes and n_{ij} = 0, 1, we have

$N_{i.}$ = Number of blocks containing the i-th treatment

(13.5) $N_{.j}$ = number of treatments in the j-th block = k

$N_{..}$ = total number of observations

These provide very little simplification of the normal equations, only the values of $N_{.j}$ may be replaced by k.

Note that (13.3) represents a system of b + t + 1 equations in b + t + 1 unknowns: $\hat{\mu}$, the $\hat{\tau}$'s, and the $\hat{\rho}$'s. It is easy to show that the last b equations sum to the first equation, and so do the t equations involving $y_{i..}$ on the right hand side. This means that the system (13.3) does not have a unique solution. We can obtain a unique solution though, if we impose two conditions on the estimates. A convenient choice is

(13.6) $\sum_{i} \hat{\tau}_i = 0$ and $\sum_{j} \hat{\rho}_j = 0$

Unfortunately, only the first equation in (13.3) simplifies under these conditions. Consequently, solutions for the $\hat{\tau}$'s and $\hat{\rho}$'s are not easily obtained (imposing any other conditions would not help much, if any.) We will only indicate the steps used in obtaining the solutions and merely state the final expression for estimating the treatment parameters. Each of the last b equations is solved for $\hat{\rho}_j$ which is then substituted into the t equations containing the sum, $\sum_{j} n_{ij}\hat{\rho}_j$. This gives the system of t equations in t unknowns:

(13.7) $(N_{i.} - \sum_{j} n^2_{ij}/N_{.j})\hat{\tau}_i - \sum_{h}\sum_{j} n_{ij}n_{hj}\hat{\tau}_h/N_{.j} = Q_i$
$\phantom{(13.7)   (N_{i.} - \sum_{j} n^2_{ij}/N_{.j})\hat{\tau}_i - \sum_{h}}_{h \neq i}$

where

(13.8) $Q_i = y_{i..} - \sum_{j} n_{ij}y_{.j.}/N_{.j}$

Because $\sum_{i} Q_i = 0$, the system (13.7) does not have a unique solution unless we impose a condition on the estimates. Usually, $\sum_{i} \hat{\tau}_i = 0$ is used.

While the system (13.7) may be difficult to solve in the general case, a number of simplifications usually can be made for specific types of designs. In particular, this will be true for the classes of designs discussed in the next two chapters. From equal block sizes and $n_{ij} = 0, 1$, we have $N_{.j} = k$ and

$$\sum_j n_{ij}^2 = \sum_j n_{ij} = N_i.$$

which may be incorporated in the equations of (13.7). This gives the following simplified form for the equations to estimate the treatment parameters:

(13.9) $$(1 - 1/k)N_{i.}\,\hat{\tau}_i - (1/k) \sum_h \sum_{\substack{j \\ h \neq i}} n_{ij}n_{hj}\hat{\tau}_h = Q_i$$

where

(13.10) $$Q_i = y_{i..} - \sum_j n_{ij}\bar{y}_{.j}.$$

For later usage let us introduce

(13.11) $$\sum_j n_{ij}n_{hj} = \lambda_{ih}$$

and

(13.12) $$B_i = \sum_j n_{ij}y_{.j}.$$

Because $n_{ij}n_{hj} = 1$ only when both treatments labeled i and h appear in the j-th block, we see that λ_{ih} represents the number of times that treatments i and h occur together in incomplete blocks. The quantity B_i represents the sum of all block totals in which the i-th treatment occurs.

Let us now discuss the quantities specified in (13.10). For each i, the value of Q_i is obtained by subtracting a certain sum from the i-th treatment total. Recalling that $n_{ij} = 1$ if the i-th treatment appears in the j-th block (and $n_{ij} = 0$ otherwise), we see that $\sum_j n_{ij}\bar{y}_{.j.} = B_i/k$ represents the sum of all block means for the blocks in which the i-th treatment occurs. This subtraction has the net effect of removing the block effects. Remember that we eliminated the $\hat{\rho}$'s from the equations on the left hand side when we set up (13.7).

Example 13.1

Let us modify the RCB design of Example 6.3 by eliminating one of the t = 4 treatments (bleaching agents) per block. This gives us k = 3 and b = 5. The resulting IB design is:

		Block					
		1	2	3	4	5	$y_{i..}$
Agent, i	1	21		25	18	22	86
	2		38	27	17		82
	3	16	25		18	21	80
	4	28	35	27		24	114
$y_{.j.}$		65	98	79	53	67	362

Table of n_{ij} and related values

Block, j		1	2	3	4	5	$N_{i.}$
Agent, i	1	1	0	1	1	1	4
	2	0	1	1	1	0	3
	3	1	1	0	1	1	4
	4	1	1	1	0	1	4
$N_{.j}$		3	3	3	3	3	15 = N..

Using $Q_i = y_{i..} - \sum_j n_{ij} y_{.j.}/3$

$$Q_1 = 86 - (65 + 79 + 53 + 67)/3 = -2$$

$$Q_2 = 82 - (98 + 79 + 53)/3 = 16/3$$

$$Q_3 = 80 - (65 + 98 + 53 + 67)/3 = -43/3$$

$$Q_4 = 114 - (65 + 98 + 79 + 67)/3 = 11$$

In the equations (13.9), the expression $\sum\limits_{j} n_{ij}n_{hj}$ is calculated
from the products of corresponding elements in rows h and i of
the "table of n_{ij} values". The resulting equations (13.9) are

$$(8/3)\hat{\tau}_1 - (2\hat{\tau}_2 + 3\hat{\tau}_3 + 3\hat{\tau}_4)/3 = -2$$

$$(6/3)\hat{\tau}_2 - (2\hat{\tau}_1 + 2\hat{\tau}_3 + 2\hat{\tau}_4)/3 = 16/3$$

$$(8/3)\hat{\tau}_3 - (3\hat{\tau}_1 + 2\hat{\tau}_2 + 3\hat{\tau}_4)/3 = -43/3$$

$$(8/3)\hat{\tau}_4 - (3\hat{\tau}_1 + 2\hat{\tau}_2 + 3\hat{\tau}_3)/3 = 11$$

Multiplying each equation by 3 to clear the fractions, and then
rearranging the terms of the left hand side leads to the system:

$$8\hat{\tau}_1 - 2\hat{\tau}_2 - 3\hat{\tau}_3 - 3\hat{\tau}_4 = -6$$

$$-2\hat{\tau}_1 + 6\hat{\tau}_2 - 2\hat{\tau}_3 - 2\hat{\tau}_4 = 16$$

$$-3\hat{\tau}_1 - 2\hat{\tau}_2 + 8\hat{\tau}_3 - 3\hat{\tau}_4 = -43$$

$$-3\hat{\tau}_1 - 2\hat{\tau}_2 - 3\hat{\tau}_3 + 8\hat{\tau}_4 = 33$$

Note that both left and right hand sides of this last system of
equations add to zero. We impose the condition $\sum\limits_{i} \hat{\tau}_i = 0$ which
allows us to replace $\hat{\tau}_4$ by $(- \hat{\tau}_1 - \hat{\tau}_2 - \hat{\tau}_3)$. We need to use only
3 (any 3) of the four equations. The reduced system is

$$11\hat{\tau}_1 + \hat{\tau}_2 \qquad = -6$$

$$8\hat{\tau}_2 \qquad = 16$$

$$\hat{\tau}_2 + 11\hat{\tau}_3 = -43$$

which may be easily solved and gives

$$\hat{\tau}_1 = -0.73 \qquad \hat{\tau}_2 = 2 \qquad \hat{\tau}_3 = -4.09$$

$$\hat{\tau}_4 = - \hat{\tau}_1 - \hat{\tau}_2 - \hat{\tau}_3 = 2.82$$

These estimates compare quite well with those found earlier in
Example 6.3. ▓

We have just spent considerable time on the estimation of treatment parameters in the general IB design. These estimates, adjusted for block effects, play a very important role in the analysis of variance. It can be shown that the sum of squares for "treatments eliminating block effects" is given by

$$(13.13) \qquad \text{SS(Trts. elim. blocks)} = \sum_i \hat{r}_i Q_i$$

For brevity, we label this quantity SS(Trts. adj.). Under the assumptions stated earlier in Section 13.2, this sum of squares is independent of the ordinary block sum of squares, denoted in IB analyses as sum of squares for blocks "ignoring treatments" and given by

$$(13.14) \qquad \text{SS(Blocks ign. trts.)} = \frac{1}{k} \sum y^2_{.j.} - C$$

where

$$(13.15) \qquad C = y^2_{...} / N_{..}$$

is the general correction term. For brevity, we will label this quantity SS(Blocks unadj). The total sum of squares is calculated according to

$$(13.16) \qquad \text{SSY} = \sum\sum\sum_{ijg} y^2_{ijg} - C$$

Finally, the difference between the total sum of squares and the two sums of squares defined in (13.13) and (13.14) is called the *intrablock error sum of squares;* that is,

$$(13.17) \qquad \text{SSE} = \text{SSY} - \text{SS(Blocks unadj.)} - \text{SS(Trts. adj.)}$$

We summarize these calculations by giving the ANOVA for the intrablock analysis of a general IB design.

Table 13.1   Intrablock ANOVA for a General IB Design

Source	df	SS	MS
Blocks (unadj.)	b-1	From (13.16)	
Treatments (adj.)	t-1	From (13.13)	MST
Intrablock Error	N..-b-t+1	SSE	MSE
Total	N..-1	SSY	

Under the assumptions discussed earlier in Section 13.2, a test of hypothesis of equal treatment parameters may be made from entries in the ANOVA table. The hypothesis is

(13.18) H_0: $\tau_1 = \tau_2 = \cdots = \tau_t$

and the test statistic is

(13.19) $F = MST/MSE$

which has the F distribution with t-1 and N..-b-t+1 degrees of freedom. The critical region corresponds to the upper tail area of this F distribution.

Example 13.2

We now complete the intrablock analysis for the IB design given in the previous example. The sums of squares are:

$$C = (362)^2/15 = 8736.3$$

$$SS(\text{Blocks unadj.}) = [(65)^2 + (98)^2 + \cdots + (67)^2)]/3 - C$$
$$= 27368/3 - 8736.3$$
$$= 9122.7 - 8736.3 = 386.4$$

$$SS(\text{Trts. adj.}) = (-0.73)(-2) + (2)(16/3)$$
$$+ (-4.09)(-43/3) + (2.82)(11)$$

$$= 101.8$$

$$SSY = (21)^2 + (16)^2 + \cdots + (24)^2 - C$$
$$= 9296 - 8736.3$$
$$= 559.7$$

$$SSE = 559.7 - 386.4 - 101.8 = 71.5$$

ANOVA for IB Design of Bleaching Agents

Source	df	SS	MS
Blocks unadj.	4	386.4	
Treatments adj.	3	101.8	33.93
Intrablock error	7	71.5	10.21
Total	14	559.7	

To test H_0: $\tau_1 = \tau_2 = \tau_3 = \tau_4 = 0$, we calculate

$$F_{3,7} = 33.93/10.21 = 3.32$$

which gives an observed significance level of $P > 0.1$. ✳

13.5   Criteria for Using Intrablock and Interblock Information

The statements in this section are applicable to any incomplete block design. Reference will be made to this material as other designs are discussed.

There are situations where even an intrablock analysis should not be undertaken. Unfortunately, one must complete most of the intrablock analysis before this determination can be made.

From the ANOVA for a general IB design, Table 13.1, the total sum of squares was subdivided (partitioned orthogonally) as follows:

(13.20)   SSY = SS(Bks. unadj)+SS(Trts. adj)+SS(Intrablock Error)

In Section 13.2, we remarked that an IB design may be viewed as a two-factor experiment without interaction. Now let the roles of the block and treatment factors be interchanged---so that "blocks would be adjusted but treatments would not". An intrablock analysis for this situation would have the following subdivision of the total sum of squares:

(13.21)   SSY = SS(Bks. adj)+SS(Trts. unadj)+SS(Intrablock error)

Equating the right hand sides of (13.20) and (13.21) gives the following basic identity for incomplete block designs:

(13.22)   SS(Blocks adj) + SS(Trts. unadj.)

$$= \text{SS(Blocks unadj.)} + \text{SS(Trts adj.)}$$

This identity allows us to easily set up the ANOVA table for a "second type" of intrablock analysis. All we need to calculate is the sum of squares for treatments unadjusted for block effects; that is, ignoring the incomplete block structure. Thus,

(13.23) $\text{SS(Trts. unadj)} = \sum_i (y_{i..}^2 / N_{i.}) - C$

Then using (13.22) to calculate SS(Blocks adj.), we have

Table 13.2 Intrablock ANOVA for Basic BIB Design

Source	df	SS	MS	EMS
Blocks adj.	b-1	From (13.22)	MSB	$\sigma^2_\epsilon + \dfrac{t(r-1)}{b-1}\sigma^2_\rho$
Treatments unadj.	t-1	From (13.23)	---	
Intrablock error	rt-b-t+1	From (13.20)	MSE	σ^2_ϵ
Total	rt-1	SSY		

When MSB < MSE, Rao [3] recommends that an intrablock analysis not be performed. Instead, the experiment should be analyzed as a CR design, or as an RCB design if the incomplete blocks are arranged in replicate groups (as in Example 10.3.)

The analysis with recovery of interblock information requires random block effects. This is necessary so that intrablock and interblock estimates can be combined according to Theorem 1.4. But such a combined analysis should be undertaken only when sufficient degrees of freedom are available for estimation of block variability, of which MSB plays a leading role (see the EMS in Table 13.2). To insure that MSB is a reliable estimator, Yates [6] and Kempthorne [1] recommend that it be based on at least 10 degrees of freedom while Rao [3] recommends at least 12.

The interblock estimates would never be used by themselves, but only in combination with intrablock estimates. Accordingly, the recovery of interblock information would be considered only after establishing that an intrablock analysis should be done.

In summary then, for any incomplete block design, an intrablock analysis should be performed only when MSB > MSE. And a combined analysis, one with recovery of interblock information, should be performed only when one has random block effects and

(13.24)
$$MSB > MSE$$
$$\text{df on MSB} > 10 \text{ (or 12)}.$$

PROBLEMS

1. A plant scientist was conducting an experiment to gather preliminary information on the yield of 5 new varieties of wheat. Originally, the design was an RCB with 4 blocks. But during the growing season, a wind storm damaged 4 plots. Yields, y, were obtained from the following plots:

	Variety				
Block	1	2	3	4	5
1	x	x		x	x
2	x		x	x	x
3	x	x	x	x	
4	x	x	x		x

(a) Give the linear model for this experiment. Explain terms used and specify ranges on subscripts.
(b) What assumptions are necessary for an intrablock analysis of this experiment?
(c) Give the sources and df of an intrablock ANOVA for this experiment.

2. An engineer performed an accelerated life test experiment on a particular electronic component made by four different companies. Six test racks were purchased for the experiment. Four components, one from each company, were randomly placed in each test rack. Then, the test racks were powered (at various locations throughout the laboratory, depending upon available space and power sources.) After 500 hours, one component had not failed on each rack. These were located at the same position of each test rack. An examination revealed that these positions malfunctioned, sometime after being powered up. All other positions were functioning properly at the end of experiment; that is, after 500 hours. Times to failure, y, were obtained as indicated:

Test Rack	Company			
(Block)	1	2	3	4
1		x	x	x
2	x	x		x
3	x		x	x
4	x	x	x	
5		x	x	x
6	x		x	x

(a) Give the linear model for this experiment. Explain terms used and specify ranges on subscripts.
(b) What assumptions are necessary for an intrablock analysis of this experiment?
(c) Give the sources and df of an intrablock ANOVA for this experiment.

3. The plant scientist, of Problem 1, recorded the following yields, y, in pounds, from the undamaged plots:

			Variety		
Block	1	2	3	4	5
1	58	65		49	55
2	51		44	42	52
3	54	60	42	47	
4	60	68	47		56

(a) Set up the table of n_{ij} and related numbers.

(b) Set up the normal equations (13.9) for a least squares analysis of this data.

4. The engineer, Problem 2, recorded the following times to failure, y, in hours:

		Company		
Rack	1	2	3	4
1		344	386	321
2	372	355		308
3	388		402	330
4	369	370	393	
5		368	379	315
6	376		391	326

(a) Set up the table of n_{ij} and related numbers.

(b) Set up the normal equations (13.9) for a least squares analysis of this data.

5. Refer to the wheat experiment of Problems 1 and 3:

(a) Set up the reduced normal equations by imposing the condition $\Sigma \; \hat{\tau}_i = 0$.

(b) Estimate the treatment parameters from the system of reduced normal equations.

6. Refer to the life test experiment of Problems 2 and 4:

(a) Set up the reduced normal equations by imposing the condition $\Sigma \; \hat{\tau}_i = 0$.

(b) Estimate the treatment parameters.

7. Refer to the wheat experiment of Problems 1, 3 and 5:

(a) Set up the intrablock ANOVA table.
(b) Make the intrablock test of zero treatment effects.

8. Refer to the life test experiment of Problems 2, 4, and 6:

(a) Set up the intrablock ANOVA table.
(b) Make the intrablock test of zero treatment effects.

9. A scientist comes to you for help in designing an experiment.
    He is investigating four treatments which have an equally-
    spaced, gradient structure.   The only thing he is interested
    in is whether the mean responses to these four treatments are
    linearly related.   (He wants maximum information on this.)
    Enviromental and climatic conditions are certain to have an
    effect on the response of interest so days must be considered
    as a blocking factor.   Unfortunately the nature of the treat-
    ments is such that only two of the treatments can occur on
    any one day.   He has no trouble obtaining up to 20 pairs of
    homogeneous EU to use for the experiment.

    For simplicity, suppose the 4 treatments are:

    T_1 1 unit

    T_2 2 units

    T_3 3 units

    T_4 4 units

    Design an experiment for this scientist.   Give one replicate
    of the treatment and indicate necessary randomization(s).

10. Use the information given in the previous problem and design
    an experiment for the scientist if his only interest is in
    knowing whether the mean responses have

    (a) a quadratic relationship
    (b) both a linear and a quadratic relationship

REFERENCES

[1] Kempthorne, O., *Experimental Design*, Wiley, New York, 1952.
[2] Pearce, S. C., Experimenting with blocks of natural size,
     Biometrics, Vol. 20, 1964.
[3] Rao, C. R., General methods of analysis for incomplete block
     designs, *Jour. Amer. Stat. Assoc.*, Vol.42, 1946.
[4] Yates, F., Incomplete randomized blocks, *Ann. Eugenics*,
     Vol. 7, 1936.
[5] Yates,F., The recovery of inter-block information in variety
     trials arranged in three-dimensional lattices, *Ann. Eugenics*
     Vol. 9, 1939.
[6] Yates, F., The recovery of inter-block information in bal-
     anced incomplete block designs, *Ann. Eugenics*, Vol. 10, 1940.

## Chapter 14

BALANCED INCOMPLETE BLOCK DESIGNS

14.1 Introduction

The previous chapter contains basic ideas and concepts of general
incomplete block designs. This material will be applied now to a
special class of designs known as *balanced incomplete block
designs*. Before formally defining them, let us recall some facts
which will be helpful in understanding the "balanced" aspect.

In completely confounded designs with factorial treatments, we
sacrificed information on certain treatment contrasts so that
full information would be available on contrasts of most interest
to us. The amount of information governs the precision (variance)
with which a contrast is estimated. Partially confounded designs
allow some information on all contrasts, provided that sufficient
replications are used. For general incomplete block designs,
there might be many classes of contrasts, each class estimated
with a different amount of information.

Two reasons may be cited for having an interest in classes of
designs having some type of balance: (i) the number of different
information measures might be kept to a minimum, and (ii) the
amount(s) of information might be specified in some manner.

Balanced incomplete block designs have the feature that the
difference of every pair of treatment parameters is estimated
with the same amount of information; that is, estimated with the
same variance.

14.2   Balanced Incomplete Block (BIB) Designs

We assume a set of t treatments are to be investigated but blocks of size t are not available.   Rather, homogeneous blocks of size k < t exist,   or can be constructed.   Thus,   an incomplete block design   is   mandatory but we may be able to choose   from   designs having   different numbers   of blocks and block   compositions   (and therefore   having different balance features.) We now define   the first special class of incomplete block designs.

Definition 14.1

A *balanced incomplete block (BIB) design* is an experimental setup   where t treatments are to be allocated to b blocks in such a way that:

(i) Each block contains k < t treatments.
(ii) Each treatment appears in exactly r blocks.
(iii) Every pair of treatments   occurs together in λ blocks; that   is,   every   pair of treatments appears   together equally often.

From   Definition 14.1,   it follows that $\lambda < r < b$ must   hold   for every BIB design, otherwise every treatment would appear in every block.

Example 14.1

For t = 4 treatments and blocks of size k = 3,   a BIB design can be constructed with $\lambda = 2$, b = 4 and r = 3. Letting treatments be denoted by 1, 2, 3, 4, the block compositions would be

Block	Treatments		
1	1	2	3
2	1	2	4
3	1	3	4
4	2	3	4

We see that each treatment appears 3 times and that each of the 6 pairs of treatments occurs twice within blocks. ※

Example 14.2

For t = 4 treatments and blocks of size k = 2,   a BIB design   can be constructed with $\lambda = 1$,   b = 6 and r = 3.   Letting treatments be denoted by 1, 2, 3, 4, the block compositions would be

Block	Treatments
1	1,2
2	3,4
3	1,3
4	2,4
5	1,4
6	2,3

Let us point out how this BIB is the same and how it differs from the one of the previous example. The number of treatments is the same but more blocks are required. The total number of EU is the same for the two designs but every pair of treatments occurs together only once in the present BIB compared to twice in the previous design. This is the price we must pay for smaller block sizes.

Note that this BIB design is the same one given in Example 10.3 where a factorial structure was assumed for the treatments. In the earlier example though, blocks were arranged in complete replicates. This is not assumed here (but see Section 14.6). ❋

Example 14.3

Suppose the following 4 incomplete blocks are used to investigate t = 6 treatments.

Block	Treatments		
1	1	2	4
2	1	3	6
3	2	3	5
4	4	5	6

Even though each treatment occurs in two blocks this is not a BIB design. Twelve pairs of treatments each occur once within the four blocks but three pairs of treatments, (1,5), (2,6), and (3,4), do not occur at all within blocks. ❋

Any BIB design can be characterized by five numerical values: t, k, b, r, and λ. Two inequalities were indicated earlier for these values; namely, k < t and λ < r < b. Various equalities also hold among these values. For any BIB design, two important equalities are

(14.1) $rt = bk$

and

(14.2) $\lambda(t-1) = r(k-1)$.

The left hand side of (14.1) represents the number of times all treatments appear in the experiment---this number must equal the totality of EU which the right hand side of (14.1) represents. To establish the equality in (14.2), consider only those blocks where a specific treatment occurs, say treatment 1. Besides treatment 1 there are $\lambda(t-1)$ other treatment units in these blocks because each of the other t-1 treatments must occur λ times with treatment 1. But treatment 1 must appear in r blocks each containing k-1 units other than the one on which treatment 1 occurs. This product, $r(k-1)$, represents the number of units available for the $\lambda(t-1)$ "other treatments".

Example 14.4

For the BIB design in Example 14.1, we had $t = 4$, $b = 4$, $k = 3$, $r = 3$ and $\lambda = 2$. From these values we have

$$rt = 12 \qquad\qquad \lambda(t-1) = 6$$
$$bk = 12 \qquad\qquad r(k-1) = 6$$

and therefore the equalities in (14.1) and (14.2) hold. ✳

Example 14.5

For the incomplete block design presented in Example 14.3, four of the numerical values are $t = 6$, $b = 4$, $k = 3$, and $r = 2$.
    The equality in (14.1) holds; that is, $rt = 12 = bk$. But because some pairs of treatments do not occur within blocks, while some pairs occur once, there is not a constant value of λ. Hence, the equality in (14.2) cannot be satisfied; this is not a BIB design. ✳

14.3   Some Design Considerations

For any values of t and k, a BIB design always exists. One simply continues to add blocks of size k until the conditions of Definition 14.1 hold. A major disadvantage is that BIB designs formed in this manner may get prohibitively large.
    By adding additional blocks, one may be able to construct a

BIB design from an existing IB design. Block compositions of the existing design might be inappropriate for conversion to a BIB design. Of course, one can always change the block compositions at the design stage, if the changes necessary to give a BIB design are readily discernible.

Example 14.6

Suppose we would like to convert the incomplete block design of Example 14.3 to a BIB design. Beginning with $t = 6$ treatments and blocks of size $k = 3$, we note that the equality in (14.1) implies $3b = 6r$, or $b = 2r$. To satisfy the equality in (14.2), we would need $\lambda = r(k-1)/(t-1) = 2r/5$. The smallest value of r giving an integer value of λ is $r = 5$. Thus, the BIB design would have $t = 6$, $k = 3$, $r = 5$, $b = 10$ and $\lambda = 2$.

The addition of the following 6 blocks would convert the IB design of Example 14.3 to a BIB design:

Block	Treatments		
5	1	2	6
6	1	3	5
7	1	4	5
8	2	3	4
9	2	5	6
10	3	4	6

The researcher would need to decide if 10 blocks of size 3 can be obtained and if an experiment of this size is practical. ※

For given values of t and k, the equalities in (14.1) and (14.2) can be used to determine possible values of b, r, and λ for a BIB design. Not all values obtained in this manner lead to BIB designs; multiples of b, r, and λ may be required. By trial and error, block compositions would be formed to construct the BIB design. Generally, this is a difficult task.

Example 14.7

An experimenter wishes to construct a BIB design for $t = 16$ treatments in blocks of $k = 6$ units each. By (14.1)

$$3b = 8r \qquad \text{or} \qquad b = 8r/3$$

and by (14.2)

$$15\lambda = 5r \qquad \text{or} \qquad \lambda = r/3$$

Thus, r must be some multiple of 3 for integer values of b and λ. Using r = 3 gives values of b = 8 and λ = 1. But a BIB design does not exist for these parameters. Instead, using r = 6 leads to a BIB design with parameters

$$t = 16, \; k = 6, \; r = 6, \; b = 16, \; \lambda = 2. \; ✳$$

Systematic methods of constructing BIB designs have been studied quite extensively by Bose [2], Fisher [4], Hussain [6] John [7], and Bhattacharya [1]. Various aspects of finite geometries are used in establishing the existence as well as construction of BIB designs. An in-depth discussion of this material is beyond the scope of this book. But to indicate some basic ideas, we will discuss several methods of constructing BIB designs which utilize Latin Square designs and their extensions (Section 8.14).

First, a BIB design for p treatments can be constructed from a pxp LS design. Simply omit any one row and ignore the row and column blockings (but see Section 14.8.) Each column of the remaining (p-1)xp array forms an incomplete block of k = p-1 treatments. Because every treatment appears once in every row of an LS design, the BIB design will contain every treatment r = p-1 times. And for the equality in (14.2) to be satisfied, every pair of treatments occurs together in incomplete blocks λ = p-2 times. In summary then, omitting any one row of an LS design leads to a BIB design with parameters

(14.3) $t = b = p, \qquad r = k = p\text{-}1, \text{ and } \qquad \lambda = p\text{-}2.$

Example 14.8

A BIB design for t = 5 treatments in blocks of size k = 4 can be constructed from the following 5 x 5 LS design:

E	C	B	A	D
A	E	D	C	B
B	D	A	E	C
C	B	E	D	A
D	A	C	B	E

Let us choose to omit the third row of the above array and place the remaining 4 treatments of each column in a block. The result is a BIB design with blocks

Block	Treatments
1	E A C D
2	C E B A
3	B D E C
4	A C D B
5	D B A E

with parameters $t = b = 5$, $r = k = 4$, and $\lambda = 3$. ✸

The above method of constructing a BIB design is most valuable for a small number of treatments. The block size, being one less than the number of treatments, quickly becomes too large.

We now consider two additional types of BIB designs that may be constructed from LS designs and their extensions: (i) for $t = p^2$ treatments in blocks of size $k = p$, and (ii) for $t = p^2 + p + 1$ treatments in blocks of size $k = p + 1$.

For the first type, an "extended" LS design of size $p \times p$ with $p-1$ languages is required. Simply number the cells with treatment numbers 1, 2, ---, p^2. Then, form $p + 1$ sets of p blocks each using: (i) rows, (ii) columns and (iii) letters of each language. A BIB design constructed in this way has parameters

$$(14.4) \qquad t = p^2, \quad r = p + 1, \quad k = p, \quad \text{and} \quad \lambda = 1 .$$

Example 14.9

A BIB design for $t = 9$ treatments in blocks of $k = 3$ units can be constructed from a 3 x 3 Graeco-Latin square design (because $k-1 = 2$ languages are required.) In the cells of the following GLS design, the numbers represent the treatments:

Aα1	Cβ2	Bγ3
Cγ4	Bα5	Aβ6
Bβ7	Aγ8	Cα9

The BIB design consists of 12 blocks ($p + 1 = 4$ sets of 3 blocks) formed from the rows, the columns, and letters of each language. The block compositions are

Block	Treatments			Block	Treatments		
1	1	2	3	7	1	6	8
2	4	5	6	8	3	5	7
3	7	8	9	9	2	4	9
4	1	4	7	10	1	5	9
5	2	5	8	11	2	6	7
6	3	6	9	12	3	4	8 ※

BIB designs for $t = p^2 + p + 1$ treatments in blocks of size k = p + 1 can be constructed in the following manner. Place any p+1 treatments in the first block and the remaining p^2 treatments in the cells of a pxp extended LS design having p-1 languages. Each treatment of the first block is used to generate a set of p additional blocks. One (treatment of the first block) is included with the treatments in each row of the LS, another with the treatments in each column of the LS, and a different one with the treatments associated with each letter of a given language of the LS. A BIB design constructed by this method has parameters

$$(14.5) \qquad t = b = p^2 + p + 1, \quad r = k = p + 1, \quad \lambda = 1$$

Example 14.10

A BIB design will be constructed for t = 13 treatments in blocks of size k = 4. Using (14.5) with p = 3, the remaining parameters are

$$b = 13, \ r = 4, \ \lambda = 1.$$

Let the first block contain treatments 1,2,3,4. We need a 3 x 3 extended LS design with p - 1 = 2 languages (that is, a basic GLS design), into whose cells we place the remaining 9 treatments:

Aα5	Cβ6	Bγ7
Cγ8	Bα9	Aβ10
Bβ11	Aγ12	Cα13

The 13 blocks for this BIB design would be:

Block	Treatments				Block	Treatments			
1	1	2	3	4	8	3	5	10	12
2	1	5	6	7	9	3	7	9	11
3	1	8	9	10	10	3	6	8	13
4	1	11	12	13	11	4	5	9	13
5	2	5	8	11	12	4	6	10	11
6	2	6	9	12	13	4	7	8	12
7	2	7	10	13					※

14.4   Linear Model and Assumptions for Basic BIB Designs

The linear model and assumptions for a general incomplete block design appear in Section 13.2. These are appropriate for basic BIB design where the incomplete blocks are arranged completely at random. We repeat the model here for reference purposes:

$$(14.6) \qquad y_{ijg} = \mu + \tau_i + \rho_j + \epsilon_{ijg}; \quad \begin{aligned} i &= 1, 2, ---, t \\ j &= 1, 2, ---, b \\ g &= n_{ij} \end{aligned}$$

where n_{ij} = 1 if the i-th treatment appears in the j-th block, and equals zero otherwise. The next 3 sections deal with basic BIB designs. In Section 14.7 we consider BIB designs where the incomplete blocks are arranged to form complete replicates of the treatments.

14.5   Intrablock Analysis for a Basic BIB Design

At this time it might be useful to review Section 13.5 where we discussed criteria (in particular, MSB > MSE) necessary for doing an intrablock analysis.

The results given for the intrablock amalysis of a general incomplete block design may be simplified for the BIB designs. The quantities in (13.5) and (13.11) are each constant for a BIB design; namely,

$$(14.7) \qquad \begin{aligned} N_{i.} &= r & N_{..} &= bk = rt \\ N_{.j} &= k & \sum_j n_{ij} n_{hj} &= \lambda \end{aligned}$$

Using these equalities, slight simplifications can be made to the normal equations (13.3). More importantly, the system of equations given in (13.7) and used for estimating the treatment parameters, takes on a particularly simple form. First we have

(14.8) $(r - r/k)\hat{\tau}_i - (\lambda/k) \sum_{h \neq i} \hat{\tau}_h = Q_i$

To solve this system uniquely, we impose the condition $\sum_h \hat{\tau}_h = 0$ from which we obtain

(14.9) $\sum_{h \neq i} \hat{\tau}_h = -\hat{\tau}_i$

Making this replacement in (14.8) gives an easily solved system:

(14.10) $(r - r/k + \lambda/k)\hat{\tau}_i = Q_i$

where

(14.11) $Q_i = y_{i..} - (1/k)B_i$

with $B_i = \sum_j n_{ij} y_{.j.}$ representing the sum of all block totals in which the i-th treatment appears. A final simplification can be made on the coefficient of $\hat{\tau}_i$ in (14.10) by using the equality $r(k-1) = \lambda(t-1)$. This gives intrablock estimates:

(14.12) $\hat{\tau}_i = (k/\lambda t)Q_i$

Example 14.11

A nutrition specialist studied the effect of six diets on weight gain of domestic rabbits. When rabbits reached four weeks of age, they were assigned to a diet. It was decided to block on litters of rabbits but from past experience about sizes of litters, it was felt that only 3 or 4 (uniform) rabbits could be selected from each available litter.
    A BIB design with t=6, k=3, b=10, r=5, and λ=2 was conducted and the following weight gains (in ounces) were obtained:

Block	Treatments and Observations			Block Totals
1	6 42.2	2 32.6	3 35.2	110.0
2	3 40.9	1 40.1	2 38.1	119.1
3	3 34.6	6 34.3	4 37.5	106.4
4	1 44.9	5 40.8	3 43.9	129.6
5	5 32.0	3 40.9	4 37.3	110.2
6	2 37.3	6 42.8	5 40.5	120.6
7	4 37.9	1 45.2	2 40.6	123.7
8	1 44.0	5 38.5	6 51.9	134.4
9	4 27.5	2 30.6	5 20.6	78.7
10	6 41.7	4 42.3	1 37.3	121.3

To estimate the treatment parameters according to (14.11) and (14.12), the following tabulations are made:

Treatment(i)	Total($y_{i..}$)	B_i	Q_i
1	211.5	628.1	2.13
2	179.2	552.1	-4.83
3	195.5	575.3	3.73
4	182.5	540.3	2.40
5	172.4	573.5	-18.77
6	212.9	592.7	15.33

Entries in the last two columns are calculated as

$$B_1 = 119.1 + 129.6 + 123.7 + 134.4 + 121.3 = 628.1$$

$$Q_1 = 211.5 - (628.1)/3 = 2.13$$

and so on. The treatment effects can now be obtained from the last column and are

$$\hat{\tau}_i = (3/12)Q_i: \quad 0.53, -1.21, 0.93, 0.60, -4.69, 3.83$$

We note that the values of Q_i as well as the estimates, $\hat{\tau}_i$, sum to zero, within rounding. ✳

The sums of squares for the intrablock analysis of any BIB design have the same expressions as the ones in (13.13) through (13.17), except that the divisor of the general correction term can be replaced by rt (or bk.) Clearly then, the ANOVA given in Table 13.1 is appropriate for the basic BIB design. And the test procedure described in the paragraph following Table 13.1 also is appropriate for testing equality of treatment parameters in the basic BIB design.

Example 14.12

We now complete the intrablock analysis for the rabbit nutrition experiment of the previous example. The overall correction term is

$$C = y^2../30 = (1154)^2/30 = 44390.5$$

The sums of squares calculations are

$$SS(\text{Blocks unadj.}) = (1/3)[(110)^2 + (119.1)^2 + \cdots + (121.3)^2] - C$$
$$= 45120.9 - 44390.5 = 730.4$$

$$SS(\text{Trts. adj.}) = 0.53(2.13) + (-1.21)(-4.83) + \cdots + 3.83(15.33)$$
$$= 158.6$$

$$SSY = (42.2)^2 + (32.6)^2 + \cdots + (37.3)^2 - C$$
$$= 45430.4 - 44390.5 = 1039.9$$

$$SS(\text{Intrablock error}) = 1039.9 - 730.4 - 158.6 = 150.9$$

Summarizing these calculations gives the intrablock ANOVA table:

ANOVA for Rabbit Nutrition Study

Source	df	SS	MS
Blocks (unadj.)	9	730.4	---
Diets (adj.)	5	158.6	31.7
Intrablock error	15	150.9	10.1
Total	29	1039.9	

To test equality of diet effects, the hypotheses are

$$H_0 : \tau_1 = \tau_2 = \cdots = \tau_6 = 0$$

$$H_1 : \text{Some } \tau\text{'s are not zero.}$$

The appropriate test statistic is the ratio of the diet and intrablock error mean squares. Under H_0 and the assumptions stated earlier, this ratio is an F variable with 5 and 15 degrees of freedom. Thus

$$F_{5,15} = 31.7/10.1 = 3.14$$

which has an observed level of significance of $P < 0.05$. Hence, we conclude that the diets produce different effects relative to growth. ▓

We now consider contrasts of the treatment parameters, say

(14.13) $\pi = \sum_i c_i \tau_i$

where $\sum_i c_i = 0$. An unbiased estimator of this contrast is

(14.14) $\hat{\pi} = \sum_i c_i \hat{\tau}_i = (k/\lambda t) \sum_i c_i Q_i$

Under the assumption that the residual components in model (14.6) are independent and identically distributed with zero mean and variance σ_ϵ^2, we can show that the variance of $\hat{\pi}$ has the form

(14.15) $\text{Var}(\hat{\pi}) = (k/\lambda t)^2 (\sum_i c_i^2) \sigma_\epsilon^2$

The unknown variance, σ^2_ϵ, would be estimated by the intrablock error mean square and used in forming an estimated variance of $\hat{\pi}$.

Inferences about a contrast π are straightforward under the assumption that the residual components are normally distributed. The Student's t distribution is appropriate. For testing

$$(14.16) \qquad H_0: \pi = 0 \qquad vs. \qquad H_1: \pi \neq 0$$

the test statistic is

$$(14.17) \qquad t = \frac{\hat{\pi}}{(k/\lambda t)\sqrt{MSE\Sigma c_i^2}}$$

which has degrees of freedom equal to those of MSE. The critical region for the alternative hypothesis stated in (14.16) would be a two-tailed region of the t distribution. For a one-sided alternative, the critical region would be the "matching" one-tailed region.

A two-sided confidence interval for the contrast $\pi = \sum_i c_i \tau_i$ having coverage probability of $1-\alpha$, has limits L and U given by

$$(14.18) \qquad L,U = (k/\lambda t)\sum_i c_i Q_i \pm t_{\alpha/2}(k/\lambda t)\sqrt{MSE\Sigma c_i^2}$$

where $t_{\alpha/2}$ is the upper $\alpha/2$ percentage point of a t distribution with degrees of freedom equal to those of MSE.

Example 14.13

For the rabbit nutrition study of Example 14.11, suppose we are told that diets 1, 2, and 3 contain a supplement which diets 4, 5, and 6 do not. Then we might be interested in comparing the first 3 diets (collectively) with the last 3 to determine if the supplement has an effect. The contrast of interest in this case would be

$$\pi = \tau_1 + \tau_2 + \tau_3 - \tau_4 - \tau_5 - \tau_6$$

and has an estimated value of

$$\hat{\pi} = (3/12)\sum_i c_i Q_i = 0.25(2.13-4.83+3.73-2.40+18.77-15.33)$$
$$= 0.52$$

To test that the supplement has no effect; that is, produces no greater weight gains, we use

$$H_0: \tau_1 + \tau_2 + \tau_3 - \tau_4 - \tau_5 - \tau_6 = 0$$

$$H_1: \tau_1 + \tau_2 + \tau_3 - \tau_4 - \tau_5 - \tau_6 \neq 0$$

$$t_{15} = 0.52/(3/12)\sqrt{10.1(6)}$$

$$= 0.27$$

Because the observed significance level is $P > 0.25$, we have no evidence that the supplement produces greater weight gains.

For illustrative purposes, let us construct a 90% CI for the contrast π. The limits are, by (14.18):

$$L,U = 0.52 \pm 1.75(3/12)\sqrt{10.1(6)}$$

$$= 0.52 \pm 3.41 \rightarrow -2.89; \; 3.93$$

The resulting CI for π contains both positive and negative values so it could well be that the supplement produces lower or greater weight gains (than no supplement). ※

14.6 Analysis with Recovery of Interblock Information: Basic BIB Designs

Conditions necessary for utilizing interblock information were discussed in Sections 13.3 and 13.5. Briefly, there should be at least 10 degrees of freedom on the mean square for "adjusted blocks", MSB, and MSB should exceed the intrablock error mean square, MSE.

As indicated previously, recovery of interblock information requires random block effects. Specifically, it is assumed that the ρ's are independent and identically distributed with zero mean and variance σ_ρ^2, and that the ρ's and ϵ's are independent. All other assumptions for the intrablock analysis are retained.

Yates [9] showed that certain contrasts among the incomplete blocks provide estimates of the treatment parameters in a BIB design. We first discuss these interblock esimates. Recalling that we used B_i to represent the total of all observations from blocks in which the i-th treatment appears, let us define

$$(14.19) \qquad \bar{B}. = (1/t) \sum_i B_i$$

Then, the interblock estimator of τ_i is

(14.20) $\qquad \tilde{\tau}_i = (B_i - \bar{B}.)/(r-\lambda)$

which is an unbiased estimator of τ_i.

Example 14.14

We assume random litter (block) effects for the rabbit nutrition study.   This would not be an unreasonable assumption unless, for example, we were using litters only from some specific genetic background and wished to make inferences about the diet effects only as they pertain to such litters.

The interblock estimates of the treatment parameters can be readily calculated from the information contained in Example 14.11.   We have   r = 5, λ = 2, and

Treatment(i)	B_i	$\tilde{\tau}_i = (B_i - \bar{B}.)/(r-\lambda)$
1	628.1	17.03
2	552.1	-8.30
3	575.3	-0.57
4	540.3	-12.23
5	573.5	-1.17
6	592.7	5.23

$$\bar{B}. = (628.1 + \cdots + 592.7)/6 = 577.0$$

We observe that these estimates, the $\tilde{\tau}_i$, are quite different from the intrablock estimates given in Example 14.7.   As we will see later, the variances of these estimates are considerably larger than for the intrablock estimates. ▨

Using all assumptions indicated at the beginning of this section, it is known (see Graybill [5] or Kempthorne [8], for example) that the intrablock and interblock estimators, given by (14.12) and (14.20) are independent, with variances

(14.21) $\qquad V_1 = \text{Var}(\hat{\tau}_i) = \dfrac{k(t-1)}{\lambda t^2}\sigma_\epsilon^2$

(14.22) $\qquad V_2 = \text{Var}(\tilde{\tau}_i) = \dfrac{k(t-1)}{t(r-\lambda)}(\sigma_\epsilon^2 + k\sigma_\rho^2)$

The conditions of Theorem 1.4 are satisfied so the best linear unbiased estimator of τ_i, using both intrablock and interblock information, is the combined estimator

$$(14.23) \qquad \hat{\tau}_i^* = \frac{V_2 \hat{\tau}_i + V_1 \tilde{\tau}_i}{V_1 + V_2}$$

Expressions (14.21) and (14.22) reveal that interblock estimators have a greater variance than the intrablock estimators. The unknown variance components, σ_ϵ^2 and σ_ρ^2, in expressions (14.21) and (14.22) are estimated from the "second" intrablock analysis, see Table 13.2. From the EMS of that table, we see that these estimates are

$$(14.24) \qquad \hat{\sigma}_\epsilon^2 = MSE \qquad \text{and} \qquad \hat{\sigma}_\rho^2 = \frac{b-1}{t(r-1)}(MSB - MSE)$$

Remember that MSB is the mean square for adjusted blocks and the earlier stated assumption MSB > MSE.

Example 14.15

Let us estimate the variance components for the rabbit nutrition experiment. From the ANOVA table in Example 14.12, we see that

$$\hat{\sigma}_\epsilon^2 = MSE = 10.1$$

To estimate σ_ρ^2, we need to calculate MSB, the mean square for "Blocks, adjusted". This can be obtained with the aid of identity (13.22) after calculating the sum of squares for "Treatments, unadjusted". We have

$$SS(\text{Trts. unadj.}) = (1/5)[(211.5)^2 + \cdots + (212.9)^2] - (1154)^2/30$$
$$= 44683.9 - 44390.5 = 293.4$$

$$SS(\text{Blocks adj.}) = 730.4 + 158.6 - 293.4$$
$$= 595.6$$

$$MSB = 595.6/9 = 66.2$$

then the estimate of σ_ρ^2 is

$$\hat{\sigma}^2_\rho = [9/6(4)](66.2\text{-}10.1) = 21.04.$$

We may now construct estimates of V_1 and V_2; namely,

$$\hat{V}_1 = [3(5)/2(36)](10.1) = 2.10$$

$$\hat{V}_2 = [3(5)/6(3)][10.1 + 3(21.04)] = 61.01$$

From these variance estimates, we see that interblock estimates are nearly 30 times as variable as the intrablock estimates. ▧

We can substitute the estimates of V_1 and V_2 into (14.23) to calculate the combined estimates. Equivalently, we can substitute for the estimates of the variance components from (14.24) into (14.23) to get the combined estimates in terms of MSB and MSE. After a fair amount of algebraic manipulation, we obtain

(14.25) $\hat{\tau}^*_i = (\bar{y}_{i..} - \bar{y}_{...}) + \omega[(t\text{-}k)(\bar{y}_{i..} - \bar{y}_{...}) - (t\text{-}1)(B_i - \bar{B}.)/r]$

where $\bar{y}_{i..}$ and $\bar{y}_{...}$ are "unadjusted" treatment and overall means, respectively, and

(14.26) $\omega = \dfrac{(b\text{-}1)(MSB - MSE)}{(t\text{-}k)(b\text{-}t)MSE + t(k\text{-}1)(b\text{-}1)MSB}$

is called a *weighting factor*.

Example 14.16

The combined estimates of treatment parameters will be calculated for the rabbit nutrition experiment. From the previous example, we have MSB = 66.2 and MSE = 10.1 giving

$$\omega = \frac{9(66.2 - 10.1)}{3(4)(10.1) + 6(2)(9)(66.2)}$$

From information given in Examples 14.11 and 14.14, we obtain $\bar{y}_{...} = 1154/30 = 38.47$ and tabulate

Treatment(i)	$\bar{y}_{i..}$	$\bar{y}_{i..} - \bar{y}_{...}$	$B_i - \bar{B}.$
1	211.5	3.83	51.1
2	179.2	-2.63	-24.9
3	195.5	0.63	- 1.7
4	182.5	-1.97	-36.7
5	172.4	-3.99	- 3.5
6	212.9	4.11	15.7

By (14.26), the combined estimates of treatment parameters are

$$\hat{\tau}_1^* = 3.83 + 0.0694[3(3.83)-5(51.1)/5]$$

$$= 3.83 + 0.0694\,(-39.61)$$

$$= 3.83 - 2.75 = 1.08$$

$$\hat{\tau}_2^* = -2.63 + 0.0694[3(-2.63)+24.9]$$

$$= -2.63 + 1.18 = -1.45$$

$$\hat{\tau}_3^* = 0.63 + 0.0694[3(0.63)+1.7]$$

$$= 0.63 + 0.25 = 0.88$$

$$\hat{\tau}_4^* = -1.97 + 0.0694[3(-1.97)+36.7]$$

$$= -1.97 + 2.14 = 0.17$$

$$\hat{\tau}_5^* = -3.99 + 0.0694[3(-3.99)+3.5]$$

$$= -3.99 - 0.59 = -4.58$$

$$\hat{\tau}_6^* = 4.11 + 0.0694[3(4.11)-15.7]$$

$$= 4.11 - 0.23 = 3.88$$

Earlier we calculated both intrablock and interblock estimators as well as estimates of their variances, so we also could have calculated the combined estimators from an expression similar to (14.23). For example, we have $\hat{\tau}_1 = 0.53$, $\tilde{\tau}_1 = 17.03$, $V_1 = 2.1$ and $V_2 = 61.01$ which gives, by (14.23),

$$\hat{\tau}_1^* = \frac{(61.01)(0.53) + (2.1)(17.03)}{2.1 + 61.01} = 1.08 \;\text{※}$$

We need to consider "adjusted treatment means" when a combined analysis is performed. The difference of any pair of combined estimates is equivalent to the difference of the corresponding adjusted treatment means. Instead of these differences, some researchers prefer to know the values of individual adjusted

treatment means.   These means may be obtained from the   combined
estimates   upon addition of $\bar{y}_{...}$,   the overall   unadjusted   mean.
Thus

$$(14.27) \qquad \text{Adj. } \bar{y}_{i..} = \hat{r}_i^* + \bar{y}_{...} = \frac{1}{r} y_{i..}^*$$

Variances   (actually standard deviations) of these adjusted means
are   of interest.   Unfortunately,   the exact variances cannot   be
determined because we used estimates of V_1 and V_2 in   calculating
the   combined   estimates   and the weighting   factor.   A   quantity
called the *effective error variance* often is used as a approximation (see Federer [3], for example.)   For the BIB designs, it is

$$(14.28) \qquad MSE^* = MSE[1 + (t-k)\omega]$$

This is just an increase of the intrablock error variation,   MSE,
to   account for variation introduced by the interblock estimates.
The intrablock error, MSE, is the appropriate variation when only
intrablock information is utilized. When based on both estimates,
the variance of any adjusted treatment mean is estimated by

$$(14.29) \qquad \hat{Var}(\text{adj. } \bar{y}_{i..}) = (1/r)MSE^*$$

We   now   discuss   a test of equal treatment   effects   when   a
combined   analysis   is performed.   The effective   error   variance
plays   a   role in this test.   Variation among treatments will   be
measured from the combined estimates,   or equivalently,   from the
adjusted means given in (14.27). The sum of squares for measuring
this variation is

$$(14.30) \qquad SS^*(\text{Trts. adj.}) = \frac{1}{r} \sum_i (y_{i..}^*)^2 - C$$

where C is the general correction term. Then, the mean square for
"treatments adjusted" under a combined analysis is

$$(14.31) \qquad MS^*(\text{Trts. adj.}) = SS^*(\text{Trts. adj.})/(t-1)$$

The test statistic for testing equal treatment effects under a combined analysis is

(14.32) $F = MS^*(\text{Trts. adj.})/MSE^*$

which has an approximate F distribution with $(t-1)$ and $(tr-t-b+1)$ degrees of freedom. The critical region corresponds to an upper tail area of this distribution. Note, this is not an exact test.

Example 14.17

Let us return to the rabbit nutrition study. From previous examples, we have $t=6$, $k=3$, $r=5$, $\lambda=2$, $MSE=10.1$, $\omega = 0.0694$, and $\bar{y}... = 38.47$. The adjusted treatment means are, using (14.27):

$$\text{Adj. } \bar{y}_{1..} = \hat{\tau}_1^* + \bar{y}... = 1.08 + 38.47 = 39.55$$

$$\text{Adj. } \bar{y}_{2..} = -1.45 + 38.47 = 37.02$$

$$\text{Adj. } \bar{y}_{3..} = 0.88 + 38.47 = 39.35$$

$$\text{Adj. } \bar{y}_{4..} = 0.17 + 38.47 = 38.64$$

$$\text{Adj. } \bar{y}_{5..} = 4.58 + 38.47 = 33.89$$

$$\text{Adj. } \bar{y}_{6..} = 3.88 + 38.47 = 42.35$$

The effective error variance is, by (14.28):

$$MSE^* = 10.1[1 + 3(0.0694)] = 12.2$$

The estimated variance of an adjusted treatment mean is

$$MSE^*/r = 2(12.2)/5 = 2.44.$$

We now make a test of equal treatment effects under a combined analysis. From the adjusted treatment means given above, we can calculate

$$SS^*(\text{Trts. adj.}) = (1/5)[(197.75)^2 + (185.1)^2 + (196.75)^2$$

$$+ (193.2)^2 + (169.45)^2 + (211.75)^2] - C$$

$$= 4459.10 - 44390.5$$

$$= 200.5$$

$$MS^*(\text{Trts. adj.}) = 200.5/5 = 40.1$$

The test statistic is an approximate F variable with 5 and 15 degrees of freedom, whose calculated value is

$$F = 40.1/12.2 = 3.29$$

which has an observed significance level of $P < 0.05$. Hence, we have evidence of different effects of the six diets on the growth of rabbits. ※

14.7.1   BIB Designs with Replicate Groups

In some experiments the incomplete blocks can be arranged in complete replicates of the treatments. This might occur if we have incomplete blocks at different locations, if the experiment is conducted over a series of days, and so on. An example will illustrate the experimental layout.

Example 14.18

Suppose a BIB design with $t=4$, $k=2$, $\lambda=1$, $b=6$ and $r=3$ had the following layout:

	Replicate 1		Replicate 2		Replicate 3	
Block 1	1	2	1	3	1	4
Block 2	3	4	2	4	2	3

Each treatment occurs within each replicate but in incomplete blocks of size 2. Compare the design with the design given earlier in Example 10.3. ※

The analysis of a BIB design in this category is analogous to that for the basic BIB with changes to reflect the presence of replicates. Now the partitioning of the total sum of squares includes sums of squares "due to Replicates" and "Blocks within Replicates". In the basic BIB design, these two sums of squares are combined as "sum of squares due to Blocks". This change is reflected likewise in the linear model. Assuming b* incomplete blocks within each replicate, we have

(14.33) $y_{mijg} = \mu + \pi_m + \tau_i + \rho_{jm} + \epsilon_{mijg}$;
$\begin{aligned} m &= 1,\ 2,\ \cdots,\ r \\ i &= 1,\ 2,\ \cdots,\ t \\ j &= 1,\ 2,\ \cdots,\ b* \\ g &= n_{ij} \end{aligned}$

where

π_m = effect due to the m-th replicate

(14.34) ρ_{jm} = effect due to the j-th block within the m-th replicate

The sums of squares to measure these effects are

(14.35) $SSR = \dfrac{1}{t} \sum_m y_{m\ldots}^2 - C$

(14.36) $SSBlocks(R) = \dfrac{1}{k} \sum_m \sum_j y_{m.j.}^2 - C - SSR$

where $C = y_{\ldots}^2/rt$ is the general correction term. Then the ANOVA table for the intrablock analysis would be

Table 14.1 Intrablock ANOVA for BIB Design with Replicate Groups

Source	df	SS
Replicates	r-1	SSR
Blocks(Reps unadj)	b-r	SSBlocks(R)
Treatments(adj.)	t-1	SST
Intrablock error	rt-b-t+1	SSE
Total	rt-1	SSY

For the recovery of interblock information, the estimates of the treatment parameters may be calculated by (14.25) with the following change in the weighting factor

(14.37) $\omega = \dfrac{r(MSB - MSE)}{k(b-r-t+1)MSE + rt(k-1)MSB}$

where MSB is the mean square for Blocks(Reps. adj.); this sum of squares may be obtained from the basic identity (13.22) with "Blocks" replaced by "Blocks(Reps)".

The formula for the effective error variance is the same as for the basic BIB design.

14.7.2 Balanced Lattices

The lattice designs represent a special subset of incomplete block designs having $t=k^2$ unstructured treatments in blocks of size k arranged in replicate groups. A balanced lattice has r = k + 1 replicates and $\lambda = 1$. They require b = k(k+1) blocks which might result in a large experiment. (For fewer replicates, see Chapter 15.)

All material of the previous section is applicable for use in analyzing a balanced lattice design. Various simplifications occur because $t=k^2$; most notably, the weighting factor reduces to

$$(14.38) \qquad \omega = (MSB - MSE)/kMSB$$

14.8 Incomplete Latin Square Designs: Youden Squares

In some situations, incomplete block designs are required for controlling 2 sources of heterogeneity among experimental units. A general class of such designs is generated from LS designs by *omitting one or more rows. These are called incomplete Latin* square designs. We noted in Section 14.3 that the omission of a single row of an LS design would give a BIB design. The omission of two or more rows may or may not lead to a BIB design (see also Chapter 15.)

We restrict attention in the remainder of this section to incomplete Latin square designs having two or more rows omitted, and which are also BIB designs. This special subset of incomplete Latin square designs is known as the Youden squares, after Youden [10] and [11] who developed them for greenhouse studies.

Because every treatment occurs once in every row of a Youden square, contrasts of treatments will contain no row effects. Some treatments are missing in each column of a Youden square so that contrasts of treatments will contain column effects. Thus, treatment effects must be adjusted for column (incomplete block) effects but not for row (complete block) effects.

The analysis of a Youden square is the same as for a basic BIB design with an additional source of variation due to column blocking. Therefore, the intrablock analysis for a Youden square having t columns (and treatments) and k rows has the following partial ANOVA.

Table 14.2   Intrablock ANOVA for Youden Square

Source	df	MS
Columns(unadj.)	$t-1$	MSC
Rows	$k-1$	MSR
Treatments(adj.)	$t-1$	MST
Intrablock error	$(t-1)(k-2)$	MSE
Total	$tk-1$	

The analysis with recovery of interblock (column) information, if appropriate, would utilize (14.25) and (14.26). The weighting factor reduces to the simple form

$$(14.39) \qquad \omega = (MSB - MSE)/t(k-1)MSB$$

where MSB is now the mean square for "Columns adjusted for treatments". The sum of squares for Columns(adj) is obtained from the basic identity (13.22) with "Columns" substituted for "Blocks".

PROBLEMS

1. An incomplete block design consist of the following blocks:

Block		Treatments		
1	1	2	3	6
2	1	2	5	7
3	1	3	4	7
4	1	4	5	6
5	2	3	4	5
6	2	4	6	7
7	3	5	6	7

(a) Verify that this is a BIB design.
(b) Specify the parameters of this design.

2. An incomplete block design consists of the following blocks:

Block	Treatments			Block	Treatments		
1	1	2	3	7	2	6	7
2	1	4	7	8	3	4	8
3	1	5	9	9	3	5	7
4	1	6	8	10	3	6	9
5	2	4	9	11	4	5	6
6	2	5	8	12	7	8	9

(a) Verify that this is a BIB design.
(b) Specify the parameters of this design.

3. An incomplete block design consists of the following blocks:

Block	Treatments					Block	Treatments				
1	1	2	3	4	5	4	1	2	4	5	6
2	1	2	3	4	6	5	2	3	4	5	6
3	1	2	3	5	6						

(a) Is this a BIB design?
(b) If this is a BIB design, give the parameters. If not, add blocks to make this a BIB (change the composition of any existing blocks, if necessary.)

4. An incomplete block design consists of the following blocks:

Block	Treatments			Block	Treatments		
1	1	2	3	6	2	3	6
2	1	2	4	7	2	5	6
3	1	3	5	8	3	4	5
4	1	4	6	9	3	4	6
5	1	5	6				

(a) Is this a BIB design?
(b) If this is a BIB design, give the parameters. If not, add blocks to make this a BIB (change the composition of any existing blocks, if necessary.)

5. An engineer is planning an experiment to investigate the strength of ceramic components made under 5 percentages of silicon. There are 6 ovens available but only 4 components can be baked in an oven at one time. If possible, construct a BIB design using up to 6 ovens with 4 components per oven. If impossible, how many additional ovens would be required?

6. Repeat Problem 5 if only 3 components are placed in an oven.

7. An educator is planning an experiment to investigate the effects of 6 methods of teaching computer concepts to third grade students. Third grade classes at a given school were taken to be blocks. A total of 8 schools are available but some schools have only 4 third grade classes. If possible, construct a BIB design using up to 8 schools with 4 classes per school. If impossible, how many additional schools would be required?

8. Repeat Problem 7 if only 3 classes are available per school.

9. Construct a BIB design for 6 treatments in blocks of size 5 using a Latin square design.

10. Construct a BIB design for 7 treatments in blocks of size 6 using a Latin square design.

11. Construct a BIB design for 16 treatments in blocks of size 4 using an extended Latin square design.

12. Construct a BIB design for 25 treatments in blocks of size 5 using an extended Latin square design.

13. Construct a BIB design for 7 treatments in blocks of size 3 using an extended Latin square design.

14. Construct a BIB design for 21 treatments in blocks of size 5 using an extended Latin square design.

15. A researcher studied 7 filters for their ability to remove particles from the exhaust of gasoline engines. One gasoline engine was used throughout the experiment. A filter was weighed and placed in the filtering system and attached to the exhaust system of the engine. After 2 hours of running, the filter was removed and the increase in "dry weight" was determined. Due to set-up and running times, only 3 filters could be tested on any one day. The filters (treatments) were tested in a BIB design over a period of 7 days (blocks.) The results were (in mg.):

Day	Treatments and Weight Increases, y		
1	5	4	7
	176	192	155
2	4	1	2
	210	171	221
3	3	1	7
	129	155	141
4	6	3	4
	228	152	206

Day	Treatments and Weight Increases, y		
5	2	7	6
	215	146	209
6	6	5	1
	212	208	158
7	2	3	5
	209	144	194

(a) Perform the intrablock ANOVA.
(b) Test equality of filter effects.
(c) Do a combined analysis, with recovery of interblock information, if appropriate. Ignore the condition on the number of degrees of freedom. Make appropriate tests.

16. An agronomist conducted a field experiment with 25 varieties of barley in a BIB design with 5 EU (thus, 5 varieties) per block. The blocks were arranged in 6 replicates. The yields, y (pounds per plot) together with the varieties (designated by #) are given below.

						Replicate						
		1		2		3		4		5		6
Block	#	y	#	y	#	y	#	y	#	y	#	y
	22	530	12	484	12	426	4	650	16	371	17	210
	21	445	22	487	6	456	25	339	24	355	1	456
1	25	453	17	380	5	464	12	381	2	446	17	210
	24	388	2	408	24	258	16	478	13	400	24	480
	23	458	7	517	18	468	8	465	16	371	8	540
	18	472	20	585	13	430	19	470	15	441	7	554
	19	435	25	480	19	475	2	441	21	381	5	495
2	20	505	10	155	25	385	15	470	4	478	16	348
	17	478	5	518	7	463	23	262	7	516	14	402
	16	490	15	436	1	433	6	470	18	300	23	302
	4	593	13	447	4	525	1	427	22	514	2	482
	2	577	23	384	17	465	14	417	19	477	25	371
3	3	487	18	405	23	391	22	412	11	426	18	450
	5	471	3	522	10	342	10	300	8	394	9	255
	1	425	8	587	11	408	18	351	5	427	11	289

					Replicate						
	1		2		3		4		5		6
Block	#	y	#	y	#	y	#	y	#	y	#	y
	13	513	16	455	8	510	21	402	3	390	4	374
	15	525	21	424	20	451	17	235	17	393	20	457
4	14	506	1	451	14	380	13	378	6	497	22	396
	12	516	11	454	2	527	5	557	25	326	13	362
	11	475	6	476	21	445	9	294	14	344	6	578
	8	489	4	687	3	538	11	364	23	335	3	370
	10	300	19	536	15	566	20	488	9	320	21	458
5	6	466	24	404	16	453	3	370	20	391	19	446
	9	472	14	394	9	454	7	300	1	421	12	498
	7	435	9	475	22	573	24	243	12	433	10	293

(Courtesy of Dr. T. Starling, Agronomy Dept., Va. Tech.)

(a) Construct the intrablock ANOVA.
(b) Test equality of varietal effects.
(c) Do a combined analysis, with recovery of interblock information, if appropriate. Make appropriate tests.

REFERENCES

[1] Bhattacharya, K.N., A new solution in symmetrical balanced incomplete block designs, *Sankhya*, Vol. 7, 1946.
[2] Bose, R.C., On the construction of balanced incomplete block designs, *Ann. Eugenics*, Vol. 9, 1939.
[3] Federer, W. T., *Experimental Design*, Macmillan, New York, 1955.
[4] Fisher, R. A., An examination of the different possible solutions of a problem in incomplete blocks, *Ann. Eugenics*, Vol. 10, 1940.
[5] Graybill, F. A., *An Introduction to Linear Statistical Models, Vol. 1*, McGraw-Hill, New York, 1961.
[6] Hussain, Q. N., On the totality of the solutions for the symmetrical incomplete block designs, *Sankhya*, Vol. 7, 1945.
[7] John, P. W., *Statistical Design and Analysis of Experiments*, Macmillan, New York, 1971.
[8] Kempthorne, O., *Design and Analysis of Experiments*, Wiley, New York, 1952.

[9] Yates, F., The recovery of inter-block information in
     balanced incomplete block designs, *Ann. Eugenics*, Vol. 10,
     1940.
[10] Youden, W. J., Use of incomplete block replications in
     estimating tobacco-mosaic virus, *Contr. Boyce Thompson Inst.*
     Vol. 9, 1937.
[11] Youden, W. J., Experimental designs to increase accuracy of
     greenhouse studies, *Contr. Boyce Thompson Inst.*, Vol. 11,
     1940.

## Chapter 15

PARTIALLY BALANCED INCOMPLETE BLOCK DESIGNS

15.1 Introduction

For some experiments, a researcher may find it impractical to adhere to the conditions of balanced incomplete block designs (Definition 14.1.) In particular, the number of blocks necessary for balanced designs often demands an unreasonable amount of time and resources. Any incomplete block design other than a BIB design lacks the full balance originally proposed by Yates and requires a more complicated estimation and analysis process.

Certain classes of incomplete block designs (which are not BIB designs) have some balance properties that can be utilized in simplifying the general analysis discussed in Chapter 13.

15.2 The Partially Balanced Designs

This special subset of incomplete block designs not satisfying the conditions of Definition 14.1 were developed by Bose and Nair [3]. Their goal was to construct incomplete block experiments with "some degree of balance," and whose analysis would not be "unreasonably complex". The name *partially balanced* was given to this special set of incomplete block designs which they proposed. Without having the full balance feature, some rather severe symmetry properties must hold. We now define this special class of incomplete block designs.

Definition 15.1

A *partially balanced incomplete block (PBIB) design* is an experimental set-up where t treatments are to be allocated to b blocks in such a way that:

(i) Each block contains k < t treatments.

(ii) Each treatment appears in exactly r blocks.

(iii) For any specified treatment, the other t-1 treatments belong to $m \geq 2$ classes; each of the n_i treatments belonging to the i-th class occurs λ_i times with the specified treatment. The numbers n_i and λ_i remain the same no matter which treatment is the specified one. Pairs of treatments that occur together λ_i times within blocks are called *i-th associates*.

(iv) Consider two treatments, say u and v, that are k-th associates. Then, the number of treatments common to the i-th associates of u and j-th associates of v,

$$\text{denoted by } p_{ij}^k,$$

must remain constant for all pairs of k-th associates.

The notation given in Definition 15.1 is commonly used in the literature of PBIB designs. The quantities $r, t, b, k, \lambda_1, \lambda_2, ---,$ λ_m, $n_1, n_2, ---, n_m$ are called *parameters of the first kind* while the collection of p_{ij}^k are called *parameters of the second kind*.

Example 15.1

An experiment with the following blocks represents a PBIB design:

1	6	8

1	2	7

2	4	9

3	5	7

4	5	8

3	6	9

For this design, r = 2, k = 3, b = 6 and t = 9. A listing of treatments and associate classes is:

Treatment	First Associates	Second Associates
1	2,6,7,8	3,4,5,9
2	1,4,7,9	3,5,6,8
3	5,6,7,9	1,2,4,8
4	2,5,8,9	1,3,6,7
5	3,4,7,8	1,2,6,9
6	1,3,8,9	2,4,5,7
7	1,2,3,5	4,6,8,9
8	1,4,5,6	2,3,7,9
9	2,3,4,6	1,5,7,8

There are $m = 2$ associate classes with $n_1 = 4$, $n_2 = 4$, $\lambda_1 = 1$, and $\lambda_2 = 0$. That is, any treatment occurs once with each of the $n_1 = 4$ treatments and does not occur with the $n_2 = 4$ remaining treatments.   From the above listing we can determine parameters of the second kind. We note that treatments u=1 and v=2 are first associates.   Only treatment 7 is common to the first associates of these 2 treatments, so $p_{11}^1 = 1$. First associates of treatment 1 and second associates of treatment 2 have treatments 6 and 8 in common, so $p_{12}^1 = 2$. This same number holds for treatments common to the *second* associates of treatment 1 and *first* associates of treatment 2 (common treatments are 4 and 9).   Finally, treatments 3 and 5 are common to the second associates of both treatments 1 and 2, so $p_{22}^1 = 2$. The reader can easily verify that the same numbers are obtained for the p_{ij}^1 if any other pair of first associates are chosen, for example, treatments 3 and 6.

Treatments 2 and 3 are second associates; from this pair of treatments we can determine the p_{ij}^2. Two treatments, 7 and 9, are common to the first associates of both 2 and 3 so $p_{11}^2 = 2$. Two treatments, 1 and 4, are common to the first associates of treatment 2 and second associates of treatment 3, so $p_{12}^2 = 2$. And, only treatment 8 is common to the second associates of both treatments 2 and 3, so $p_{22}^2 = 1$. Again, the reader can easily verify that these same numbers are obtained for the p_{ij}^2 if any other pair of second associates are chosen, for example, when the treatments 5 and 9 are used. ※

For convenience and clarity, it is often helpful to display the parameters of the second kind as a series of matrices, one

for each associate class.   For two associate classes, these are
given by

(15.1) $P_1 = \begin{pmatrix} p_{11}^1 & p_{12}^1 \\ p_{21}^1 & p_{22}^1 \end{pmatrix}$ $P_2 = \begin{pmatrix} p_{11}^2 & p_{12}^2 \\ p_{21}^2 & p_{22}^2 \end{pmatrix}$

Example 15.2

For the previous example,   the matrices for the parameters of the
second kind are

$$P_1 = \begin{pmatrix} 1 & 2 \\ 2 & 2 \end{pmatrix}   P_2 = \begin{pmatrix} 2 & 2 \\ 2 & 1 \end{pmatrix} \text{\small\textbf{※}}$$

    The   collection of parameters of the first and   second   kind
will completely specify a PBIB design.   As with the BIB designs,
certain conditions hold among the parameters.   Specifically

(15.2) $n_1 + n_2 + \cdots + n_m = t - 1$

(15.3) $n_1 \lambda_1 + n_2 \lambda_2 + \cdots + n_m \lambda_m = r(k-1)$

(15.4) $p_{ij}^k = p_{ji}^k$ for $k = 1, 2, \cdots, m$

(15.5) $n_k p_{ij}^k = n_i p_{jk}^i = n_j p_{ik}^j$

(15.6) $\sum_j p_{ij}^k = n_i - 1$ if $i = k$

$= n_i$ if $i \neq k$

The condition in (15.2) should be easy to see: for any treatment,
the remaining t-1 treatments are distributed among the set of m
associate classes.   To verify the condition in (15.3),   consider
all blocks in which any specified treatment occurs, say treatment
1.   In these blocks,   there are r(k-1) experimental units for the
treatments other than treatment 1.   But each of the n_1 treatments

in the first associate class must occur λ_1 times with treatment 1, the n_2 treatments in the second associate class must occur λ_2 times with treatment 1, and so on through the m-th associate class. This gives the left-hand side of (15.3).

Those equalities given in (15.4) and (15.5) are symmetry properties that we alluded to earlier. We shall not try to justify these conditions but merely point out two useful results. A number of simplifications can be made in the general incomplete block analysis when the symmetry conditions can be incorporated. And these conditions are used in constructing PBIB designs; that is, forming the block compositions. Note that the conditions specified in (15.4) allow us to interchange any two treatment designations without disturbing the balance features.

The conditions (15.6) are concerned with row sums of the P matrices defined in (15.1). Specifically, each row of P_i must sum to the i-th associate number, n_i, except the i-th row which must sum to $n_i - 1$.

Example 15.3

Using the PBIB design discussed in Examples 15.1 and 15.2, let us show that the conditions in (15.2) through (15.6) hold. We have: $r = 2$, $b = 6$, $t = 9$, $k = 3$, $m = 2$, $n_1 = 4$, $n_2 = 4$, $\lambda_1 = 1$, and $\lambda_2 = 0$. The first condition is easily seen for $n_1 + n_2 = 8 = t - 1$. To show the second condition, we have $n_1\lambda_1 + n_2\lambda_2 = 4 = r(k-1)$.

The symmetry of matrices P_1 and P_2 appearing in Example 15.2 is a result of the condition in (15.4). The equalities given by (15.5) reduce to

$$n_1 P_{12}^1 = n_2 P_{11}^2$$

$$n_1 P_{22}^1 = n_2 P_{12}^2$$

and for our design, each of these is seen to be $4(2) = 4(2)$.

A check of the matrices in Example 15.2 reveals that for P_1, the entries in the two rows sum to 3 and 4, respectively. These numbers are $n_1 - 1$ and n_2, as they should be. And for P_2, the entries is the two rows sum to 4 and 3, respectively, which are n_1 and $n_2 - 1$. ▓

Example 15.4

Now let us use the conditions in (15.2) through (15.6) to show how a PBIB design may be constructed. Suppose a researcher has t = 10 treatments to investigate and can obtain blocks of k = 4 units each. He is willing to consider m = 2 associate classes with λ_1 = 2 and λ_2 = 1.

Two conditions that must hold are

$$n_1 + n_2 = 9$$

$$2n_1 + n_2 = 3r$$

Solving the first of these equations for n_2 and substituting into the second gives $n_1 = 3(r-3)$. This implies that r must be at least 4 but no more than 6 (because n_1 cannot exceed 9). If we consider r = 4, we have $n_1 = 3$ and $n_2 = 6$. Then conditions (15.2) through (15.6) lead to the following equalities

$$p_{11}^1 + p_{12}^1 = 2$$

$$p_{12}^1 + p_{22}^1 = 6$$

$$p_{11}^2 + p_{12}^2 = 3$$

$$p_{12}^2 + p_{22}^2 = 5$$

$$3p_{12}^1 = 6p_{11}^2$$

$$3p_{22}^1 = 6p_{12}^2$$

From the first of these equations, and the fact that the p_{ij}^k must be integers, there are only 3 possibilities to consider; namely,

p_{11}^1	p_{12}^1
0	2
1	1
2	0

The fifth equality reduces to $p_{12}^1 = 2p_{11}^2$ which means that p_{12}^1

cannot be an odd number so the second possibility above can be eliminated. Letting $p^1_{11} = 0$ and $p^1_{12} = 2$, we can determine the remaining parameters with the aid of equalities (15.2) thru (15.5). The P matrices are

$$P_1 = \begin{pmatrix} 0 & 2 \\ 2 & 4 \end{pmatrix} \qquad P_2 = \begin{pmatrix} 1 & 2 \\ 2 & 3 \end{pmatrix}$$

The entries of these two matrices are the parameters of the second kind. The parameters of the first kind are: $t = 10$, $k = 4$, $b = 10$, $r = 4$, $n_1 = 3$, $n_2 = 6$, $\lambda_1 = 2$, and $\lambda_2 = 1$.

A table of treatment associates can be constructed with the aid of the P matrices. Recall that their entries specify the number of treatments common to first and second associate pairs. If (arbitrarily) we let treatments 2,3, and 4 be first associates of treatment 1, then one possible table is:

Treatment	First Associates	Second Associates
1	2,3,4	5,6,7,8,9,10
2	1,5,6	3,4,7,8,9,10
3	1,7,8	2,4,5,6,9,10
4	1,9,10	2,3,5,6,7,8
5	2,7,9	1,3,4,6,8,10
6	2,8,10	1,3,4,5,7,9
7	3,5,10	1,2,4,6,8,9
8	3,6,9	1,2,4,5,7,10
9	4,5,8	1,2,3,6,7,10
10	4,6,7	1,2,3,5,8,9

The ten blocks can be constructed easily by placing each treatment and its first associates together in a block. Thus, (1,2,3,4), (1,2,5,6), ... , (4,6,7,10). ✳

15.3 Some Design Considerations

PBIB designs having two associate classes are the most common. Considerable attention has been given to these PBIB designs, by Bose [1] and Bose and Shimamoto [4], and also by Bose, et. al [2] and Clatworthy [5] who give extensive catalogs of existing designs. These catalogs give parameters of the first kind, some constants used in the analysis, and various efficiency factors. The listings are useful for checking the existence of a proposed design. In some cases the experimenter may have the flexibility

of adding   or deleting a few treatments thereby finding a   more
manageable design than the one originally proposed.

We   have   shown how to construct the block compositions of a
PBIB design with 2 associate classes in Example 15.4.   As might
be expected, the construction becomes more tedious with increases
in   the   number   of treatments or associate classes.   For each
additional   associate   class there are additional   parameters of
both the first and second kind.

One   can   always construct a PBIB design from   a BIB   design
$(t,r,b,k, \quad \lambda = 1)$ by deleting all blocks containing one specified
treatment, Rao [6]. The remaining blocks will form a 2-associate
class   PBIB design with $\lambda_1 = 1$ and $\lambda_2 = 0$. It is fairly easy to
show that this PBIB design has treatments, blocks, and replicates
given by

(15.7) $t* = t-1$ $b* = b-r$ $r* = r(t-k)/(t-1)$

We point out that a PBIB obtained in this manner may not be   the
most   desirable;   the   number of blocks may be quite   large,   for
example.

Another   way   to construct a PBIB design with 2   associate
classes is from an existing PBIB design with 2 associate classes.
For   each existing block,   construct a new block   containing   all
treatments   not   appearing in the existing block,   Bose and   Nair
[3]. If the existing PBIB has parameters of the first   kind   of
t,b,k,r,λ_1, and λ_2, the new design has parameters of the first
kind given by

$$t* = t; \quad r* = b-r; \quad k* = t-k$$
(15.8)
$$b* = b; \quad \lambda_1^* = \lambda_1 + b - 2r; \quad \lambda_2^* = \lambda_2 + b - 2r$$

If the number of treatments is large,   a new PBIB design obtained
in this manner may not be appropriate as the block size, k*, may
be too large.

15.4   General Comments About Analyses

The   comments   made in Section 13.5 for general   IB   designs   are
appropriate here. To briefly summarize: One may consider either
an intrablock analysis or an analysis with recovery of interblock
information.   The   latter requires random blocks and sufficient
degrees   of   freedom   for estimating MSB,   the   mean   square   for
"Blocks adjusted for treatment effects". And if MSB is less than

MSE, the intrablock error mean square, one would not even perform an intrablock analysis. Instead, a completely randomized or randomized complete block analysis would be done, depending upon the arrangement of the incomplete blocks.

As expected, the number of calculations required for analyzing a PBIB design are greater than for a BIB design of "comparable size". For estimation of treatment effects, we generally need to make an adjustment for each associate class. These adjusted quantities are used then in calculating sums of squares for testing purposes. Thus, it is clear that calculations increase with increases in the number of associate classes.

In the BIB designs, the differences of all pairs of treatment means have a common variance. But this is not true in the PBIB designs. We need a different variance for each associate class. This is so because pairs which form different associate classes occur together different numbers of times in the experiment. In other words, all pairs of means in a BIB design are compared with the same precision, but in a PBIB design, the precision with which pairs of means are compared depends upon which associates they are.

15.5   Analyses of PBIB Designs

We will discuss the analyses of PBIB designs with 2 associate classes. For 3 associate classes, the reader should consult the paper by Rao [6].

We first consider the intrablock analysis. Treatment effects will be adjusted for block effects, as was done in BIB designs. Now the adjustments consist of 2 components because of the 2 associate classes. Other than a few more calculations, the analysis closely parallels that of the BIB designs.

The notation and computations are those used most frequently in the literature. We begin by listing various constants that appear in the computations; these are

$$A_{11} = \lambda_1 + r(k-1) \qquad\qquad B_{11} = \lambda_1 - \lambda_2$$

$$A_{12} = \lambda_2 + r(k-1) \qquad\qquad B_{12} = \lambda_2 - \lambda_1$$

$$(15.9) \qquad B_{21} = A_{11} + B_{11}(p_{22}^2 - p_{22}^1) \qquad A_{21} = B_{11}p_{12}^1$$

$$B_{22} = A_{12} + B_{11}(p_{11}^2 - p_{11}^1) \qquad A_{22} = B_{12}p_{12}^2$$

$$\Delta = A_{11}B_{21} - A_{21}B_{11}$$

Then the following treatment calculations can be made

(15.10) $y_{i.}$ = observed total for i-th treatment

(15.11) B_i = sum of blocks containing i-th treatment

(15.12) $Q_i = y_{i.} - B_i/k$

(15.13) $\sum_i Q_{i1}$ = sum of Q_i for the first associates of the i-th treatment

(15.14) $\hat{\tau}_i = k(B_{22}Q_i - B_{12} \sum_i Q_{i1})/\Delta$

The estimates given by (15.14) are the adjusted treatment effects from which the adjusted treatment means can be obtained upon addition of $\bar{y}..$, the observed overall mean. All of these are intrablock estimates, of course, without recovery of interblock information.

The ANOVA table for the intrablock analysis may be completed now. The format is identical to Table 13.1 for the intrablock analysis of a general IB design. All sums of squares of Section 13.4 are appropriate here if the quantities in (15.10) thru (15.14) are substituted. We illustrate these computations in the following example.

Example 15.5

Twelve filters were tested for their ability to collect air pollutants. Due to the time required to make a run, change the filters, and "clean" the system, only 5 filters could be tested on a given day. Environmental conditions were expected to affect the filtering abilities so days were taken to be blocks with random effects. The experiment was conducted on 12 days as a PBIB design with parameters

$$t = b = 12 \qquad r = k = 5$$

$$\lambda_1 = 1 \qquad \lambda_2 = 2 \qquad n_1 = 2 \qquad n_2 = 9$$

$$P_1 = \begin{pmatrix} 1 & 0 \\ 0 & 9 \end{pmatrix} \qquad P_2 = \begin{pmatrix} 0 & 2 \\ 2 & 6 \end{pmatrix}$$

The constants (15.9) needed for an intrablock analysis are

$$A_{11} = 1 + 5(4) = 21 \qquad A_{21} = (1-2)(0) = 0$$

$$A_{12} = 2 + 5(4) = 22 \qquad A_{22} = (2-1)(2) = 2$$

$$B_{11} = 1-2 = -1 \qquad B_{21} = 21 + (-1)(6-9) = 24$$

$$B_{12} = 2-1 = 1 \qquad B_{22} = 22 + (-1)(0-1) = 23$$

$$\Delta = 22(23)-2(1) = 504$$

The variable of interest was the increase in weight (in grams) of the filter after 40 minutes of operation.

Block	Treatments and Observations					Block Totals $y_{.j}$
1	1 18.3	2 12.2	5 21.9	8 17.7	11 24.0	94.1
2	1 11.9	3 22.3	6 14.5	9 4.5	12 18.5	71.7
3	1 11.4	4 11.2	7 15.8	11 17.3	12 13.1	68.8
4	1 25.7	4 8.9	8 21.7	9 9.9	10 13.5	79.7
5	1 15.3	5 18.8	6 8.4	7 14.1	10 15.6	72.2
6	2 8.8	3 12.2	4 22.8	7 16.7	10 12.4	72.9
7	2 14.1	4 19.0	5 8.5	9 13.3	12 14.0	68.9
8	2 13.7	6 10.1	7 9.6	8 11.6	12 16.5	61.5

Block	Treatments and Observations					Block Totals $y_{.j}$
9	2 9.3	6 9.6	9 14.2	10 11.7	11 18.9	63.7
10	3 14.8	4 7.6	6 12.6	8 9.5	11 17.8	62.3
11	3 18.5	5 15.4	7 18.6	9 12.9	11 15.1	80.5
12	3 7.7	5 13.6	8 12.1	10 4.6	12 16.2	54.2

$$y.. = 850.5$$

To facilitate calculations, the following table is constructed:

Treatment Number	Total ($y_{i.}$)	B_i	Q_i	First Assoc.	$\Sigma\, Q_{i1}$	$\hat{\tau}_i$
1	82.6	386.5	5.3	2,3	-7.0	1.28
2	58.1	361.4	-14.2	1,3	12.5	-3.36
3	75.5	341.6	7.2	1,2	-8.9	1.73
4	69.5	352.6	-1.0	5,6	-6.8	-0.16
5	78.2	369.9	4.2	4,6	-12.0	1.08
6	55.2	331.1	-11.0	4,5	3.2	-2.54
7	74.8	355.9	3.6	8,9	-15.9	0.98
8	72.6	351.8	2.2	7,9	-14.5	0.65
9	54.8	364.5	-18.1	7,8	5.8	-4.07
10	57.8	342.7	-10.7	11,12	32.5	-2.76
11	93.1	369.4	19.2	10,12	2.6	4.36
12	78.3	325.1	13.3	10,11	8.5	2.95
Checks	850.5	4252.5	0		0	0.14

(The check value for the last column indicates that additional decimals should have been used in the calculations.) The sums of squares calculations are:

Correction: $C = (850.5)^2/60 = 12,055.84$

Total: SS $= (18.3)^2 + \cdots + (16.2)^2 - C = 1258.97$

Blocks (ignoring trts.): SSB $= [(94.1)^2 + \cdots + (54.2)^2]/5 - C$

$$= 243.12$$

Trts.(elim. Blocks): $\sum\limits_{i} \hat{r}_i Q_i = (1.28)(5.3) + \cdots + (2.95)(13.3)$

$$= 330.69$$

Intrablock Error: SSE = 1258.97 - 243.12 - 330.69 = 685.16

Intrablock ANOVA for Air Pollutant Filtering

Source	df	SS	MS
Blocks(ign.trts.)	11	243.12	
Treatments(adj.)	11	330.69	30.06 = MST
Intrablock Error	37	685.16	18.52 = MSE
Total	59	1258.97	※

Under the customary assumption that the error components, the ϵ's, are independent, identically distributed normal variables with zero mean, the intrablock ANOVA provides entries to make an F test of no treatment effects. And from the completed ANOVA we obtain an estimate of intrablock error, MSE, from which we may estimate the variances for differences of adjusted treatment means. For differences of 2 first associate means, the estimated variance is

$$(15.15) \qquad \hat{V}_1 = 2kB_{21}MSE/\Delta$$

and for differences of 2 second associate means, the estimated variance is

$$(15.16) \qquad \hat{V}_2 = 2kB_{22}MSE/\Delta$$

Appropriate standard deviations are obtained from square roots of the expressions in (15.15) and (15.16).

Example 15.6

For our earlier example, a test of

$$H_0: \tau_1 = \tau_2 = \cdots = \tau_{12} = 0$$

would be made by calculating

$$F = MST/MSE = 30.06/18.52 = 1.62$$

which gives an observed significance level of $P > 0.10$.
  Estimated variances for pairs of adjusted mean differences are

$$\hat{V}_1 = 2(5)(24)(18.52)/504 = 8.82$$

$$\hat{V}_2 = 2(5)(23)(18.52)/504 = 8.45$$

where $\hat{V}_1$ is for first associate pairs (1,2 or 7,8, etc.)   and $\hat{V}_2$ is for second associate pairs (1,4 or 7,10, etc.). ▓

As indicated earlier, the analysis with recovery of interblock
information is done under the assumption of random block effects.
This "combined analysis" is fairly complicated for PBIB designs
and, as usually recommended, is not performed without sufficient
degrees of freedom available for estimating the block variation,
σ^2_ρ. Generally, 10 or fewer degrees of freedom are considered
inadequate with 12 or more considered sufficient.
   To estimate the block variation, we need the sum of squares
for blocks adjusted for treatment effects.   This sum of squares
can be obtained from the basic identity [see also (13.22)]

(15.17)   SS Blocks(unadj.) + SS Trts.(adj.) =

                            SS Blocks(adj.) + SS Trts.(unadj.)

The two sums of squares on the left hand side of this equation
appear in an intrablock ANOVA table. The second term on the right
hand side is calculated easily from unadjusted treatment totals:

(15.18) $SS \; Trts(unadj.) = \dfrac{1}{r} \sum_i y^2_{i.} - C$

The estimate of the block variation can be obtained now from the
"adjusted block mean square"

(15.19) $\hat{\sigma}^2_\rho = MSB = SS \; Blocks(adj.)/(b-1)$

Then, to combine intrablock and interblock information, we use the estimated variance components, MSE and MSB, to calculate the following weights

(15.20) $\quad w = 1/MSE$

(15.21) $\quad w' = \dfrac{(bk-t)}{k(b-1)MSB-(t-k)MSE}$

Another set of constants is required to complete the analysis. These are

$$R = r[w + w'/(k-1)]$$

$$\Lambda_1 = \lambda_1(w - w')$$

$$\Lambda_2 = \lambda_2(w - w')$$

$$A'_{11} = R(k-1) + \Lambda_1$$

$$A'_{12} = R(k-1) + \Lambda_2$$

$$B'_{11} = \Lambda_1 - \Lambda_2$$

(15.22) $\quad B'_{12} = \Lambda_2 - \Lambda_1$

$$A'_{21} = B'_{11}p^1_{12}$$

$$A'_{22} = -B'_{11}p^2_{12}$$

$$B'_{21} = A'_{11} + B'_{11}(p^2_{22} - p^1_{22})$$

$$B'_{22} = A'_{12} + B'_{11}(p^2_{11} - p^1_{11})$$

$$\Delta' = A'_{11}B'_{21} - A'_{21}B'_{11}$$

With these constants, the following treatment quantities can be calculated

(15.23) $\quad P_i = kwQ_i + w'B_i - rkw'\bar{y}..$

(15.24) $\quad \sum\limits_i P_{i1} =$ Sum of P_i for the first associates of the i-th treatment

(15.25) $\hat{\tau}_i^* = [B'_{22}P_i - B'_{12} \sum_i P_{i1}]/\Delta'$

The last expression gives estimated treatment effects utilizing both intrablock and interblock information. Adding the overall mean, $\bar{y}..$, to these estimates gives the adjusted treatment means with recovery of interblock information. Differences of these means have estimated variances given by

(15.26)
$$\hat{V}_1 = 2kB'_{21}/\Delta'$$
$$\hat{V}_2 = 2kB'_{22}/\Delta'$$

for each pair of means which are first and second associates, respectively.

The overall efficiency factor for a PBIB design is

(15.27) $EF = \dfrac{(t-1)\Delta}{rk[(t-1)B_{22} + n_1 B_{12}]}$

Example 15.7

For the PBIB design analyzed in Example 15.5, we note there are only 11 degrees of freedom available for estimation of MSB. This is a borderline situation making the recovery of interblock information questionable. For illustrative purposes, we will do a complete analysis utilizing both types of information.

From the table appearing in Example 15.5, we calculate the sum of squares for treatments unadjusted (that is, ignoring blocks):

$$SS \; Trts.(unadj.) = [(82.6)^2 + \cdots + (78.3)^2]/5 - C$$
$$= 12379.87 - 12055.84$$
$$= 324.0$$

Then, by (15.17) and the ANOVA table in Example 15.5

$$SS \; Blocks(adj.) = 243.12 + 330.69 - 324.0$$
$$= 249.81$$

giving

$$MSB = 249.81/11 = 22.71$$

which exceeds MSE = 18.52.   Therefore,   the intrablock   analysis performed earlier is appropriate.

The   weights for combining the two types of information   are, by (15.20) and (15.21),

$$w = 1/MSE = 1/18.52 = 0.054$$

$$w' = 48[5(11)(22.71) - 7(18.52)]$$

$$= 48/1119.41$$

$$= 0.043$$

The constants specified in (15.22) are

$$R = 5[0.054 + 0.043/4] = 0.32375$$

$$\Lambda_1 = 0.054 - 0.043 = 0.011$$

$$\Lambda_2 = 2(0.054 - 0.043) = 0.022$$

$$A'_{11} = 0.32375(4) + 0.011 = 1.306$$

$$A'_{12} = 0.32375(4) + 0.022 = 1.317$$

$$B'_{11} = 0.011 - 0.022 = -0.011$$

$$B'_{12} = 0.011$$

$$A'_{21} = 0.011(0) = 0$$

$$A'_{22} = 0.011(2) = 0.022$$

$$B'_{21} = 1.306 - 0.011(6-9) = 1.339$$

$$B'_{22} = 1.317 - 0.011(0-1) = 1.328$$

$$\Delta' = 1.306(1.339) - 0 = 1.749$$

Combined   estimates may be calculated by forming the   tabulations below. A portion of the table in Example 15.5 is reproduced here to aid in the calculations. From the above constants and (15.23) thru (15.25),   the entries in the fourth and seventh columns are given by

$$P_i = 5(0.054)Q_i + 0.043B_i - 5(5)(0.043)\bar{y}..$$

$$= 0.27Q_i + 0.043B_i - 15.238$$

$$\hat{\tau}_i^* = [1.328P_i - 0.011 \sum_i P_{i1}]/1.749$$

Trt.	Q_i	B_i	P_i	First Associates	$\sum P_{i1}$	$\hat{\tau}_i^*$
1	5.3	386.5	2.81	2,3	-2.14	2.12
2	-14.2	361.4	-3.53	1,3	4.20	-2.71
3	7.2	341.6	1.39	1,2	-0.72	1.06
4	-1.0	352.6	-0.35	5,6	-2.17	-0.25
5	4.2	369.9	1.80	4,6	-4.32	1.39
6	-11.0	331.1	-3.97	4,5	1.45	-3.02
7	3.6	355.9	1.04	8,9	-3.97	0.76
8	2.2	351.8	0.48	7,9	-3.41	0.39
9	-18.1	364.5	-4.45	7,8	1.52	-3.39
10	-10.7	342.7	-3.39	11,12	8.16	-2.63
11	19.2	369.4	5.83	10.12	-1.06	4.42
12	13.3	325.1	2.33	10,11	2.44	1.75
Checks	0		-0.01		-0.02	-0.11

(The check value for the last column indicates that additional decimals should have been used in the calculations.) Estimates of treatment effects utilizing both intrablock and interblock information appear in the last column. By adding $\bar{y}.. = 14.175$ to each, the adjusted treatment means are obtained. Differences of these means have estimated variances of

$$\hat{V}_1 = 2(5)(1.339)/1.749 = 7.66$$
$$\hat{V}_2 = 2(5)(1.328)/1.749 = 7.59$$

respectively, for pairs that are first and second associates.
The overall efficiency factor for this design is

$$EF = 11(504)/25[11(23 + 2(1)] = 0.87$$

It is interesting to note that this value of EF is given in [5] for a PBIB having these parameters of the first and second kind.※

PROBLEMS

1. A partially balanced incomplete block design consists of 6 blocks with the following compositions:

Block	Treatments			
1	1	2	7	8
2	1	3	6	8
3	1	4	5	8
4	2	3	6	7
5	2	4	5	7
6	3	4	5	6

(a) Give a table of treatments and associate classes.
(b) Give the parameters of this design.

2. A partially balanced incomplete block design consists of 9 blocks with the following compositions:

Block	Treatments			Block	Treatments			Block	Treatments		
1	1	2	8	4	2	3	7	7	4	5	7
2	1	3	5	5	2	4	6	8	5	8	9
3	1	4	6	6	3	6	9	9	6	7	8

(a) Give a table of treatments and associate classes.
(b) Give the parameters of this design.

3. The following BIB design appeared in Problem 1 of Chapter 14:

Block	Treatments				Block	Treatments			
1	1	2	3	6	5	2	3	4	5
2	1	2	5	7	6	2	4	6	7
3	1	3	4	7	7	3	5	6	7
4	1	4	5	6					

Show that when all blocks containing treatment 7 are eliminated, the remaining blocks form a PBIB design and give

(a) a table of treatments and associate classes.
(b) the parameters of the design.

4. The following BIB design appeared in Problem 2 of Chapter 14:

Block	Treatments			Block	Treatments		
1	1	2	3	7	2	6	7
2	1	4	7	8	3	4	8
3	1	5	9	9	3	5	7
4	1	6	8	10	3	6	9
5	2	4	9	11	4	5	6
6	2	5	8	12	7	8	9

Show that when all blocks containing treatment 5 are eliminated, the remaining blocks form a PBIB design and give

(a) a table of treatments and associate classes.
(b) the parameters of the design.

5. Refer to the PBIB design obtained in Problem 3 above. For each existing block of this design, construct a new block containing all "other treatments". Show that this new set of blocks gives a PBIB design with parameters of the first kind as indicated in (15.8).

6. Refer to the PBIB design obtained in Problem 4 above. For each existing block of this design, construct a new block containing all "other treatments". Show that this new set of blocks gives a PBIB design with parameters of the first kind as indicated in (15.8).

7. A person conducts the following incomplete block design:

Block	Treatments		
1	1	2	3
2	2	3	5
3	3	4	5
4	1	2	4
5	1	4	5

Is this a PBIB design? Why or Why not?

8. A person conducts the following incomplete block design:

Block	Treatment			
1	1	2	5	6
2	1	3	4	5
3	1	2	4	6
4	2	3	5	6
5	2	4	5	6
6	3	4	5	6

Is this a PBIB design? Why or why not?

9. A chemist conducted an experiment to investigate the yield of a chemical under eight treatments (different temperatures, catalysts, and so on.) Due to equipment and work involved, only 3 treatments could be investigated on a single day. Therefore, the experiment was conducted as a PBIB design with the following yields (in grams):

Day	Treatments and Yields			Day	Treatments and Yields		
1	6	5	8	5	5	7	4
	14.2	13.6	11.5		12.8	16.1	13.6
2	5	3	2	6	7	1	6
	10.8	14.6	12.9		14.9	13.1	13.7
3	1	4	2	7	4	6	3
	15.3	13.8	13.7		14.4	15.0	14.9
4	3	8	1	8	2	7	8
	12.9	10.7	14.0		12.7	15.5	11.3

(a) Give a table of treatments and their associates.
(b) Give the parameters of this design.

10. A plant breeder was interested in performing a yield study on
36 varieties of wheat. He could set up blocks of 6 plots (EU)
each, but didn't have enough land to do a BIB design. He
decided to conduct a PBIB design with 18 blocks. The
yields, y (in pounds per plot) together with the varieties
(designated by #) are given below.

Block

1		2		3		4		5		6	
#	y	#	y	#	y	#	y	#	y	#	y
26	559	21	520	16	452	20	463	36	459	7	554
29	438	20	470	9	561	34	638	31	567	10	658
28	583	24	643	30	581	13	562	32	608	11	613
30	602	19	557	31	575	27	651	34	646	12	670
25	585	23	693	2	519	5	600	35	530	9	630
27	530	22	539	23	711	12	603	33	580	8	542

7		8		9		10		11		12	
#	y	#	y	#	y	#	y	#	y	#	y
13	641	6	584	5	564	26	507	12	600	7	548
15	661	4	517	35	473	2	467	36	491	31	507
14	492	5	480	17	517	14	487	30	540	19	563
16	587	1	555	23	675	32	512	24	642	13	659
17	483	2	595	11	600	8	426	6	518	25	561
18	629	3	536	29	363	20	409	18	645	1	558

13		14		15		16		17		18	
#	y	#	y	#	y	#	y	#	y	#	y
29	461	10	537	19	542	7	529	15	554	34	555
15	632	17	483	26	555	14	523	33	511	28	451
22	531	32	587	18	681	6	619	9	509	22	479
36	565	3	643	33	597	21	525	21	531	16	465
8	462	24	681	11	650	35	458	3	597	10	625
1	517	25	601	4	588	28	470	27	642	4	570

(Courtesy of Dr. T. Starling, Agronomy Dept., Va. Tech.)

(a) Give a table of treatments and their associates.
(b) Give the parameters of this design.

11. Refer to the chemical experiment of Problem 9.

(a) Give the intrablock ANOVA.
(b) Test equality of mean yields under the 8 treatments.
(c) Give standard deviations for differences of any pair of first associate means and second associate means.

12. Refer to the wheat experiment of Problem 10.

(a) Give the intrablock ANOVA.
(b) Test equality of mean yield for the varieties.
(c) Give standard deviations for differences of any pair of first associate means and second associate means.

13. Refer to the chemical experiment, Problems 9 and 11.

(a) Estimate the treatment effects under a combined analysis, with recovery of interblock information, if MSB > MSE. (Ignore the condition on degrees of freedom.)
(b) Give standard deviations for differences of any pair of first associate means and second associate means.

14. Refer to the wheat experiment, Problems 10 and 12.

(a) Estimate the treatment effects under a combined analysis, with recovery of interblock information, if MSB > MSE.
(b) Give standard deviations for differences of any pair of first associate means and second associate means.

REFERENCES

[1] Bose, R. C., Partially balanced incomplete block designs
     with two associate classes involving only two replications,
     Calcutta Stat. Assoc. Bulletin, Vol. 3, 1951.
[2] Bose, R.C., W.H. Clatworthy and S.S. Shrikhande, Tables of
     partially balanced designs with two associate classes, North
     Carolina Agr. Expt. Sta. Tech. Bulletin No. 107, 1954.
[3] Bose, R. C., and K. R. Nair, Partially balanced incomplete
     block designs, *Sankhya*, Vol. 4, 1939.
[4] Bose, R. C., and T. Shimamoto, Classification and analysis
     of partially balanced incomplete block designs with two
     associate classes, *Jour. Amer. Stat. Assoc.*, Vol. 47, 1952.
[5] Clatworthy, W. H., Contributions on partially balanced
     incomplete block designs with two associate classes, Nat.
     Bur. Standards, App. Math. Series 47, 1956.
[6] Rao, C.R., General method of analysis for incomplete block
     designs, *Jour. Amer. Stat. Assoc.*, Vol. 42, 1947.

STATISTICAL TABLES

TABLE A.1   RANDOM DIGITS

One hundred rows and one hundred columns, numbered 00 through 99.

Column

Row	00	10	20	30	40	50	60	70	80	90
00	12210 78708	26642 30263	18817 83563	43508 30067	00814 52666	51919 45569	91860 41304	59222 93948	88948 47086	49362 86003
01	08207 84246	54377 04106	27748 65845	68959 24971	48296 06660	30127 45385	30234 98810	06260 75947	57456 97761	88838 22960
02	02389 16728	53183 18029	47943 18426	54760 01032	05582 45145	07971 82846	07027 25831	90061 94692	29094 57091	93837 33066
03	45957 89199	48812 49184	72609 12674	63480 17832	59507 47912	96576 24488	49578 79055	25783 99998	65379 66222	99221 75840
04	60472 69188	47417 62238	21912 70058	58128 67803	06484 90108	91080 82938	36243 71823	31278 22173	75359 39710	33339 08461
10	17706 78400	39590 91343	68952 86244	32101 41970	41949 83142	01800 09706	56313 23110	83305 10671	39703 92305	42191 31407
11	68412 01172	53330 63158	79023 67487	92224 44119	80203 47668	67121 75971	31359 34596	62816 69180	38355 61796	84294 73953
12	38380 85484	78758 48901	15416 54075	60035 92619	53423 55260	64857 48245	27859 09608	82797 71769	50174 90089	19431 39101
13	33691 53551	82856 22508	02838 18926	48821 80106	93755 96540	01323 39760	58859 52056	75559 46496	19928 06313	10797 15985
14	30028 36404	38583 56854	70045 94757	29048 97631	76683 95703	95758 43737	85846 31846	00058 21348	49757 35588	53949 64442
20	86840 68495	31789 12190	85453 55099	32460 81494	25440 93376	56922 80775	57973 38831	01846 47130	32641 16458	17645 22255
21	74290 85310	48915 74363	21216 40536	69311 31228	69523 49379	50792 02799	89744 78554	44158 19541	80735 58381	74031 76384
22	57658 02695	07529 81325	38551 00341	70933 31770	11103 50566	61693 27308	04102 55206	05527 17939	56908 57411	01570 47340
23	57241 21735	05496 49587	25762 56783	72103 44942	35027 55472	87707 76310	27595 73615	35403 88844	81980 24229	18319 54932
24	81346 95718	52962 95122	33641 98424	94558 49682	00395 90835	36276 26357	45389 39517	68745 55408	95320 82078	99118 57326

Row

Column

Table of Random Digits (columns 00–90, rows 00–74)

```
Row |    00          10          20          30          40          50          60          70          80          90
----+--------------------------------------------------------------------------------------------------------------------------
 00 | 26203 95922  34009 69651  01356 64264  61445 26597  04005 38318  27637 32241  27585 73829  66220 95383  65705 36807  53387 59555
 01 | 75437 58382  36108 44359  98592 33148  41686 63480  84801 64312  02991 40662  83302 59049  15457 69414  18618 04209  30085 31223
 02 | 73126 65038  02219 80983  66429 26432  92263 02342  72437 74278  51623 06344  29473 77424  67785 08562  34115 83221  87065 18095
 03 | 22732 98009  62337 35191  44256 97591  31050 09614  32011 75975  34346 02632  45899 88753  63170 06616  30465 86803  98791 13673
 04 | 85893 00162  29357 03746  70544 45658  43214 ?????  32926 52149  47282 57598  28118 04994  08887 00094  95344 58673  42795 29893
    |
 10 | 13774 55916  86636 60615  20977 97863  19044 94411  61029 15045  35888 79003  29220 08406  13662 85055  77148 56712  66746 49616
 11 | 88184 67131  39604 46018  53367 12097  85278 97970  37989 73168  40108 67694  50651 11401  39889 13902  99292 57869  24045 93778
 12 | 06781 68222  29334 46006  06793 64226  47956 40942  67801 94233  15601 22929  28247 68373  97439 17025  78341 43384  92373 60253
 13 | 95685 15658  90989 95500  60094 24631  39331 66466  98200 57210  78227 30094  04876 58572  25850 20846  05034 55569  37733 30390
 14 | 95483 11390  79500 95929  21018 80234  11967 62100  15249 75466  23825 28632  43270 93860  49685 12500  10387 82803  80673 58194
    |
 20 | 98447 86088  38236 52711  97289 05929  34306 41298  20159 16429  27027 25989  20185 53716  37541 88748  13514 27498  84272 58443
 21 | 86732 24965  57529 65692  45102 94170  72991 34499  25469 88295  47440 24940  61890 78572  36545 73937  57257 22363  28977 11005
 22 | 00017 14428  13950 22867  96012 00020  06657 36040  32013 73593  35224 99122  15964 10246  26843 65489  06682 56489  89942 08153
 23 | 19577 23379  32314 88972  80943 00810  24438 02910  26327 11837  87393 77921  80075 14476  91900 93900  75792 44368  00075 38150
 24 | 99949 02938  04327 68140  96504 90920  43025 42032  59946 39301  45360 03895  59431 95393  47330 26818  81437 42951  64196 97629
    |
 30 | 07888 07495  49523 62903  72418 67769  73333 50603  40895 79813  90874 30321  72812 18341  09729 97336  59725 00131  09502 06343
 31 | 71122 74229  02785 51996  17100 75967  17346 28091  15781 12987  40168 85185  63532 63131  04938 45549  24277 10902  77056 54461
 32 | 66431 12160  32367 29420  02365 41955  78345 64987  69707 19315  35340 07477  80054 70824  74395 16461  53571 01803  68476 50648
 33 | 07508 74254  15399 78034  41472 74756  36962 91117  24407 87084  21756 36715  18357 52825  35973 54937  02781 65717  05299 40908
 34 | 08235 04811  09261 23772  48732 38675  23262 05142  21310 52588  24609 83910  15782 00844  54962 56301  92097 60495  84012 88934
    |
 40 | 48891 10539  77297 06672  60476 03209  56167 45544  78258 21422  24810 61498  24935 86366  58262 44565  91426 86742  61747 79346
 41 | 73871 02158  64630 46343  47332 15481  71394 96315  73568 80645  75555 42967  02810 16754  08813 40079  62385 21488  38665 94197
 42 | 34170 02175  09960 18993  17807 66444  32730 21212  46076 25920  49718 45037  73341 17414  91868 24102  76123 67138  43728 43627
 43 | 71583 99318  45044 48166  87077 55556  63659 33391  85205 13101  04194 83554  98004 14744  63132 75018  75167 24090  02458 78215
 44 | 01108 73834  09699 74820  30068 86921  78110 28980  24355 68185  66155 24222  91229 63841  03271 56726  36817 51182  94336 20894
    |
 50 | 20522 12904  93684 21725  57411 71846  80425 00453  19748 21457  10801 76783  05312 30807  40006 04465  70163 17305  46414 76468
 51 | 79608 42670  40653 95662  12532 49174  48503 67040  69216 27753  40422 75576  12054 17028  58495 87538  23570 66469  46900 95568
 52 | 68090 48226  92873 37646  30421 74228  35929 44178  38348 91363  95572 12125  88937 08659  46840 83231  48694 85960  89763 58305
 53 | 19084 66561  09518 45488  31864 35233  62351 49727  02447 07250  57276 49498  83787 87636  89511 64735  75611 94911  91224 67928
 54 | 07646 32173  72941 15268  61602 62354  27853 58697  90386 37514  80566 94963  43951 51326  33274 54833  86699 66988  91224 72484
    |
 60 | 63196 74349  39550 70915  83488 07746  17894 50004  91058 41346  44517 55108  17435 33109  60343 46193  66019 43713  24097 52921
 61 | 86398 28564  87112 86288  97693 26056  17915 92559  35381 75027  55424 87650  13896 90005  99458 20153  86688 13650  75201 79447
 62 | 63642 03881  57412 19351  27350 31510  45854 15923  02510 43820  03646 54402  97762 16434  62430 28438  13602 63236  81431 75641
 63 | 71121 53702  40309 92143  19131 65330  63967 53721  85795 53721  14537 53791  75413 39128  82975 73849  27269 73444  26120 06824
 64 | 13701 53089  57760 40693  34610 98000  45016 71126  59165 50190  ????? ?????  ????? ?????  ????? ?????  80802 96976  04878 35832
    |
 70 | 71134 32621  80057 67356  89035 65907  57480 04850  92464 32201  01644 33630  71247 59273  07811 33546  88628 06469  86257 39298
 71 | 33767 36215  39791 72287  29800 67640  97975 74758  46021 73423  79387 94217  77995 93285  13354 84980  83590 63494  06036 18502
 72 | 33146 61482  78422 55628  18394 48569  51299 86316  19323 41580  74962 49488  54662 93588  50466 86195  62458 06195  07995 71054
 73 | 36625 32972  51316 64412  82599 79023  38452 03155  74595 13130  21165 45577  46383 38855  21561 89332  94248 09703  78397 38770
 74 | 63621 36578  05384 34537  25673 58665  29591 66953  98274 34545  58519 95396  73607 72106  76597 85596  99075 39195  99605 66179
```

Row

TABLE A.1 (Continued)

Column / Row

Row	00		10		20		30		40		50		60		70		80		90	
70	22624	74007	67711	37494	37526	00412	47081	69740	38591	44042	46982	79519	22294	15676	83484	98279	79200	02640	22501	43073
71	56577	81154	72712	59626	77020	40221	70415	21555	13815	15986	58463	67619	18006	05028	32441	83599	28915	05362	21612	64681
72	68561	00765	43531	76052	72134	75017	93077	40022	29505	76950	43055	00020	39254	68439	27399	24259	04641	50935	07112	55117
73	29278	37732	91524	28083	49045	23812	81820	66774	33620	64598	84073	38387	14337	90766	60436	65757	57590	17880	13776	35810
74	05725	26121	68788	96440	70878	63396	59123	86114	46700	98438	93542	37270	09361	62404	74056	52964	67372	81398	01482	97589
75	12415	12423	39879	62843	88286	04293	49225	20626	61318	10872	54467	20234	52813	85296	14542	73241	74848	39001	97598	76641
76	56651	39944	91948	87327	20492	92155	94288	75984	73002	78618	43608	42832	93917	67031	50220	94089	64858	27691	16719	99870
77	48493	98522	25685	40367	43342	52382	28333	30879	59352	86497	64808	01692	46424	64722	87162	05582	01452	14980	17397	07403
78	25123	62925	33514	71567	65553	97229	55058	40209	62112	18160	78703	93006	59651	48404	82284	66405	89818	00989	56112	78144
79	85607	00506	63005	26044	87342	39708	30757	75972	54177	31864	14886	70359	32158	30401	20829	22534	88848	07669	25100	48602
80	16373	13846	52495	45645	10162	39115	45771	25100	16356	37750	69280	61856	78974	91485	01583	11620	53740	32705	80391	56749
81	40801	68181	06980	50698	26657	94736	02366	84979	18942	84572	99680	99636	54107	79588	90845	21652	58875	13171	68531	18550
82	32100	76474	58798	77166	31884	50951	29955	94471	53066	46180	01662	21554	63836	41530	21864	81711	68921	61749	36051	78024
83	76932	16642	68834	38684	07709	97540	70822	32654	85262	92296	67852	69123	14280	11647	65125	82427	61594	32015	93473	05627
84	57106	64646	95365	10663	32804	72795	25166	79078	26789	51843	13911	67691	97854	89950	40963	06697	82660	69097	65284	49808
85	05418	98621	75977	52630	27675	96279	13152	44711	46961	45757	95822	09552	65950	34875	64250	41385	80133	70818	09286	30769
86	37221	69111	23417	17803	11075	66454	95453	60328	68808	94125	44068	24928	27345	34235	44124	06435	06281	43723	97380	76080
87	23823	99934	93432	05912	93150	84781	95233	03767	24838	94955	99222	66415	71069	62293	77467	35751	22548	23799	96272	58777
88	84321	03554	23836	64544	27469	23266	57244	28275	82170	07575	08442	61287	72421	35777	61079	42462	17761	94518	98114	74031
89	02344	23656	61016	09765	19867	21790	46457	25411	03500	46762	14967	60637	32097	28122	87708	19378	93372	23225	38453	80331
90	70325	88171	23685	97217	76585	88384	51221	31272	17338	80422	27864	15358	16499	91903	62987	98198	15036	93293	68241	44450
91	22987	01461	15898	25857	74145	84101	90051	36721	45737	65918	99678	16125	52978	79815	85990	18659	00113	93253	49186	25165
92	83159	50055	54172	05122	07096	55938	95953	33901	54679	64178	89143	79403	22324	54261	97830	42630	48494	09999	69961	39421
93	68181	68525	53187	92984	58323	41389	66592	56584	14326	07823	34135	31532	42025	83214	83730	28249	25629	11494	70726	45051
94	65843	69659	40222	34435	50890	24504	70304	43227	29870	44698	72117	97579	36071	29261	89937	78208	23747	56756	37453	51344
95	54251	76480	23860	30082	69132	08840	98017	54350	66249	75725	19725	76199	08620	22662	52907	25194	84597	93419	95762	14991
96	42504	70709	53223	89394	06872	79919	06862	25977	42529	78046	96997	66390	27609	41570	17749	23185	24475	54451	91471	33969
97	56201	69520	52851	03783	60280	27101	07911	12147	97079	52817	44158	67618	15572	95162	95842	08301	11906	68081	40436	58735
98	25284	80305	76573	14254	56770	83548	44466	06063	04232	62838	33839	40750	18898	61650	09970	47651	41205	65020	33537	01022
99	00702	10030	47850	44094	84878	51686	39037	19661	73877	24708	53070	61630	84434	05732	18094	71669	41033	82402	16415	83958

TABLE A.2   RANDOM PERMUTATIONS OF 7

```
7 1 3 1 4   6 1 7 6 4   2 6 6 5 3   3 1 1 6 4   6 7 4 4 7   6 2 5 7 7
6 7 1 7 1   7 5 6 7 5   7 1 5 3 6   2 4 5 4 5   5 1 3 5 4   1 4 6 5 3
5 3 4 5 5   5 7 3 3 2   6 4 3 4 5   7 5 7 3 7   3 2 5 3 6   7 7 4 1 4
2 5 2 2 7   3 3 5 2 6   3 3 1 6 7   6 7 3 2 2   4 6 7 1 3   3 5 1 2 6
3 2 7 6 2   2 6 2 1 3   1 2 7 7 4   1 2 4 5 3   1 5 1 2 1   5 6 7 6 5
4 4 6 4 3   1 2 1 4 7   5 5 4 2 2   5 6 6 7 1   2 3 6 6 2   4 1 3 3 1
1 6 5 3 6   4 4 4 5 1   4 7 2 1 1   4 3 2 1 6   7 4 2 7 5   2 3 2 4 2

4 6 2 3 6   4 3 6 5 4   5 4 7 7 1   2 6 2 7 5   4 1 3 2 1   3 3 5 2 2
5 5 6 2 4   6 4 3 6 7   1 5 3 5 2   7 2 3 6 2   3 3 7 6 5   6 7 1 7 7
7 7 1 5 3   7 1 1 2 3   6 1 2 2 4   1 1 7 5 7   5 2 1 1 3   4 4 7 1 4
1 1 4 7 1   3 2 4 4 5   4 7 6 4 3   3 4 1 2 6   1 4 5 4 6   7 2 3 5 5
2 4 7 6 2   2 6 5 3 6   2 6 4 1 6   5 3 6 1 1   2 5 2 3 7   1 5 6 6 1
3 3 3 1 7   1 7 7 7 1   7 2 1 6 7   6 7 4 4 3   7 6 4 7 4   5 6 4 3 6
6 2 5 4 5   5 5 2 1 2   3 3 5 3 5   4 5 5 3 4   6 7 6 5 2   2 1 2 4 3

7 3 5 5 1   1 5 5 2 1   5 1 6 1 1   4 4 2 6 3   2 3 5 3 3   3 6 3 1 6
3 5 1 3 6   6 3 1 6 6   7 6 7 4 4   6 3 4 3 6   7 5 1 6 7   1 1 6 2 4
4 6 2 6 7   3 7 2 3 2   4 3 1 2 5   1 7 5 7 4   3 4 4 1 6   4 4 7 7 3
2 2 4 4 2   5 1 7 4 7   6 5 3 5 6   2 2 6 2 1   5 6 7 4 2   2 5 2 3 1
5 4 7 2 3   7 4 6 5 4   1 4 4 6 3   5 1 1 5 2   4 2 3 2 5   6 2 1 5 2
1 7 6 7 4   2 6 4 1 3   3 7 2 7 7   7 5 7 1 7   6 7 6 7 1   7 3 5 6 7
6 1 3 1 5   4 2 3 7 5   2 2 5 3 2   3 6 3 4 5   1 1 2 5 4   5 7 4 4 5

5 2 4 7 1   6 4 5 6 7   7 1 1 6 1   6 7 4 4 1   2 3 5 6 4   7 1 7 6 6
2 6 5 1 4   7 1 1 1 3   4 7 7 2 2   3 5 2 3 7   5 7 7 5 7   5 7 4 4 7
4 1 2 6 7   1 2 7 5 1   3 5 6 1 6   4 6 7 2 3   4 6 4 2 3   1 5 5 1 3
3 5 7 3 6   4 3 2 7 5   1 4 2 5 5   1 3 1 1 2   6 5 1 4 6   6 4 6 5 1
6 4 3 5 5   3 5 4 3 6   2 3 4 3 3   5 2 6 7 4   3 4 6 1 5   2 6 2 7 2
1 3 6 4 2   5 6 6 2 4   5 2 3 7 7   2 4 5 6 5   1 2 2 7 2   4 3 1 3 5
7 7 1 2 3   2 7 3 4 2   6 6 5 4 4   7 1 3 5 6   7 1 3 3 1   3 2 3 2 4

1 4 4 5 2   3 2 1 6 1   4 1 2 6 7   1 7 5 7 4   5 4 1 2 3   4 6 7 1 2
3 3 5 1 7   6 1 7 7 4   1 6 6 3 1   7 4 6 2 3   3 5 6 4 5   7 7 3 3 7
4 7 6 3 4   1 6 2 2 7   5 3 3 2 4   2 6 1 5 5   6 6 2 2 3   1 5 1 2 4
7 2 1 2 5   4 5 4 4 5   2 7 7 1 5   4 3 3 1 6   1 3 3 7 6   5 1 2 7 5
5 5 2 7 3   2 3 5 5 6   6 2 1 5 3   3 5 2 3 2   2 7 7 5 1   6 4 6 6 6
6 1 3 6 1   5 4 3 3 2   7 5 5 7 6   6 2 4 4 7   7 1 5 1 7   2 3 5 5 1
2 6 7 4 6   7 7 6 1 3   3 4 4 4 2   5 1 7 6 1   4 2 4 6 2   3 2 4 4 3

7 3 5 6 6   1 3 2 3 1   4 2 7 4 4   7 6 3 4 6   5 4 1 5 5   6 5 7 7 2
4 5 6 4 1   2 7 5 2 5   5 7 4 7 3   3 2 7 7 2   3 5 6 2 7   5 2 3 5 7
3 7 7 7 4   4 5 4 4 4   6 4 2 5 2   5 7 5 2 7   1 6 3 3 1   7 7 1 3 3
5 6 4 1 5   6 1 7 5 7   3 1 1 3 1   6 4 2 1 1   7 7 5 6 4   3 6 5 2 6
2 1 2 2 3   7 6 1 6 6   2 5 3 2 5   2 5 4 5 4   6 2 2 4 3   4 3 4 4 4
6 4 3 5 7   5 4 6 7 2   1 3 6 6 7   4 1 6 6 5   4 1 4 7 2   2 4 2 1 5
1 2 1 3 2   3 2 3 1 3   7 6 5 1 6   1 3 1 3 3   2 3 7 1 2   1 1 6 6 1

2 2 6 5 4   3 2 7 1 5   6 1 3 2 3   6 5 7 4 4   3 1 7 2 3   3 6 5 5 3
7 5 5 7 7   1 3 1 3 2   7 4 7 4 4   2 2 5 6 1   6 7 3 4 6   2 4 2 7 7
5 1 7 2 3   4 6 5 4 7   4 6 4 7 2   1 4 3 1 5   2 6 5 1 2   6 3 4 4 5
6 3 1 6 1   7 5 3 7 6   5 3 5 5 6   3 7 4 3 2   1 2 1 3 5   7 7 6 1 4
1 6 3 3 2   2 1 4 6 4   3 7 6 3 1   7 3 6 5 3   7 5 4 6 4   5 1 7 6 1
3 7 2 4 6   5 7 6 2 1   1 2 2 1 7   5 1 2 7 7   4 3 6 5 1   4 2 1 3 6
4 4 4 1 5   6 4 2 5 3   2 5 1 6 5   4 6 1 2 6   5 4 2 7 7   1 5 3 2 2
```

Computed by the authors

TABLE A.2   (Continued)

```
2 3 4 3 4    2 2 3 5 1    5 6 4 6 5    1 5 1 3 2    3 2 1 3 1    1 1 4 7
4 6 3 2 7    3 4 1 3 2    2 2 7 4 6    2 6 7 2 4    1 4 3 1 3    7 7 1 3
7 5 2 5 2    4 6 6 2 4    3 1 1 3 4    5 7 3 7 3    5 6 2 5 5    2 6 5 2
1 1 5 7 1    7 7 4 6 5    7 4 2 7 2    7 1 4 4 1    6 5 7 2 6    3 4 3 5
5 4 7 6 3    6 5 2 4 6    6 3 6 1 7    3 3 6 1 7    4 7 4 4 7    6 3 6 4
6 2 6 1 5    5 1 5 1 7    1 5 3 5 1    6 2 5 6 5    7 1 5 6 2    5 5 2 1
3 7 1 4 6    1 3 7 7 3    4 7 5 2 3    4 4 2 5 6    2 3 6 7 4    4 2 7 6

1 2 6 2 1    5 4 6 6 1    7 2 3 1 7    5 3 2 6 3    7 2 5 2 2    6 1 7 1
3 4 5 5 5    6 2 1 1 6    5 7 6 7 6    7 5 1 5 4    4 3 4 7 3    7 6 3 6
5 3 4 3 3    4 1 3 7 5    6 4 7 3 3    3 4 5 4 1    5 6 6 1 4    5 2 4 3
4 6 7 6 7    3 5 5 3 2    3 3 2 5 5    1 6 6 2 6    1 1 2 3 5    2 4 2 7
6 5 1 1 2    2 7 4 4 4    4 1 4 6 4    4 7 4 3 2    3 4 7 6 6    3 3 6 2
7 7 3 4 4    7 3 2 5 3    2 6 1 2 1    2 1 3 1 5    6 7 1 5 1    1 5 1 4
2 1 2 7 6    1 6 7 2 7    1 5 5 4 2    6 2 7 7 7    2 5 3 4 7    4 7 5 5

3 4 2 2 7    4 7 5 5 2    6 2 5 7 7    1 5 4 3 6    2 6 1 2 6    6 3 5 2
1 5 7 3 1    1 3 4 1 5    4 3 4 4 5    4 7 5 5 2    4 5 3 6 7    4 1 2 4
2 1 5 6 4    2 2 3 7 3    5 5 3 1 3    3 2 3 6 4    1 4 7 3 2    1 4 3 3
4 7 3 7 5    7 5 2 6 6    7 7 6 6 6    6 1 6 4 1    3 3 5 1 4    7 2 4 1
5 3 4 4 3    5 6 1 2 7    1 6 7 2 2    2 3 1 1 7    5 2 6 5 5    5 5 6 6
6 2 1 1 2    3 1 7 4 4    2 1 2 5 4    7 4 7 2 5    7 1 4 7 1    2 6 7 7
7 6 6 5 6    6 4 6 3 1    3 4 1 3 1    5 6 2 7 3    6 7 2 4 3    3 7 1 5

3 6 3 6 5    5 6 5 1 4    7 5 1 5 2    4 1 1 1 2    4 7 1 2 5    5 5 1 3
1 2 5 1 3    6 1 6 5 2    2 2 6 2 6    7 3 4 6 4    5 3 5 3 7    3 4 2 4
5 1 2 2 1    3 7 7 7 3    4 4 2 3 4    6 4 5 2 1    2 4 6 4 1    2 1 5 5
7 4 1 7 4    1 4 1 2 6    6 3 3 1 5    3 6 6 3 5    6 6 4 5 4    6 2 7 7
2 3 6 3 7    4 5 4 4 5    5 6 5 7 7    5 5 2 4 7    1 2 7 6 2    7 7 6 2
4 7 4 5 6    7 3 2 6 7    3 1 4 4 1    2 7 7 7 3    7 5 2 7 3    1 6 4 1
6 5 7 4 2    2 2 3 3 1    1 7 7 6 3    1 2 3 5 6    3 1 3 1 6    4 3 3 6

7 2 2 6 4    6 6 4 7 7    3 7 4 6 7    3 1 4 6 3    1 6 7 7 5    5 4 7 2
5 7 7 4 1    4 3 2 6 4    2 6 1 1 6    7 3 6 1 1    2 5 1 5 3    3 2 2 7
6 4 1 2 5    3 1 1 4 6    7 4 6 5 5    5 7 2 2 4    5 7 3 3 6    1 6 3 6
4 6 4 3 6    7 5 6 3 1    4 1 5 7 3    6 2 7 4 2    7 2 4 6 1    6 3 4 3
3 3 3 1 3    2 4 5 1 5    6 3 7 3 4    4 5 1 5 7    4 4 2 4 7    7 5 5 1
2 5 6 5 7    5 2 7 2 3    5 5 3 2 1    2 6 5 3 6    3 3 6 1 2    2 7 1 4
1 1 5 7 2    1 7 3 5 2    1 2 2 4 2    1 4 3 7 5    6 1 5 2 4    4 1 6 5

6 7 6 3 3    5 4 6 3 6    2 3 7 7 6    1 1 5 3 5    1 3 7 6 7    6 3 7 1
4 5 2 4 6    7 6 5 4 3    4 7 5 5 4    5 7 2 4 1    2 5 2 7 3    7 4 1 6
1 3 5 6 4    3 7 2 5 4    6 6 1 4 1    2 2 4 1 3    6 6 1 3 1    5 7 6 4
2 1 3 2 5    6 3 1 2 1    3 2 6 2 3    3 5 7 5 4    3 2 4 4 2    2 6 4 7
3 4 7 5 2    2 2 3 7 2    5 5 2 6 7    6 4 6 2 6    4 7 6 2 6    3 2 3 5
5 2 4 1 1    4 1 7 6 7    7 4 3 3 5    4 3 3 6 2    7 4 5 1 4    1 5 5 2
7 6 1 7 7    1 5 4 1 5    1 1 4 1 2    7 6 1 7 7    5 1 3 5 5    4 1 2 3

7 1 7 6 2    2 7 3 2 1    3 3 6 2 1    2 5 1 4 5    2 5 5 5 3    4 3 2 4
2 5 1 4 1    3 6 4 6 3    6 5 2 1 4    5 6 7 2 3    7 1 1 7 4    5 7 1 7
4 4 2 7 3    5 6 7 1 6    2 7 7 6 3    3 4 3 3 7    1 4 3 2 6    3 5 5 1
5 6 5 5 5    1 1 1 7 5    5 6 1 7 6    4 7 6 6 1    3 2 6 4 7    1 2 6 5
6 3 4 2 7    7 4 5 4 2    4 4 4 3 2    7 1 4 1 2    4 7 2 3 1    2 1 7 2
1 2 6 3 4    4 5 2 5 4    1 1 5 5 5    6 3 5 7 6    6 3 4 6 2    6 4 3 6
3 7 3 1 6    3 2 6 3 7    7 2 3 4 7    1 2 2 5 4    5 6 7 1 5    7 6 4 3

7 1 7 4 5    6 6 6 7 1    3 6 3 6 1    3 7 6 3 4    2 1 4 4 6    4 7 5 3
2 7 5 2 4    7 5 7 1 2    5 7 7 7 7    1 1 7 1 2    6 6 1 5 3    6 1 2 4
1 5 4 6 1    3 1 2 4 7    4 5 2 2 6    7 5 1 4 3    4 2 6 3 4    2 3 5 5
3 6 6 7 7    2 2 3 2 6    7 4 4 4 2    5 6 4 7 7    7 5 3 6 5    3 3 6 7
5 2 1 3 3    4 7 1 3 5    1 2 1 1 3    4 4 5 2 1    3 3 7 2 7    1 6 4 2
6 3 2 5 2    5 3 5 5 4    6 1 5 5 5    6 3 2 5 5    1 4 5 1 2    5 4 1 6
4 4 3 1 6    1 4 4 6 3    2 3 6 3 4    2 2 3 6 6    5 7 2 7 1    2 5 7 1
```

TABLE A.3   RANDOM PERMUTATIONS OF 12

3	10	10	10	11		1	5	4	4	10		12	6	11	2	12		12	4	8	10	6
9	6	8	4	8		9	10	6	5	6		2	2	4	11	3		6	6	7	1	3
10	12	6	8	4		11	12	2	6	7		11	5	7	7	10		5	10	11	2	9
11	9	12	12	1		10	11	5	8	4		10	11	10	3	5		7	5	4	4	12
6	8	3	6	10		5	3	10	2	1		1	8	12	9	1		11	11	10	9	10
4	1	4	1	6		4	8	1	7	12		3	1	5	5	7		9	12	3	8	7
2	4	5	11	9		6	7	12	10	9		6	7	3	1	11		4	3	6	11	8
1	3	9	3	12		7	6	3	3	5		5	9	9	10	4		2	7	2	5	4
7	11	1	9	5		3	9	9	9	2		7	4	6	4	2		3	2	5	12	11
8	2	11	5	2		8	1	7	1	11		9	3	8	8	6		10	1	1	6	5
5	5	7	2	7		12	2	8	11	8		4	12	2	12	8		8	9	12	7	1
12	7	2	7	3		2	4	11	12	3		8	10	1	6	9		1	8	9	3	2
6	12	10	11	6		8	4	11	12	5		8	5	12	7	5		7	8	5	11	12
11	9	5	7	7		5	11	7	6	9		6	10	9	5	4		10	7	6	10	2
3	8	2	1	5		3	8	6	11	3		2	7	3	10	2		6	6	1	7	6
1	6	9	10	1		2	3	10	4	1		9	2	8	9	12		4	4	8	4	1
10	5	12	9	2		12	12	5	3	4		11	8	10	8	7		11	5	11	12	10
9	10	8	5	12		7	9	3	5	2		3	4	1	11	10		9	2	10	5	4
2	4	11	4	8		1	1	4	9	12		4	3	7	4	3		1	12	2	8	3
8	2	6	8	3		6	7	8	8	7		10	12	2	12	11		5	11	9	2	11
4	3	4	3	11		9	2	9	7	11		7	6	5	2	9		3	1	7	9	5
12	1	7	6	4		4	6	1	2	10		12	9	11	6	8		12	3	3	1	9
7	11	1	2	9		10	5	2	10	8		1	11	4	3	6		2	9	12	3	8
5	7	3	12	10		11	10	12	1	6		5	1	6	1	1		8	10	4	6	7
12	12	7	12	5		9	10	9	4	8		7	5	11	8	2		5	1	2	11	2
5	8	8	9	2		1	2	3	7	11		5	12	4	10	11		11	3	1	9	12
8	11	1	1	4		8	6	7	2	6		6	1	2	6	3		12	12	4	4	9
6	10	11	4	8		5	4	8	9	10		3	4	10	12	7		2	4	5	1	11
11	1	5	6	10		3	5	1	11	9		2	11	9	7	4		7	11	12	2	5
9	9	4	2	9		11	11	4	3	5		11	7	1	3	8		4	6	11	10	4
10	7	3	10	6		10	3	2	5	2		8	10	6	9	10		10	10	9	3	10
7	5	10	11	11		2	8	10	8	7		9	6	3	11	5		8	5	8	12	8
4	3	12	3	7		6	7	12	12	3		12	2	12	1	1		6	2	3	8	7
3	2	6	7	1		7	12	6	6	4		10	3	7	2	12		1	8	7	6	6
2	6	2	5	3		12	1	5	1	1		4	8	8	5	6		3	9	6	7	3
1	4	9	8	12		4	9	11	10	12		1	9	5	4	9		9	7	10	5	1
12	9	11	3	1		8	8	5	9	3		3	11	11	6	2		1	3	10	6	9
8	5	12	1	9		10	1	3	7	9		5	3	5	11	3		7	4	12	1	4
4	7	8	9	3		4	6	8	3	5		10	2	8	10	10		11	9	5	8	5
7	4	5	11	4		2	4	6	5	1		12	1	2	8	7		2	11	8	9	2
1	2	2	4	10		9	11	10	11	10		2	5	4	2	8		8	5	11	3	1
6	10	9	12	2		6	10	1	2	8		8	8	3	7	12		9	2	4	7	11
11	8	3	5	8		3	3	11	10	6		11	12	6	3	11		4	1	2	5	10
5	12	1	2	11		7	9	2	12	7		6	10	9	9	1		3	10	3	12	3
9	1	6	8	7		1	7	4	8	4		4	6	10	4	4		6	6	6	4	8
3	3	4	10	6		5	2	7	4	11		7	4	1	5	9		12	12	7	11	12
10	6	10	7	12		12	12	9	6	12		9	7	12	1	5		10	7	1	2	6
2	11	7	6	5		11	5	12	1	2		1	9	7	12	6		5	8	9	10	7

Computed by the authors

TABLE A.3   (Continued)

5	11	2	11	8	11	5	6	10	12	10	7	11	9	3	8	1	12	2	11
4	3	11	2	1	8	2	1	11	3	2	11	2	11	12	5	10	8	6	10
8	1	6	6	3	4	3	4	5	9	4	5	7	6	4	12	4	7	12	2
3	12	5	3	5	2	9	3	12	8	3	1	10	2	10	9	9	4	9	3
10	6	4	9	2	5	7	2	8	10	8	3	9	7	8	4	3	6	4	12
11	10	1	12	12	1	12	9	2	4	12	10	1	12	2	7	7	9	11	9
7	5	3	8	11	3	11	10	4	11	7	12	12	10	7	2	12	3	10	1
2	4	12	1	9	6	10	12	7	2	5	2	6	3	9	11	6	11	8	7
9	8	8	5	7	12	8	11	6	1	9	8	3	4	6	3	2	5	3	5
1	2	7	10	10	10	6	7	3	5	1	9	5	8	11	10	5	10	1	4
12	9	10	7	6	7	1	5	9	7	11	4	4	1	5	1	11	2	7	8
6	7	9	4	4	9	4	8	1	6	6	6	8	5	1	6	8	1	5	6
8	6	11	7	9	8	9	1	12	11	8	11	3	10	11	5	12	2	12	4
3	12	9	1	11	7	3	9	2	1	7	5	4	12	2	3	1	8	9	11
2	8	4	11	8	5	2	12	8	2	11	7	2	6	12	10	5	7	4	2
5	4	6	3	4	6	11	11	4	5	12	3	7	9	6	6	4	5	6	5
10	2	3	6	7	1	1	7	10	8	6	1	9	11	1	4	10	10	11	9
4	1	2	4	1	9	10	5	3	7	2	9	8	8	10	1	2	4	8	6
12	11	8	10	5	2	6	10	5	9	9	4	12	5	9	9	7	1	2	8
1	7	7	9	12	10	7	3	1	3	4	12	6	2	4	7	3	9	1	10
9	9	5	12	2	11	4	4	7	10	10	2	10	4	8	8	6	12	3	12
6	10	10	2	6	3	8	6	9	4	3	10	5	3	5	12	11	11	5	1
11	3	1	8	3	4	5	2	6	6	1	8	11	1	7	11	9	3	7	3
7	5	12	5	10	12	12	8	11	12	5	6	1	7	3	2	8	6	10	7
12	7	7	11	4	8	2	2	4	2	1	5	3	1	3	9	10	9	12	7
4	10	3	3	12	11	9	11	10	9	11	7	8	7	6	4	1	10	9	12
5	2	5	6	10	10	3	1	1	1	4	4	7	3	12	3	6	4	1	10
1	11	1	12	6	7	5	4	8	6	10	6	9	2	10	12	8	7	4	1
11	3	4	4	8	9	11	8	9	4	8	10	10	10	7	8	5	11	10	3
2	5	11	7	9	3	4	12	3	12	6	11	2	5	11	5	12	12	11	8
3	8	8	10	2	4	7	10	12	7	12	3	12	8	9	6	9	5	7	6
10	9	9	2	3	12	10	3	5	11	2	9	11	12	1	2	4	8	5	5
9	4	10	8	11	6	12	9	6	5	7	8	6	11	5	7	7	2	6	2
8	1	12	9	1	2	8	5	11	3	5	12	1	4	2	1	11	6	3	4
7	6	2	1	7	1	1	6	7	8	9	2	5	9	4	11	3	1	8	9
6	12	6	5	5	5	6	7	2	10	3	1	4	6	8	10	2	3	2	11
2	6	12	8	6	7	10	4	11	1	2	6	7	1	7	5	4	12	7	5
1	1	1	9	10	11	11	10	7	8	8	4	10	8	5	7	12	6	3	3
4	5	6	4	11	2	2	1	8	10	7	8	4	6	12	10	7	2	1	1
6	4	8	3	4	5	4	3	9	11	5	9	8	3	10	11	1	3	6	9
9	9	2	2	2	4	12	5	6	4	1	1	6	7	2	2	6	9	4	10
8	12	10	7	1	1	3	12	4	9	10	12	3	4	4	4	11	11	10	12
11	7	9	1	9	12	7	2	5	7	3	7	12	12	1	1	3	1	9	11
7	2	4	10	5	9	9	7	10	2	11	5	1	10	3	8	8	4	2	7
3	11	7	6	3	3	6	9	12	12	4	10	11	9	11	3	5	10	8	4
10	8	3	12	8	6	8	8	3	5	12	11	5	11	8	12	2	5	5	2
12	3	5	11	7	10	1	6	1	6	6	3	9	5	6	9	10	7	12	8
5	10	11	5	12	8	5	11	2	3	9	2	2	2	9	6	9	8	11	6
7	3	1	12	7	10	5	9	10	11	11	7	5	1	1	7	2	1	12	4
3	10	9	3	11	5	6	5	7	5	6	2	10	11	10	3	9	2	3	8
1	11	7	7	4	8	12	12	8	7	2	9	1	9	8	6	12	12	1	9
12	2	6	1	8	7	4	8	2	3	5	12	4	7	9	12	3	8	10	12
5	8	8	5	2	4	3	6	9	1	4	3	7	5	11	4	6	4	9	3
4	4	5	8	6	11	1	3	5	10	12	4	9	4	4	10	1	7	2	10
6	7	11	10	9	6	9	4	11	2	10	11	3	12	3	11	11	9	5	6
10	5	12	6	10	2	8	10	12	9	1	6	2	10	6	8	5	3	4	11
11	1	4	2	5	3	7	7	6	8	3	8	11	3	2	9	10	6	8	5
2	6	10	9	1	12	2	1	4	6	9	5	6	6	12	2	8	10	6	7
9	12	3	4	3	9	10	11	3	4	7	10	12	8	5	5	7	11	7	1
8	9	2	11	12	1	11	2	1	12	8	1	8	2	7	1	4	5	11	2

TABLE A.4   LATIN SQUARES

2 x 2

A B
B A

3 x 3

A B C
B C A
C A B

4 x 4

A B C D	A B C D	A B C D	A B C D
B C D A	B D A C	B A D C	B A D C
C D A B	C A D B	C D A B	C D B A
D A B C	D C B A	D C B A	D C A B

5 x 5

A B C D E
B D E A C
C A D E B
D E B C A
E C A B D

6 x 6

A B C D E F
B F D C A E
C A E B F D
D E B F C A
E C F A D B
F D A E B C

7 x 7

A B C D E F G
B E G A C D F
C A D B F G E
D F A E G C B
E C F G B A D
F G B C D E A
G D E F A B C

8 x 8

A B C D E F G H
B D A H G C E F
C E G F A B H D
D A B C H G F E
E C H A F D B G
F G D E B H A C
G H F B C E D A
H F E G D A C B

9 x 9

A B C D E F G H I
B H F E C A I D G
C A H I D B E G F
D I B C G E A F H
E C I F B G H A D
F G E A H I D B C
G D A B I H F C E
H E D G F C B I A
I F G H A D C E B

10 x 10

A B C D E F G H I J
B F J I A C E D G H
C A G E D J I B H F
D H I A C E B J F G
E C A F J G H I B D
F D E B G H A C J I
G J D C H I F A E B
H I B G F A J E D C
I G H J B D C F A E
J E F H I B D G C A

TABLE A.5   THE STANDARD NORMAL DISTRIBUTION

Entries are $P(Z \leq z_\gamma) = \gamma$ for values of z_γ from 0.00 to 3.99.

z_γ	.00	.01	.02	.03	.04
0.0	0.50000	0.50399	0.50798	0.51197	0.51596
0.1	0.53983	0.54380	0.54776	0.55172	0.55567
0.2	0.57926	0.58317	0.58707	0.59096	0.59484
0.3	0.61791	0.62172	0.62552	0.62930	0.63307
0.4	0.65542	0.65910	0.66276	0.66640	0.67003
0.5	0.69146	0.69498	0.69847	0.70194	0.70540
0.6	0.72575	0.72907	0.73237	0.73565	0.73891
0.7	0.75804	0.76115	0.76424	0.76731	0.77035
0.8	0.78815	0.79103	0.79389	0.79673	0.79955
0.9	0.81594	0.81859	0.82121	0.82381	0.82639
1.0	0.84135	0.84375	0.84614	0.84850	0.85083
1.1	0.86433	0.86650	0.86864	0.87076	0.87286
1.2	0.88493	0.88686	0.88877	0.89065	0.89251
1.3	0.90320	0.90490	0.90658	0.90824	0.90988
1.4	0.91924	0.92073	0.92220	0.92364	0.92507
1.5	0.93319	0.93448	0.93574	0.93699	0.93822
1.6	0.94520	0.94630	0.94738	0.94845	0.94950
1.7	0.95543	0.95637	0.95728	0.95818	0.95907
1.8	0.96407	0.96485	0.96562	0.96638	0.96712
1.9	0.97128	0.97193	0.97257	0.97320	0.97381
2.0	0.97725	0.97778	0.97831	0.97882	0.97932
2.1	0.98214	0.98257	0.98300	0.98341	0.98382
2.2	0.98610	0.98645	0.98679	0.98713	0.98745
2.3	0.98928	0.98956	0.98983	0.99010	0.99036
2.4	0.99180	0.99202	0.99224	0.99245	0.99266
2.5	0.99379	0.99396	0.99413	0.99430	0.99446
2.6	0.99534	0.99547	0.99560	0.99573	0.99585
2.7	0.99653	0.99664	0.99674	0.99683	0.99693
2.8	0.99744	0.99752	0.99760	0.99767	0.99774
2.9	0.99813	0.99819	0.99825	0.99831	0.99836
3.0	0.99865	0.99869	0.99874	0.99878	0.99882
3.1	0.99903	0.99906	0.99910	0.99913	0.99916
3.2	0.99931	0.99934	0.99936	0.99938	0.99940
3.3	0.99952	0.99953	0.99955	0.99957	0.99958
3.4	0.99966	0.99968	0.99969	0.99970	0.99971
3.5	0.99977	0.99978	0.99978	0.99979	0.99980
3.6	0.99984	0.99985	0.99985	0.99986	0.99986
3.7	0.99989	0.99990	0.99990	0.99990	0.99991
3.8	0.99993	0.99993	0.99993	0.99994	0.99994
3.9	0.99995	0.99995	0.99996	0.99996	0.99996

TABLE A.5   (Continued)

.05	.06	.07	.08	.09	z_γ
0.51994	0.52392	0.52790	0.53188	0.53586	0.0
0.55962	0.56356	0.56750	0.57143	0.57535	0.1
0.59871	0.60257	0.60642	0.61026	0.61409	0.2
0.63683	0.64058	0.64431	0.64803	0.65173	0.3
0.67365	0.67724	0.68082	0.68439	0.68793	0.4
0.70884	0.71226	0.71566	0.71904	0.72241	0.5
0.74215	0.74537	0.74857	0.75175	0.75490	0.6
0.77337	0.77637	0.77935	0.78231	0.78524	0.7
0.80234	0.80511	0.80785	0.81057	0.81327	0.8
0.82894	0.83147	0.83398	0.83646	0.83891	0.9
0.85314	0.85543	0.85769	0.85993	0.86214	1.0
0.87493	0.87698	0.87900	0.88100	0.88298	1.1
0.89435	0.89617	0.89796	0.89973	0.90147	1.2
0.91149	0.91309	0.91466	0.91621	0.91774	1.3
0.92647	0.92786	0.92922	0.93056	0.93189	1.4
0.93943	0.94062	0.94179	0.94295	0.94408	1.5
0.95053	0.95154	0.95254	0.95352	0.95449	1.6
0.95994	0.96080	0.96164	0.96246	0.96327	1.7
0.96784	0.96856	0.96926	0.96995	0.97062	1.8
0.97441	0.97500	0.97558	0.97615	0.97670	1.9
0.97982	0.98030	0.98077	0.98124	0.98169	2.0
0.98422	0.98461	0.98500	0.98537	0.98574	2.1
0.98778	0.98809	0.98840	0.98870	0.98899	2.2
0.99061	0.99086	0.99111	0.99134	0.99158	2.3
0.99286	0.99305	0.99324	0.99343	0.99361	2.4
0.99461	0.99477	0.99492	0.99506	0.99520	2.5
0.99598	0.99609	0.99621	0.99632	0.99643	2.6
0.99702	0.99711	0.99720	0.99728	0.99736	2.7
0.99781	0.99788	0.99795	0.99801	0.99807	2.8
0.99841	0.99846	0.99851	0.99856	0.99861	2.9
0.99886	0.99889	0.99893	0.99896	0.99900	3.0
0.99918	0.99921	0.99924	0.99926	0.99929	3.1
0.99942	0.99944	0.99946	0.99948	0.99950	3.2
0.99960	0.99961	0.99962	0.99964	0.99965	3.3
0.99972	0.99973	0.99974	0.99975	0.99976	3.4
0.99981	0.99981	0.99982	0.99983	0.99983	3.5
0.99987	0.99987	0.99988	0.99988	0.99989	3.6
0.99991	0.99992	0.99992	0.99992	0.99992	3.7
0.99994	0.99994	0.99995	0.99995	0.99995	3.8
0.99996	0.99996	0.99996	0.99997	0.99997	3.9

Computed by the authors

TABLE A.6   PERCENTAGE POINTS OF T DISTRIBUTIONS

Upper tail area

ν	0.25	0.10	0.05	0.025	0.01	0.005
1	1.000	3.078	6.314	12.706	31.821	63.657
2	0.816	1.886	2.920	4.303	6.965	9.925
3	0.765	1.638	2.353	3.182	4.541	5.841
4	0.741	1.533	2.132	2.776	3.747	4.604
5	0.727	1.476	2.015	2.571	3.365	4.032
6	0.718	1.440	1.943	2.447	3.143	3.707
7	0.711	1.415	1.895	2.365	2.998	3.499
8	0.706	1.397	1.860	2.306	2.896	3.355
9	0.703	1.383	1.833	2.262	2.821	3.250
10	0.700	1.372	1.812	2.228	2.764	3.169
11	0.697	1.363	1.796	2.201	2.718	3.106
12	0.695	1.356	1.782	2.179	2.681	3.055
13	0.694	1.350	1.771	2.160	2.650	3.012
14	0.692	1.345	1.761	2.145	2.624	2.977
15	0.691	1.341	1.753	2.131	2.602	2.947
16	0.690	1.337	1.746	2.120	2.583	2.921
17	0.689	1.333	1.740	2.110	2.567	2.898
18	0.688	1.330	1.734	2.101	2.552	2.878
19	0.688	1.328	1.729	2.093	2.539	2.861
20	0.687	1.325	1.725	2.086	2.528	2.845
21	0.686	1.323	1.721	2.080	2.518	2.831
22	0.686	1.321	1.717	2.074	2.508	2.819
23	0.685	1.319	1.714	2.069	2.500	2.807
24	0.685	1.318	1.711	2.064	2.492	2.797
25	0.684	1.316	1.708	2.060	2.485	2.787
26	0.684	1.315	1.706	2.056	2.479	2.779
27	0.684	1.314	1.703	2.052	2.473	2.771
28	0.683	1.313	1.701	2.048	2.467	2.763
29	0.683	1.311	1.699	2.045	2.462	2.756
30	0.683	1.310	1.697	2.042	2.457	2.750
40	0.681	1.303	1.684	2.021	2.423	2.704
60	0.679	1.296	1.671	2.000	2.390	2.660
120	0.677	1.289	1.658	1.980	2.358	2.617
∞	0.674	1.282	1.645	1.960	2.326	2.576

TABLE A.7   PERCENTAGE POINTS OF CHI-SQUARED DISTRIBUTIONS

Let χ^2_ν denote a chi-squared variable with ν degrees of freedom.

Entries are $\chi^2_{\nu;\gamma}$ where $P(\chi^2_\nu \leq \chi^2_{\nu;\gamma}) = \gamma$.

ν	.005	.01	.025	.05	.10	.90	.95	.975	.99	.995
1	0.00	0.00	0.00	0.00	0.02	2.71	3.84	5.02	6.63	7.88
2	0.01	0.02	0.05	0.10	0.21	4.61	5.99	7.38	9.21	10.60
3	0.07	0.11	0.22	0.35	0.58	6.25	7.81	9.35	11.34	12.84
4	0.21	0.30	0.48	0.71	1.06	7.78	9.49	11.14	13.28	14.86
5	0.41	0.55	0.83	1.15	1.61	9.24	11.07	12.83	15.09	16.75
6	0.68	0.87	1.24	1.64	2.20	10.64	12.59	14.45	16.81	18.55
7	0.99	1.24	1.69	2.17	2.83	12.02	14.07	16.01	18.48	20.28
8	1.34	1.65	2.18	2.73	3.49	13.36	15.51	17.53	20.09	21.96
9	1.73	2.09	2.70	3.33	4.17	14.68	16.92	19.02	21.67	23.59
10	2.16	2.56	3.25	3.94	4.87	15.99	18.31	20 48	23.21	25.19
11	2.60	3.05	3.82	4.57	5.58	17.28	19.68	21.92	24.73	26.76
12	3.07	3.57	4.40	5.23	6.30	18.55	21.03	23.34	26.22	28.30
13	3.57	4.11	5.01	5.89	7.04	19.81	22.36	24.74	27.69	29.82
14	4.07	4.66	5.63	6.57	7.79	21.06	23.68	26.12	29.14	31.32
15	4.60	5.23	6.26	7.26	8.55	22.31	25.00	27.49	30.58	32.80
16	5.14	5.81	6.91	7.96	9.31	23.54	26.30	28.85	32.00	34.27
17	5.70	6.41	7.56	8.67	10.09	24.77	27.59	30.19	33.41	35.72
18	6.26	7.01	8.23	9.39	10.86	25.99	28.87	31.53	34.81	37.16
19	6.84	7.63	8.91	10.12	11.65	27.20	30.14	32.85	36.19	38.58
20	7.43	8.26	9.59	10.85	12.44	28.41	31.41	34.17	37.57	40.00
21	8.03	8.90	10.28	11.59	13.24	29.62	32.67	35.48	38.93	41.40
22	8.64	9.54	10.98	12.34	14.04	30.81	33.92	36.78	40.29	42.80
23	9.26	10.20	11.69	13.09	14.85	32.01	35.17	38.08	41.64	44.18
24	9.89	10.86	12.40	13.85	15.66	33.20	36.42	39.36	42.98	45.56
25	10.52	11.52	13.12	14.61	16.47	34.38	37.65	40.65	44.31	46.93
30	13.79	14.95	16.79	18.49	20.60	40.26	43.77	46.98	50.89	53.67
40	20.71	22.16	24.43	26.51	29.05	51.81	55.76	59.34	63.69	66.77
50	27.99	29.71	32.36	34.76	37.69	63.17	67.50	71.42	76.15	79.49
60	35.53	37.48	40.48	43.19	46.46	74.40	79.08	83.30	88.38	91.95
80	51.17	53.54	57.15	60.39	64.28	96.58	101.9	106.6	112.3	116.3
100	67.33	70.06	74.22	77.93	82.36	118.5	124.3	129.6	135.8	140.2

TABLE A.8   PERCENTAGE POINTS OF THE MAXIMUM F DISTRIBUTIONS

For a set of k independent variances (mean squares) each
based on ν degrees of freedom.

Upper 5% points

ν \ k	2	3	4	5	6	7	8	9	10	11	12
2	39·0	87·5	142	202	266	333	403	475	550	626	704
3	15·4	27·8	39·2	50·7	62·0	72·9	83·5	93·9	104	114	124
4	9·60	15·5	20·6	25·2	29·5	33·6	37·5	41·1	44·6	48·0	51·
5	7·15	10·8	13·7	16·3	18·7	20·8	22·9	24·7	26·5	28·2	29·
6	5·82	8·38	10·4	12·1	13·7	15·0	16·3	17·5	18·6	19·7	20·
7	4·99	6·94	8·44	9·70	10·8	11·8	12·7	13·5	14·3	15·1	15·
8	4·43	6·00	7·18	8·12	9·03	9·78	10·5	11·1	11·7	12·2	12·
9	4·03	5·34	6·31	7·11	7·80	8·41	8·95	9·45	9·91	10·3	10·
10	3·72	4·85	5·67	6·34	6·92	7·42	7·87	8·28	8·66	9·01	9·
12	3·28	4·16	4·79	5·30	5·72	6·09	6·42	6·72	7·00	7·25	7·
15	2·86	3·54	4·01	4·37	4·68	4·95	5·19	5·40	5·59	5·77	5·
20	2·46	2·95	3·29	3·54	3·76	3·94	4·10	4·24	4·37	4·49	4·
30	2·07	2·40	2·61	2·78	2·91	3·02	3·12	3·21	3·29	3·36	3·
60	1·67	1·85	1·96	2·04	2·11	2·17	2·22	2·26	2·30	2·33	2·
∞	1·00	1·00	1·00	1·00	1·00	1·00	1·00	1·00	1·00	1·00	1·

Upper 1% points

ν \ k	2	3	4	5	6	7	8	9	10	11	12
2	199	448	729	1036	1362	1705	2063	2432	2813	3204	3605
3	47·5	85	120	151	184	21(6)	24(9)	28(1)	31(0)	33(7)	36(
4	23·2	37	49	59	69	79	89	97	106	113	120
5	14·9	22	28	33	38	42	46	50	54	57	60
6	11·1	15·5	19·1	22	25	27	30	32	34	36	37
7	8·89	12·1	14·5	16·5	18·4	20	22	23	24	26	27
8	7·50	9·9	11·7	13·2	14·5	15·8	16·9	17·9	18·9	19·8	21
9	6·54	8·5	9·9	11·1	12·1	13·1	13·9	14·7	15·3	16·0	16
10	5·85	7·4	8·6	9·6	10·4	11·1	11·8	12·4	12·9	13·4	13
12	4·91	6·1	6·9	7·6	8·2	8·7	9·1	9·5	9·9	10·2	10
15	4·07	4·9	5·5	6·0	6·4	6·7	7·1	7·3	7·5	7·8	8
20	3·32	3·8	4·3	4·6	4·9	5·1	5·3	5·5	5·6	5·8	5
30	2·63	3·0	3·3	3·4	3·6	3·7	3·8	3·9	4·0	4·1	4
60	1·96	2·2	2·3	2·4	2·4	2·5	2·5	2·6	2·6	2·7	2
∞	1·00	1·0	1·0	1·0	1·0	1·0	1·0	1·0	1·0	1·0	1

TABLE A.9   ORTHOGONAL POLYNOMIAL COEFFICIENTS

t	Degree	ξ_{ij}									$\Sigma_i \xi^2_{ij}$
3	1	-1	0	1							2
	2	1	-2	1							6
4	1	-3	-1	1	3						20
	2	1	-1	-1	1						4
	3	-1	3	-3	1						20
5	1	-2	-1	0	1	2					10
	2	2	-1	-2	-1	2					14
	3	-1	2	0	-2	1					10
	4	1	-4	6	-4	1					70
6	1	-5	-3	-1	1	3	5				70
	2	5	-1	-4	-4	-1	5				84
	3	-5	7	4	-4	-7	5				180
	4	1	-3	2	2	-3	1				28
7	1	-3	-2	-1	0	1	2	3			28
	2	5	0	-3	-4	-3	0	5			84
	3	-1	1	1	0	-1	-1	1			6
	4	3	-7	1	6	1	-7	3			154
8	1	-7	-5	-3	-1	1	3	5	7		168
	2	7	1	-3	-5	-5	-3	1	7		168
	3	-7	5	7	3	-3	-7	-5	7		264
	4	7	-13	-3	9	9	-3	-13	7		616
9	1	-4	-3	-2	-1	0	1	2	3	4	60
	2	28	7	-8	-17	-20	-17	-8	7	28	2772
	3	-14	7	13	9	0	-9	-13	-7	14	990
	4	14	-21	-11	9	18	9	-11	-21	14	2002
10	1	-9	-7	-5	-3	-1	1	3	5	7	9	330
	2	6	2	-1	-3	-4	-4	-3	-1	2	6	132
	3	-42	14	35	31	12	-12	-31	-35	-14	42	8580
	4	18	-22	-17	3	18	18	3	-17	-22	18	2860

Computed by the authors

TABLE A.10   PERCENTAGE POINTS OF F DISTRIBUTIONS

Let F_{ν_1,ν_2} denote an F variable having ν_1 and ν_2 degrees of freedom for the numerator and denominator, respectively. Entries are $F_{\nu_1,\nu_2;\gamma}$ where $P(F_{\nu_1,\nu_2} \leq F_{\nu_1,\nu_2;\gamma}) = \gamma$.

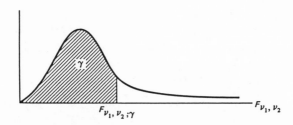

For $\nu_1 = 10$ and $\nu_2 = 17$, the table gives:

$$F_{10,17;0.005} = 0.186; \qquad F_{10,17;0.10} = 0.450$$
$$F_{10,17;0.975} = 2.92; \qquad F_{10,17;0.99} = 3.59$$

Computed by the authors

TABLE A.10   (Continued)

v_2	v_1 / γ	1	2	3	4	5	6	7	8	9	10	v_1 / γ
1	0.005	0.000	0.005	0.018	0.032	0.044	0.054	0.062	0.068	0.073	0.078	0.005
	0.010	0.000	0.010	0.029	0.047	0.062	0.073	0.082	0.089	0.095	0.100	0.010
	0.025	0.002	0.026	0.057	0.082	0.100	0.113	0.124	0.132	0.139	0.144	0.025
	0.050	0.006	0.054	0.099	0.130	0.151	0.167	0.179	0.188	0.195	0.201	0.050
	0.100	0.025	0.117	0.181	0.220	0.246	0.265	0.279	0.289	0.298	0.304	0.100
	0.900	39.86	49.50	53.59	55.83	57.24	58.20	58.91	59.44	59.86	60.19	0.900
	0.950	161.4	199.5	215.7	224.6	230.2	234.0	236.8	238.9	240.5	241.9	0.950
	0.975	647.8	799.5	864.2	899.6	921.8	937.1	948.2	956.7	963.3	968.6	0.975
	0.990	4052.	5000.	5403.	5625.	5764.	5859.	5928.	5981.	6022.	6056.	0.990
	0.995	16211	20000	21615	22500	23056	23437	23715	23926	24091	24225	0.995
2	0.005	0.000	0.005	0.020	0.038	0.055	0.069	0.081	0.091	0.099	0.106	0.005
	0.010	0.000	0.010	0.032	0.056	0.075	0.092	0.105	0.116	0.125	0.132	0.010
	0.025	0.001	0.026	0.062	0.094	0.119	0.138	0.153	0.165	0.175	0.183	0.025
	0.050	0.005	0.053	0.105	0.144	0.173	0.194	0.211	0.224	0.235	0.244	0.050
	0.100	0.020	0.111	0.183	0.231	0.265	0.289	0.307	0.321	0.333	0.342	0.100
	0.900	8.53	9.00	9.16	9.24	9.29	9.33	9.35	9.37	9.38	9.39	0.900
	0.950	18.51	19.00	19.16	19.25	19.30	19.33	19.35	19.37	19.38	19.40	0.950
	0.975	38.51	39.00	39.17	39.25	39.30	39.33	39.36	39.37	39.39	39.40	0.975
	0.990	98.50	99.00	99.17	99.25	99.30	99.33	99.36	99.37	99.39	99.40	0.990
	0.995	198.5	199.0	199.2	199.2	199.3	199.3	199.4	199.4	199.4	199.4	0.995
3	0.005	0.000	0.005	0.021	0.041	0.060	0.077	0.092	0.104	0.115	0.124	0.005
	0.010	0.000	0.010	0.034	0.060	0.083	0.102	0.118	0.132	0.143	0.153	0.010
	0.025	0.001	0.026	0.065	0.100	0.129	0.152	0.170	0.185	0.197	0.207	0.025
	0.050	0.005	0.052	0.108	0.152	0.185	0.210	0.230	0.246	0.259	0.270	0.050
	0.100	0.019	0.109	0.185	0.239	0.276	0.304	0.325	0.342	0.356	0.367	0.100
	0.900	5.54	5.46	5.39	5.34	5.31	5.29	5.27	5.25	5.24	5.23	0.900
	0.950	10.13	9.55	9.28	9.12	9.01	8.94	8.89	8.85	8.81	8.79	0.950
	0.975	17.44	16.04	15.44	15.10	14.89	14.74	14.63	14.55	14.48	14.43	0.975
	0.990	34.12	30.82	29.47	28.72	28.25	27.93	27.70	27.52	27.38	27.27	0.990
	0.995	55.55	49.80	47.50	46.24	45.45	44.91	44.52	44.23	44.00	43.82	0.995
4	0.005	0.000	0.005	0.022	0.043	0.064	0.083	0.099	0.114	0.126	0.136	0.005
	0.010	0.000	0.010	0.035	0.063	0.088	0.109	0.127	0.143	0.156	0.167	0.010
	0.025	0.001	0.025	0.066	0.104	0.135	0.161	0.181	0.198	0.212	0.224	0.025
	0.050	0.004	0.052	0.110	0.157	0.193	0.221	0.243	0.261	0.275	0.288	0.050
	0.100	0.018	0.108	0.187	0.243	0.284	0.314	0.338	0.356	0.371	0.384	0.100
	0.900	4.54	4.32	4.19	4.11	4.05	4.01	3.98	3.95	3.94	3.92	0.900
	0.950	7.71	6.94	6.59	6.39	6.26	6.16	6.09	6.04	6.00	5.96	0.950
	0.975	12.22	10.65	9.98	9.60	9.36	9.20	9.07	8.98	8.90	8.84	0.975
	0.990	21.20	18.00	16.69	15.98	15.52	15.21	14.98	14.80	14.66	14.55	0.990
	0.995	31.33	26.28	24.26	23.15	22.46	21.97	21.62	21.35	21.14	20.97	0.995
5	0.005	0.000	0.005	0.022	0.045	0.067	0.087	0.105	0.120	0.134	0.146	0.005
	0.010	0.000	0.010	0.035	0.064	0.091	0.114	0.134	0.151	0.165	0.177	0.010
	0.025	0.001	0.025	0.067	0.107	0.140	0.167	0.189	0.208	0.223	0.236	0.025
	0.050	0.004	0.052	0.111	0.160	0.198	0.228	0.252	0.271	0.287	0.301	0.050
	0.100	0.017	0.108	0.188	0.247	0.290	0.322	0.347	0.367	0.383	0.397	0.100
	0.900	4.06	3.78	3.62	3.52	3.45	3.40	3.37	3.34	3.32	3.30	0.900
	0.950	6.61	5.79	5.41	5.19	5.05	4.95	4.88	4.82	4.77	4.73	0.950
	0.975	10.01	8.43	7.76	7.39	7.15	6.98	6.85	6.76	6.68	6.62	0.975
	0.990	16.26	13.27	12.06	11.39	10.97	10.67	10.46	10.29	10.16	10.05	0.990
	0.995	22.78	18.31	16.53	15.56	14.94	14.51	14.20	13.96	13.77	13.62	0.995

TABLE A.10 (Continued)

v_2	v_1 γ	1	2	3	4	5	6	7	8	9	10	v_1 γ
6	0.005	0.000	0.005	0.022	0.046	0.069	0.090	0.109	0.126	0.140	0.153	0.005
	0.010	0.000	0.010	0.036	0.066	0.094	0.118	0.139	0.157	0.172	0.186	0.010
	0.025	0.001	0.025	0.068	0.109	0.143	0.172	0.195	0.215	0.231	0.246	0.025
	0.050	0.004	0.052	0.112	0.162	0.202	0.233	0.259	0.279	0.296	0.311	0.050
	0.100	0.017	0.107	0.189	0.249	0.294	0.327	0.354	0.375	0.392	0.406	0.100
	0.900	3.78	3.46	3.29	3.18	3.11	3.05	3.01	2.98	2.96	2.94	0.900
	0.950	5.99	5.14	4.76	4.53	4.39	4.28	4.21	4.15	4.10	4.06	0.950
	0.975	8.81	7.26	6.60	6.23	5.99	5.82	5.70	5.60	5.52	5.46	0.975
	0.990	13.75	10.92	9.78	9.15	8.75	8.47	8.26	8.10	7.98	7.87	0.990
	0.995	18.63	14.54	12.92	12.03	11.46	11.07	10.79	10.57	10.39	10.25	0.995
7	0.005	0.000	0.005	0.022	0.046	0.070	0.093	0.113	0.130	0.145	0.159	0.005
	0.010	0.000	0.010	0.036	0.067	0.096	0.121	0.143	0.162	0.178	0.192	0.010
	0.025	0.001	0.025	0.068	0.110	0.146	0.176	0.200	0.221	0.238	0.253	0.025
	0.050	0.004	0.052	0.113	0.164	0.205	0.238	0.264	0.286	0.304	0.319	0.050
	0.100	0.017	0.107	0.190	0.251	0.297	0.332	0.359	0.381	0.399	0.414	0.100
	0.900	3.59	3.26	3.07	2.96	2.88	2.83	2.78	2.75	2.72	2.70	0.900
	0.950	5.59	4.74	4.35	4.12	3.97	3.87	3.79	3.73	3.68	3.64	0.950
	0.975	8.07	6.54	5.89	5.52	5.29	5.12	4.99	4.90	4.82	4.76	0.975
	0.990	12.25	9.55	8.45	7.85	7.46	7.19	6.99	6.84	6.72	6.62	0.990
	0.995	16.24	12.40	10.88	10.05	9.52	9.16	8.89	8.68	8.51	8.38	0.995
8	0.005	0.000	0.005	0.023	0.047	0.072	0.095	0.115	0.133	0.149	0.164	0.005
	0.010	0.000	0.010	0.036	0.068	0.097	0.123	0.146	0.166	0.183	0.198	0.010
	0.025	0.001	0.025	0.069	0.111	0.148	0.179	0.204	0.226	0.244	0.259	0.025
	0.050	0.004	0.052	0.113	0.166	0.208	0.241	0.268	0.291	0.310	0.326	0.050
	0.100	0.017	0.107	0.190	0.253	0.299	0.335	0.363	0.386	0.405	0.421	0.100
	0.900	3.46	3.11	2.92	2.81	2.73	2.67	2.62	2.59	2.56	2.54	0.900
	0.950	5.32	4.46	4.07	3.84	3.69	3.58	3.50	3.44	3.39	3.35	0.950
	0.975	7.57	6.06	5.42	5.05	4.82	4.65	4.53	4.43	4.36	4.30	0.975
	0.990	11.26	8.65	7.59	7.01	6.63	6.37	6.18	6.03	5.91	5.81	0.990
	0.995	14.69	11.04	9.60	8.81	8.30	7.95	7.69	7.50	7.34	7.21	0.995
9	0.005	0.000	0.005	0.023	0.047	0.073	0.096	0.117	0.136	0.153	0.168	0.005
	0.010	0.000	0.010	0.037	0.068	0.098	0.125	0.149	0.169	0.187	0.202	0.010
	0.025	0.001	0.025	0.069	0.112	0.150	0.181	0.207	0.230	0.248	0.265	0.025
	0.050	0.004	0.052	0.113	0.167	0.210	0.244	0.272	0.295	0.315	0.331	0.050
	0.100	0.017	0.107	0.191	0.254	0.302	0.338	0.367	0.390	0.410	0.426	0.100
	0.900	3.36	3.01	2.81	2.69	2.61	2.55	2.51	2.47	2.44	2.42	0.900
	0.950	5.12	4.26	3.86	3.63	3.48	3.37	3.29	3.23	3.18	3.14	0.950
	0.975	7.21	5.71	5.08	4.72	4.48	4.32	4.20	4.10	4.03	3.96	0.975
	0.990	10.56	8.02	6.99	6.42	6.06	5.80	5.61	5.47	5.35	5.26	0.990
	0.995	13.61	10.11	8.72	7.96	7.47	7.13	6.88	6.69	6.54	6.42	0.995
10	0.005	0.000	0.005	0.023	0.048	0.073	0.098	0.119	0.139	0.156	0.171	0.005
	0.010	0.000	0.010	0.037	0.069	0.099	0.127	0.151	0.172	0.190	0.206	0.010
	0.025	0.001	0.025	0.069	0.113	0.151	0.183	0.210	0.233	0.252	0.269	0.025
	0.050	0.004	0.052	0.114	0.168	0.211	0.246	0.275	0.299	0.319	0.336	0.050
	0.100	0.017	0.106	0.191	0.255	0.303	0.340	0.370	0.394	0.414	0.431	0.100
	0.900	3.29	2.92	2.73	2.61	2.52	2.46	2.41	2.38	2.35	2.32	0.900
	0.950	4.96	4.10	3.71	3.48	3.33	3.22	3.14	3.07	3.02	2.98	0.950
	0.975	6.94	5.46	4.83	4.47	4.24	4.07	3.95	3.85	3.78	3.72	0.975
	0.990	10.04	7.56	6.55	5.99	5.64	5.39	5.20	5.06	4.94	4.85	0.990
	0.995	12.83	9.43	8.08	7.34	6.87	6.54	6.30	6.12	5.97	5.85	0.995

TABLE A.10   (Continued)

v_2	v_1 γ	1	2	3	4	5	6	7	8	9	10	v_1 γ
11	0.005	0.000	0.005	0.023	0.048	0.074	0.099	0.121	0.141	0.158	0.174	0.005
	0.010	0.000	0.010	0.037	0.069	0.100	0.128	0.153	0.174	0.193	0.210	0.010
	0.025	0.001	0.025	0.070	0.114	0.152	0.185	0.212	0.236	0.256	0.273	0.025
	0.050	0.004	0.052	0.114	0.168	0.213	0.248	0.278	0.302	0.322	0.340	0.050
	0.100	0.017	0.106	0.191	0.256	0.305	0.343	0.373	0.397	0.417	0.434	0.100
	0.900	3.23	2.86	2.66	2.54	2.45	2.39	2.34	2.30	2.27	2.25	0.900
	0.950	4.84	3.98	3.59	3.36	3.20	3.09	3.01	2.95	2.90	2.85	0.950
	0.975	6.72	5.26	4.63	4.28	4.04	3.88	3.76	3.66	3.59	3.53	0.975
	0.990	9.65	7.21	6.22	5.67	5.32	5.07	4.89	4.74	4.63	4.54	0.990
	0.995	12.23	8.91	7.60	6.88	6.42	6.10	5.86	5.68	5.54	5.42	0.995
12	0.005	0.000	0.005	0.023	0.048	0.075	0.100	0.122	0.143	0.161	0.177	0.005
	0.010	0.000	0.010	0.037	0.070	0.101	0.130	0.155	0.176	0.196	0.213	0.010
	0.025	0.001	0.025	0.070	0.114	0.153	0.186	0.214	0.238	0.259	0.276	0.025
	0.050	0.004	0.052	0.114	0.169	0.214	0.250	0.280	0.305	0.325	0.343	0.050
	0.100	0.016	0.106	0.192	0.257	0.306	0.344	0.375	0.400	0.420	0.438	0.100
	0.900	3.18	2.81	2.61	2.48	2.39	2.33	2.28	2.24	2.21	2.19	0.900
	0.950	4.75	3.89	3.49	3.26	3.11	3.00	2.91	2.85	2.80	2.75	0.950
	0.975	6.55	5.10	4.47	4.12	3.89	3.73	3.61	3.51	3.44	3.37	0.975
	0.990	9.33	6.93	5.95	5.41	5.06	4.82	4.64	4.50	4.39	4.30	0.990
	0.995	11.75	8.51	7.23	6.52	6.07	5.76	5.52	5.35	5.20	5.09	0.995
13	0.005	0.000	0.005	0.023	0.049	0.075	0.101	0.124	0.144	0.163	0.179	0.005
	0.010	0.000	0.010	0.037	0.070	0.102	0.131	0.156	0.178	0.198	0.215	0.010
	0.025	0.001	0.025	0.070	0.115	0.154	0.188	0.216	0.240	0.261	0.279	0.025
	0.050	0.004	0.051	0.115	0.170	0.215	0.251	0.282	0.307	0.328	0.346	0.050
	0.100	0.016	0.106	0.192	0.257	0.307	0.346	0.377	0.402	0.423	C.441	0.100
	0.900	3.14	2.76	2.56	2.43	2.35	2.28	2.23	2.20	2.16	2.14	0.900
	0.950	4.67	3.81	3.41	3.18	3.03	2.92	2.83	2.77	2.71	2.67	0.950
	0.975	6.41	4.97	4.35	4.00	3.77	3.60	3.48	3.39	3.31	3.25	0.975
	0.990	9.07	6.70	5.74	5.21	4.86	4.62	4.44	4.30	4.19	4.10	0.990
	0.995	11.37	8.19	6.93	6.23	5.79	5.48	5.25	5.08	4.94	4.82	0.995
14	0.005	0.000	0.005	0.023	0.049	0.076	0.101	0.125	0.146	0.164	0.181	0.005
	0.010	0.000	0.010	0.037	0.070	0.102	0.131	0.157	0.180	0.200	0.217	0.010
	0.025	0.001	0.025	0.070	0.115	0.155	0.189	0.218	0.242	0.263	0.282	0.025
	0.050	0.004	0.051	0.115	0.170	0.216	0.253	0.283	0.309	0.331	0.349	0.050
	0.100	0.016	0.106	0.192	0.258	0.308	0.347	0.378	0.404	0.425	0.443	0.100
	0.900	3.10	2.73	2.52	2.39	2.31	2.24	2.19	2.15	2.12	2.10	0.900
	0.950	4.60	3.74	3.34	3.11	2.96	2.85	2.76	2.70	2.65	2.60	0.950
	0.975	6.30	4.86	4.24	3.89	3.66	3.50	3.38	3.29	3.21	3.15	0.975
	0.990	8.86	6.51	5.56	5.04	4.69	4.46	4.28	4.14	4.03	3.94	0.990
	0.995	11.06	7.92	6.68	6.00	5.56	5.26	5.03	4.86	4.72	4.60	0.995
15	0.005	0.000	0.005	0.023	0.049	0.076	0.102	0.126	0.147	0.166	0.183	0.005
	0.010	0.000	0.010	0.037	0.070	0.103	0.132	0.158	0.181	0.202	0.219	0.010
	0.025	0.001	0.025	0.070	0.116	0.156	0.190	0.219	0.244	0.265	0.284	0.025
	0.050	0.004	0.051	0.115	0.171	0.217	0.254	0.285	0.311	0.333	0.351	0.050
	0.100	0.016	0.106	0.192	0.258	0.309	0.348	0.380	0.406	0.427	0.446	0.100
	0.900	3.07	2.70	2.49	2.36	2.27	2.21	2.16	2.12	2.09	2.06	0.900
	0.950	4.54	3.68	3.29	3.06	2.90	2.79	2.71	2.64	2.59	2.54	0.950
	0.975	6.20	4.77	4.15	3.80	3.58	3.41	3.29	3.20	3.12	3.06	0.975
	0.990	8.68	6.36	5.42	4.89	4.56	4.32	4.14	4.00	3.89	3.80	0.990
	0.995	10.80	7.70	6.48	5.80	5.37	5.07	4.85	4.67	4.54	4.42	0.995

TABLE A.10   (Continued)

v_2	v_1 / γ	1	2	3	4	5	6	7	8	9	10	v_1 / γ
16	0.005	0.000	0.005	0.023	0.049	0.076	0.102	0.126	0.148	0.167	0.184	0.005
	0.010	0.000	0.010	0.037	0.071	0.103	0.133	0.159	0.183	0.203	0.221	0.010
	0.025	0.001	0.025	0.070	0.116	0.156	0.191	0.220	0.245	0.267	0.286	0.025
	0.050	0.004	0.051	0.115	0.171	0.217	0.255	0.286	0.312	0.335	0.354	0.050
	0.100	0.016	0.106	0.192	0.259	0.310	0.349	0.381	0.407	0.429	0.448	0.100
	0.900	3.05	2.67	2.46	2.33	2.24	2.18	2.13	2.09	2.06	2.03	0.900
	0.950	4.49	3.63	3.24	3.01	2.85	2.74	2.66	2.59	2.54	2.49	0.950
	0.975	6.12	4.69	4.08	3.73	3.50	3.34	3.22	3.12	3.05	2.99	0.975
	0.990	8.53	6.23	5.29	4.77	4.44	4.20	4.03	3.89	3.78	3.69	0.990
	0.995	10.58	7.51	6.30	5.64	5.21	4.91	4.69	4.52	4.38	4.27	0.995
17	0.005	0.000	0.005	0.023	0.049	0.077	0.103	0.127	0.149	0.168	0.186	0.005
	0.010	0.000	0.010	0.037	0.071	0.104	0.134	0.160	0.184	0.204	0.223	0.010
	0.025	0.001	0.025	0.070	0.116	0.157	0.192	0.221	0.247	0.269	0.288	0.025
	0.050	0.004	0.051	0.115	0.171	0.218	0.256	0.287	0.314	0.336	0.356	0.050
	0.100	0.016	0.106	0.192	0.259	0.310	0.350	0.382	0.409	0.431	0.450	0.100
	0.900	3.03	2.64	2.44	2.31	2.22	2.15	2.10	2.06	2.03	2.00	0.900
	0.950	4.45	3.59	3.20	2.96	2.81	2.70	2.61	2.55	2.49	2.45	0.950
	0.975	6.04	4.62	4.01	3.66	3.44	3.28	3.16	3.06	2.98	2.92	0.975
	0.990	8.40	6.11	5.18	4.67	4.34	4.10	3.93	3.79	3.68	3.59	0.990
	0.995	10.38	7.35	6.16	5.50	5.07	4.78	4.56	4.39	4.25	4.14	0.995
18	0.005	0.000	0.005	0.023	0.049	0.077	0.103	0.128	0.150	0.170	0.187	0.005
	0.010	0.000	0.010	0.037	0.071	0.104	0.134	0.161	0.185	0.206	0.224	0.010
	0.025	0.001	0.025	0.070	0.116	0.157	0.192	0.222	0.248	0.270	0.290	0.025
	0.050	0.004	0.051	0.115	0.172	0.218	0.257	0.288	0.315	0.338	0.357	0.050
	0.100	0.016	0.106	0.193	0.260	0.311	0.351	0.384	0.410	0.432	0.451	0.100
	0.900	3.01	2.62	2.42	2.29	2.20	2.13	2.08	2.04	2.00	1.98	0.900
	0.950	4.41	3.55	3.16	2.93	2.77	2.66	2.58	2.51	2.46	2.41	0.950
	0.975	5.98	4.56	3.95	3.61	3.38	3.22	3.10	3.01	2.93	2.87	0.975
	0.990	8.29	6.01	5.09	4.58	4.25	4.01	3.84	3.71	3.60	3.51	0.990
	0.995	10.22	7.21	6.03	5.37	4.96	4.66	4.44	4.28	4.14	4.03	0.995
19	0.005	0.000	0.005	0.023	0.049	0.077	0.104	0.128	0.151	0.171	0.188	0.005
	0.010	0.000	0.010	0.037	0.071	0.104	0.135	0.162	0.186	0.207	0.226	0.010
	0.025	0.001	0.025	0.070	0.117	0.158	0.193	0.223	0.249	0.271	0.291	0.025
	0.050	0.004	0.051	0.115	0.172	0.219	0.257	0.289	0.316	0.339	0.359	0.050
	0.100	0.016	0.106	0.193	0.260	0.311	0.352	0.385	0.411	0.434	0.453	0.100
	0.900	2.99	2.61	2.40	2.27	2.18	2.11	2.06	2.02	1.98	1.96	0.900
	0.950	4.38	3.52	3.13	2.90	2.74	2.63	2.54	2.48	2.42	2.38	0.950
	0.975	5.92	4.51	3.90	3.56	3.33	3.17	3.05	2.96	2.88	2.82	0.975
	0.990	8.18	5.93	5.01	4.50	4.17	3.94	3.77	3.63	3.52	3.43	0.990
	0.995	10.07	7.09	5.92	5.27	4.85	4.56	4.34	4.18	4.04	3.93	0.995
20	0.005	0.000	0.005	0.023	0.050	0.078	0.104	0.129	0.151	0.171	0.190	0.005
	0.010	0.000	0.010	0.037	0.071	0.105	0.135	0.162	0.187	0.208	0.227	0.010
	0.025	0.001	0.025	0.070	0.117	0.158	0.193	0.224	0.250	0.273	0.293	0.025
	0.050	0.004	0.051	0.115	0.172	0.219	0.258	0.290	0.317	0.341	0.360	0.050
	0.100	0.016	0.106	0.193	0.260	0.312	0.353	0.385	0.412	0.435	0.454	0.100
	0.900	2.97	2.59	2.38	2.25	2.16	2.09	2.04	2.00	1.96	1.94	0.900
	0.950	4.35	3.49	3.10	2.87	2.71	2.60	2.51	2.45	2.39	2.35	0.950
	0.975	5.87	4.46	3.86	3.51	3.29	3.13	3.01	2.91	2.84	2.77	0.975
	0.990	8.10	5.85	4.94	4.43	4.10	3.87	3.70	3.56	3.46	3.37	0.990
	0.995	9.94	6.99	5.82	5.17	4.76	4.47	4.26	4.09	3.96	3.85	0.995

TABLE A.10   (Continued)

ν_2	ν_1 / γ	1	2	3	4	5	6	7	8	9	10	ν_1 / γ
21	0.005	0.000	0.005	0.023	0.050	0.078	0.105	0.129	0.152	0.172	0.191	0.005
	0.010	0.000	0.010	0.037	0.071	0.105	0.136	0.163	0.187	0.209	0.228	0.010
	0.025	0.001	0.025	0.071	0.117	0.158	0.194	0.225	0.251	0.274	0.294	0.025
	0.050	0.004	0.051	0.115	0.173	0.220	0.259	0.291	0.318	0.342	0.362	0.050
	0.100	0.016	0.106	0.193	0.260	0.312	0.353	0.386	0.413	0.436	0.456	0.100
	0.900	2.96	2.57	2.36	2.23	2.14	2.08	2.02	1.98	1.95	1.92	0.900
	0.950	4.32	3.47	3.07	2.84	2.68	2.57	2.49	2.42	2.37	2.32	0.950
	0.975	5.83	4.42	3.82	3.48	3.25	3.09	2.97	2.87	2.80	2.73	0.975
	0.990	8.02	5.78	4.87	4.37	4.04	3.81	3.64	3.51	3.40	3.31	0.990
	0.995	9.83	6.89	5.73	5.09	4.68	4.39	4.18	4.01	3.88	3.77	0.995
22	0.005	0.000	0.005	0.023	0.050	0.078	0.105	0.130	0.153	0.173	0.192	0.005
	0.010	0.000	0.010	0.037	0.072	0.105	0.136	0.164	0.188	0.210	0.229	0.010
	0.025	0.001	0.025	0.071	0.117	0.159	0.195	0.225	0.252	0.275	0.295	0.025
	0.050	0.004	0.051	0.116	0.173	0.220	0.259	0.292	0.319	0.343	0.363	0.050
	0.100	0.016	0.106	0.193	0.261	0.313	0.354	0.387	0.414	0.437	0.457	0.100
	0.900	2.95	2.56	2.35	2.22	2.13	2.06	2.01	1.97	1.93	1.90	0.900
	0.950	4.30	3.44	3.05	2.82	2.66	2.55	2.46	2.40	2.34	2.30	0.950
	0.975	5.79	4.38	3.78	3.44	3.22	3.05	2.93	2.84	2.76	2.70	0.975
	0.990	7.95	5.72	4.82	4.31	3.99	3.76	3.59	3.45	3.35	3.26	0.990
	0.995	9.73	6.81	5.65	5.02	4.61	4.32	4.11	3.94	3.81	3.70	0.995
23	0.005	0.000	0.005	0.023	0.050	0.078	0.105	0.130	0.153	0.174	0.192	0.005
	0.010	0.000	0.010	0.037	0.072	0.105	0.136	0.164	0.189	0.211	0.230	0.010
	0.025	0.001	0.025	0.071	0.117	0.159	0.195	0.226	0.253	0.276	0.296	0.025
	0.050	0.004	0.051	0.116	0.173	0.221	0.260	0.293	0.320	0.344	0.364	0.050
	0.100	0.016	0.106	0.193	0.261	0.313	0.354	0.388	0.415	0.438	0.458	0.100
	0.900	2.94	2.55	2.34	2.21	2.11	2.05	1.99	1.95	1.92	1.89	0.900
	0.950	4.28	3.42	3.03	2.80	2.64	2.53	2.44	2.37	2.32	2.27	0.950
	0.975	5.75	4.35	3.75	3.41	3.18	3.02	2.90	2.81	2.73	2.67	0.975
	0.990	7.88	5.66	4.76	4.26	3.94	3.71	3.54	3.41	3.30	3.21	0.990
	0.995	9.63	6.73	5.58	4.95	4.54	4.26	4.05	3.88	3.75	3.64	0.995
24	0.005	0.000	0.005	0.023	0.050	0.078	0.106	0.131	0.154	0.175	0.193	0.005
	0.010	0.000	0.010	0.037	0.072	0.106	0.137	0.165	0.189	0.211	0.231	0.010
	0.025	0.001	0.025	0.071	0.117	0.159	0.195	0.227	0.253	0.277	0.297	0.025
	0.050	0.004	0.051	0.116	0.173	0.221	0.260	0.293	0.321	0.345	0.365	0.050
	0.100	0.016	0.106	0.193	0.261	0.313	0.355	0.388	0.416	0.439	0.459	0.100
	0.900	2.93	2.54	2.33	2.19	2.10	2.04	1.98	1.94	1.91	1.88	0.900
	0.950	4.26	3.40	3.01	2.78	2.62	2.51	2.42	2.36	2.30	2.25	0.950
	0.975	5.72	4.32	3.72	3.38	3.15	2.99	2.87	2.78	2.70	2.64	0.975
	0.990	7.82	5.61	4.72	4.22	3.90	3.67	3.50	3.36	3.26	3.17	0.990
	0.995	9.55	6.66	5.52	4.89	4.49	4.20	3.99	3.83	3.69	3.59	0.995
25	0.005	0.000	0.005	0.023	0.050	0.078	0.106	0.131	0.154	0.175	0.194	0.005
	0.010	0.000	0.010	0.037	0.072	0.106	0.137	0.165	0.190	0.212	0.232	0.010
	0.025	0.001	0.025	0.071	0.118	0.160	0.196	0.227	0.254	0.278	0.298	0.025
	0.050	0.004	0.051	0.116	0.173	0.221	0.261	0.294	0.322	0.346	0.366	0.050
	0.100	0.016	0.106	0.193	0.261	0.314	0.355	0.389	0.417	0.440	0.460	0.100
	0.900	2.92	2.53	2.32	2.18	2.09	2.02	1.97	1.93	1.89	1.87	0.900
	0.950	4.24	3.39	2.99	2.76	2.60	2.49	2.40	2.34	2.28	2.24	0.950
	0.975	5.69	4.29	3.69	3.35	3.13	2.97	2.85	2.75	2.68	2.61	0.975
	0.990	7.77	5.57	4.68	4.18	3.85	3.63	3.46	3.32	3.22	3.13	0.990
	0.995	9.48	6.60	5.46	4.84	4.43	4.15	3.94	3.78	3.64	3.54	0.995

TABLE A.10   (Continued)

ν_2	ν_1 / γ	1	2	3	4	5	6	7	8	9	10	ν_1 / γ
30	0.005	0.000	0.005	0.023	0.050	0.079	0.107	0.133	0.156	0.178	0.197	0.005
	0.010	0.000	0.010	0.037	0.072	0.107	0.138	0.167	0.192	0.215	0.235	0.010
	0.025	0.001	0.025	0.071	0.118	0.161	0.197	0.229	0.257	0.281	0.302	0.025
	0.050	0.004	0.051	0.116	0.174	0.222	0.263	0.296	0.325	0.349	0.370	0.050
	0.100	0.016	0.106	0.193	0.262	0.315	0.357	0.391	0.420	0.444	0.464	0.100
	0.900	2.88	2.49	2.28	2.14	2.05	1.98	1.93	1.88	1.85	1.82	0.900
	0.950	4.17	3.32	2.92	2.69	2.53	2.42	2.33	2.27	2.21	2.16	0.950
	0.975	5.57	4.18	3.59	3.25	3.03	2.87	2.75	2.65	2.57	2.51	0.975
	0.990	7.56	5.39	4.51	4.02	3.70	3.47	3.30	3.17	3.07	2.98	0.990
	0.995	9.18	6.35	5.24	4.62	4.23	3.95	3.74	3.58	3.45	3.34	0.995
40	0.005	0.000	0.005	0.023	0.051	0.080	0.108	0.135	0.159	0.181	0.201	0.005
	0.010	0.000	0.010	0.037	0.073	0.108	0.140	0.169	0.195	0.219	0.240	0.010
	0.025	0.001	0.025	0.071	0.119	0.162	0.200	0.232	0.260	0.285	0.307	0.025
	0.050	0.004	0.051	0.116	0.175	0.224	0.265	0.299	0.329	0.354	0.376	0.050
	0.100	0.016	0.106	0.194	0.263	0.317	0.360	0.394	0.423	0.448	0.469	0.100
	0.900	2.84	2.44	2.23	2.09	2.00	1.93	1.87	1.83	1.79	1.76	0.900
	0.950	4.08	3.23	2.84	2.61	2.45	2.34	2.25	2.18	2.12	2.08	0.950
	0.975	5.42	4.05	3.46	3.13	2.90	2.74	2.62	2.53	2.45	2.39	0.975
	0.990	7.31	5.18	4.31	3.83	3.51	3.29	3.12	2.99	2.89	2.80	0.990
	0.995	8.83	6.07	4.98	4.37	3.99	3.71	3.51	3.35	3.22	3.12	0.995
60	0.005	0.000	0.005	0.023	0.051	0.081	0.110	0.137	0.162	0.185	0.206	0.005
	0.010	0.000	0.010	0.037	0.073	0.109	0.142	0.172	0.199	0.223	0.245	0.010
	0.025	0.001	0.025	0.071	0.120	0.163	0.202	0.235	0.264	0.290	0.313	0.025
	0.050	0.004	0.051	0.116	0.176	0.226	0.267	0.303	0.333	0.359	0.382	0.050
	0.100	0.016	0.106	0.194	0.264	0.318	0.362	0.398	0.428	0.453	0.475	0.100
	0.900	2.79	2.39	2.18	2.04	1.95	1.87	1.82	1.77	1.74	1.71	0.900
	0.950	4.00	3.15	2.76	2.53	2.37	2.25	2.17	2.10	2.04	1.99	0.950
	0.975	5.29	3.93	3.34	3.01	2.79	2.63	2.51	2.41	2.33	2.27	0.975
	0.990	7.08	4.98	4.13	3.65	3.34	3.12	2.95	2.82	2.72	2.63	0.990
	0.995	8.49	5.79	4.73	4.14	3.76	3.49	3.29	3.13	3.01	2.90	0.995
120	0.005	0.000	0.005	0.024	0.051	0.082	0.111	0.139	0.165	0.189	0.211	0.005
	0.010	0.000	0.010	0.038	0.074	0.110	0.144	0.174	0.202	0.227	0.250	0.010
	0.025	0.001	0.025	0.072	0.120	0.165	0.204	0.238	0.268	0.295	0.319	0.025
	0.050	0.004	0.051	0.117	0.177	0.227	0.270	0.306	0.337	0.364	0.388	0.050
	0.100	0.016	0.105	0.194	0.265	0.320	0.365	0.401	0.432	0.458	0.480	0.100
	0.900	2.75	2.35	2.13	1.99	1.90	1.82	1.77	1.72	1.68	1.65	0.900
	0.950	3.92	3.07	2.68	2.45	2.29	2.17	2.09	2.02	1.96	1.91	0.950
	0.975	5.15	3.80	3.23	2.89	2.67	2.52	2.39	2.30	2.22	2.16	0.975
	0.990	6.85	4.79	3.95	3.48	3.17	2.96	2.79	2.66	2.56	2.47	0.990
	0.995	8.18	5.54	4.50	3.92	3.55	3.28	3.09	2.93	2.81	2.71	0.995
∞	0.005	0.000	0.005	0.024	0.052	0.082	0.113	0.141	0.168	0.193	0.216	0.005
	0.010	0.000	0.010	0.038	0.074	0.111	0.145	0.177	0.206	0.232	0.256	0.010
	0.025	0.001	0.025	0.072	0.121	0.166	0.206	0.241	0.272	0.300	0.325	0.025
	0.050	0.004	0.051	0.117	0.178	0.229	0.273	0.310	0.342	0.369	0.394	0.050
	0.100	0.016	0.105	0.195	0.266	0.322	0.367	0.405	0.436	0.463	0.487	0.100
	0.900	2.71	2.30	2.08	1.94	1.85	1.77	1.72	1.67	1.63	1.60	0.900
	0.950	3.84	3.00	2.60	2.37	2.21	2.10	2.01	1.94	1.88	1.83	0.950
	0.975	5.02	3.69	3.12	2.79	2.57	2.41	2.29	2.19	2.11	2.05	0.975
	0.990	6.63	4.61	3.78	3.32	3.02	2.80	2.64	2.51	2.41	2.32	0.990
	0.995	7.88	5.30	4.28	3.72	3.35	3.09	2.90	2.74	2.63	2.52	0.995

TABLE A.10   (Continued)

ν_2	ν_1 / γ	11	12	13	14	15	16	17	18	19	20	ν_1 / γ
1	0.005	0.082	0.085	0.088	0.090	0.093	0.095	0.096	0.098	0.099	0.101	0.005
	0.010	0.104	0.107	0.110	0.113	0.115	0.117	0.119	0.121	0.122	0.124	0.010
	0.025	0.149	0.153	0.156	0.159	0.161	0.164	0.166	0.167	0.169	0.170	0.025
	0.050	0.206	0.211	0.214	0.217	0.220	0.223	0.225	0.227	0.228	0.230	0.050
	0.100	0.310	0.315	0.319	0.322	0.325	0.328	0.330	0.333	0.334	0.336	0.100
	0.900	60.47	60.71	60.90	61.07	61.22	61.35	61.46	61.57	61.66	61.74	0.900
	0.950	243.0	243.9	244.7	245.4	246.0	246.5	246.9	247.3	247.7	248.0	0.950
	0.975	973.0	976.7	979.8	982.5	984.9	986.9	988.7	990.3	991.8	993.1	0.975
	0.990	6083.	6106.	6126.	6143.	6157.	6170.	6181.	6192.	6201.	6209.	0.990
	0.995	24334	24426	24505	24572	24631	24682	24727	24767	24804	24836	0.995
2	0.005	0.112	0.118	0.122	0.126	0.130	0.133	0.136	0.139	0.141	0.143	0.005
	0.010	0.139	0.144	0.149	0.153	0.157	0.161	0.164	0.166	0.169	0.171	0.010
	0.025	0.190	0.196	0.201	0.206	0.210	0.213	0.217	0.219	0.222	0.224	0.025
	0.050	0.251	0.257	0.263	0.267	0.272	0.275	0.278	0.281	0.284	0.286	0.050
	0.100	0.350	0.356	0.362	0.367	0.371	0.375	0.378	0.381	0.384	0.386	0.100
	0.900	9.40	9.41	9.41	9.42	9.42	9.43	9.43	9.44	9.44	9.44	0.900
	0.950	19.40	19.41	19.42	19.42	19.43	19.43	19.44	19.44	19.44	19.45	0.950
	0.975	39.41	39.41	39.42	39.43	39.43	39.44	39.44	39.44	39.45	39.45	0.975
	0.990	99.41	99.42	99.42	99.43	99.43	99.44	99.44	99.44	99.45	99.45	0.990
	0.995	199.4	199.4	199.4	199.4	199.4	199.4	199.4	199.4	199.4	199.4	0.995
3	0.005	0.132	0.138	0.144	0.150	0.154	0.159	0.162	0.166	0.169	0.172	0.005
	0.010	0.161	0.168	0.174	0.180	0.185	0.189	0.193	0.196	0.200	0.203	0.010
	0.025	0.216	0.224	0.230	0.236	0.241	0.245	0.249	0.253	0.256	0.259	0.025
	0.050	0.279	0.287	0.293	0.299	0.304	0.309	0.313	0.316	0.320	0.323	0.050
	0.100	0.376	0.384	0.391	0.396	0.402	0.406	0.410	0.414	0.417	0.420	0.100
	0.900	5.22	5.22	5.21	5.21	5.20	5.20	5.19	5.19	5.19	5.19	0.900
	0.950	8.77	8.75	8.73	8.72	8.71	8.70	8.69	8.68	8.67	8.67	0.950
	0.975	14.38	14.35	14.32	14.29	14.27	14.25	14.23	14.22	14.20	14.19	0.975
	0.990	27.18	27.11	27.04	26.99	26.94	26.91	26.88	26.85	26.82	26.80	0.990
	0.995	43.68	43.56	43.46	43.38	43.31	43.26	43.22	43.17	43.14	43.12	0.995
4	0.005	0.145	0.153	0.160	0.167	0.172	0.177	0.182	0.186	0.190	0.193	0.005
	0.010	0.176	0.185	0.192	0.199	0.204	0.210	0.214	0.218	0.222	0.226	0.010
	0.025	0.234	0.243	0.250	0.257	0.263	0.268	0.273	0.277	0.281	0.285	0.025
	0.050	0.298	0.307	0.315	0.321	0.327	0.333	0.337	0.342	0.345	0.349	0.050
	0.100	0.394	0.403	0.411	0.418	0.423	0.429	0.433	0.437	0.441	0.445	0.100
	0.900	3.91	3.90	3.89	3.88	3.87	3.86	3.86	3.85	3.85	3.84	0.900
	0.950	5.94	5.91	5.89	5.87	5.86	5.84	5.83	5.82	5.81	5.80	0.950
	0.975	8.79	8.75	8.71	8.68	8.66	8.63	8.61	8.59	8.58	8.56	0.975
	0.990	14.45	14.37	14.31	14.25	14.20	14.15	14.11	14.08	14.05	14.02	0.990
	0.995	20.82	20.71	20.60	20.52	20.44	20.37	20.31	20.26	20.21	20.17	0.995
5	0.005	0.156	0.165	0.173	0.180	0.186	0.192	0.197	0.202	0.206	0.210	0.005
	0.010	0.188	0.197	0.206	0.213	0.220	0.225	0.231	0.235	0.240	0.244	0.010
	0.025	0.247	0.257	0.265	0.273	0.280	0.286	0.291	0.296	0.300	0.304	0.025
	0.050	0.312	0.322	0.331	0.338	0.345	0.351	0.356	0.361	0.365	0.369	0.050
	0.100	0.408	0.418	0.426	0.433	0.440	0.446	0.451	0.455	0.460	0.463	0.100
	0.900	3.28	3.27	3.26	3.25	3.24	3.23	3.22	3.22	3.21	3.21	0.900
	0.950	4.70	4.68	4.66	4.64	4.62	4.60	4.59	4.58	4.57	4.56	0.950
	0.975	6.57	6.52	6.49	6.46	6.43	6.40	6.38	6.36	6.34	6.33	0.975
	0.990	9.96	9.89	9.82	9.77	9.72	9.68	9.64	9.61	9.58	9.55	0.990
	0.995	13.49	13.38	13.29	13.21	13.15	13.08	13.03	12.98	12.94	12.90	0.995

TABLE A.10 (Continued)

v_2	v_1 / γ	11	12	13	14	15	16	17	18	19	20	v_1 / γ
6	0.005	0.164	0.174	0.182	0.190	0.197	0.204	0.209	0.214	0.219	0.224	0.005
	0.010	0.197	0.207	0.216	0.224	0.232	0.238	0.244	0.249	0.254	0.258	0.010
	0.025	0.258	0.268	0.277	0.286	0.293	0.299	0.305	0.310	0.315	0.320	0.025
	0.050	0.323	0.334	0.343	0.351	0.358	0.365	0.371	0.376	0.380	0.385	0.050
	0.100	0.419	0.429	0.438	0.446	0.453	0.459	0.465	0.470	0.474	0.478	0.100
	0.900	2.92	2.90	2.89	2.88	2.87	2.86	2.85	2.85	2.84	2.84	0.900
	0.950	4.03	4.00	3.98	3.96	3.94	3.92	3.91	3.90	3.88	3.87	0.950
	0.975	5.41	5.37	5.33	5.30	5.27	5.24	5.22	5.20	5.18	5.17	0.975
	0.990	7.79	7.72	7.66	7.60	7.56	7.52	7.48	7.45	7.42	7.40	0.990
	0.995	10.13	10.03	9.95	9.88	9.81	9.76	9.71	9.66	9.62	9.59	0.995
7	0.005	0.171	0.181	0.190	0.199	0.206	0.213	0.219	0.225	0.230	0.235	0.005
	0.010	0.205	0.216	0.225	0.234	0.241	0.248	0.255	0.260	0.266	0.270	0.010
	0.025	0.266	0.277	0.287	0.296	0.304	0.311	0.317	0.323	0.328	0.333	0.025
	0.050	0.332	0.343	0.353	0.362	0.369	0.376	0.383	0.388	0.393	0.398	0.050
	0.100	0.427	0.438	0.448	0.456	0.463	0.470	0.476	0.481	0.486	0.490	0.100
	0.900	2.68	2.67	2.65	2.64	2.63	2.62	2.61	2.61	2.60	2.59	0.900
	0.950	3.60	3.57	3.55	3.53	3.51	3.49	3.48	3.47	3.46	3.44	0.950
	0.975	4.71	4.67	4.63	4.60	4.57	4.54	4.52	4.50	4.48	4.47	0.975
	0.990	6.54	6.47	6.41	6.36	6.31	6.27	6.24	6.21	6.18	6.16	0.990
	0.995	8.27	8.18	8.10	8.03	7.97	7.91	7.87	7.83	7.79	7.75	0.995
8	0.005	0.176	0.187	0.197	0.206	0.214	0.221	0.228	0.234	0.239	0.245	0.005
	0.010	0.211	0.222	0.232	0.242	0.250	0.257	0.264	0.270	0.275	0.281	0.010
	0.025	0.273	0.285	0.295	0.304	0.313	0.320	0.327	0.333	0.338	0.343	0.025
	0.050	0.339	0.351	0.361	0.371	0.379	0.386	0.392	0.398	0.404	0.409	0.050
	0.100	0.434	0.446	0.456	0.464	0.472	0.479	0.485	0.491	0.496	0.500	0.100
	0.900	2.52	2.50	2.49	2.48	2.46	2.45	2.45	2.44	2.43	2.42	0.900
	0.950	3.31	3.28	3.26	3.24	3.22	3.20	3.19	3.17	3.16	3.15	0.950
	0.975	4.24	4.20	4.16	4.13	4.10	4.08	4.05	4.03	4.02	4.00	0.975
	0.990	5.73	5.67	5.61	5.56	5.52	5.48	5.44	5.41	5.38	5.36	0.990
	0.995	7.10	7.01	6.94	6.87	6.81	6.76	6.72	6.68	6.64	6.61	0.995
9	0.005	0.181	0.192	0.203	0.212	0.220	0.228	0.235	0.241	0.247	0.253	0.005
	0.010	0.216	0.228	0.239	0.248	0.257	0.265	0.272	0.278	0.284	0.289	0.010
	0.025	0.279	0.291	0.302	0.312	0.320	0.328	0.335	0.341	0.347	0.353	0.025
	0.050	0.345	0.358	0.368	0.378	0.386	0.394	0.401	0.407	0.413	0.418	0.050
	0.100	0.440	0.452	0.462	0.471	0.479	0.487	0.493	0.499	0.504	0.509	0.100
	0.900	2.40	2.38	2.36	2.35	2.34	2.33	2.32	2.31	2.30	2.30	0.900
	0.950	3.10	3.07	3.05	3.03	3.01	2.99	2.97	2.96	2.95	2.94	0.950
	0.975	3.91	3.87	3.83	3.80	3.77	3.74	3.72	3.70	3.68	3.67	0.975
	0.990	5.18	5.11	5.05	5.01	4.96	4.92	4.89	4.86	4.83	4.81	0.990
	0.995	6.31	6.23	6.15	6.09	6.03	5.98	5.94	5.90	5.86	5.83	0.995
10	0.005	0.185	0.197	0.207	0.217	0.226	0.234	0.241	0.248	0.254	0.260	0.005
	0.010	0.220	0.233	0.244	0.254	0.263	0.271	0.278	0.285	0.291	0.297	0.010
	0.025	0.284	0.296	0.308	0.318	0.327	0.335	0.342	0.349	0.355	0.361	0.025
	0.050	0.350	0.363	0.374	0.384	0.393	0.401	0.408	0.415	0.421	0.426	0.050
	0.100	0.445	0.457	0.468	0.477	0.486	0.493	0.500	0.506	0.511	0.516	0.100
	0.900	2.30	2.28	2.27	2.26	2.24	2.23	2.22	2.22	2.21	2.20	0.900
	0.950	2.94	2.91	2.89	2.86	2.84	2.83	2.81	2.80	2.79	2.77	0.950
	0.975	3.66	3.62	3.58	3.55	3.52	3.50	3.47	3.45	3.44	3.42	0.975
	0.990	4.77	4.71	4.65	4.60	4.56	4.52	4.49	4.46	4.43	4.41	0.990
	0.995	5.75	5.66	5.59	5.53	5.47	5.42	5.38	5.34	5.31	5.27	0.995

TABLE A.10   (Continued)

ν_2	ν_1 γ	11	12	13	14	15	16	17	18	19	20	ν_1 γ
11	0.005	0.188	0.200	0.212	0.222	0.231	0.239	0.247	0.254	0.260	0.266	0.005
	0.010	0.224	0.237	0.248	0.259	0.268	0.277	0.284	0.291	0.298	0.304	0.010
	0.025	0.288	0.301	0.313	0.323	0.332	0.341	0.348	0.355	0.362	0.368	0.025
	0.050	0.355	0.368	0.380	0.390	0.399	0.407	0.414	0.421	0.427	0.433	0.050
	0.100	0.449	0.462	0.473	0.482	0.491	0.499	0.506	0.512	0.518	0.523	0.100
	0.900	2.23	2.21	2.19	2.18	2.17	2.16	2.15	2.14	2.13	2.12	0.900
	0.950	2.82	2.79	2.76	2.74	2.72	2.70	2.69	2.67	2.66	2.65	0.950
	0.975	3.47	3.43	3.39	3.36	3.33	3.30	3.28	3.26	3.24	3.23	0.975
	0.990	4.46	4.40	4.34	4.29	4.25	4.21	4.18	4.15	4.12	4.10	0.990
	0.995	5.32	5.24	5.16	5.10	5.05	5.00	4.96	4.92	4.89	4.86	0.995
12	0.005	0.191	0.204	0.215	0.226	0.235	0.244	0.252	0.259	0.266	0.272	0.005
	0.010	0.227	0.241	0.253	0.263	0.273	0.281	0.289	0.297	0.303	0.309	0.010
	0.025	0.292	0.305	0.317	0.328	0.337	0.346	0.354	0.361	0.368	0.374	0.025
	0.050	0.359	0.372	0.384	0.395	0.404	0.412	0.420	0.427	0.433	0.439	0.050
	0.100	0.453	0.466	0.477	0.487	0.496	0.504	0.511	0.517	0.523	0.528	0.100
	0.900	2.17	2.15	2.13	2.12	2.10	2.09	2.08	2.08	2.07	2.06	0.900
	0.950	2.72	2.69	2.66	2.64	2.62	2.60	2.58	2.57	2.56	2.54	0.950
	0.975	3.32	3.28	3.24	3.21	3.18	3.15	3.13	3.11	3.09	3.07	0.975
	0.990	4.22	4.16	4.10	4.05	4.01	3.97	3.94	3.91	3.88	3.86	0.990
	0.995	4.99	4.91	4.84	4.77	4.72	4.67	4.63	4.59	4.56	4.53	0.995
13	0.005	0.194	0.207	0.219	0.229	0.239	0.248	0.256	0.264	0.271	0.277	0.005
	0.010	0.230	0.244	0.256	0.267	0.277	0.286	0.294	0.302	0.308	0.315	0.010
	0.025	0.295	0.309	0.321	0.332	0.342	0.351	0.359	0.366	0.373	0.379	0.025
	0.050	0.362	0.376	0.388	0.399	0.408	0.417	0.425	0.432	0.439	0.445	0.050
	0.100	0.456	0.469	0.481	0.491	0.500	0.508	0.515	0.522	0.528	0.533	0.100
	0.900	2.12	2.10	2.08	2.07	2.05	2.04	2.03	2.02	2.01	2.01	0.900
	0.950	2.63	2.60	2.58	2.55	2.53	2.51	2.50	2.48	2.47	2.46	0.950
	0.975	3.20	3.15	3.12	3.08	3.05	3.03	3.00	2.98	2.96	2.95	0.975
	0.990	4.02	3.96	3.91	3.86	3.82	3.78	3.75	3.72	3.69	3.66	0.990
	0.995	4.72	4.64	4.57	4.51	4.46	4.41	4.37	4.33	4.30	4.27	0.995
14	0.005	0.196	0.209	0.222	0.233	0.243	0.252	0.260	0.268	0.275	0.281	0.005
	0.010	0.233	0.247	0.259	0.270	0.281	0.290	0.298	0.306	0.313	0.320	0.010
	0.025	0.298	0.312	0.324	0.336	0.346	0.355	0.363	0.371	0.378	0.384	0.025
	0.050	0.365	0.379	0.392	0.403	0.412	0.421	0.429	0.437	0.443	0.449	0.050
	0.100	0.459	0.472	0.484	0.494	0.504	0.512	0.519	0.526	0.532	0.538	0.100
	0.900	2.07	2.05	2.04	2.02	2.01	2.00	1.99	1.98	1.97	1.96	0.900
	0.950	2.57	2.53	2.51	2.48	2.46	2.44	2.43	2.41	2.40	2.39	0.950
	0.975	3.09	3.05	3.01	2.98	2.95	2.92	2.90	2.88	2.86	2.84	0.975
	0.990	3.86	3.80	3.75	3.70	3.66	3.62	3.59	3.56	3.53	3.51	0.990
	0.995	4.51	4.43	4.36	4.30	4.25	4.20	4.16	4.12	4.09	4.06	0.995
15	0.005	0.198	0.212	0.224	0.235	0.246	0.255	0.264	0.272	0.279	0.286	0.005
	0.010	0.235	0.249	0.262	0.274	0.284	0.293	0.302	0.310	0.317	0.324	0.010
	0.025	0.300	0.315	0.328	0.339	0.349	0.359	0.367	0.375	0.382	0.389	0.025
	0.050	0.368	0.382	0.395	0.406	0.416	0.425	0.433	0.441	0.448	0.454	0.050
	0.100	0.461	0.475	0.487	0.498	0.507	0.515	0.523	0.530	0.536	0.542	0.100
	0.900	2.04	2.02	2.00	1.99	1.97	1.96	1.95	1.94	1.93	1.92	0.900
	0.950	2.51	2.48	2.45	2.42	2.40	2.38	2.37	2.35	2.34	2.33	0.950
	0.975	3.01	2.96	2.92	2.89	2.86	2.84	2.81	2.79	2.77	2.76	0.975
	0.990	3.73	3.67	3.61	3.56	3.52	3.49	3.45	3.42	3.40	3.37	0.990
	0.995	4.33	4.25	4.18	4.12	4.07	4.02	3.98	3.95	3.91	3.88	0.995

TABLE A.10   (Continued)

ν_2	ν_1 / γ	11	12	13	14	15	16	17	18	19	20	ν_1 / γ
16	0.005	0.200	0.214	0.227	0.238	0.249	0.258	0.267	0.275	0.282	0.289	0.005
	0.010	0.237	0.252	0.265	0.276	0.287	0.297	0.305	0.313	0.321	0.328	0.010
	0.025	0.303	0.317	0.330	0.342	0.353	0.362	0.371	0.379	0.386	0.393	0.025
	0.050	0.370	0.385	0.398	0.409	0.419	0.429	0.437	0.445	0.451	0.458	0.050
	0.100	0.464	0.478	0.490	0.500	0.510	0.519	0.526	0.533	0.540	0.546	0.100
	0.900	2.01	1.99	1.97	1.95	1.94	1.93	1.92	1.91	1.90	1.89	0.900
	0.950	2.46	2.42	2.40	2.37	2.35	2.33	2.32	2.30	2.29	2.28	0.950
	0.975	2.93	2.89	2.85	2.82	2.79	2.76	2.74	2.72	2.70	2.68	0.975
	0.990	3.62	3.55	3.50	3.45	3.41	3.37	3.34	3.31	3.28	3.26	0.990
	0.995	4.18	4.10	4.03	3.97	3.92	3.87	3.83	3.80	3.76	3.73	0.995
17	0.005	0.202	0.216	0.229	0.240	0.251	0.261	0.270	0.278	0.286	0.293	0.005
	0.010	0.239	0.254	0.267	0.279	0.290	0.299	0.308	0.317	0.324	0.331	0.010
	0.025	0.305	0.320	0.333	0.345	0.356	0.365	0.374	0.382	0.390	0.396	0.025
	0.050	0.372	0.387	0.400	0.412	0.422	0.432	0.440	0.448	0.455	0.462	0.050
	0.100	0.466	0.480	0.492	0.503	0.513	0.522	0.529	0.537	0.543	0.549	0.100
	0.900	1.98	1.96	1.94	1.93	1.91	1.90	1.89	1.88	1.87	1.86	0.900
	0.950	2.41	2.38	2.35	2.33	2.31	2.29	2.27	2.26	2.24	2.23	0.950
	0.975	2.87	2.82	2.79	2.75	2.72	2.70	2.67	2.65	2.63	2.62	0.975
	0.990	3.52	3.46	3.40	3.35	3.31	3.27	3.24	3.21	3.19	3.16	0.990
	0.995	4.05	3.97	3.90	3.84	3.79	3.75	3.71	3.67	3.64	3.61	0.995
18	0.005	0.203	0.218	0.231	0.243	0.253	0.263	0.272	0.281	0.289	0.296	0.005
	0.010	0.241	0.256	0.269	0.281	0.292	0.302	0.311	0.320	0.327	0.335	0.010
	0.025	0.307	0.322	0.335	0.347	0.358	0.368	0.377	0.385	0.393	0.400	0.025
	0.050	0.374	0.389	0.403	0.414	0.425	0.434	0.443	0.451	0.458	0.465	0.050
	0.100	0.468	0.482	0.494	0.505	0.515	0.524	0.532	0.539	0.546	0.552	0.100
	0.900	1.95	1.93	1.92	1.90	1.89	1.87	1.86	1.85	1.84	1.84	0.900
	0.950	2.37	2.34	2.31	2.29	2.27	2.25	2.23	2.22	2.20	2.19	0.950
	0.975	2.81	2.77	2.73	2.70	2.67	2.64	2.62	2.60	2.58	2.56	0.975
	0.990	3.43	3.37	3.32	3.27	3.23	3.19	3.16	3.13	3.10	3.08	0.990
	0.995	3.94	3.86	3.79	3.73	3.68	3.64	3.60	3.56	3.53	3.50	0.995
19	0.005	0.205	0.219	0.233	0.245	0.256	0.266	0.275	0.283	0.291	0.299	0.005
	0.010	0.243	0.258	0.271	0.283	0.294	0.305	0.314	0.322	0.330	0.338	0.010
	0.025	0.308	0.324	0.337	0.350	0.361	0.371	0.380	0.388	0.396	0.403	0.025
	0.050	0.376	0.391	0.405	0.417	0.427	0.437	0.446	0.454	0.461	0.468	0.050
	0.100	0.469	0.484	0.496	0.508	0.518	0.527	0.535	0.542	0.549	0.555	0.100
	0.900	1.93	1.91	1.89	1.88	1.86	1.85	1.84	1.83	1.82	1.81	0.900
	0.950	2.34	2.31	2.28	2.26	2.23	2.21	2.20	2.18	2.17	2.16	0.950
	0.975	2.76	2.72	2.68	2.65	2.62	2.59	2.57	2.55	2.53	2.51	0.975
	0.990	3.36	3.30	3.24	3.19	3.15	3.12	3.08	3.05	3.03	3.00	0.990
	0.995	3.84	3.76	3.70	3.64	3.59	3.54	3.50	3.46	3.43	3.40	0.995
20	0.005	0.206	0.221	0.234	0.246	0.258	0.268	0.277	0.286	0.294	0.301	0.005
	0.010	0.244	0.259	0.273	0.285	0.297	0.307	0.316	0.325	0.333	0.340	0.010
	0.025	0.310	0.325	0.339	0.352	0.363	0.373	0.382	0.391	0.399	0.406	0.025
	0.050	0.378	0.393	0.407	0.419	0.430	0.439	0.448	0.456	0.464	0.471	0.050
	0.100	0.471	0.486	0.498	0.510	0.520	0.529	0.537	0.544	0.551	0.557	0.100
	0.900	1.91	1.89	1.87	1.86	1.84	1.83	1.82	1.81	1.80	1.79	0.900
	0.950	2.31	2.28	2.25	2.22	2.20	2.18	2.17	2.15	2.14	2.12	0.950
	0.975	2.72	2.68	2.64	2.60	2.57	2.55	2.52	2.50	2.48	2.46	0.975
	0.990	3.29	3.23	3.18	3.13	3.09	3.05	3.02	2.99	2.96	2.94	0.990
	0.995	3.76	3.68	3.61	3.55	3.50	3.46	3.42	3.38	3.35	3.32	0.995

TABLE A.10   (Continued)

v_2	v_1 / γ	11	12	13	14	15	16	17	18	19	20	v_1 / γ
21	0.005	0.207	0.222	0.236	0.248	0.259	0.270	0,279	0.288	0.296	0.304	0.005
	0.010	0.245	0.261	0.275	0.287	0.299	0.309	0.319	0.327	0.335	0.343	0.010
	0.025	0.311	0.327	0.341	0.354	0.365	0.375	0.385	0.393	0.401	0.408	0.025
	0.050	0.379	0.395	0.409	0.421	0.432	0.442	0.451	0.459	0.466	0.473	0.050
	0.100	0.472	0.487	0.500	0.511	0.522	0.531	0.539	0.547	0.554	0.560	0.100
	0.900	1.90	1.87	1.86	1.84	1.83	1.81	1.80	1.79	1.78	1.78	0.900
	0.950	2.28	2.25	2.22	2.20	2.18	2.16	2.14	2.12	2.11	2.10	0.950
	0.975	2.68	2.64	2.60	2.56	2.53	2.51	2.48	2.46	2.44	2.42	0.975
	0.990	3.24	3.17	3.12	3.07	3.03	2.99	2.96	2.93	2.90	2.88	0.990
	0.995	3.68	3.60	3.54	3.48	3.43	3.38	3.34	3.31	3.27	3.24	0.995
22	0.005	0.208	0.223	0.237	0.250	0.261	0.272	0.281	0.290	0.299	0.306	0.005
	0.010	0.247	0.262	0.276	0.289	0.300	0.311	0.321	0.330	0.338	0.345	0.010
	0.025	0.313	0.329	0.343	0.355	0.367	0.377	0.387	0.395	0.404	0.411	0.025
	0.050	0.381	0.396	0.410	0.423	0.434	0.444	0.453	0.461	0.469	0.476	0.050
	0.100	0.474	0.489	0.502	0.513	0.523	0.533	0.541	0.549	0.556	0.562	0.100
	0.900	1.88	1.86	1.84	1.83	1.81	1.80	1.79	1.78	1.77	1.76	0.900
	0.950	2.26	2.23	2.20	2.17	2.15	2.13	2.11	2.10	2.08	2.07	0.950
	0.975	2.65	2.60	2.56	2.53	2.50	2.47	2.45	2.43	2.41	2.39	0.975
	0.990	3.18	3.12	3.07	3.02	2.98	2.94	2.91	2.88	2.85	2.83	0.990
	0.995	3.61	3.54	3.47	3.41	3.36	3.31	3.27	3.24	3.21	3.18	0.995
23	0.005	0.209	0.225	0.238	0.251	0.263	0.273	0.283	0.292	0.301	0.308	0.005
	0.010	0.248	0.263	0.277	0.290	0.302	0.313	0.323	0.332	0.340	0.348	0.010
	0.025	0.314	0.330	0.344	0.357	0.369	0.379	0.389	0.398	0.406	0.413	0.025
	0.050	0.382	0.398	0.412	0.424	0.435	0.446	0.455	0.463	0.471	0.478	0.050
	0.100	0.475	0.490	0.503	0.515	0.525	0.534	0.543	0.551	0.558	0.564	0.100
	0.900	1.87	1.84	1.83	1.81	1.80	1.78	1.77	1.76	1.75	1.74	0.900
	0.950	2.24	2.20	2.18	2.15	2.13	2.11	2.09	2.08	2.06	2.05	0.950
	0.975	2.62	2.57	2.53	2.50	2.47	2.44	2.42	2.39	2.37	2.36	0.975
	0.990	3.14	3.07	3.02	2.97	2.93	2.89	2.86	2.83	2.80	2.78	0.990
	0.995	3.55	3.47	3.41	3.35	3.30	3.25	3.21	3.18	3.15	3.12	0.995
24	0.005	0.210	0.226	0.240	0.252	0.264	0.275	0.285	0.294	0.302	0.310	0.005
	0.010	0.249	0.265	0.279	0.292	0.304	0.314	0.324	0.333	0.342	0.350	0.010
	0.025	0.315	0.331	0.346	0.359	0.370	0.381	0.391	0.400	0.408	0.415	0.025
	0.050	0.383	0.399	0.413	0.426	0.437	0.447	0.457	0.465	0.473	0.480	0.050
	0.100	0.476	0.491	0.504	0.516	0.527	0.536	0.545	0.552	0.560	0.566	0.100
	0.900	1.85	1.83	1.81	1.80	1.78	1.77	1.76	1.75	1.74	1.73	0.900
	0.950	2.22	2.18	2.15	2.13	2.11	2.09	2.07	2.05	2.04	2.03	0.950
	0.975	2.59	2.54	2.50	2.47	2.44	2.41	2.39	2.36	2.35	2.33	0.975
	0.990	3.09	3.03	2.98	2.93	2.89	2.85	2.82	2.79	2.76	2.74	0.990
	0.995	3.50	3.42	3.35	3.30	3.25	3.20	3.16	3.12	3.09	3.06	0.995
25	0.005	0.211	0.227	0.241	0.254	0.266	0.276	0.286	0.296	0.304	0.312	0.005
	0.010	0.250	0.266	0.280	0.293	0.305	0.316	0.326	0.335	0.344	0.352	0.010
	0.025	0.316	0.332	0.347	0.360	0.372	0.383	0.392	0.401	0.410	0.417	0.025
	0.050	0.384	0.400	0.415	0.427	0.439	0.449	0.458	0.467	0.475	0.482	0.050
	0.100	0.477	0.492	0.506	0.517	0.528	0.538	0.546	0.554	0.561	0.568	0.100
	0.900	1.84	1.82	1.80	1.79	1.77	1.76	1.75	1.74	1.73	1.72	0.900
	0.950	2.20	2.16	2.14	2.11	2.09	2.07	2.05	2.04	2.02	2.01	0.950
	0.975	2.56	2.51	2.48	2.44	2.41	2.38	2.36	2.34	2.32	2.30	0.975
	0.990	3.06	2.99	2.94	2.89	2.85	2.81	2.78	2.75	2.72	2.70	0.990
	0.995	3.45	3.37	3.30	3.25	3.20	3.15	3.11	3.08	3.04	3.01	0.995

TABLE A.10 (Continued)

v_2	v_1 / γ	11	12	13	14	15	16	17	18	19	20	v_1 / γ
	0.005	0.215	0.231	0.246	0.259	0.271	0.283	0.293	0.303	0.312	0.320	0.005
	0.010	0.254	0.270	0.285	0.299	0.311	0.323	0.333	0.343	0.352	0.360	0.010
	0.025	0.321	0.337	0.352	0.366	0.378	0.389	0.400	0.409	0.418	0.426	0.025
	0.050	0.389	0.405	0.420	0.433	0.445	0.456	0.466	0.475	0.483	0.490	0.050
	0.100	0.482	0.497	0.511	0.523	0.534	0.544	0.553	0.561	0.568	0.575	0.100
30												
	0.900	1.79	1.77	1.75	1.74	1.72	1.71	1.70	1.69	1.68	1.67	0.900
	0.950	2.13	2.09	2.06	2.04	2.01	1.99	1.98	1.96	1.95	1.93	0.950
	0.975	2.46	2.41	2.37	2.34	2.31	2.28	2.26	2.23	2.21	2.20	0.975
	0.990	2.91	2.84	2.79	2.74	2.70	2.66	2.63	2.60	2.57	2.55	0.990
	0.995	3.25	3.18	3.11	3.06	3.01	2.96	2.92	2.89	2.85	2.82	0.995
	0.005	0.220	0.237	0.252	0.266	0.279	0.291	0.302	0.312	0.322	0.331	0.005
	0.010	0.259	0.276	0.292	0.306	0.319	0.331	0.342	0.353	0.362	0.371	0.010
	0.025	0.327	0.344	0.360	0.374	0.387	0.399	0.409	0.419	0.429	0.437	0.025
	0.050	0.395	0.412	0.428	0.441	0.454	0.465	0.475	0.485	0.493	0.502	0.050
	0.100	0.487	0.504	0.518	0.530	0.542	0.552	0.562	0.570	0.578	0.585	0.100
40												
	0.900	1.74	1.71	1.70	1.68	1.66	1.65	1.64	1.62	1.61	1.61	0.900
	0.950	2.04	2.00	1.97	1.95	1.92	1.90	1.89	1.87	1.85	1.84	0.950
	0.975	2.33	2.29	2.25	2.21	2.18	2.15	2.13	2.11	2.09	2.07	0.975
	0.990	2.73	2.66	2.61	2.56	2.52	2.48	2.45	2.42	2.39	2.37	0.990
	0.995	3.03	2.95	2.89	2.83	2.78	2.74	2.70	2.66	2.63	2.60	0.995
	0.005	0.225	0.243	0.259	0.274	0.287	0.300	0.312	0.323	0.333	0.343	0.005
	0.010	0.265	0.283	0.299	0.314	0.328	0.341	0.353	0.364	0.374	0.383	0.010
	0.025	0.333	0.351	0.368	0.383	0.396	0.409	0.420	0.431	0.441	0.450	0.025
	0.050	0.402	0.419	0.435	0.450	0.463	0.475	0.486	0.496	0.505	0.514	0.050
	0.100	0.494	0.510	0.525	0.538	0.550	0.561	0.571	0.580	0.589	0.596	0.100
60												
	0.900	1.68	1.66	1.64	1.62	1.60	1.59	1.58	1.56	1.55	1.54	0.900
	0.950	1.95	1.92	1.89	1.86	1.84	1.82	1.80	1.78	1.76	1.75	0.950
	0.975	2.22	2.17	2.13	2.09	2.06	2.03	2.01	1.98	1.96	1.94	0.975
	0.990	2.56	2.50	2.44	2.39	2.35	2.31	2.28	2.25	2.22	2.20	0.990
	0.995	2.82	2.74	2.68	2.62	2.57	2.53	2.49	2.45	2.42	2.39	0.995
	0.005	0.231	0.249	0.266	0.282	0.297	0.310	0.323	0.335	0.346	0.356	0.005
	0.010	0.271	0.290	0.307	0.323	0.338	0.352	0.364	0.376	0.387	0.397	0.010
	0.025	0.340	0.359	0.376	0.392	0.406	0.420	0.432	0.443	0.454	0.464	0.025
	0.050	0.409	0.427	0.444	0.459	0.473	0.486	0.497	0.508	0.518	0.527	0.050
	0.100	0.500	0.518	0.533	0.547	0.560	0.571	0.582	0.591	0.600	0.609	0.100
120												
	0.900	1.62	1.60	1.58	1.56	1.54	1.53	1.52	1.50	1.49	1.48	0.900
	0.950	1.87	1.83	1.80	1.78	1.75	1.73	1.71	1.69	1.67	1.66	0.950
	0.975	2.10	2.05	2.01	1.98	1.94	1.92	1.89	1.87	1.84	1.82	0.975
	0.990	2.40	2.34	2.28	2.23	2.19	2.15	2.12	2.09	2.06	2.03	0.990
	0.995	2.62	2.54	2.48	2.42	2.37	2.33	2.29	2.25	2.22	2.19	0.995
	0.005	0.237	0.256	0.274	0.291	0.307	0.321	0.335	0.348	0.360	0.372	0.005
	0.010	0.278	0.298	0.316	0.333	0.349	0.363	0.377	0.390	0.402	0.413	0.010
	0.025	0.347	0.367	0.385	0.402	0.417	0.432	0.445	0.457	0.469	0.480	0.025
	0.050	0.416	0.436	0.453	0.469	0.484	0.498	0.510	0.522	0.532	0.543	0.050
	0.100	0.507	0.525	0.542	0.556	0.570	0.582	0.593	0.604	0.613	0.622	0.100
∞												
	0.900	1.57	1.55	1.52	1.50	1.49	1.47	1.46	1.44	1.43	1.42	0.900
	0.950	1.79	1.75	1.72	1.69	1.67	1.64	1.62	1.60	1.59	1.57	0.950
	0.975	1.99	1.94	1.90	1.87	1.83	1.80	1.78	1.75	1.73	1.71	0.975
	0.990	2.25	2.18	2.13	2.08	2.04	2.00	1.97	1.93	1.90	1.88	0.990
	0.995	2.43	2.36	2.29	2.24	2.19	2.14	2.10	2.06	2.03	2.00	0.995

TABLE A.10   (Continued)

v_2	v_1 γ	21	22	23	24	25	30	40	60	120	∞	v_1 γ
1	0.005	0.102	0.103	0.104	0.105	0.106	0.109	0.113	0.118	0.122	0.127	0.005
	0.010	0.125	0.126	0.127	0.128	0.129	0.132	0.137	0.141	0.146	0.151	0.010
	0.025	0.172	0.173	0.174	0.175	0.176	0.180	0.184	0.189	0.194	0.199	0.025
	0.050	0.231	0.233	0.234	0.235	0.236	0.240	0.245	0.250	0.255	0.260	0.050
	0.100	0.338	0.339	0.340	0.342	0.343	0.347	0.353	0.358	0.364	0.370	0.100
	0.900	61.81	61.88	61.95	62.00	62.05	62.26	62.53	62.79	63.06	63.33	0.900
	0.950	248.3	248.6	248.8	249.1	249.3	250.1	251.1	252.2	253.3	254.3	0.950
	0.975	994.3	995.4	996.3	997.3	998.1	1001.	1006.	1010.	1014.	1018.	0.975
	0.990	6216.	6223.	6229.	6235.	6240.	6261.	6287.	6313.	6340.	6366.	0.990
	0.995	24866	24893	24917	24940	24961	25044	25148	25256	25360	25465	0.995
2	0.005	0.145	0.147	0.149	0.150	0.152	0.157	0.165	0.173	0.181	0.189	0.005
	0.010	0.173	0.175	0.177	0.178	0.180	0.186	0.193	0.201	0.209	0.217	0.010
	0.025	0.226	0.228	0.230	0.232	0.233	0.239	0.247	0.255	0.263	0.271	0.025
	0.050	0.288	0.290	0.292	0.294	0.295	0.302	0.309	0.317	0.326	0.334	0.050
	0.100	0.388	0.390	0.392	0.394	0.396	0.402	0.410	0.418	0.426	0.434	0.100
	0.900	9.44	9.45	9.45	9.45	9.45	9.46	9.47	9.47	9.48	9.49	0.900
	0.950	19.45	19.45	19.45	19.45	19.46	19.46	19.47	19.48	19.49	19.50	0.950
	0.975	39.45	39.45	39.45	39.46	39.46	39.46	39.47	39.48	39.49	39.50	0.975
	0.990	99.45	99.45	99.46	99.46	99.46	99.47	99.47	99.48	99.49	99.50	0.990
	0.995	199.5	199.5	199.5	199.5	199.5	199.5	199.5	199.5	199.5	199.5	0.995
3	0.005	0.175	0.177	0.179	0.181	0.183	0.191	0.201	0.211	0.222	0.234	0.005
	0.010	0.205	0.208	0.210	0.212	0.214	0.222	0.232	0.242	0.253	0.264	0.010
	0.025	0.262	0.264	0.267	0.269	0.271	0.279	0.289	0.299	0.310	0.321	0.025
	0.050	0.325	0.328	0.330	0.332	0.334	0.342	0.352	0.363	0.373	0.384	0.050
	0.100	0.423	0.425	0.428	0.430	0.432	0.439	0.449	0.459	0.470	0.480	0.100
	0.900	5.18	5.18	5.18	5.18	5.18	5.17	5.17	5.16	5.15	5.13	0.900
	0.950	8.66	8.66	8.65	8.65	8.64	8.63	8.61	8.61	8.56	8.53	0.950
	0.975	14.18	14.17	14.16	14.16	14.15	14.12	14.10	14.10	13.96	13.90	0.975
	0.990	26.78	26.77	26.75	26.73	26.72	26.70	26.67	26.76	26.27	26.13	0.990
	0.995	43.10	43.08	43.08	43.05	43.04	43.05	43.09	43.64	42.15	41.84	0.995
4	0.005	0.196	0.199	0.202	0.205	0.207	0.216	0.229	0.242	0.255	0.269	0.005
	0.010	0.229	0.232	0.235	0.237	0.239	0.249	0.261	0.274	0.287	0.301	0.010
	0.025	0.288	0.291	0.293	0.296	0.298	0.308	0.320	0.332	0.346	0.359	0.025
	0.050	0.352	0.355	0.358	0.360	0.362	0.372	0.384	0.396	0.409	0.422	0.050
	0.100	0.448	0.451	0.453	0.456	0.458	0.467	0.478	0.490	0.502	0.514	0.100
	0.900	3.84	3.84	3.83	3.83	3.83	3.82	3.80	3.79	3.78	3.76	0.900
	0.950	5.79	5.79	5.78	5.77	5.77	5.75	5.72	5.69	5.66	5.63	0.950
	0.975	8.55	8.53	8.52	8.51	8.50	8.46	8.41	8.36	8.31	8.26	0.975
	0.990	13.99	13.97	13.95	13.93	13.91	13.84	13.75	13.66	13.57	13.46	0.990
	0.995	20.13	20.09	20.06	20.03	20.01	19.89	19.76	19.61	19.50	19.33	0.995
5	0.005	0.214	0.217	0.220	0.223	0.226	0.237	0.251	0.266	0.282	0.299	0.005
	0.010	0.247	0.251	0.254	0.257	0.259	0.270	0.285	0.300	0.315	0.331	0.010
	0.025	0.308	0.311	0.314	0.317	0.320	0.330	0.344	0.359	0.374	0.390	0.025
	0.050	0.372	0.376	0.379	0.382	0.384	0.395	0.408	0.422	0.437	0.452	0.050
	0.100	0.467	0.470	0.473	0.476	0.478	0.488	0.501	0.514	0.527	0.541	0.100
	0.900	3.20	3.20	3.19	3.19	3.19	3.17	3.16	3.14	3.12	3.11	0.900
	0.950	4.55	4.54	4.53	4.53	4.52	4.50	4.46	4.43	4.40	4.37	0.950
	0.975	6.31	6.30	6.29	6.28	6.27	6.23	6.17	6.12	6.06	6.02	0.975
	0.990	9.53	9.50	9.48	9.47	9.45	9.38	9.29	9.20	9.09	9.02	0.990
	0.995	12.87	12.83	12.81	12.78	12.75	12.65	12.52	12.39	12.23	12.14	0.995

TABLE A.10   (Continued)

ν_2	γ	21	22	23	24	25	30	40	60	120	∞	γ
6	0.005	0.228	0.231	0.235	0.238	0.241	0.253	0.269	0.286	0.304	0.323	0.005
	0.010	0.262	0.266	0.270	0.273	0.276	0.288	0.304	0.321	0.338	0.357	0.010
	0.025	0.324	0.327	0.331	0.334	0.337	0.349	0.364	0.381	0.398	0.415	0.025
	0.050	0.389	0.392	0.396	0.399	0.402	0.413	0.428	0.444	0.460	0.477	0.050
	0.100	0.482	0.485	0.488	0.491	0.494	0.505	0.519	0.533	0.548	0.564	0.100
	0.900	2.83	2.83	2.82	2.82	2.81	2.80	2.78	2.76	2.74	2.72	0.900
	0.950	3.86	3.86	3.85	3.84	3.83	3.81	3.77	3.74	3.70	3.67	0.950
	0.975	5.15	5.14	5.13	5.12	5.11	5.06	5.01	4.96	4.90	4.85	0.975
	0.990	7.37	7.35	7.33	7.31	7.30	7.23	7.14	7.05	6.96	6.88	0.990
	0.995	9.56	9.53	9.50	9.47	9.45	9.36	9.24	9.12	8.98	8.88	0.995
7	0.005	0.239	0.243	0.247	0.251	0.254	0.267	0.285	0.304	0.324	0.345	0.005
	0.010	0.275	0.279	0.283	0.286	0.289	0.303	0.320	0.339	0.358	0.379	0.010
	0.025	0.337	0.341	0.345	0.348	0.351	0.364	0.381	0.399	0.418	0.437	0.025
	0.050	0.402	0.406	0.409	0.413	0.416	0.428	0.445	0.462	0.479	0.498	0.050
	0.100	0.494	0.498	0.501	0.504	0.507	0.519	0.534	0.550	0.566	0.583	0.100
	0.900	2.59	2.58	2.58	2.58	2.57	2.56	2.54	2.51	2.49	2.47	0.900
	0.950	3.43	3.43	3.42	3.41	3.40	3.38	3.34	3.30	3.27	3.23	0.950
	0.975	4.45	4.44	4.43	4.41	4.40	4.36	4.31	4.25	4.20	4.14	0.975
	0.990	6.13	6.11	6.09	6.07	6.06	5.99	5.91	5.82	5.73	5.65	0.990
	0.995	7.72	7.69	7.67	7.64	7.62	7.53	7.42	7.31	7.18	7.08	0.995
8	0.005	0.249	0.254	0.258	0.261	0.265	0.279	0.299	0.319	0.341	0.364	0.005
	0.010	0.285	0.290	0.294	0.297	0.301	0.315	0.334	0.354	0.376	0.398	0.010
	0.025	0.348	0.352	0.356	0.360	0.363	0.377	0.395	0.415	0.435	0.456	0.025
	0.050	0.413	0.417	0.421	0.425	0.428	0.441	0.459	0.477	0.496	0.516	0.050
	0.100	0.505	0.508	0.512	0.515	0.518	0.531	0.547	0.563	0.581	0.599	0.100
	0.900	2.42	2.41	2.41	2.40	2.40	2.38	2.36	2.34	2.32	2.29	0.900
	0.950	3.14	3.13	3.12	3.12	3.11	3.08	3.04	3.01	2.97	2.93	0.950
	0.975	3.98	3.97	3.96	3.95	3.94	3.89	3.84	3.78	3.73	3.67	0.975
	0.990	5.34	5.32	5.30	5.28	5.26	5.20	5.12	5.03	4.94	4.86	0.990
	0.995	6.58	6.55	6.53	6.50	6.48	6.40	6.29	6.18	6.06	5.95	0.995
9	0.005	0.258	0.262	0.267	0.271	0.274	0.290	0.310	0.332	0.356	0.381	0.005
	0.010	0.294	0.299	0.303	0.307	0.311	0.326	0.346	0.368	0.391	0.415	0.010
	0.025	0.357	0.362	0.366	0.370	0.374	0.388	0.408	0.428	0.450	0.473	0.025
	0.050	0.423	0.427	0.431	0.435	0.438	0.452	0.471	0.490	0.511	0.532	0.050
	0.100	0.513	0.517	0.521	0.525	0.528	0.541	0.558	0.575	0.594	0.613	0.100
	0.900	2.29	2.29	2.28	2.28	2.27	2.25	2.23	2.21	2.18	2.16	0.900
	0.950	2.93	2.92	2.91	2.90	2.89	2.86	2.83	2.79	2.75	2.71	0.950
	0.975	3.65	3.64	3.63	3.61	3.60	3.56	3.51	3.45	3.39	3.33	0.975
	0.990	4.79	4.76	4.75	4.73	4.71	4.65	4.57	4.48	4.40	4.31	0.990
	0.995	5.80	5.78	5.75	5.73	5.71	5.62	5.52	5.41	5.30	5.19	0.995
10	0.005	0.265	0.270	0.275	0.279	0.283	0.299	0.321	0.344	0.370	0.397	0.005
	0.010	0.302	0.307	0.311	0.316	0.320	0.336	0.357	0.380	0.405	0.431	0.010
	0.025	0.366	0.370	0.375	0.379	0.383	0.398	0.419	0.440	0.464	0.488	0.025
	0.050	0.431	0.435	0.440	0.444	0.447	0.462	0.481	0.502	0.523	0.546	0.050
	0.100	0.521	0.525	0.529	0.533	0.536	0.550	0.567	0.586	0.605	0.625	0.100
	0.900	2.19	2.19	2.18	2.18	2.17	2.16	2.13	2.11	2.08	2.06	0.900
	0.950	2.76	2.75	2.75	2.74	2.73	2.70	2.66	2.62	2.58	2.54	0.950
	0.975	3.40	3.39	3.38	3.37	3.35	3.31	3.26	3.20	3.14	3.08	0.975
	0.990	4.38	4.36	4.34	4.33	4.31	4.25	4.17	4.08	4.00	3.91	0.990
	0.995	5.25	5.22	5.20	5.17	5.15	5.07	4.97	4.86	4.75	4.64	0.995

TABLE A.10   (Continued)

ν_2	ν_1 / γ	21	22	23	24	25	30	40	60	120	∞	ν_1 / γ
11	0.005	0.272	0.277	0.282	0.286	0.290	0.307	0.330	0.355	0.382	0.411	0.005
	0.010	0.309	0.314	0.319	0.323	0.327	0.344	0.367	0.391	0.417	0.445	0.010
	0.025	0.373	0.378	0.382	0.387	0.391	0.407	0.428	0.451	0.476	0.502	0.025
	0.050	0.438	0.443	0.447	0.451	0.455	0.470	0.491	0.512	0.535	0.559	0.050
	0.100	0.528	0.532	0.536	0.540	0.543	0.557	0.576	0.595	0.615	0.637	0.100
	0.900	2.12	2.11	2.11	2.10	2.10	2.08	2.05	2.03	2.00	1.97	0.900
	0.950	2.64	2.63	2.62	2.61	2.60	2.57	2.53	2.49	2.45	2.40	0.950
	0.975	3.21	3.20	3.18	3.17	3.16	3.12	3.06	3.00	2.94	2.88	0.975
	0.990	4.08	4.06	4.04	4.02	4.01	3.94	3.86	3.78	3.69	3.60	0.990
	0.995	4.83	4.80	4.78	4.76	4.74	4.65	4.55	4.44	4.34	4.23	0.995
12	0.005	0.278	0.283	0.288	0.292	0.297	0.315	0.339	0.365	0.393	0.424	0.005
	0.010	0.315	0.320	0.325	0.330	0.334	0.352	0.375	0.401	0.428	0.458	0.010
	0.025	0.379	0.384	0.389	0.394	0.398	0.415	0.437	0.461	0.487	0.514	0.025
	0.050	0.444	0.449	0.454	0.458	0.462	0.478	0.499	0.522	0.545	0.571	0.050
	0.100	0.533	0.538	0.542	0.546	0.549	0.564	0.583	0.603	0.625	0.647	0.100
	0.900	2.05	2.05	2.04	2.04	2.03	2.01	1.99	1.96	1.93	1.90	0.900
	0.950	2.53	2.52	2.51	2.51	2.50	2.47	2.43	2.38	2.34	2.30	0.950
	0.975	3.06	3.04	3.03	3.02	3.01	2.96	2.91	2.85	2.79	2.72	0.975
	0.990	3.84	3.82	3.80	3.78	3.76	3.70	3.62	3.54	3.45	3.36	0.990
	0.995	4.50	4.48	4.45	4.43	4.41	4.33	4.23	4.12	4.01	3.90	0.995
13	0.005	0.283	0.288	0.293	0.298	0.303	0.321	0.346	0.374	0.403	0.436	0.005
	0.010	0.321	0.326	0.331	0.336	0.340	0.359	0.383	0.410	0.438	0.470	0.010
	0.025	0.385	0.390	0.395	0.400	0.404	0.422	0.445	0.470	0.497	0.526	0.025
	0.050	0.450	0.455	0.460	0.464	0.468	0.485	0.507	0.530	0.555	0.581	0.050
	0.100	0.539	0.543	0.547	0.551	0.555	0.570	0.590	0.611	0.633	0.656	0.100
	0.900	2.00	1.99	1.99	1.98	1.98	1.96	1.93	1.90	1.88	1.85	0.900
	0.950	2.45	2.44	2.43	2.42	2.41	2.38	2.34	2.30	2.25	2.21	0.950
	0.975	2.93	2.92	2.91	2.89	2.88	2.84	2.78	2.72	2.66	2.60	0.975
	0.990	3.64	3.62	3.60	3.59	3.57	3.51	3.43	3.34	3.25	3.17	0.990
	0.995	4.24	4.22	4.19	4.17	4.15	4.07	3.97	3.87	3.76	3.65	0.995
14	0.005	0.288	0.293	0.298	0.303	0.308	0.327	0.353	0.382	0.413	0.447	0.005
	0.010	0.326	0.331	0.336	0.341	0.346	0.365	0.390	0.418	0.448	0.480	0.010
	0.025	0.390	0.395	0.401	0.405	0.410	0.428	0.452	0.478	0.506	0.536	0.025
	0.050	0.455	0.460	0.465	0.470	0.474	0.491	0.513	0.538	0.563	0.591	0.050
	0.100	0.543	0.548	0.552	0.556	0.560	0.576	0.596	0.618	0.640	0.665	0.100
	0.900	1.96	1.95	1.94	1.94	1.93	1.91	1.89	1.86	1.83	1.80	0.900
	0.950	2.38	2.37	2.36	2.35	2.34	2.31	2.27	2.22	2.18	2.13	0.950
	0.975	2.83	2.81	2.80	2.79	2.78	2.73	2.67	2.61	2.55	2.49	0.975
	0.990	3.48	3.46	3.44	3.43	3.41	3.35	3.27	3.18	3.09	3.00	0.990
	0.995	4.03	4.01	3.98	3.96	3.94	3.86	3.76	3.66	3.55	3.44	0.995
15	0.005	0.292	0.298	0.303	0.308	0.313	0.333	0.360	0.389	0.421	0.457	0.005
	0.010	0.330	0.336	0.341	0.346	0.351	0.370	0.397	0.425	0.456	0.491	0.010
	0.025	0.395	0.400	0.405	0.410	0.415	0.433	0.458	0.485	0.514	0.546	0.025
	0.050	0.460	0.465	0.470	0.474	0.479	0.496	0.520	0.545	0.571	0.600	0.050
	0.100	0.547	0.552	0.557	0.561	0.565	0.581	0.602	0.624	0.647	0.672	0.100
	0.900	1.92	1.91	1.90	1.90	1.89	1.87	1.85	1.82	1.79	1.76	0.900
	0.950	2.32	2.31	2.30	2.29	2.28	2.25	2.20	2.16	2.11	2.07	0.950
	0.975	2.74	2.73	2.71	2.70	2.69	2.64	2.58	2.52	2.46	2.40	0.975
	0.990	3.35	3.33	3.31	3.29	3.28	3.21	3.13	3.05	2.96	2.87	0.990
	0.995	3.86	3.83	3.81	3.79	3.77	3.69	3.58	3.48	3.37	3.26	0.995

TABLE A.10   (Continued)

ν_2	ν_1 γ	21	22	23	24	25	30	40	60	120	∞	ν_1 γ
16	0.005	0.296	0.302	0.307	0.312	0.317	0.338	0.365	0.396	0.430	0.467	0.005
	0.010	0.334	0.340	0.346	0.351	0.355	0.375	0.403	0.432	0.464	0.500	0.010
	0.025	0.399	0.405	0.410	0.415	0.419	0.439	0.464	0.492	0.522	0.555	0.025
	0.050	0.464	0.469	0.474	0.479	0.483	0.501	0.525	0.551	0.579	0.608	0.050
	0.100	0.551	0.556	0.561	0.565	0.569	0.585	0.607	0.629	0.654	0.680	0.100
	0.900	1.88	1.88	1.87	1.87	1.86	1.84	1.81	1.78	1.75	1.72	0.900
	0.950	2.26	2.25	2.24	2.24	2.23	2.19	2.15	2.11	2.06	2.01	0.950
	0.975	2.67	2.65	2.64	2.63	2.61	2.57	2.51	2.45	2.38	2.32	0.975
	0.990	3.24	3.22	3.20	3.18	3.16	3.10	3.02	2.93	2.84	2.75	0.990
	0.995	3.71	3.68	3.66	3.64	3.62	3.54	3.44	3.33	3.22	3.11	0.995
17	0.005	0.299	0.305	0.311	0.316	0.321	0.342	0.371	0.402	0.437	0.476	0.005
	0.010	0.338	0.344	0.349	0.355	0.360	0.380	0.408	0.438	0.472	0.509	0.010
	0.025	0.403	0.409	0.414	0.419	0.424	0.443	0.470	0.498	0.529	0.563	0.025
	0.050	0.468	0.473	0.478	0.483	0.487	0.506	0.530	0.557	0.585	0.616	0.050
	0.100	0.555	0.560	0.564	0.569	0.573	0.589	0.611	0.635	0.659	0.686	0.100
	0.900	1.85	1.85	1.84	1.84	1.83	1.81	1.78	1.75	1.72	1.69	0.900
	0.950	2.22	2.21	2.20	2.19	2.18	2.15	2.10	2.06	2.01	1.96	0.950
	0.975	2.60	2.59	2.57	2.56	2.55	2.50	2.44	2.38	2.32	2.25	0.975
	0.990	3.14	3.12	3.10	3.08	3.07	3.00	2.92	2.83	2.75	2.65	0.990
	0.995	3.58	3.56	3.53	3.51	3.49	3.41	3.31	3.21	3.10	2.98	0.995
18	0.005	0.303	0.309	0.315	0.320	0.325	0.347	0.376	0.408	0.444	0.484	0.005
	0.010	0.341	0.347	0.353	0.359	0.364	0.385	0.413	0.444	0.479	0.517	0.010
	0.025	0.406	0.412	0.418	0.423	0.428	0.448	0.475	0.504	0.536	0.571	0.025
	0.050	0.471	0.477	0.482	0.487	0.491	0.510	0.535	0.562	0.592	0.624	0.050
	0.100	0.558	0.563	0.568	0.572	0.576	0.593	0.615	0.639	0.665	0.693	0.100
	0.900	1.83	1.82	1.82	1.81	1.80	1.78	1.75	1.72	1.69	1.66	0.900
	0.950	2.18	2.17	2.16	2.15	2.14	2.11	2.06	2.02	1.97	1.92	0.950
	0.975	2.54	2.53	2.52	2.50	2.49	2.44	2.38	2.32	2.26	2.19	0.975
	0.990	3.05	3.03	3.02	3.00	2.98	2.92	2.84	2.75	2.66	2.57	0.990
	0.995	3.47	3.45	3.42	3.40	3.38	3.30	3.20	3.10	2.99	2.87	0.995
19	0.005	0.306	0.312	0.318	0.323	0.329	0.351	0.381	0.414	0.451	0.492	0.005
	0.010	0.344	0.351	0.357	0.362	0.367	0.389	0.418	0.450	0.485	0.525	0.010
	0.025	0.409	0.416	0.421	0.426	0.431	0.452	0.479	0.509	0.542	0.578	0.025
	0.050	0.474	0.480	0.485	0.490	0.495	0.514	0.540	0.567	0.597	0.630	0.050
	0.100	0.561	0.566	0.571	0.575	0.579	0.597	0.619	0.644	0.670	0.698	0.100
	0.900	1.81	1.80	1.79	1.79	1.78	1.76	1.73	1.70	1.67	1.63	0.900
	0.950	2.14	2.13	2.12	2.11	2.11	2.07	2.03	1.98	1.93	1.88	0.950
	0.975	2.49	2.48	2.46	2.45	2.44	2.39	2.33	2.27	2.20	2.13	0.975
	0.990	2.98	2.96	2.94	2.92	2.91	2.84	2.76	2.67	2.58	2.49	0.990
	0.995	3.37	3.35	3.33	3.31	3.29	3.21	3.11	3.00	2.89	2.78	0.995
20	0.005	0.308	0.315	0.321	0.327	0.332	0.354	0.385	0.419	0.457	0.500	0.005
	0.010	0.347	0.354	0.360	0.365	0.370	0.392	0.422	0.455	0.492	0.532	0.010
	0.025	0.412	0.419	0.424	0.430	0.435	0.456	0.484	0.514	0.548	0.585	0.025
	0.050	0.477	0.483	0.488	0.493	0.498	0.518	0.544	0.572	0.603	0.637	0.050
	0.100	0.563	0.569	0.573	0.578	0.582	0.600	0.623	0.648	0.675	0.704	0.100
	0.900	1.79	1.78	1.77	1.77	1.76	1.74	1.71	1.68	1.64	1.61	0.900
	0.950	2.11	2.10	2.09	2.08	2.07	2.04	1.99	1.95	1.90	1.84	0.950
	0.975	2.45	2.43	2.42	2.41	2.40	2.35	2.29	2.22	2.16	2.09	0.975
	0.990	2.92	2.90	2.88	2.86	2.84	2.78	2.69	2.61	2.52	2.42	0.990
	0.995	3.29	3.27	3.24	3.22	3.20	3.12	3.02	2.92	2.81	2.69	0.995

TABLE A.10   (Continued)

ν_2	ν_1 / γ	21	22	23	24	25	30	40	60	120	∞	ν_1 / γ
21	0.005	0.311	0.318	0.324	0.329	0.335	0.358	0.389	0.424	0.463	0.507	0.005
	0.010	0.350	0.356	0.363	0.368	0.374	0.396	0.426	0.460	0.497	0.539	0.010
	0.025	0.415	0.421	0.427	0.433	0.438	0.459	0.488	0.519	0.554	0.592	0.025
	0.050	0.480	0.486	0.491	0.496	0.501	0.521	0.548	0.577	0.608	0.643	0.050
	0.100	0.566	0.571	0.576	0.581	0.585	0.603	0.626	0.652	0.679	0.709	0.100
	0.900	1.77	1.76	1.75	1.75	1.74	1.72	1.69	1.66	1.62	1.59	0.900
	0.950	2.08	2.07	2.06	2.05	2.05	2.01	1.96	1.92	1.87	1.81	0.950
	0.975	2.41	2.39	2.38	2.37	2.36	2.31	2.25	2.18	2.11	2.04	0.975
	0.990	2.86	2.84	2.82	2.80	2.79	2.72	2.64	2.55	2.46	2.36	0.990
	0.995	3.22	3.19	3.17	3.15	3.13	3.05	2.95	2.84	2.73	2.61	0.995
22	0.005	0.313	0.320	0.326	0.332	0.338	0.361	0.393	0.428	0.468	0.514	0.005
	0.010	0.352	0.359	0.365	0.371	0.376	0.399	0.430	0.464	0.503	0.546	0.010
	0.025	0.418	0.424	0.430	0.436	0.441	0.462	0.491	0.523	0.559	0.598	0.025
	0.050	0.482	0.488	0.494	0.499	0.504	0.524	0.551	0.581	0.613	0.649	0.050
	0.100	0.568	0.573	0.578	0.583	0.588	0.606	0.630	0.655	0.683	0.714	0.100
	0.900	1.75	1.74	1.74	1.73	1.73	1.70	1.67	1.64	1.60	1.57	0.900
	0.950	2.06	2.05	2.04	2.03	2.02	1.98	1.94	1.89	1.84	1.78	0.950
	0.975	2.37	2.36	2.34	2.33	2.32	2.27	2.21	2.14	2.08	2.00	0.975
	0.990	2.81	2.78	2.77	2.75	2.73	2.67	2.58	2.50	2.40	2.31	0.990
	0.995	3.15	3.12	3.10	3.08	3.06	2.98	2.88	2.77	2.66	2.55	0.995
23	0.005	0.316	0.322	0.329	0.335	0.340	0.364	0.396	0.433	0.474	0.521	0.005
	0.010	0.355	0.362	0.368	0.374	0.379	0.402	0.434	0.469	0.508	0.552	0.010
	0.025	0.420	0.427	0.433	0.438	0.443	0.465	0.495	0.528	0.564	0.604	0.025
	0.050	0.485	0.491	0.496	0.502	0.507	0.527	0.555	0.585	0.617	0.654	0.050
	0.100	0.570	0.576	0.581	0.585	0.590	0.608	0.633	0.659	0.687	0.719	0.100
	0.900	1.74	1.73	1.72	1.72	1.71	1.69	1.66	1.62	1.59	1.55	0.900
	0.950	2.04	2.02	2.01	2.01	2.00	1.96	1.91	1.86	1.81	1.76	0.950
	0.975	2.34	2.33	2.31	2.30	2.29	2.24	2.18	2.11	2.04	1.97	0.975
	0.990	2.76	2.74	2.72	2.70	2.69	2.62	2.54	2.45	2.35	2.26	0.990
	0.995	3.09	3.06	3.04	3.02	3.00	2.92	2.82	2.71	2.60	2.48	0.995
24	0.005	0.318	0.325	0.331	0.337	0.343	0.367	0.400	0.437	0.479	0.527	0.005
	0.010	0.357	0.364	0.370	0.376	0.382	0.405	0.437	0.473	0.513	0.558	0.010
	0.025	0.422	0.429	0.435	0.441	0.446	0.468	0.498	0.531	0.568	0.610	0.025
	0.050	0.487	0.493	0.499	0.504	0.509	0.530	0.558	0.588	0.622	0.659	0.050
	0.100	0.572	0.578	0.583	0.588	0.592	0.611	0.635	0.662	0.691	0.723	0.100
	0.900	1.72	1.71	1.71	1.70	1.70	1.67	1.64	1.61	1.57	1.53	0.900
	0.950	2.01	2.00	1.99	1.98	1.97	1.94	1.89	1.84	1.79	1.73	0.950
	0.975	2.31	2.30	2.28	2.27	2.26	2.21	2.15	2.08	2.01	1.94	0.975
	0.990	2.72	2.70	2.68	2.66	2.64	2.58	2.49	2.40	2.31	2.21	0.990
	0.995	3.04	3.01	2.99	2.97	2.95	2.87	2.77	2.66	2.55	2.43	0.995
25	0.005	0.320	0.327	0.333	0.339	0.345	0.369	0.403	0.441	0.483	0.533	0.005
	0.010	0.359	0.366	0.372	0.378	0.384	0.408	0.440	0.477	0.517	0.564	0.010
	0.025	0.424	0.431	0.437	0.443	0.448	0.471	0.501	0.535	0.573	0.615	0.025
	0.050	0.489	0.495	0.501	0.506	0.511	0.532	0.561	0.592	0.626	0.664	0.050
	0.100	0.574	0.580	0.585	0.590	0.594	0.613	0.638	0.665	0.694	0.727	0.100
	0.900	1.71	1.70	1.70	1.69	1.68	1.66	1.63	1.59	1.56	1.52	0.900
	0.950	2.00	1.98	1.97	1.96	1.96	1.92	1.87	1.82	1.77	1.71	0.950
	0.975	2.28	2.27	2.26	2.24	2.23	2.18	2.12	2.05	1.98	1.91	0.975
	0.990	2.68	2.66	2.64	2.62	2.60	2.54	2.45	2.36	2.27	2.17	0.990
	0.995	2.99	2.96	2.94	2.92	2.90	2.82	2.72	2.61	2.50	2.38	0.995

TABLE A.10   (Continued)

v_2	v_1 / γ	21	22	23	24	25	30	40	60	120	∞	v_1 / γ
30	0.005	0.328	0.335	0.342	0.349	0.355	0.381	0.416	0.457	0.504	0.559	0.005
	0.010	0.368	0.375	0.382	0.388	0.394	0.419	0.454	0.493	0.538	0.589	0.010
	0.025	0.433	0.440	0.447	0.453	0.458	0.482	0.515	0.551	0.592	0.639	0.025
	0.050	0.497	0.504	0.510	0.516	0.521	0.543	0.573	0.606	0.643	0.685	0.050
	0.100	0.582	0.588	0.593	0.598	0.603	0.622	0.649	0.678	0.710	0.745	0.100
	0.900	1.66	1.65	1.64	1.64	1.63	1.61	1.57	1.54	1.50	1.46	0.900
	0.950	1.92	1.91	1.90	1.89	1.88	1.84	1.79	1.74	1.68	1.62	0.950
	0.975	2.18	2.16	2.15	2.14	2.12	2.07	2.01	1.94	1.87	1.79	0.975
	0.990	2.53	2.51	2.49	2.47	2.45	2.39	2.30	2.21	2.11	2.01	0.990
	0.995	2.80	2.77	2.75	2.73	2.71	2.63	2.52	2.42	2.30	2.18	0.995
40	0.005	0.339	0.347	0.355	0.362	0.368	0.396	0.436	0.481	0.534	0.599	0.005
	0.010	0.379	0.387	0.394	0.401	0.408	0.435	0.473	0.517	0.567	0.628	0.010
	0.025	0.445	0.453	0.459	0.466	0.472	0.498	0.533	0.573	0.620	0.674	0.025
	0.050	0.509	0.516	0.522	0.529	0.534	0.558	0.591	0.627	0.669	0.717	0.050
	0.100	0.592	0.598	0.604	0.610	0.615	0.636	0.664	0.696	0.731	0.772	0.100
	0.900	1.60	1.59	1.58	1.57	1.57	1.54	1.51	1.47	1.42	1.38	0.900
	0.950	1.83	1.81	1.80	1.79	1.78	1.74	1.69	1.64	1.58	1.51	0.950
	0.975	2.05	2.03	2.02	2.01	1.99	1.94	1.88	1.80	1.72	1.64	0.975
	0.990	2.35	2.33	2.31	2.29	2.27	2.20	2.11	2.02	1.92	1.80	0.990
	0.995	2.57	2.55	2.52	2.50	2.48	2.40	2.30	2.18	2.06	1.93	0.995
60	0.005	0.352	0.361	0.369	0.376	0.383	0.414	0.458	0.510	0.572	0.652	0.005
	0.010	0.392	0.401	0.409	0.416	0.423	0.453	0.495	0.545	0.604	0.679	0.010
	0.025	0.458	0.466	0.474	0.481	0.487	0.515	0.555	0.600	0.654	0.720	0.025
	0.050	0.522	0.529	0.536	0.543	0.549	0.575	0.611	0.652	0.700	0.759	0.050
	0.100	0.604	0.610	0.616	0.622	0.628	0.650	0.682	0.717	0.757	0.807	0.100
	0.900	1.53	1.53	1.52	1.51	1.50	1.48	1.44	1.40	1.35	1.29	0.900
	0.950	1.73	1.72	1.71	1.70	1.69	1.65	1.59	1.53	1.47	1.39	0.950
	0.975	1.93	1.91	1.90	1.88	1.87	1.82	1.74	1.67	1.58	1.48	0.975
	0.990	2.17	2.15	2.13	2.12	2.10	2.03	1.94	1.84	1.73	1.60	0.990
	0.995	2.36	2.33	2.31	2.29	2.27	2.19	2.08	1.96	1.83	1.69	0.995
120	0.005	0.366	0.376	0.384	0.393	0.401	0.435	0.485	0.545	0.623	0.733	0.005
	0.010	0.407	0.416	0.425	0.433	0.441	0.474	0.522	0.579	0.652	0.755	0.010
	0.025	0.473	0.482	0.490	0.498	0.505	0.536	0.580	0.633	0.698	0.788	0.025
	0.050	0.536	0.544	0.552	0.559	0.565	0.594	0.634	0.682	0.740	0.819	0.050
	0.100	0.616	0.623	0.630	0.636	0.642	0.667	0.702	0.742	0.791	0.856	0.100
	0.900	1.47	1.46	1.46	1.45	1.44	1.41	1.37	1.32	1.26	1.19	0.900
	0.950	1.64	1.63	1.62	1.61	1.60	1.55	1.50	1.43	1.35	1.25	0.950
	0.975	1.81	1.79	1.77	1.76	1.75	1.69	1.61	1.53	1.43	1.31	0.975
	0.990	2.01	1.99	1.97	1.95	1.93	1.86	1.76	1.66	1.53	1.38	0.990
	0.995	2.16	2.13	2.11	2.09	2.07	1.98	1.87	1.75	1.61	1.43	0.995
∞	0.005	0.383	0.393	0.403	0.412	0.421	0.460	0.518	0.592	0.699	1.000	0.005
	0.010	0.424	0.434	0.443	0.452	0.461	0.498	0.554	0.624	0.724	1.000	0.010
	0.025	0.490	0.499	0.508	0.517	0.525	0.560	0.611	0.675	0.763	1.000	0.025
	0.050	0.552	0.561	0.569	0.577	0.584	0.616	0.663	0.720	0.798	1.000	0.050
	0.100	0.630	0.638	0.646	0.652	0.659	0.687	0.727	0.774	0.839	1.000	0.100
	0.900	1.41	1.40	1.39	1.38	1.38	1.34	1.29	1.24	1.17	1.00	0.900
	0.950	1.56	1.54	1.53	1.52	1.51	1.46	1.39	1.32	1.22	1.00	0.950
	0.975	1.69	1.67	1.66	1.64	1.63	1.57	1.48	1.39	1.27	1.00	0.975
	0.990	1.85	1.83	1.81	1.79	1.77	1.70	1.59	1.47	1.32	1.00	0.990
	0.995	1.97	1.95	1.92	1.90	1.88	1.79	1.67	1.53	1.36	1.00	0.995

TABLE A.11   STUDENTIZED RANGE VALUES

Values for order k and ν degrees of freedom

Upper 5% points

k	2	3	4	5	6	7	8	9	10
ν 1	17.97	26.98	32.82	37.08	40.41	43.12	45.40	47.36	49.07
2	6.08	8.33	9.80	10.88	11.74	12.44	13.03	13.54	13.99
3	4.50	5.91	6.82	7.50	8.04	8.48	8.85	9.18	9.46
4	3.93	5.04	5.76	6.29	6.71	7.05	7.35	7.60	7.83
5	3.64	4.60	5.22	5.67	6.03	6.33	6.58	6.80	6.99
6	3.46	4.34	4.90	5.30	5.63	5.90	6.12	6.32	6.49
7	3.34	4.16	4.68	5.06	5.36	5.61	5.82	6.00	6.16
8	3.26	4.04	4.53	4.89	5.17	5.40	5.60	5.77	5.92
9	3.20	3.95	4.41	4.76	5.02	5.24	5.43	5.59	5.74
10	3.15	3.88	4.33	4.65	4.91	5.12	5.30	5.46	5.60
11	3.11	3.82	4.26	4.57	4.82	5.03	5.20	5.35	5.49
12	3.08	3.77	4.20	4.51	4.75	4.95	5.12	5.27	5.39
13	3.06	3.73	4.15	4.45	4.69	4.88	5.05	5.19	5.32
14	3.03	3.70	4.11	4.41	4.64	4.83	4.99	5.13	5.25
15	3.01	3.67	4.08	4.37	4.59	4.78	4.94	5.08	5.20
16	3.00	3.65	4.05	4.33	4.56	4.74	4.90	5.03	5.15
17	2.98	3.63	4.02	4.30	4.52	4.70	4.86	4.99	5.11
18	2.97	3.61	4.00	4.28	4.49	4.67	4.82	4.96	5.07
19	2.96	3.59	3.98	4.25	4.47	4.65	4.79	4.92	5.04
20	2.95	3.58	3.96	4.23	4.45	4.62	4.77	4.90	5.01
24	2.92	3.53	3.90	4.17	4.37	4.54	4.68	4.81	4.92
30	2.89	3.49	3.85	4.10	4.30	4.46	4.60	4.72	4.82
40	2.86	3.44	3.79	4.04	4.23	4.39	4.52	4.63	4.73
60	2.83	3.40	3.74	3.98	4.16	4.31	4.44	4.55	4.65
120	2.80	3.36	3.68	3.92	4.10	4.24	4.36	4.47	4.56
∞	2.77	3.31	3.63	3.86	4.03	4.17	4.29	4.39	4.47

Extracted with permission of the Biometrika Trustees from
The Biometrika Tables for Statisticians,   Vol. 1, 3rd. ed.,   1966,
edited by E. S. Pearson and H. O. Hartley.

TABLE A.11   (Continued)

Values for order k and ν degrees of freedom

Upper 1% points

k	2	3	4	5	6	7	8	9	10
ν 1	90.03	135.0	164.3	185.6	202.2	215.8	227.2	237.0	245.6
2	14.04	19.02	22.29	24.72	26.63	28.20	29.53	30.68	31.69
3	8.26	10.62	12.17	13.33	14.24	15.00	15.64	16.20	16.69
4	6.51	8.12	9.17	9.96	10.58	11.10	11.55	11.93	12.27
5	5.70	6.98	7.80	8.42	8.91	9.32	9.67	9.97	10.24
6	5.24	6.33	7.03	7.56	7.97	8.32	8.61	8.87	9.10
7	4.95	5.92	6.54	7.01	7.37	7.68	7.94	8.17	8.37
8	4.75	5.64	6.20	6.62	6.96	7.24	7.47	7.68	7.86
9	4.60	5.43	5.96	6.35	6.66	6.91	7.13	7.33	7.49
10	4.48	5.27	5.77	6.14	6.43	6.67	6.87	7.05	7.21
11	4.39	5.15	5.62	5.97	6.25	6.48	6.67	6.84	6.99
12	4.32	5.05	5.50	5.84	6.10	6.32	6.51	6.67	6.81
13	4.26	4.96	5.40	5.73	5.98	6.19	6.37	6.53	6.67
14	4.21	4.89	5.32	5.63	5.88	6.08	6.26	6.41	6.54
15	4.17	4.84	5.25	5.56	5.80	5.99	6.16	6.31	6.44
16	4.13	4.79	5.19	5.49	5.72	5.92	6.08	6.22	6.35
17	4.10	4.74	5.14	5.43	5.66	5.85	6.01	6.15	6.27
18	4.07	4.70	5.09	5.38	5.60	5.79	5.94	6.08	6.20
19	4.05	4.67	5.05	5.33	5.55	5.73	5.89	6.02	6.14
20	4.02	4.64	5.02	5.29	5.51	5.69	5.84	5.97	6.09
24	3.96	4.55	4.91	5.17	5.37	5.54	5.69	5.81	5.92
30	3.89	4.45	4.80	5.05	5.24	5.40	5.54	5.65	5.76
40	3.82	4.37	4.70	4.93	5.11	5.26	5.39	5.50	5.60
60	3.76	4.28	4.59	4.82	4.99	5.13	5.25	5.36	5.45
120	3.70	4.20	4.50	4.71	4.87	5.01	5.12	5.21	5.30
∞	3.64	4.12	4.40	4.60	4.76	4.88	4.99	5.08	5.16

TABLE A.12   VALUES FOR DUNCAN'S NEW MULTIPLE RANGE TEST (NMRT)

$$\alpha = 0.05$$

p, the rank order of means

ν	2	3	4	5	6	7	8	9	10
2	6.085	6.085	6.085	6.085	6.085	6.085	6.085	6.085	6.085
3	4.501	4.516	4.516	4.516	4.516	4.516	4.516	4.516	4.516
4	3.927	4.013	4.033	4.033	4.033	4.033	4.033	4.033	4.033
5	3.635	3.749	3.797	3.814	3.814	3.814	3.814	3.814	3.814
6	3.461	3.587	3.649	3.680	3.694	3.697	3.697	3.697	3.697
7	3.344	3.477	3.548	3.588	3.611	3.622	3.626	3.626	3.626
8	3.261	3.399	3.475	3.521	3.549	3.566	3.575	3.579	3.579
9	3.199	3.339	3.420	3.470	3.502	3.523	3.536	3.544	3.547
10	3.151	3.293	3.376	3.430	3.465	3.489	3.505	3.516	3.522
11	3.113	3.256	3.342	3.397	3.435	3.462	3.480	3.493	3.501
12	3.082	3.225	3.313	3.370	3.410	3.439	3.459	3.474	3.484
13	3.055	3.200	3.289	3.348	3.389	3.419	3.442	3.458	3.470
14	3.033	3.178	3.268	3.329	3.372	3.403	3.426	3.444	3.457
15	3.014	3.160	3.250	3.312	3.356	3.389	3.413	3.432	3.446
16	2.998	3.144	3.235	3.298	3.343	3.376	3.402	3.422	3.437
17	2.984	3.130	3.222	3.285	3.331	3.366	3.392	3.412	3.429
18	2.971	3.118	3.210	3.274	3.321	3.356	3.383	3.405	3.421
19	2.960	3.107	3.199	3.264	3.311	3.347	3.375	3.397	3.415
20	2.950	3.097	3.190	3.255	3.303	3.339	3.368	3.391	3.409
24	2.919	3.066	3.160	3.226	3.276	3.315	3.345	3.370	3.390
30	2.888	3.035	3.131	3.199	3.250	3.290	3.322	3.349	3.371
40	2.858	3.006	3.102	3.171	3.224	3.266	3.300	3.328	3.352
60	2.829	2.976	3.073	3.143	3.198	3.241	3.277	3.307	3.333
120	2.800	2.947	3.045	3.116	3.172	3.217	3.254	3.287	3.314
∞	2.772	2.918	3.017	3.089	3.146	3.193	3.232	3.265	3.294

TABLE A.12   (Continued)

$$\alpha = 0.05$$

p, the rank order of means

ν	11	12	13	14	15	16	18	20	30
2	6.085	6.085	6.085	6.085	6.085	6.085	6.085	6.085	6.085
3	4.516	4.516	4.516	4.516	4.516	4.516	4.516	4.516	4.516
4	4.033	4.033	4.033	4.033	4.033	4.033	4.033	4.033	4.033
5	3.814	3.814	3.814	3.814	3.814	3.814	3.814	3.814	3.814
6	3.697	3.697	3.697	3.697	3.697	3.697	3.697	3.697	3.697
7	3.626	3.626	3.626	3.626	3.626	3.626	3.626	3.626	3.626
8	3.579	3.579	3.579	3.579	3.579	3.579	3.579	3.579	3.579
9	3.547	3.547	3.547	3.547	3.547	3.547	3.547	3.547	3.547
10	3.525	3.526	3.526	3.526	3.526	3.526	3.526	3.526	3.526
11	3.506	3.509	3.510	3.510	3.510	3.510	3.510	3.510	3.510
12	3.491	3.496	3.498	3.499	3.499	3.499	3.499	3.499	3.499
13	3.478	3.484	3.488	3.490	3.490	3.490	3.490	3.490	3.490
14	3.467	3.474	3.479	3.482	3.484	3.484	3.485	3.485	3.485
15	3.457	3.465	3.471	3.476	3.478	3.480	3.481	3.481	3.481
16	3.449	3.458	3.465	3.470	3.473	3.477	3.478	3.478	3.478
17	3.441	3.451	3.459	3.465	3.469	3.473	3.476	3.476	3.476
18	3.435	3.445	3.454	3.460	3.465	3.470	3.474	3.474	3.474
19	3.429	3.440	3.449	3.456	3.462	3.467	3.472	3.474	3.474
20	3.424	3.436	3.445	3.453	3.459	3.464	3.470	3.473	3.474
24	3.406	3.420	3.432	3.441	3.449	3.456	3.465	3.471	3.477
30	3.389	3.405	3.418	3.430	3.439	3.447	3.460	3.470	3.486
40	3.373	3.390	3.405	3.418	3.429	3.439	3.456	3.469	3.500
60	3.355	3.374	3.391	3.406	3.419	3.431	3.451	3.467	3.515
120	3.337	3.359	3.377	3.394	3.409	3.423	3.446	3.466	3.532
∞	3.320	3.343	3.363	3.382	3.399	3.414	3.442	3.466	3.550

TABLE A.12   (Continued)

$$\alpha = 0.01$$

p, the rank order of means

ν	2	3	4	5	6	7	8	9	10
2	14.04	14.04	14.04	14.04	14.04	14.04	14.04	14.04	14.04
3	8.261	8.321	8.321	8.321	8.321	8.321	8.321	8.321	8.321
4	6.512	6.677	6.740	6.756	6.756	6.756	6.756	6.756	6.756
5	5.702	5.893	5.989	6.040	6.065	6.074	6.074	6.074	6.074
6	5.243	5.439	5.549	5.614	5.655	5.680	5.694	5.701	5.703
7	4.949	5.145	5.260	5.334	5.383	5.416	5.439	5.454	5.464
8	4.746	4.939	5.057	5.135	5.189	5.227	5.256	5.276	5.291
9	4.596	4.787	4.906	4.986	5.043	5.086	5.118	5.142	5.160
10	4.482	4.671	4.790	4.871	4.931	4.975	5.010	5.037	5.058
11	4.392	4.579	4.697	4.780	4.841	4.887	4.924	4.952	4.975
12	4.320	4.504	4.622	4.706	4.767	4.815	4.852	4.883	4.907
13	4.260	4.442	4.560	4.644	4.706	4.755	4.793	4.824	4.850
14	4.210	4.391	4.508	4.591	4.654	4.704	4.743	4.775	4.802
15	4.168	4.347	4.463	4.547	4.610	4.660	4.700	4.733	4.760
16	4.131	4.309	4.425	4.509	4.572	4.622	4.663	4.696	4.724
17	4.099	4.275	4.391	4.475	4.539	4.589	4.630	4.664	4.693
18	4.071	4.246	4.362	4.445	4.509	4.560	4.601	4.635	4.664
19	4.046	4.220	4.335	4.419	4.483	4.534	4.575	4.610	4.639
20	4.024	4.197	4.312	4.395	4.459	4.510	4.552	4.587	4.617
24	3.956	4.126	4.239	4.322	4.386	4.437	4.480	4.516	4.546
30	3.889	4.056	4.168	4.250	4.314	4.366	4.409	4.445	4.477
40	3.825	3.988	4.098	4.180	4.244	4.296	4.339	4.376	4.408
60	3.762	3.922	4.031	4.111	4.174	4.226	4.270	4.307	4.340
120	3.702	3.858	3.965	4.044	4.107	4.158	4.202	4.239	4.272
∞	3.643	3.796	3.900	3.978	4.040	4.091	4.135	4.172	4.205

TABLE A.12 (Continued)

$$\alpha = 0.01$$

p, the rank order of means

ν	11	12	13	14	15	16	18	20	30
2	14.04	14.04	14.04	14.04	14.04	14.04	14.04	14.04	14.04
3	8.321	8.321	8.321	8.321	8.321	8.321	8.321	8.321	8.321
4	6.756	6.756	6.756	6.756	6.756	6.756	6.756	6.756	6.756
5	6.074	6.074	6.074	6.074	6.074	6.074	6.074	6.074	6.074
6	5.703	5.703	5.703	5.703	5.703	5.703	5.703	5.703	5.703
7	5.470	5.472	5.472	5.472	5.472	5.472	5.472	5.472	5.472
8	5.302	5.309	5.314	5.316	5.317	5.317	5.317	5.317	5.317
9	5.174	5.185	5.193	5.199	5.203	5.205	5.206	5.206	5.206
10	5.074	5.088	5.098	5.106	5.112	5.117	5.122	5.124	5.124
11	4.994	5.009	5.021	5.031	5.039	5.045	5.054	5.059	5.061
12	4.927	4.944	4.958	4.969	4.978	4.986	4.998	5.006	5.011
13	4.872	4.889	4.904	4.917	4.928	4.937	4.950	4.960	4.972
14	4.824	4.843	4.859	4.872	4.884	4.894	4.910	4.921	4.940
15	4.783	4.803	4.820	4.834	4.846	4.857	4.874	4.887	4.914
16	4.748	4.768	4.786	4.800	4.813	4.825	4.844	4.858	4.890
17	4.717	4.738	4.756	4.771	4.785	4.797	4.816	4.832	4.869
18	4.689	4.711	4.729	4.745	4.759	4.772	4.792	4.808	4.850
19	4.665	4.686	4.705	4.722	4.736	4.749	4.771	4.788	4.833
20	4.642	4.664	4.684	4.701	4.716	4.729	4.751	4.769	4.818
24	4.573	4.596	4.616	4.634	4.651	4.665	4.690	4.710	4.770
30	4.504	4.528	4.550	4.569	4.586	4.601	4.628	4.650	4.721
40	4.436	4.461	4.483	4.503	4.521	4.537	4.566	4.591	4.671
60	4.368	4.394	4.417	4.438	4.456	4.474	4.504	4.530	4.620
120	4.301	4.327	4.351	4.372	4.392	4.410	4.442	4.469	4.568
∞	4.235	4.261	4.285	4.307	4.327	4.345	4.379	4.408	4.514

TABLE 1.13   VALUES FOR DUNNETT'S COMPARISONS WITH A CONTROL

For two-sided comparisons made with $\alpha = 0.05$.

k, the number of means, excluding the control

ν	1	2	3	4	5	6	7	8	9
5	2.57	3.03	3.29	3.48	3.62	3.73	3.82	3.90	3.97
6	2.45	2.86	3.10	3.26	3.39	3.49	3.57	3.64	3.71
7	2.36	2.75	2.97	3.12	3.24	3.33	3.41	3.47	3.53
8	2.31	2.67	2.88	3.02	3.13	3.22	3.29	3.35	3.41
9	2.26	2.61	2.81	2.95	3.05	3.14	3.20	3.26	3.32
10	2.23	2.57	2.76	2.89	2.99	3.07	3.14	3.19	3.24
11	2.20	2.53	2.72	2.84	2.94	3.02	3.08	3.14	3.19
12	2.18	2.50	2.68	2.81	2.90	2.98	3.04	3.09	3.14
13	2.16	2.48	2.65	2.78	2.87	2.94	3.00	3.06	3.10
14	2.14	2.46	2.63	2.75	2.84	2.91	2.97	3.02	3.07
15	2.13	2.44	2.61	2.73	2.82	2.89	2.95	3.00	3.04
16	2.12	2.42	2.59	2.71	2.80	2.87	2.92	2.97	3.02
17	2.11	2.41	2.58	2.69	2.78	2.85	2.90	2.95	3.00
18	2.10	2.40	2.56	2.68	2.76	2.83	2.89	2.94	2.98
19	2.09	2.39	2.55	2.66	2.75	2.81	2.87	2.92	2.96
20	2.09	2.38	2.54	2.65	2.73	2.80	2.86	2.90	2.95
24	2.06	2.35	2.51	2.61	2.70	2.76	2.81	2.86	2.90
30	2.04	2.32	2.47	2.58	2.66	2.72	2.77	2.82	2.86
40	2.02	2.29	2.44	2.54	2.62	2.68	2.73	2.77	2.81
60	2.00	2.27	2.41	2.51	2.58	2.64	2.69	2.73	2.77
120	1.98	2.24	2.38	2.47	2.55	2.60	2.65	2.69	2.73
∞	1.96	2.21	2.35	2.44	2.51	2.57	2.61	2.65	2.69

Reproduced from: (1) C. W. Dunnett, "New Tables for Multiple Comparisons with a Control." *Biometrics*, 20: 482-491. 1964. With permission of the author and The Biometric Society. (2)C. W. Dunnett, "A Multiple Comparison Procedure for Comparing Several Treatments with a Control." *Jour. Amer. Stat. Assoc.*, 50: 1112-1118. 1955. With permission of the author and The American Statistical Association.

TABLE 1.13   (Continued)

For two-sided comparisons made with $\alpha = 0.01$.

k, the number of means, excluding the control

ν	1	2	3	4	5	6	7	8	9
5	4.03	4.63	4.98	5.22	5.41	5.56	5.69	5.80	5.89
6	3.71	4.21	4.51	4.71	4.87	5.00	5.10	5.20	5.28
7	3.50	3.95	4.21	4.39	4.53	4.64	4.74	4.82	4.89
8	3.36	3.77	4.00	4.17	4.29	4.40	4.48	4.56	4.62
9	3.25	3.63	3.85	4.01	4.12	4.22	4.30	4.37	4.43
10	3.17	3.53	3.74	3.88	3.99	4.08	4.16	4.22	4.28
11	3.11	3.45	3.65	3.79	3.89	3.98	4.05	4.11	4.16
12	3.05	3.39	3.58	3.71	3.81	3.89	3.96	4.02	4.07
13	3.01	3.33	3.52	3.65	3.74	3.82	3.89	3.94	3.99
14	2.98	3.29	3.47	3.59	3.69	3.76	3.83	3.88	3.93
15	2.95	3.25	3.43	3.55	3.64	3.71	3.78	3.83	3.88
16	2.92	3.22	3.39	3.51	3.60	3.67	3.73	3.78	3.83
17	2.90	3.19	3.36	3.47	3.56	3.63	3.69	3.74	3.79
18	2.88	3.17	3.33	3.44	3.53	3.60	3.66	3.71	3.75
19	2.86	3.15	3.31	3.42	3.50	3.57	3.63	3.68	3.72
20	2.85	3.13	3.29	3.40	3.48	3.55	3.60	3.65	3.69
24	2.80	3.07	3.22	3.32	3.40	3.47	3.52	3.57	3.61
30	2.75	3.01	3.15	3.25	3.33	3.39	3.44	3.49	3.52
40	2.70	2.95	3.09	3.19	3.26	3.32	3.37	3.41	3.44
60	2.66	2.90	3.03	3.12	3.19	3.25	3.29	3.33	3.37
120	2.62	2.85	2.97	3.06	3.12	3.18	3.22	3.26	3.29
∞	2.58	2.79	2.92	3.00	3.06	3.11	3.15	3.19	3.22

TABLE 1.13 (Continued)

For one-sided comparisons made with $\alpha = 0.05$.

k, the number of means, excluding the control

ν	1	2	3	4	5	6	7	8	9
5	2.02	2.44	2.68	2.85	2.98	3.08	3.16	3.24	3.30
6	1.94	2.34	2.56	2.71	2.83	2.92	3.00	3.07	3.12
7	1.89	2.27	2.48	2.62	2.73	2.82	2.89	2.95	3.01
8	1.86	2.22	2.42	2.55	2.66	2.74	2.81	2.87	2.92
9	1.83	2.18	2.37	2.50	2.60	2.68	2.75	2.81	2.86
10	1.81	2.15	2.34	2.47	2.56	2.64	2.70	2.76	2.81
11	1.80	2.13	2.31	2.44	2.53	2.60	2.67	2.72	2.77
12	1.78	2.11	2.29	2.41	2.50	2.58	2.64	2.69	2.74
13	1.77	2.09	2.27	2.39	2.48	2.55	2.61	2.66	2.71
14	1.76	2.08	2.25	2.37	2.46	2.53	2.59	2.64	2.69
15	1.75	2.07	2.24	2.36	2.44	2.51	2.57	2.62	2.67
16	1.75	2.06	2.23	2.34	2.43	2.50	2.56	2.61	2.65
17	1.74	2.05	2.22	2.33	2.42	2.49	2.54	2.59	2.64
18	1.73	2.05	2.21	2.32	2.41	2.48	2.53	2.58	2.62
19	1.73	2.03	2.20	2.31	2.40	2.47	2.52	2.57	2.61
20	1.72	2.03	2.19	2.30	2.39	2.46	2.51	2.56	2.60
24	1.71	2.01	2.17	2.28	2.36	2.43	2.48	2.53	2.57
30	1.70	1.99	2.15	2.25	2.33	2.40	2.45	2.50	2.54
40	1.68	1.97	2.13	2.23	2.31	2.37	2.42	2.47	2.51
60	1.67	1.95	2.10	2.21	2.28	2.35	2.39	2.44	2.48
120	1.66	1.93	2.08	2.18	2.26	2.32	2.37	2.41	2.45
∞	1.64	1.92	2.06	2.16	2.23	2.29	2.34	2.38	2.42

TABLE 1.13   (Continued)

For one-sided comparisons made with $\alpha = 0.01$.

k, the number of means, excluding the control

ν	1	2	3	4	5	6	7	8	9
5	3.37	3.90	4.21	4.43	4.60	4.73	4.85	4.94	5.03
6	3.14	3.61	3.88	4.07	4.21	4.33	4.43	4.51	4.59
7	3.00	3.42	3.66	3.83	3.96	4.07	4.15	4.23	4.30
8	2.90	3.29	3.51	3.67	3.79	3.88	3.96	4.03	4.09
9	2.82	3.19	3.40	3.55	3.66	3.75	3.82	3.89	3.94
10	2.76	3.11	3.31	3.45	3.56	3.64	3.71	3.78	3.83
11	2.72	3.06	3.25	3.38	3.48	3.56	3.63	3.69	3.74
12	2.68	3.01	3.19	3.32	3.42	3.50	3.56	3.62	3.67
13	2.65	2.97	3.15	3.27	3.37	3.44	3.51	3.56	3.61
14	2.62	2.94	3.11	3.23	3.32	3.40	3.46	3.51	3.56
15	2.60	2.91	3.08	3.20	3.29	3.36	3.42	3.47	3.52
16	2.58	2.88	3.05	3.17	3.26	3.33	3.39	3.44	3.48
17	2.57	2.86	3.03	3.14	3.23	3.30	3.36	3.41	3.45
18	2.55	2.84	3.01	3.12	3.21	3.27	3.33	3.38	3.42
19	2.54	2.83	2.99	3.10	3.18	3.25	3.31	3.36	3.40
20	2.53	2.81	2.97	3.08	3.17	3.23	3.29	3.34	3.38
24	2.49	2.77	2.92	3.03	3.11	3.17	3.22	3.27	3.31
30	2.46	2.72	2.87	2.97	3.05	3.11	3.16	3.21	3.24
40	2.42	2.68	2.82	2.92	2.99	3.05	3.10	3.14	3.18
60	2.39	2.64	2.78	2.87	2.94	3.00	3.04	3.08	3.12
120	2.36	2.60	2.73	2.82	2.89	2.94	2.99	3.03	3.06
∞	2.33	2.56	2.68	2.77	2.84	2.89	2.93	2.97	3.00

TABLE 1.14   VALUES FOR BONFERRONI'S   T

m = number of contrasts; α = 0.05; ν = degrees of freedom

ν	5	7	10	12	15	20	24	30	40	60	120	∞
2	3.17	2.84	2.64	2.56	2.49	2.42	2.39	2.36	2.33	2.30	2.27	2.24
3	3.54	3.13	2.87	2.78	2.69	2.61	2.58	2.54	2.50	2.47	2.43	2.39
4	3.81	3.34	3.04	2.94	2.84	2.75	2.70	2.66	2.62	2.58	2.54	2.50
5	4.04	3.50	3.17	3.06	2.95	2.85	2.80	2.75	2.71	2.66	2.62	2.58
6	4.22	3.64	3.28	3.15	3.04	2.93	2.88	2.83	2.78	2.73	2.68	2.64
7	4.38	3.76	3.37	3.24	3.11	3.00	2.94	2.89	2.84	2.79	2.74	2.69
8	4.53	3.86	3.45	3.31	3.18	3.06	3.00	2.94	2.89	2.84	2.79	2.74
9	4.66	3.95	3.52	3.37	3.24	3.11	3.05	2.90	2.93	2.88	2.83	2.77
10	4.78	4.03	3.58	3.43	3.29	3.16	3.09	3.03	2.97	2.92	2.86	2.81
15	5.25	4.36	3.83	3.65	3.48	3.33	3.26	3.19	3.12	3.06	2.99	2.94
20	5.60	4.59	4.01	3.80	3.62	3.46	3.38	3.30	3.23	3.16	3.09	3.02
25	5.89	4.78	4.15	3.93	3.74	3.55	3.47	3.39	3.31	3.24	3.16	3.09
30	6.15	4.95	4.27	4.04	3.82	3.63	3.54	3.46	3.38	3.30	3.22	3.15
35	6.36	5.09	4.37	4.13	3.90	3.70	3.61	3.52	3.43	3.34	3.27	3.19
40	6.56	5.21	4.45	4.20	3.97	3.76	3.66	3.57	3.48	3.39	3.31	3.23
45	6.70	5.31	4.53	4.26	4.02	3.80	3.70	3.61	3.51	3.42	3.34	3.26
50	6.86	5.40	4.59	4.32	4.07	3.85	3.74	3.65	3.55	3.46	3.37	3.29

m = number of contrasts; α = 0.01; ν = degrees of freedom

ν	5	7	10	12	15	20	24	30	40	60	120	∞
2	4.78	4.03	3.58	3.43	3.29	3.16	3.09	3.03	2.97	2.92	2.86	2.81
3	5.25	4.36	3.83	3.65	3.48	3.33	3.26	3.19	3.12	3.06	2.99	2.94
4	5.60	4.59	4.01	3.80	3.62	3.46	3.38	3.30	3.23	3.16	3.09	3.02
5	5.89	4.78	4.15	3.93	3.74	3.55	3.47	3.39	3.31	3.24	3.16	3.09
6	6.15	4.95	4.27	4.04	3.82	3.63	3.54	3.46	3.38	3.30	3.22	3.15
7	6.36	5.09	4.37	4.13	3.90	3.70	3.61	3.52	3.43	3.34	3.27	3.19
8	6.56	5.21	4.45	4.20	3.97	3.76	3.66	3.57	3.48	3.39	3.31	3.23
9	6.70	5.31	4.53	4.26	4.02	3.80	3.70	3.61	3.51	3.42	3.34	3.26
10	6.86	5.40	4.59	4.32	4.07	3.85	3.74	3.65	3.55	3.46	3.37	3.29
15	7.51	5.79	4.86	4.56	4.29	4.03	3.91	3.80	3.70	3.59	3.50	3.40
20	8.00	6.08	5.06	4.73	4.42	4.15	4.04	3.90	3.79	3.69	3.58	3.48
25	8.37	6.30	5.20	4.86	4.53	4.25	4.1*	3.98	3.88	3.76	3.64	3.54
30	8.68	6.49	5.33	4.95	4.61	4.33	4.2*	4.13	3.93	3.81	3.69	3.59
35	8.95	6.67	5.44	5.04	4.71	4.39	4.3*	4.26	3.97	3.84	3.73	3.63
40	9.19	6.83	5.52	5.12	4.78	4.46	4.3*	4.1*	4.01	3.89	3.77	3.66
45	9.41	6.93	5.60	5.20	4.84	4.52	4.3*	4.2*	4.1*	3.93	3.80	3.69
50	9.68	7.06	5.70	5.27	4.90	4.56	4.4*	4.2*	4.1*	3.97	3.83	3.72

* Obtained by graphical interpolation

Reproduced from: O. J. Dunn, "Multiple Comparisons Among Means", *Jour. Amer. Stat. Assoc.*, 56: 52-64, 1961. With permission of the author and The American Statistical Association.

TABLE A.15   VALUES OF THE ARCSIN TRANSFORMATION

$$X = \sin^{-1}\sqrt{p} \quad \text{for proportion p}$$

p	X	p	X	p	X	p	X	p	X	p	X
.001	.032	.04	.201	.34	.623	.64	0.927	.94	1.323	.97	1.397
.002	.045	.05	.226	.35	.633	.65	0.938	.941	1.325	.971	1.4
.003	.055	.06	.247	.36	.644	.66	0.948	.942	1.328	.972	1.403
.004	.063	.07	.268	.37	.654	.67	0.959	.943	1.33	.973	1.406
.005	.071	.08	.287	.38	.664	.68	0.97	.944	1.332	.974	1.409
.006	.078	.09	.305	.39	.674	.69	0.98	.945	1.334	.975	1.412
.007	.084	.1	.322	.4	.685	.7	0.991	.946	1.336	.976	1.415
.008	.09	.11	.338	.41	.695	.71	1.002	.947	1.338	.977	1.419
.009	.095	.12	.354	.42	.705	.72	1.013	.948	1.341	.978	1.422
.01	.1	.13	.369	.43	.715	.73	1.024	.949	1.343	.979	1.425
.011	.105	.14	.383	.44	.725	.74	1.036	.95	1.345	.98	1.429
.012	.11	.15	.398	.45	.735	.75	1.047	.951	1.348	.981	1.433
.013	.114	.16	.412	.46	.745	.76	1.059	.952	1.35	.982	1.436
.014	.119	.17	.425	.47	.755	.77	1.071	.953	1.352	.983	1.44
.015	.123	.18	.438	.48	.765	.78	1.083	.954	1.355	.984	1.444
.016	.127	.19	.451	.49	.775	.79	1.095	.955	1.357	.985	1.448
.017	.131	.2	.464	.5	.785	.8	1.107	.956	1.359	.986	1.452
.018	.135	.21	.476	.51	.795	.81	1.12	.957	1.362	.987	1.457
.019	.138	.22	.488	.52	.805	.82	1.133	.958	1.364	.988	1.461
.02	.142	.23	.5	.53	.815	.83	1.146	.959	1.367	.989	1.466
.021	.145	.24	.512	.54	.825	.84	1.159	.96	1.369	.99	1.471
.022	.149	.25	.524	.55	.835	.85	1.173	.961	1.372	.991	1.476
.023	.152	.26	.535	.56	.846	.86	1.187	.962	1.375	.992	1.481
.024	.156	.27	.546	.57	.856	.87	1.202	.963	1.377	.993	1.487
.025	.159	.28	.558	.58	.866	.88	1.217	.964	1.38	.994	1.493
.026	.162	.29	.569	.59	.876	.89	1.233	.965	1.383	.995	1.5
.027	.165	.3	.58	.6	.886	.9	1.249	.966	1.385	.996	1.508
.028	.168	.31	.591	.61	.896	.91	1.266	.967	1.388	.997	1.516
.029	.171	.32	.601	.62	.907	.92	1.284	.968	1.391	.998	1.526
.03	.174	.33	.612	.63	.917	.93	1.303	.969	1.394	.999	1.539

Computed by the authors

Chapter 1

1. (a) normal, mean of 30, variance of 8
   (b) normal, mean of 30, variance of 16/3.

3. $\sqrt{2.5}$ 5. (a) 43; 50   (b) -12; 160   (c) 78; 105   (d) -8; 85

7. (a) 38; 72   (b) 2; 48   (c) -6; 56

Chapter 2

3. (a) Use pairs of rows for EU with pairs of buffer rows between
   EU.   There are two rows left over so one may use 3 buffer
   rows on each end of the parcel.
   (b) Use the middle 10 feet from each pair of treated rows. The
   yield from the "20 feet" would be the observation of
   interest.
   (c) $y_{ij} = \mu + \tau_i + \epsilon_{ij}$; $\begin{matrix} i = 1, 2, \ldots, 6 \\ j = 1, 2, \ldots, 5 \end{matrix}$

5. (a) $y_{ij} = \mu + \tau_i + \epsilon_{ij}$; $\begin{matrix} i = 1, 2, \ldots, 6 \\ j = 1, 2, \ldots, 5 \end{matrix}$

   (b)

Source	df	MS	EMS
Treatments	5	24	$\sigma_\epsilon^2 + 5\kappa_\tau^2$
Exp. Error	24	6	σ_ϵ^2

   (c) Calculated $F = 24/6 = 4$ with 5 and 24 degrees of freedom
   gives an observed significance level of $0.005 < P < 0.01$.
   (d) 2.4   (e) 2 ± 3.2

7. (a)

df	MS	EMS
4	18.9	$\sigma_\epsilon^2 + \Sigma\, r_i \tau_i^2/4$
47	7.1	σ_ϵ^2

   (b) Calculated $F = 2.66$ with 4 and 47 df.
   Observed significance level: $0.025 < P < 0.05$.
   (c) The researcher can be 99% confident that the difference in
   mean transpiration for varieties 1 and 4 is between -5.9
   and +0.9.
   (d) The ϵ's must be i.i.d. with mean zero.
   (e) The ϵ's must be i.i.d. $N(0, \sigma_\epsilon^2)$

9. (a)

df	EMS
3	$\sigma_\epsilon^2 + 10\kappa_\tau^2$
36	σ_ϵ^2

(b) Calculated $F = 12.4$ with 3 and 36 df.
Observed significance level: $P < 0.005$.

(c) 1.54

11. Calculated value of Hartley's max. $F = 1.97$.
Observed significance level: $P > 0.05$

13. (a) Gallon of water in the city's water tank.

(b) $y_{ij} = \mu + \tau_i + \epsilon_{ij}$; $\begin{array}{l} i = 1, 2, 3, 4 \\ j = 1, 2, \ldots, 10 \end{array}$

   μ = overall mean

   τ_i = filtering effect of i-th filter

   ϵ_{ij} = residual component

(c)

Source	df	SS	MS	EMS
Filters	3	275.82	91.94	$\sigma_\epsilon^2 + 10\kappa_\tau^2$
Exp.Error	36	21.65	0.601	σ_ϵ^2

(d) Calculated value of F is $91.94/0.601 = 153$ with 3 and 36 df; observed significance level is $P < 0.005$.

(e) For part (c), the ϵ's must be independent, identically distributed with zero mean. For part (d), an additional assumption of normality is required.

(f)

i	h	$\bar{y}_{i.} - \bar{y}_{h.}$
1	2	6.13
1	3	6.12
1	4	5.72
2	3	-0.01
2	4	-0.41
3	4	-0.40

The estimated standard deviation for the difference of any pair of means is $\sqrt{0.601(2/10)} = 0.35$

15. (a) $y_{ij} = \mu + \tau_i + \epsilon_{ij}$; $\begin{array}{l} i = 1, 2, 3, 4, 5 \\ j = 1, 2, 3, 4 \end{array}$

   Assumptions: $\Sigma \tau_i = 0$ and ϵ's i.i.d. $N(0, \sigma_\epsilon^2)$

(b)

Source	df	SS	MS	EMS
Models	4	589.3	147.3	$\sigma^2_\epsilon + 4\kappa^2_\tau$
Exp.Error	15	125.2	8.35	σ^2_ϵ

(c) Calculated value of F is $147.3/8.35 = 17.65$ with 4 and 15 df; observed significance level is $P < 0.005$.

(d)

i	h	$\bar{y}_{i.} - \bar{y}_{h.}$	i	h	$\bar{y}_{i.} - \bar{y}_{h.}$
1	2	-5.75	2	4	12.75
1	3	9.75	2	5	11.75
1	4	7.00	3	4	-2.75
1	5	6.00	3	5	-3.75
2	3	15.50	4	5	-1.00

The estimated standard deviation for the difference of any pair of means is $\sqrt{8.35(2/4)} = 2.04$

17. (a)

Source	df	SS	MS	EMS
Intensities	3	190.02	64.34	$\sigma^2_\epsilon + \Sigma\, r_i \tau^2_i / 3$
Exp.Error	16	11.96	0.75	σ^2_ϵ
Total	19	201.98		

(b) Calculated value of F is 84.5 with 3 and 16 df; observed significance level is $P < 0.005$.

(c)

i	h	$\bar{y}_{i.} - \bar{y}_{h.}$	$s_{\bar{y}_{i.} - \bar{y}_{h.}}$
1	2	-1.86	0.524
1	3	-5.14	0.559
1	4	-7.80	0.524
2	3	-3.29	0.581
2	4	-5.94	0.548
3	4	-2.66	0.581

(d) The psychologist can be 95% confident that the difference in mean time under intensities 1 and 4 is between 6.7 and 8.9 seconds

19. (a) $y_{ij} = \mu + \tau_i + \epsilon_{ij} + \delta_{ijk}$; $\begin{aligned} i &= 1, 2, 3, 4 \\ j &= 1, 2, \ldots, 12 \\ k &= 1, 2, 3, 4 \end{aligned}$

μ = overall mean

τ_i = effect due to the i-th condition

ϵ_{ij} = random component for j-th volunteer of i-th condition

δ_{ijk} = random component for k-th repeat of the (i,j)-th volunteer

(b)

df	SS	EMS
3	2449.2	$\sigma_\delta^2 + 4\sigma_\epsilon^2 + 48\kappa_\tau^2$
44	15906.0	$\sigma_\delta^2 + 4\sigma_\epsilon^2$
144	11966.4	σ_δ^2

(c) Calculated value of F is $816.4/361.5 = 2.26$ with 3 and 44 df; observed significance level is $0.05 < P < 0.1$.

(d) $\sqrt{361.5(2/48)} = 3.88$

21. (a)

Source	df	SS	MS	EMS
Methods	3	4.55	1.52	$\sigma_\epsilon^2 + 10\kappa_\tau^2$
Exp.Error	36	2.52	0.07	σ_ϵ^2

(b) Calculated value of F is $1.52/0.07 = 21.7$ with 3 and 36 degrees of freedom; observed significance level is $P < 0.005$.

23. (a)

Source	df	SS	MS	EMS
Hormones	4	1019.92	254.98	$\sigma_\epsilon^2 + 8\kappa_\tau^2$
Exp.Error	15	325.55	21.70	σ_ϵ^2

(b) Calculated value of F is $254.98/21.70 = 11.75$ with 4 and 15 df; observed significance level is $P < 0.005$.

Chapter 3

1. (a) All except (1,3)   (b) All except (1,3)
   (c) All except (1,3)   (d) (2,3),(3,4)

3. (a) (A,B), (A,C), (B,D), (B,E), (C,E)
   (b) (A,B), (B,E), (C,E)
   (c) (A,B), (A,C), (A,D)

5. Collection containing mean largest diameter: A,D,E.
   SCI for paired differences with largest mean diameter:
   (0,0.83), (0,2.07), (0,1.61), (0,1.33), (0,0.75

7. (a) Diets 2 and 3 differ from the control.
    (b) Collection containing smallest mean: 1,4
        SCI for paired differences with smallest fatty acid mean:
        (0,4.2), (0,11.5), (0,8.6), (0,2.4)

9. (a) (1,2), (1,3), (1,4)   (b) (1,2), (1,3), (1,4)
    (c) (1,2), (1,3), (1,4)

11. Collection C containing smallest mean: #1
    SCI for paired differences with smallest mean impurities:
    (0,0), (0.1.22), 0,1.09), (0,1.28)

13. (a) Pairs (1,2), (1,4), (2,3)   (b) Only treatment 2.

15. Only pair (3,4)

17. (a) Pairs (1,2), (1,3), (1,4)
    (b) Pairs (1,2), (1,3), (1,4).

19. (a) All pairs are significantly different
    (b) All pairs are significantly different

21. Collection C containing largest mean: #4
    SCI for differences of pairs with largest mean:
    (0,9.03), (0,7.17), (0,3.89), (0,0).

23. (a) No pairs significantly different by Tukey's procedure.
    (b) Only pair (3,4)

25. All pairs except (2,3).

27. Pairs significant by Duncan's: (1,5), (2,3), (3,4), (3,5)

Chapter 4

1. (a) Yes   (b)   No

3. Orthogonal Pairs: (1,2), (1,3), (1,4), (1,6), (1,7), (2,6)

5. (a) Contrast

	D_1	D_2	C
1	+1	-1	0
2	+1	+1	-2

    (b) Contrast 1: Drug 1 vs. Drug 2
         Contrast 2: Control vs. mean of Antibiotic drugs
    (c) Contrast 1 $H_0: \mu_1 = \mu_2$

         Contrast 2 $H_0: \mu_1 + \mu_2 - 2\mu_3 = 0$

7. (a) Contrast

	D_1	D_2	C
1	+1	-1	0
2	5	6	-11

(b) Contrast 1 : Drug 1 vs Drug 2
   Contrast 2 : Control vs "weighted mean" of antibiotic
                                 drugs.
(c) Contrast 1 H_0: $\mu_1 = \mu_2$

   Contrast 2 H_0: $\mu_3 = (5/11)\mu_1 + (6/11)\mu_2$

9. TYY = 40.416;   Contrast SS are: 1.286, 1.176, 3.24, 34.714

11. (a) C_1 = 14.6 - 20.3 = -5.7
   C_2 = 14.6 + 20.3 - 2(30.7) = -26.5
   $SS[C_1]$ = 97.47; $SS[C_2]$ = 702.25
(b) Contrast 1: Calculated F = 97.47/135.4 = 0.72;
   observed significance of, P > 0.1
   Contrast 2: Calculated F = 702.25/135.4 = 5.19;
   observed significance of 0.025 < P 0.05.

13. (a) $SS[C_1]$ = 88.61; $SS[C_2]$ = 580.12

(b) Contrast 1: Calculated value of F is 88.61/135.4 = 0.65;
   observed significance of P > 0.10.
   Contrast 2: Calculated value of F is 580.12/135.4 = 4.28;
   observed significance of 0.05 < P < 0.10.

15. (a) Any one of the mean separation procedures of Chapter 3.
   For a 5% Tukey's procedure, for example, the critical
   difference is 1.54; 3 pairs are significantly different:
   (A,B), (A,C), and (C,D).
(b) Contrasts (Orthogonal or Non-orthogonal)

Contrast	SS	F	
Effect of Blue Light (B,D vs A,C)	0.05	< 1	P > 0.1
Effect of Green Light (A,B vs C,D)	6.05	8.34	.01 < P < .025
Blue vs. Green Light, Darkness (B vs C)	1.25	1.72	P > 0.1
No light vs. Light (A vs B,C,D)	22.82	31.47	P < .005

plus others

(c) Dunnett's, using "Natural light" as a control. For 5%
   two-sided tests, the critical difference is 1.39;
   Blue(B) and Green(C) differ from Natural(A).

17.

Contrast	f-m	f-n	j-m	j-n	g-m	g-n
C_1	+1	+1	-1	-1	0	0
C_2	+1	+1	+1	+1	-2	-2
C_3	+1	-1	+1	-1	0	0
C_4	0	0	0	0	+1	-1

plus others

C_1: Freshmen, sophomores vs. juniors, seniors
C_2: Undergrads vs. Grads.
C_3: Undergrad members vs. Undergrad non-members
C_4: Grad members vs. Grad non-members.

19. (a) 17.38
    (b)

Contrast	A	B	C	D	E	F	G	H
1	+1	+1	+1	+1	-1	-1	-1	-1
2	+1	+1	-1	-1	0	0	0	0
3	0	0	0	0	+1	+1	-1	-1
4	+1	-1	0	0	0	0	0	0

Contrast 1 : Hard   vs.   Soft Woods
         2 : Pine   vs.   Poplar
         3 : Maple vs.   Oak
         4 : Paint I vs. Paint II, on pine only.

21. (a)

Contrast	Treatment						
	1	2	3	4	5	6	7
1	+6	-1	-1	-1	-1	-1	-1
2	0	+1	+1	+1	-1	-1	-1
3	0	-1	0	+1	0	0	0
4	0	+1	-2	+1	0	0	0
5	0	0	0	0	-1	0	+1
6	0	0	0	0	+1	-2	+1

Contrast 1: Truban vs. all others
         2: Nurelle at inoculation vs. Nurelle after
                                      inoculation

3: Linear within Nurelle at inoculation
4: Quadratic within Nurelle at inoculation
5: Linear within Nurelle after inoculation
6: Quadratic within Nurelle after inoculation

(b) H_0: $6\mu_1 = \mu_2 + \mu_3 + \mu_4 + \mu_5 + \mu_6 + \mu_7$
H_0: $\mu_2 + \mu_3 + \mu_4 = \mu_5 + \mu_6 + \mu_7$
and so on.

23.

Trend	SS	F	
Linear	20.74	5.09	$0.025 < P < 0.05$
Quadratic	13.52	3.32	$0.05 < P < 0.10$
Cubic	17.42	4.28	$P \approx 0.05$

Chapter 5

1. (a) $y_{ijk} = \mu + \alpha_i + \beta_j + (\alpha\beta)_{ij} + \epsilon_{ijk}$; $i = 1, 2, 3$
$j = 1, 2, 3, 4, 5$
$k = 1, 2, 3, 4, 5, 6$

μ = overall mean

α_i = effect of i-th level of A

β_j = effect of j-th level of B

$(\alpha\beta)_{ij}$ = interaction component for i-th level of A and j-th level of B

ϵ_{ij} = random component explaining all extraneous variation of (i,j,k)-th observation.

(b) The ϵ's are i.i.d. $N(0, \sigma_\epsilon^2)$

$\sum_i \alpha_i = 0$; $\sum_j \beta_j = 0$; $\sum_i (\alpha\beta)_{ij} = 0$; $\sum_j (\alpha\beta)_{ij} = 0$

(c)   ANOVA for 3x5 Factorial

Source	df	EMS
A	2	$\sigma_\epsilon^2 + 30\kappa_A^2$
B	4	$\sigma_\epsilon^2 + 18\kappa_B^2$
AxB	8	$\sigma_\epsilon^2 + 6\kappa_{AB}^2$
Exp.Error	75	σ_ϵ^2

3. (a) $y_{ijk} = \mu + \alpha_i + \beta_j + (\alpha\beta)_{ij} + \gamma_k + (\alpha\gamma)_{ik}$

$$+ (\beta\gamma)_{jk} + (\alpha\beta\gamma)_{ijk} + \epsilon_{ijkm};\ \begin{array}{l} i = 1,\ 2,\ 3 \\ j = 1,\ 2 \\ k = 1,\ 2,\ 3,\ 4 \\ m = 1,\ 2,\ \ldots,\ 5 \end{array}$$

μ = overall mean

α_i = effect of i-th level of A

β_j = effect of j-th level of B

$(\alpha\beta)_{ij}$ = interaction component for i-th level of A and j-th level of B

γ_k = effect of k-th level of C

$(\alpha\gamma)_{ik}$ = interaction component for i-th level of A and k-th level of C

$(\beta\gamma)_{jk}$ = interaction component for j-th level of B and k-th level of C

$(\alpha\beta\gamma)_{ijk}$ = interaction component for i-th level of A, j-th level of B and k-th level of C.

ϵ_{ijkm} = random component explaining all extraneous variation of (i,j,k,m)-th observation.

(b) The ϵ's are i.i.d. $N(0, \sigma_\epsilon^2)$

Each treatment component sums to zero over each of its indices.

(c)   ANOVA for 3x2x4 Factorial

Source	df	EMS
A	2	$\sigma_\epsilon^2 + 40\kappa_A^2$
B	1	$\sigma_\epsilon^2 + 60\kappa_B^2$
AB	2	$\sigma_\epsilon^2 + 20\kappa_{AB}^2$
C	3	$\sigma_\epsilon^2 + 30\kappa_C^2$
AC	6	$\sigma_\epsilon^2 + 10\kappa_{AC}^2$
BC	3	$\sigma_\epsilon^2 + 15\kappa_{BC}^2$
ABC	6	$\sigma_\epsilon^2 + 5\kappa_{ABC}^2$
Exp.Error	96	σ_ϵ^2

5. (a) $y_{ijk} = \mu + \alpha_i + \beta_j + (\alpha\beta)_{ij} + \epsilon_{ijk}$; $\quad\begin{aligned} i &= 1, 2 \\ j &= 1, 2, 3 \\ k &= 1, 2, \ldots, 8 \end{aligned}$

(b)

Source	df	EMS
A	1	$\sigma^2_\epsilon + 24\kappa^2_A$
B	2	$\sigma^2_\epsilon + 16\kappa^2_B$
AB	2	$\sigma^2_\epsilon + 8\kappa^2_{AB}$
Exp.Error	42	σ^2_ϵ

7. (a) $y_{ijk} = \mu + \alpha_i + \beta_j + (\alpha\beta)_{ij} + \gamma_k + (\alpha\gamma)_{ik}$

$\qquad + (\beta\gamma)_{jk} + (\alpha\beta\gamma)_{ijk} + \epsilon_{ijkm}$; $\quad\begin{aligned} i &= 1, 2, 3, 4 \\ j &= 1, 2, 3 \\ k &= 1, 2 \\ m &= 1, 2, 3 \end{aligned}$

(b)

Source	df	EMS
A	3	$\sigma^2_\epsilon + 18\kappa^2_A$
B	2	$\sigma^2_\epsilon + 24\kappa^2_B$
AB	6	$\sigma^2_\epsilon + 36\kappa^2_{AB}$
C	1	$\sigma^2_\epsilon + 6\kappa^2_C$
AC	3	$\sigma^2_\epsilon + 9\kappa^2_{AC}$
BC	2	$\sigma^2_\epsilon + 12\kappa^2_{BC}$
ABC	6	$\sigma^2_\epsilon + 3\kappa^2_{ABC}$
Exp.Error	48	σ^2_ϵ

9. (a) 7.9 , 2.4   (b) -0.8, -6.3   (c) 5.15
   (d) 3.55   (e) -5.5

11. (a) SSA = 132.6 , SSB = 63.0 , SSAB = 37.8
    (b) For A :   F = 4.45,
        For B :   F = 2.11,
        For AB: F = 1.27,

13. (a) NS , Sig, Sig   (b) Sig , Sig, Sig

15. (a) NS , NS , Sig , Does not exist, NS , Does not exist,
        NS , Does not exist.
    (b) Sig , Sig , Sig

17. (a)

Source	df	SS	MS	EMS
N	1	7.59	7.59	$\sigma_\epsilon^2 + 12\kappa_N^2$
S	3	3.15	1.05	$\sigma_\epsilon^2 + 6\kappa_S^2$
NxS	3	3.84	1.28	$\sigma_\epsilon^2 + 3\kappa_{NS}^2$
Exp.Error	16	0.08	0.005	σ_ϵ^2

(b) For N : $F = 1518$, $P < 0.005$
For S : $F = 210$, $P < 0.005$
For NxS: $F = 256$, $P < 0.005$

(c) NxS means: 4.54   4.64   5.27   5.92
  5.73   7.05   5.81   6.30

Standard deviation for difference of any pair of these means is 0.06.

19. (a)

Source	df	SS	MS	EMS
F	1	713.2	713.2	$\sigma_\epsilon^2 + 9\kappa_F^2$
M	2	389.3	194.7	$\sigma_\epsilon^2 + 6\kappa_M^2$
FM	2	34.2	17.1	$\sigma_\epsilon^2 + 3\kappa_{FM}^2$
Exp.Error	12	9.28	0.77	σ_ϵ^2

(b) For F: $F = 926$, $P < 0.005$
For M: $F = 253$, $P < 0.005$
For FM: $F = 22.2$, $P < 0.005$

(c) FxM means: 17.6   21.3   31.2
  5.9   11.6   14.9

Standard deviation for difference of any pair of these means is 0.72

21. With the gradient structure on Sulfur and significance of NS interaction, a trend analysis on the interaction is most appropriate.   From the means in the NS table, it is evident that the trends are different for the 2 nitrogens.

For 0 nitrogen: the dominant trend is linear, SS = 3.41, $P < 0.005$.

For 20 nitrogen: the dominant trend is cubic, SS = 2.76, $P < 0.005$.

23. Because of the significant interaction and the gradient structure on magnesium, a trend analysis on the interaction is most appropriate.

   Linear*Linear: SS = 15.87, P < 0.005

   Linear*Quadratic: SS = 18.49, P < 0.005

25. (a) A_L = 3.33, B_L = -10.5, B_Q = -25.5

   $A_L x B_L$ = 3, $A_L x B_Q$ = 1

   (b) $SS[A_L]$ = 50.0, $SS[B_L]$ = 330.75, $SS[B_Q]$ = 650.25

   $SS[A_L x B_L]$ = 6.75, $SS[A_L x B_Q]$ = 0.25

27. (a) A_L = 30.66, A_Q = -2, A_C = -8.67

   B_L = 0.75, B_Q = 0.25

   $A_L x B_L$ = 45, $A_Q x B_L$ = 7, $A_C x B_L$ = -5

   $A_L x B_Q$ = -25, $A_Q x B_Q$ = -3, $A_C x B_Q$ = 25

   (b) SS for A trends: 564, 12, 45.1
       SS for B trends: 4.5, 0.17
       SS for AB trends: 202.5, 24.5, 2.5
                          20.8, 1.5, 20.8

29. (a)

Source	df	SS	MS
T	4	3568.5	892.12
Q	1	691.2	691.2
TQ	4	511.1	127.77
Exp.Error	20	1442	72.1

   (b) For TQ: F = 1.77, P < 0.10; For Time: F = 12.37, P < .005
       For Quenching: F = 9.59, 0.005 < P < 0.01

31. (a)

Source	df	SS	MS
B	3	98,543	32,848
E	2	3,110	1,555
BE	6	1,552	259
Exp.Error	12	14,246	1187

   (b) For BE: F < 1 , P > 0.10; For Breeds: F = 27.7, P < 0.005
       For Energy: F = 1.3, P > 0.10
       The average daily intakes appear to differ only with respect to Breeds. No significant differences in average daily intakes were observed for the 3 levels of energy used. There is no evidence of an interaction of these factors.

33.

Source	df	EMS
A	1	$\sigma_\epsilon^2 + 9\sigma_{AC}^2 + 45\sigma_A^2$
B	2	$\sigma_\epsilon^2 + 3\sigma_{ABC}^2 + 15\sigma_{AB}^2 + 6\sigma_{BC}^2 + 30\kappa_B^2$
AB	2	$\sigma_\epsilon^2 + 3\sigma_{ABC}^2 + 15\sigma_{AB}^2$
C	4	$\sigma_\epsilon^2 + 9\sigma_{AC}^2 + 18\sigma_C^2$
AC	4	$\sigma_\epsilon^2 + 9\sigma_{AC}^2$
BC	8	$\sigma_\epsilon^2 + 3\sigma_{ABC}^2 + 6\sigma_{BC}^2$
ABC	8	$\sigma_\epsilon^2 + 3\sigma_{ABC}^2$
Exp.Error	60	σ_ϵ^2

35. (a) 324   (b) Each pallet made from each combination of levels of the 4 factors.

(c)

Source	df	EMS
A	5	$\sigma_\epsilon^2 + 432\kappa_A^2$
B	5	$\sigma_\epsilon^2 + 432\kappa_B^2$
AB	25	$\sigma_\epsilon^2 + 72\kappa_{AB}^2$
C	2	$\sigma_\epsilon^2 + 864\kappa_C^2$
AC	10	$\sigma_\epsilon^2 + 144\kappa_{AC}^2$
BC	10	$\sigma_\epsilon^2 + 144\kappa_{BC}^2$
ABC	50	$\sigma_\epsilon^2 + 24\kappa_{ABC}^2$
D	2	$\sigma_\epsilon^2 + 864\kappa_D^2$
AD	10	$\sigma_\epsilon^2 + 144\kappa_{AD}^2$
BD	10	$\sigma_\epsilon^2 + 144\kappa_{BD}^2$
ABD	5	$\sigma_\epsilon^2 + 24\kappa_{ABD}^2$
CD	4	$\sigma_\epsilon^2 + 288\kappa_{CD}^2$
ACD	20	$\sigma_\epsilon^2 + 48\kappa_{ACD}^2$
BCD	20	$\sigma_\epsilon^2 + 48\kappa_{BCD}^2$
ABCD	100	$\sigma_\epsilon^2 + 8\kappa_{ABCD}^2$
Exp.Error	2268	σ_ϵ^2

37. (a)

Source	df	SS	MS
(Treatments	6	6104.9)	
Type	1	48.1	48.1
Amount	2	492.1	246.1
TxA	2	503.0	251.5
Control vs. Rest	1	5061.7	5061.7
Exp.Error	28	2229.0	79.6

(b) For T : $F < 1$, $P > 0.10$
For A : $F = 3.09$, $0.05 < P < 0.10$
For TxA : $F = 3.16$, $0.05 < P < 0.10$
For Control vs Rest : $F = 63.6$, $P < 0.005$

39. (a)

Source	df	SS	MS
(Treatments	10	359.8)	
Type	1	76.8	76.8
Temp.	4	229.5	57.4
Type x Temp.	4	18.2	4.6
CrNi vs Rest	1	35.3	35.3
Exp.Error	22	351.7	16.0

(b) For Type: $F = 4.8$, $0.025 < P < 0.05$
For Temp: $F = 3.6$, $0.01 < P < 0.025$
For Type x Temp: $F < 1$, $P > 0.10$
For CrNi vs. Rest: $F = 2.21$, $P > 0.10$

41.

Source	EMS
A	$\sigma_\epsilon^2 + 6\sigma_{ABC}^2 + 12\sigma_{AB}^2 + 24\sigma_{AC}^2 + 48\kappa_A^2$
B	$\sigma_\epsilon^2 + 12\sigma_{BC}^2 + 24\sigma_B^2$
AB	$\sigma_\epsilon^2 + 6\sigma_{ABC}^2 + 12\sigma_{AB}^2$
AC	$\sigma_\epsilon^2 + 6\sigma_{ABC}^2 + 24\sigma_{AC}^2$
BC	$\sigma_\epsilon^2 + 12\sigma_{BC}^2$
ABC	$\sigma_\epsilon^2 + 6\sigma_{ABC}^2$
AD	$\sigma_\epsilon^2 + 2\sigma_{ABCD}^2 + 4\sigma_{ABD}^2 + 8\sigma_{ACD}^2 + 16\kappa_{AD}^2$

43. (a)

Source	df	MS	Source	df	MS
A	1	0.45	AD	1	0.36
B	1	0.12	BD	1	10.12
AB	1	1.81	ABD	1	0.00
C	1	4.20	CD	1	3.12
AC	1	2.42	ACD	1	0.24
BC	1	0.10	BCD	1	1.05
ABC	1	1.71	ABCD	1	0.91
D	1	52.53	Exp.Error	16	2.00

(b) Only 2 treatment components are significant: Factor D, hardening temperature and the BD interaction (hardening temperature x temper-annealing temperature). Thus, only factors B and D are influential for the austenite grain size of this high speed steel.

Chapter 6

1. (a) $y_{ij} = \mu + \tau_i + \rho_j + \epsilon_{ij}$; $\begin{array}{l} i = 1, 2, \ldots, 6 \\ j = 1, 2, \ldots, 8 \end{array}$

   μ = overall mean

   τ_i = effect of i-th treatment

   ρ_j = effect of j-th block

   ϵ_{ij} = random component explaining all extraneous variation of (i,j)-th observation.

   (b) The ϵ's are i.i.d. $N(0, \sigma_\epsilon^2)$; $\sum_i \tau_i = 0$; $\sum_j \rho_j = 0$

3. (a) $y_{ijk} = \mu + \alpha_i + \beta_j + (\alpha\beta)_{ij} + \rho_j + \epsilon_{ij}$; $\begin{array}{l} i = 1, 2, 3 \\ j = 1, 2 \\ k = 1, 2, \ldots, 8 \end{array}$

   μ = overall mean

   α_i = effect of i-th level of A

   β_j = effect of j-th level of B

   $(\alpha\beta)_{ij}$ = interaction component for the i-th level of A and the j-th level of B

   ρ_j = effect of j-th block

   ϵ_{ijk} = random component explaining all extraneous variation of (i,j,k)-th observation.

(b)   The ϵ's are i.i.d. $N(0, \sigma^2_\epsilon)$

$$\underset{i}{\Sigma}\ \alpha_i = \underset{j}{\Sigma}\ \beta_j = \underset{i}{\Sigma}(\alpha\beta)_{ij} = \underset{j}{\Sigma}(\alpha\beta)_{ij} = \underset{k}{\Sigma}\ \rho_k = 0$$

5.   (a)

Source	df	SS	MS	EMS
Blocks	5	200	40	- - -
Trts	4	150	37.5	$\sigma^2_\epsilon + 6\kappa^2_\tau$
Exp.Error	20	200	10	σ^2_ϵ

(b)   Calculated value of F = 37.5/10 = 3.75;
       Observed significance:

(c) $\sqrt{2(10/6)}$ = 1.83

7.   (a)

Source	df	SS	MS	EMS
Blocks	5	430	86	- - -
Trts	3	308	102.7	$\sigma^2_\epsilon + 6\sigma^2_\tau$
Exp.Error	15	240	16	σ^2_ϵ

(b)   Calculated value of F = 102.7/16 = 6.42;
       Observed significance: $P < 0.005$

(c) $\hat{\sigma}^2_\epsilon = 16$; $\hat{\sigma}^2_\tau = \dfrac{102.7 - 16}{6} = 14.45$

9.   (a) $y_{ij} = \mu + \tau_i + \rho_j + \epsilon_{ij}$; $i = 1, 2, 3, 4$
                                                 $j = 1, 2, 3, 4$

μ = overall mean

τ_i = effect of the i-th filter

ρ_j = effect of the j-th lab

ϵ_{ij} = random component explaining all extraneous variation of (i,j)-th observation.

(b)   The ϵ's are i.i.d. $N(0, \sigma^2_\epsilon)$

The ρ's have a $N(0, \sigma^2_\rho)$ distribution

The ϵ's and ρ's are distributed independently of each other.
The τ's are fixed parameters with $\Sigma\ \tau_i = 0$

(c)

Source	df	SS	MS	EMS
Labs(Blocks)	3	2.81	0.94	- - -
Filters	3	28.17	9.39	$\sigma^2_\epsilon + 4\kappa^2_\tau$
Exp.Error	9	1.15	0.13	σ^2_ϵ

(d) $F = 9.39/0.13 = 72.2$, $P < 0.005$
(e) 0.255

11. (a) $y_{ij} = \mu + \tau_i + \rho_j + \epsilon_{ij}$; $i = 1, 2, \ldots, 9$
$j = 1, 2, 3, 4$

μ = overall mean

τ_i = effect of the i-th defoliage scheme

ρ_j = effect of the j-th location

ϵ_{ij} = random component explaining all extraneous variation of (i,j)-th observation.

(b)

Source	df	SS	MS	EMS
Locations	3	1.30	0.43	- - -
Defol.Schemes	8	188.6	23.58	$\sigma^2_\epsilon + 4\kappa^2_\tau$
Exp.Error	24	64.9	2.71	σ^2_ϵ

(c) $F = 23.58/2.71 = 8.70$, $P < 0.005$
(d) 1.16

13. (a) $y_{ij} = \mu + \tau_i + \rho_j + \epsilon_{ij}$; $i = 1, 2, 3, 4$
$j = 1, 2, 3, 4$

μ = overall mean

τ_i = effect of the i-th brand of gasoline

ρ_j = effect of the j-th driver

ϵ_{ij} = random component explaining all extraneous variation of (i,j)-th observation.

(b)

Source	df	SS	MS	EMS
Drivers	3	152.0	50.7	- - -
Brands	3	8.5	2.8	$\sigma^2_\epsilon + 4\kappa^2_\tau$
Exp.Error	9	2.6	0.29	σ^2_ϵ

(c) F = 9.66, P < 0.005   (d)   0.38
(e) Car effects along with drivers effects are part of the conglomerate of effects called block effects.

15. (a)

Source	df	SS	MS	EMS
Blocks	3	42.8	14.3	---
Fungicides	4	216.7	54.2	$\sigma^2_\epsilon + 4\kappa^2_\tau$
Exp.Error	12	41.7	3.5	σ^2_ϵ

(b) F = 54.2/3.5 = 15.5, P < 0.005   (c)   1.32

17. (a)

Source	df	SS	MS	EMS
Blocks	1	3.0	3.0	---
A	1	65.3	65.3	$\sigma^2_\epsilon + 6\kappa^2_A$
B	2	39.5	19.8	$\sigma^2_\epsilon + 4\kappa^2_B$
AB	2	15.2	7.6	$\sigma^2_\epsilon + 2\kappa^2_{AB}$
Exp.Error	5	31.0	6.2	σ^2_ϵ

(b) For AB: F = 7.6/6.2 = 1.2, P > 0.10
     For A: F = 65.3/6.2 = 10.5, P ≈ 0.025
     For B: F = 19.8/6.2 = 3.2, P > 0.10
(d) 1.76   (e)   2.49

19. (a)

Source	df	SS	MS	EMS
Labs(Blocks)	3	2.81	0.94	---
Manuf.	1	4.41	4.41	$\sigma^2_\epsilon + 8\kappa^2_M$
Amount	1	0.72	0.72	$\sigma^2_\epsilon + 8\kappa^2_A$
M x A	1	23.04	23.04	$\sigma^2_\epsilon + 4\kappa^2_{MA}$
Exp.Error	9	1.15	0.13	σ^2_ϵ

(b) For M x A: F = 23.04/0.13 = 177.2, P < 0.005
     For Manuf.(A): F = 4.41/0.13 = 33.9, P < 0.005
     For Amount(B): F = 0.72/0.13 = 5.5, P ≈ 0.05

21. RE(RCB to CR) = [3(0.43) + 4(8)(2.71)]/35(2.71) = 0.93

23. RE(RCB to CR) = [3(0.94) + 4(3)(0.13)]/15(0.13) = 2.25

25. Because there might be different effects exerted at the different locations (due to the fan, door, and so on) one should take the locations to be blocks. One can use 10 locations, 4 plants per location. Randomly assign the 4 growth regulators to the 4 plants at each location. A new randomization would be made at each location. Random permutations from Table A.2 could be used.

27. The engineer should consider an RCB design with subsampling, where the 10 days are blocks, the treatments are a factorial set, and the 5 stepping stones per batch are the sampling units. The linear model for such an experiment would be

$$y_{ijk} = \mu + \alpha_i + \beta_j + (\alpha\beta)_{ij} + \rho_j + \epsilon_{ij} + \delta_{ijkm}; \quad i = 1, 2$$

$$j = 1, 2, 3$$
$$k = 1, 2, \ldots, 10$$
$$m = 1, 2, \ldots, 5$$

The partial ANOVA is

Source	df
Days	9
Sizes	1
Mixtures	2
S x M	2
Exp.Error	45
Sampling Error	240

Batches made by the 6 formulations would be randomized anew each day. Random permutations from Table A.2 can be used for this.

29. The use of blocks is not warranted here. All conditions within the greenhouse are fairly uniform, especially when all EU are located in one area. Thus, without further information, one should use a CR design with each treatment assigned completely at random to 4 EU.

31. A first concern would be the size of the blocks; homogeneity of 36 plots could be a problem in a 7-acre area. Depending on the equipment used to form the soil base and topsoil, one might consider plots of a smaller size. Another way to reduce the block size would be to reduce the number of levels of one or more factors. Presumably, different locations, different terrains, and so on define the blocks.

33. This is an RCB design with subsampling.

(a) $y_{ijk} = \mu + \tau_i + \rho_j + \epsilon_{ij} + \delta_{ijk};$ $i = 1, 2, 3, 4$
  $j = 1, 2, \ldots, 6$
  $k = 1, 2, 3$

(b)

Source	df	EMS
Days	5	- - -
Recipes	3	$\sigma_\delta^2 + 3\sigma_\epsilon^2 + 18\kappa_\tau^2$
Exp.Error	15	$\sigma_\delta^2 + 3\sigma_\epsilon^2$
Sampling	48	σ_δ^2

35. (a)

Source	df	MS	EMS
Days	5	25.6	- - -
Recipes	3	55.3	$\sigma_\delta^2 + 3\sigma_\epsilon^2 + 18\kappa_\tau^2$
Exp.Error	15	11.5	$\sigma_\delta^2 + 3\sigma_\epsilon^2$
Samplings	48	4.4	σ_δ^2

(b) $F = 55.3/11.5 = 4.8,\ 0.01 < P < 0.025$
(c) $\sqrt{2(11.5)/18} = 1.13$

37. (a) $y_{ijk} = \mu + \tau_i + \rho_j + (\epsilon_{ij} + \delta_{ijk};$ $i = 1, 2, 3, 4$
  $j = 1, 2, \ldots, 5$
  $k = 1, 2, 3$

   μ = overall mean

   τ_i = effect of the i-th process

   ρ_j = effect of the j-th supplier

   ϵ_{ij} = random component for Exp. Error variation

   δ_{ijk} = random component explaining variation among observations within EU.

(b)

Source	df	SS	MS	EMS
Suppliers(Blocks)	4	4.47	1.12	- - -
Processes	3	55.95	18.65	$\sigma_\delta^2 + 3\sigma_\epsilon^2 + 18\kappa_\tau^2$
Exp.Error	12	22.42	1.87	$\sigma_\delta^2 + 3\sigma_\epsilon^2$
Sampling	40	2.77	0.07	σ_δ^2

(c)   F = 18.65/1.87 = 9.97, P < 0.005.
      There is evidence of differences in the mean thickness
      of electroplating for the 4 processes.

(d) $\sqrt{2(1.87)/15}$ = 0.5

Chapter 7

1. (a) $y_{ijk} = \mu + \alpha_i + \beta_{ij} + \epsilon_{ijk}$; $\begin{array}{l} i = 1, 2, 3, 4 \\ j = 1, 2, \ldots, 5 \\ k = 1, 2, 3 \end{array}$

    μ = overall mean

    α_i = effect of the i-th coolant

    β_{ij} = effect of the j-th engine receiving the
         i-th coolant

    ϵ_{ijk} = random component explaining all extraneous
          variation

    (b)

Source	df	EMS
Coolants	3	$\sigma^2_\epsilon + 3\sigma^2_\beta + 15\kappa^2_\alpha$
Engines(C)	16	$\sigma^2_\epsilon + 3\sigma^2_\beta$
Residual	40	σ^2_ϵ

    (c) The ϵ's are i.i.d. $N(0, \sigma^2_\epsilon)$

         The β's are i.i.d. $N(0, \sigma^2_\beta)$

         The ϵ's and β's are distributed independently
         of each other.

         The α's are fixed parameters with $\sum_i \alpha_i = 0$

3. (a) $y_{ijk} = \mu + \alpha_i + \beta_{ij} + \epsilon_{ijk}$; $\begin{array}{l} i = 1, 2, \ldots, 12 \\ j = 1, 2, 3 \\ k = 1, 2, 3 \end{array}$

    μ = overall mean

    α_i = random component measuring variation among plants

    β_{ij} = random component measuring variation among
         branches on the i-th plant

    ϵ_{ijk} = random component explaining all extraneous
          variation

(b)

Source	df	EMS
Plants	11	$\sigma_\epsilon^2 + 3\sigma_\beta^2 + 9\sigma_\alpha^2$
Branches(P)	24	$\sigma_\epsilon^2 + 3\sigma_\beta^2$
Residual	72	σ_ϵ^2

(c) The ϵ's are i.i.d. $N(0, \sigma_\epsilon^2)$

The β's are i.i.d. $N(0, \sigma_\beta^2)$

The α's are i.i.d. $N(0, \sigma_\alpha^2)$

The α's, β's, and ϵ's are distributed independently of each other.

5. (a) $y_{ijkm} = \mu + \alpha_i + \beta_{ij} + \gamma_{ijk} + \epsilon_{ijkm}$;
$$\begin{aligned} i &= 1, 2, 3, 4 \\ j &= 1, 2, \ldots, 5 \\ k &= 1, 2, \ldots, 5 \\ m &= 1, 2, 3, 4 \end{aligned}$$

μ = overall mean

α_i = effect due to the i-th breed

β_{ij} = effect of the j-th ranch raising the i-th breed

γ_{ijk} = effect of the k-th sheep on the (i,j)-th ranch

ϵ_{ijk} = random component explaining all extraneous variation

(b)

Source	df	EMS
Breeds	3	$\sigma_\epsilon^2 + 4\sigma_\gamma^2 + 20\sigma_\beta^2 + 100\sigma_\alpha^2$
Ranches(B)	16	$\sigma_\epsilon^2 + 4\sigma_\gamma^2 + 20\sigma_\beta^2$
Sheep(R)(B)	80	$\sigma_\epsilon^2 + 4\sigma_\gamma^2$
Residual	300	σ_ϵ^2

(c) The ϵ's are i.i.d. $N(0, \sigma_\epsilon^2)$

The γ's are i.i.d. $N(0, \sigma_\gamma^2)$

The β's are i.i.d. $N(0, \sigma_\beta^2)$

The β's, γ's, and ϵ's are distributed independently of each other.

The α's are fixed parameters with $\sum_i \alpha_i = 0$

7. (a)

Source	df	MS	EMS
Coolants	3	59.34	$\sigma_\epsilon^2 + 3\sigma_\beta^2 + 15\kappa_\alpha^2$
Engines(C)	16	7.84	$\sigma_\epsilon^2 + 3\sigma_\beta^2$
Residual	40	3.72	σ_ϵ^2

(b) $F = 59.34/7.84 = 7.57$, $P < 0.005$

(c) $\hat{\sigma}_\epsilon^2 = 3.72$; $\hat{\sigma}_\beta^2 = 1.37$

9. (a)

Source	df	MS	EMS
Seedlings	11	0.220	$\sigma_\epsilon^2 + 3\sigma_\beta^2 + 9\sigma_\alpha^2$
Branches(S)	24	0.075	$\sigma_\epsilon^2 + 3\sigma_\beta^2$
Residual	72	0.033	σ_ϵ^2

(b) $\hat{\sigma}_\epsilon^2 = 0.033$, $\hat{\sigma}_\beta^2 = 0.014$, $\hat{\sigma}_\alpha^2 = 0.016$

(c) Seedlings: $F = 0.22/0.075 = 2.93$, $0.01 < P < 0.025$
Branches(S): $F = 0.075/0.033 = 2.27$, $P \approx 0.005$

11. (a)

Source	df	SS	MS	EMS
Brands	2	1094.1	547.06	$\sigma_\epsilon^2 + 3\sigma_\beta^2 + 18\kappa_\alpha^2$
Units(B)	15	128.1	8.54	$\sigma_\epsilon^2 + 3\sigma_\beta^2$
Residual	36	146.7	4.07	σ_ϵ^2

(b) $F = 547.06/8.54 = 64.1$, $P < 0.005$

(c) $\hat{\sigma}_\epsilon^2 = 4.07$; $\hat{\sigma}_\beta^2 = \dfrac{8.54 - 4.07}{3} = 1.49$

Chapter 8

1. (a) $y_{ijk} = \mu + \rho_i + \gamma_j + \tau_k + \epsilon_{ijk}$; $\begin{array}{l} i = 1, 2, \ldots, 6 \\ j = 1, 2, \ldots, 6 \\ k = 1, 2, \ldots, 6 \end{array}$

μ = overall mean

ρ_i = effect of i-th day

γ_j = effect of j-th operator

τ_k = effect of the k-th treatment

ϵ_{ijk} = random component explaining all extraneous variation of (i,j)-th observation.

(b)

Source	df	EMS
Days	5	---
Operators	5	---
Treatments	5	$\sigma_\epsilon^2 + 6\kappa_\tau^2$
Exp.Error	20	σ_ϵ^2

(c) The ϵ's are i.i.d. $N(0, \sigma_\epsilon^2)$

The γ's are i.i.d. $N(0, \sigma_\gamma^2)$

The ρ's are i.i.d. $N(0, \sigma_\rho^2)$

The ρ's, γ's, and ϵ's are distributed independently of each other.

The τ's are fixed parameters with $\sum_k \tau_k = 0$

3. (a) $y_{ijkm} = \mu + \rho_i + \gamma_j + \alpha_k + \beta_m + (\alpha\beta)_{km} + \epsilon_{ijkm}$

$$\text{for } \begin{array}{l} i = 1, 2, \ldots, 6 \\ j = 1, 2, \ldots, 6 \\ k = 1, 2, 3 \\ m = 1, 2 \end{array}$$

All components as described is 1.(a) except:

α_k = effect due to k-th level of A

β_m = effect due to m-th level of B

$(\alpha\beta)_{km}$ = component measuring the interaction of the k-th level of A with the m-th level of B.

(b)

Source	df	EMS
Days	5	---
Operators	5	---
A	2	$\sigma_\epsilon^2 + 12\kappa_A^2$
B	1	$\sigma_\epsilon^2 + 18\kappa_B^2$
AB	2	$\sigma_\epsilon^2 + 6\kappa_{AB}^2$
Exp.Error	20	σ_ϵ^2

(c)   All assumptions as in 1.(c) except:

The α's are fixed parameters with $\Sigma \, \alpha_k = 0$

The β's are fixed parameters with $\Sigma \, \beta_m = 0$

The $(\alpha\beta)$'s are fixed parameters with

$$\Sigma_k (\alpha\beta)_{km} = \Sigma_m (\alpha\beta)_{km} = 0$$

5.   (a) $y_{ijkm} = \mu + \rho_i + \gamma_j + \tau_k + \epsilon_{ijk} + \delta_{ijkm};$

$\quad$ $i = 1, \, 2, \, \dots, \, 6$
$\quad$ $j = 1, \, 2, \, \dots, \, 6$
$\quad$ $k = 1, \, 2, \, \dots, \, 6$
$\quad$ $m = 1, \, 2, \, 3$

All components as described in 1.(a) except

δ_{ijkm} = random component measuring the variation among observations on the same EU.

(b)

Source	df	EMS
Days	5	---
Operators	5	---
Treatments	5	$\sigma_\delta^2 + 3\sigma_\epsilon^2 + 18\kappa_\tau^2$
Exp. Error	20	$\sigma_\delta^2 + 3\sigma_\epsilon^2$
Sampling	72	σ_δ^2

(c)   All assumptions as in 1.(c) except

The δ's are i.i.d. $N(0, \, \sigma_\delta^2)$

The ϵ's, δ's, ρ's, and γ's are distributed independently of each other.

7.   (a)

Source	df	MS	EMS
Weeks	3	12.4	---
Locations	3	9.9	---
Displays	3	17.7	$\sigma_\epsilon^2 + 4\kappa_\tau^2$
Exp. Error	9	2.53	σ_ϵ^2

(b) $F = 17.7/2.53 = 7, \, P \approx 0.01$ (c) 1.12

9. RE(LS to CR) = 2.36

11. (a)

Source	df	SS	MS	EMS
Depths	2	402.9	201.4	---
Heights	2	89.5	44.9	---
Proteins	2	574.9	287.4	$\sigma_\epsilon^2 + 3\kappa_\tau^2$
Exp.Error	2	16.9	8.45	σ_ϵ^2

   (b)   F = 287.4/8.45 = 34.0, 0.025 < P < 0.05

13. RE(LS to CR) = 7.79

15. (a) $y_{ijk} = \mu + \rho_i + \gamma_j + \tau_k + \epsilon_{ijk}$; i = 1, 2, 3, 4
$$j = 1, 2, \ldots, 12$$
$$k = 1, 2, 3, 4$$

   μ = overall mean

   ρ_i = effect of i-th operator

   γ_j = effect of j-th day

   τ_k = effect of the k-th treatment

   ϵ_{ijk} = random component explaining all extraneous variation of (i,j,k)-th observation.

   (b)

Source	df	EMS
Operators	3	---
Day	11	---
Treatments	3	$\sigma_\epsilon^2 + 12\kappa_\tau^2$
Exp.Error	30	σ_ϵ^2

17. (a) $y_{ijkm} = \mu + \rho_i + \gamma_j + \alpha_k + \beta_m + (\alpha\beta)_{km} + \epsilon_{ijkm}$

for i = 1, 2, 3, 4
$$j = 1, 2, \ldots, 12$$
$$k = 1, 2$$
$$m = 1, 2$$

(b)   Source   df   EMS

Source	df	EMS
Operators	3	---
Days	11	---
A	1	$\sigma_\epsilon^2 + 24\kappa_A^2$
B	1	$\sigma_\epsilon^2 + 24\kappa_B^2$
AB	1	$\sigma_\epsilon^2 + 12\kappa_{AB}^2$
Exp.Error	30	σ_ϵ^2

19. (a)

Source	df	SS	MS	EMS
Heights	2	17.33	8.7	---
Depths	5	118.5	23.7	---
Proteins	2	694.3	347.2	$\sigma_\epsilon^2 + 6\kappa_\tau^2$
Exp.Error	8	54.4	6.8	σ_ϵ^2

(b) $F = 347.2/6.8 = 51.1$, $P < 0.005$ (c)   1.51

Chapter 9

1. (a) $y_{ij} = \mu + \tau_i + \rho_j + \beta(x_{ij} - \bar{x}..) + \epsilon_{ij}$; $\begin{array}{l} i = 1, 2, \ldots, 6 \\ j = 1, 2, \ldots, 6 \end{array}$

μ = overall mean

τ_i = effect due to the i-th variety

ρ_j = effect due to the j-th block

β = slope for the regression of yield y on number of mature plants x

ϵ_{ij} = random component explaining all extraneous variation

(b) The ϵ's are i.i.d. $N(0, \sigma_\epsilon^2)$

The regression of yield y on the number of mature plants x is linear with a common slope for all block-treatment combinations.

The ρ's are fixed with $\sum_j \rho_j = 0$

The τ's are fixed with $\sum_i \tau_i = 0$

3. (a) $y_{ij} = \mu + \tau_i + \rho_j + \beta(x_{ij} - \bar{x}..) + \epsilon_{ij}$; $\begin{array}{l} i = 1, 2, \ldots, 4 \\ j = 1, 2, \ldots, 8 \end{array}$

μ = overall mean

τ_i = effect due to the i-th training procedure

ρ_j = effect due to the j-th school

β = slope for the regression of post-training score y on the pre-training score x.

ϵ_{ij} = random component explaining all extraneous variation

(b) The ϵ's are i.i.d. $N(0, \sigma_\epsilon^2)$

The regression of post-training scores, y, on the pre-scores, x, is linear with a common slope β for all school-training procedure combinations.

The ρ's are fixed with $\sum_j \rho_j = 0$

The τ's are fixed with $\sum_i \tau_i = 0$

5. (a) $\hat{\beta} = 3778/2200 = 1.72$; (b) $F = 6487.9/40.7 = 159$, $P < 0.005$
   (c) $F = 1.62$, $P > 0.10$;   (d) $F = 219.6/40.7 = 5.4$, $P < 0.005$

7. (a) $\hat{\beta} = 1971/1343 = 1.47$; (b) $F = 2892.7/23.3 = 124$, $P < 0.005$

   (c) The covariate x, being a pre-training score cannot be affected by the training procedures (treatments)

   (d) $F = 77.4/23.3 = 3.32$, $0.025 < P < 0.05$

9. (a) $\hat{\beta} = 892/617 = 1.45$; (b) $F = 1289.6/3.83 = 336.6$, $P < 0.005$
   (c) The covariate x, being "natural" wind speed cannot be affected by treatments.
   (d) $F = 29.4/3.83 = 7.7$, $P < 0.005$.

Chapter 10

1. (a) Block 1 → (1), a, bc, abc;   Block 2 → b, c, ab, ac

   (b) Block 1 → (1), c, ab, abc;   Block 2 → a, b, ac, bc

3. (a) Block 1 → (1), ab;   Block 2 → a, b

   Block 3 → c, abc;   Block 4 → ac, bc

(b) Block 1 → (1), abc;   Block 2 → a, bc

Block 3 → b, ac;   Block 4 → c, ab

5. (a) AC   (b) B

7. (a) The generalized interaction is ABCD and one complete replicate would be

Block 1	Block 2	Block 3	Block 4
(1)	a	b	e
ac	c	d	ab
bd	de	ae	ad
abe	be	ce	bc
ade	abd	abc	cd
bce	bcd	acd	ace
cde	abce	abde	bde
abcd	acde	bcde	abcde

(b) The generalized interactions are BE,BCD,ACE, and ABDE. One complete replicate would be

Block 1	Block 2	Block 3	Block 4
(1)	a	b	c
ace	ce	cd	ae
abde	bde	ade	bd
bcd	abcd	abce	abcde

Block 5	Block 6	Block 7	Block 8
d	e	ab	ad
bc	ac	de	be
abe	abd	acd	abc
acde	bcde	bce	cde

9. (a)

Source	df	Source	df
Reps	3	D	1
Blocks(Reps)	4	AD	1
A	1	BD	1
B	1	ABD	1
AB	1	CD	1
C	1	ACD	1
AC	1	BCD	1*
BC	1*	ABCD	1*
ABC	1*	Exp.Error	41

(b) 3/4 information on ABC and BCD
   Full information on ABD and ACD

11. (a)

Source	df	Source	df	Source	df
Reps	2	ABD	1	ABCE	1*
Blocks(Reps)	9	CD	1	DE	1
A	1	ACD	1*	ADE	1*
B	1	BCD	1	BDE	1*
AB	1	ABCD	1*	ABDE	1
C	1	E	1	CDE	1*
AC	1	AE	1	ACDE	1
BC	1	BE	1	BCDE	1*
ABC	1*	ABE	1*	ABCDE	1
D	1	CE	1	BCDE	1*
AD	1	ACE	1	ABCDE	1
BD	1	BCE	1	Exp.Error	53

(b) 2/3 information on ABC,ABE,ACD,ADE,BDE,CDE
   Full information on ABD,ACE,BCD,BCE

13. (a) BC, AC, ABC

(b)

Source	df	SS	MS
Reps	2	2.3	1.2
Blocks(Reps)	3	77.5	25.8
A	1	704.2	704.2
B	1	192.7	192.7
AB	1	4.2	4.2
C	1	24.0	24.0
AC	1*	10.7	10.7
BC	1*	4.2	4.2
ABC	1*	16.7	16.7
Exp.Error	11	98.5	8.95

(c) For A: $F = 704.2/8.95 = 78.7$, $P < 0.005$
   For B: $F = 192.7/8.95 = 21.5$, $P < 0.005$
   For all other treatment components, $P > 0.10$

15. (a)

0000	1000	0100	0010
0011	1011	0111	0001
1101	0101	1001	1111
1110	0110	1010	1100

(b)

0000	1000	0100	0001
1010	0010	1110	1011
0111	1111	0011	0110
1101	0101	1001	1100

17. (a) 0000   (b) 0000
        0110   0111
        1011   1010
        1101   1101

19. (a) AB²C, AC²   (b) AB²C², A

21. (a) 000   (b) 000
        121   012
        212   021

23. (a)

000	100	010	001	200	020	002	102	012
111	211	121	112	011	101	110	210	120
222	022	202	220	122	212	221	021	201

    (b)

000	100	010	001	200	020	002	102	022
121	221	101	122	021	111	120	220	110
212	012	222	210	112	202	211	011	201

Chapter 11

1. (a) $y_{ijk} = \mu + \rho_i + \alpha_j + \delta_{ij} + \beta_k + (\alpha\beta)_{jk} + \epsilon_{ijk}$

$$\text{for } i = 1, 2, 3$$
$$j = 1, 2, \ldots, 5$$
$$k = 1, 2, 3, 4$$

(b)

Source	df	EMS
Reps	2	- - -
A	4	$\sigma_\epsilon^2 + 4\sigma_\delta^2 + 12\kappa_A^2$
Error 1	8	$\sigma_\epsilon^2 + 4\sigma_\delta^2$
B	3	$\sigma_\epsilon^2 + 15\kappa_B^2$
AB	12	$\sigma_\epsilon^2 + 3\kappa_{AB}^2$
Error 2	30	σ_ϵ^2

(c) The ϵ's are i.i.d. $N(0, \sigma_\epsilon^2)$

The δ's are i.i.d. $N(0, \sigma_\delta^2)$

The ϵ's and δ's are distributed independently of each other.

The ρ's, α's, β's, and $(\alpha\beta)$'s are fixed parameters each summing to zero.

3. (a) $y_{ijkm} = \mu + \rho_i + \alpha_j + \beta_k + (\alpha\beta)_{jk} + \delta_{ijk} + \gamma_m$

$\qquad\qquad + (\alpha\gamma)_{jm} + (\beta\gamma)_{km} + (\alpha\beta\gamma)_{jkm} + \epsilon_{ijkm}$

$$\begin{aligned}
\text{for } i &= 1, 2\\
j &= 1, 2\\
k &= 1, 2, 3\\
m &= 1, 2, 3, 4
\end{aligned}$$

(b)

Source	df	EMS
Reps	1	- - -
A	1	$\sigma_\epsilon^2 + 4\sigma_\delta^2 + 24\kappa_A^2$
B	2	$\sigma_\epsilon^2 + 4\sigma_\delta^2 + 16\kappa_B^2$
AB	2	$\sigma_\epsilon^2 + 4\sigma_\delta^2 + 8\kappa_{AB}^2$
Error 1	5	$\sigma_\epsilon^2 + 4\sigma_\delta^2$
C	3	$\sigma_\epsilon^2 + 12\kappa_C^2$
AC	3	$\sigma_\epsilon^2 + 6\kappa_{AC}^2$
BC	6	$\sigma_\epsilon^2 + 4\kappa_{BC}^2$
ABC	6	$\sigma_\epsilon^2 + 2\kappa_{ABC}^2$
Error 2	18	σ_ϵ^2

5. (a) $y_{ijk} = \mu + \rho_i + \alpha_j + \delta_{ij} + \beta_k + (\alpha\beta)_{jk} + \epsilon_{ijk}$

$$\begin{aligned}
\text{for } i &= 1, 2, 3, 4\\
j &= 1, 2, 3\\
k &= 1, 2, 3, 4
\end{aligned}$$

μ = overall mean

ρ_i = effect due to the i-th day

α_j = effect due to the j-th percentage of carbon

δ_{ij} = whole plot error component

β_k = effect due to the k-th cooling technique

$(\alpha\beta)_{jk}$ = component for the interaction of the j-th percentage of carbon with the k-th cooling technique

ϵ_{ijk} = split plot error component

(b) The ϵ's are i.i.d. $N(0, \sigma_\epsilon^2)$

The δ's are i.i.d. $N(0, \sigma_\delta^2)$

The ϵ's and δ's are distributed independently of each other.

The ρ's, α's, β's, and $(\alpha\beta)$'s are fixed parameters each summing to zero.

(c)

Source	df	EMS
Days	3	---
Percentages	2	$\sigma_\epsilon^2 + 4\sigma_\delta^2 + 16\kappa_P^2$
Error 1	6	$\sigma_\epsilon^2 + 4\sigma_\delta^2$
Coolings	3	$\sigma_\epsilon^2 + 12\kappa_C^2$
P x C	6	$\sigma_\epsilon^2 + 4\kappa_{PC}^2$
Error 2	27	σ_ϵ^2

7. (a) $y_{ijk} = \mu + \alpha_i + \delta_{ij} + \beta_k + (\alpha\beta)_{jk} + \epsilon_{ijk}$; $i = 1, 2, 3$
$j = 1, 2, 3, 4$
$k = 1, 2, 3, 4$

(b)

Source	df	EMS
Rations	2	$\sigma_\epsilon^2 + 4\sigma_\delta^2 + 16\kappa_R^2$
Error 1	9	$\sigma_\epsilon^2 + 4\sigma_\delta^2$
Tenderizers	3	$\sigma_\epsilon^2 + 12\kappa_T^2$
R x T	6	$\sigma_\epsilon^2 + 4\kappa_{RT}^2$
Error 2	27	σ_ϵ^2

9. (a)

Source	df	MS	EMS
Days	11	37.9	---
Baking T.	4	26.9	$\sigma_\epsilon^2 + 3\sigma_\delta^2 + 36\kappa_A^2$
Error 1	44	7.5	$\sigma_\epsilon^2 + 3\sigma_\delta^2$
Milk T	2	44.6	$\sigma_\epsilon^2 + 60\kappa_B^2$
BT x MT	8	0.2	$\sigma_\epsilon^2 + 12\kappa_{AB}^2$
Error 2	110	5.8	σ_ϵ^2

(b) For BT x MT, F < 1
For Milk T., F = 44.6/5.8 = 7.7, P < 0.005
For Baking T., F = 26.9/7.5 = 3.6, 0.01 < P < 0.025
(c) 0.65   (d) 1.03

11. (a)

Source	df	MS	EMS
Reps	4	296.6	- - -
M	5	100.6	$\sigma_\epsilon^2 + 4\sigma_\delta^2 + 20\kappa_M^2$
Error 1	20	7.94	$\sigma_\epsilon^2 + 4\sigma_\delta^2$
C	3	126.1	$\sigma_\epsilon^2 + 30\kappa_C^2$
MC	15	23.3	$\sigma_\epsilon^2 + 5\kappa_{MC}^2$
Error 2	72	2.66	σ_ϵ^2

(b) For MC: F = 23.3/2.66 = 8.7, P < 0.005
For C : F = 126.1/2.66 = 47.4, P < 0.005
For M : F = 100.6/7.94 = 12.7, P < 0.005
(c) 0.89   (d) 1.26

13. (a)

Source	df	SS	MS	EMS
Reps	2	449.1	224.6	- - -
Spacings	2	564.5	282.2	$\sigma_\epsilon^2 + 4\sigma_\delta^2 + 12\kappa_S^2$
Error 1	4	190.4	47.6	$\sigma_\epsilon^2 + 4\sigma_\delta^2$
Fertilizers	3	41.6	13.9	$\sigma_\epsilon^2 + 9\kappa_F^2$
S x F	6	437.0	72.8	$\sigma_\epsilon^2 + 3\kappa_{SF}^2$
Error 2	18	855.4	47.5	σ_ϵ^2

(b) For S x F: F = 72.8/47.5 = 1.53, P > 0.10
For F: F < 1, P > 0.10
For S: F = 282.2/47.6 = 5.93, 0.05 < P < 0.10
(c) 2.82   (d) 5.63

15. (a) $y_{ijkm} = \mu + \rho_i + \alpha_j + \beta_k + (\alpha\beta)_{jk} + \delta_{ijk} + \gamma_m + \eta_{im}$

$$+ (\alpha\gamma)_{jm} + (\beta\gamma)_{km} + (\alpha\beta\gamma)_{jkm} + \epsilon_{ijkm}$$

for i = 1, 2, 3
j = 1, 2
k = 1, 2, 3
m = 1, 2, 3, 4

(b)

Source	df	EMS
Reps	2	---
A	1	$\sigma_\epsilon^2 + 4\sigma_\delta^2 + 36\kappa_A^2$
B	2	$\sigma_\epsilon^2 + 4\sigma_\delta^2 + 24\kappa_B^2$
AB	2	$\sigma_\epsilon^2 + 4\sigma_\delta^2 + 12\kappa_A^2$
Error 1	10	$\sigma_\epsilon^2 + 4\sigma_\delta^2$
C	2	$\sigma_\epsilon^2 + 6\sigma_\eta^2 + 18\kappa_C^2$
Error 2	4	$\sigma_\epsilon^2 + 6\sigma_\eta^2$
AC	2	$\sigma_\epsilon^2 + 9\kappa_{AC}^2$
BC	4	$\sigma_\epsilon^2 + 6\kappa_{BC}^2$
ABC	4	$\sigma_\epsilon^2 + 3\kappa_{ABC}^2$
Error 3	20	σ_ϵ^2

17. (a) $y_{ijkm} = \mu + \rho_i + \alpha_j + \delta_{ijk} + \beta_k + (\alpha\beta)_{jk} + \eta_{im}$

$$+ \gamma_m + (\alpha\gamma)_{jm} + (\beta\gamma)_{km} + (\alpha\beta\gamma)_{jkm} + \epsilon_{ijkm}$$

$$\text{for } i = 1, 2$$
$$j = 1, 2, 3$$
$$k = 1, 2$$
$$m = 1, 2, \ldots, 6$$

(b)

Source	df	EMS
Reps	1	---
A	2	$\sigma_\epsilon^2 + 6\sigma_\eta^2 + 12\sigma_\delta^2 + 24\kappa_A^2$
Error 1	2	$\sigma_\epsilon^2 + 6\sigma_\eta^2 + 12\sigma_\delta^2$
B	1	$\sigma_\epsilon^2 + 6\sigma_\eta^2 + 36\kappa_B^2$
AB	2	$\sigma_\epsilon^2 + 6\sigma_\eta^2 + 12\kappa_{AB}^2$
Error 2	3	$\sigma_\epsilon^2 + 6\sigma_\eta^2$
C	5	$\sigma_\epsilon^2 + 12\kappa_C^2$
AC	10	$\sigma_\epsilon^2 + 4\kappa_{AC}^2$
BC	5	$\sigma_\epsilon^2 + 6\kappa_{BC}^2$
ABC	10	$\sigma_\epsilon^2 + 2\kappa_{ABC}^2$
Error 3	30	σ_ϵ^2

19. (a)

Source	df	SS	MS	EMS
Reps	2	0.88	0.44	---
Irrigations	1	0.04	0.04	$\sigma^2_\epsilon + 4\sigma^2_\eta + 8\sigma^2_\delta + 24\kappa^2_I$
Error 1	2	3.85	1.93	$\sigma^2_\epsilon + 4\sigma^2_\eta + 8\sigma^2_\delta$
Varieties	1	9.99	9.99	$\sigma^2_\epsilon + 4\sigma^2_\eta + 16\kappa^2_V$
I x V	1	2.95	2.95	$\sigma^2_\epsilon + 4\sigma^2_\eta + 8\kappa^2_{IV}$
Error 2	4	2.32	0.58	$\sigma^2_\epsilon + 4\sigma^2_\eta$
Fertilizers	3	2.80	0.93	$\sigma^2_\epsilon + 12\kappa^2_F$
I x F	3	2.46	0.82	$\sigma^2_\epsilon + 6\kappa^2_{IF}$
V x F	3	0.94	0.31	$\sigma^2_\epsilon + 6\kappa^2_{VF}$
I x V x F	3	0.81	0.27	$\sigma^2_\epsilon + 3\kappa^2_{IVF}$
Error 3	24	12.06	0.50	σ^2_ϵ

(b) Only 2 treatment components show significance:
IxV; $F = 2.95/0.58 = 5.09,\ 0.05 < P < 0.10$
V; $F = 9.99/0.58 = 17.22,\ 0.01 < P < 0.025$
(c) 0.40   (d) 0.22   (e) 0.29

Chapter 12

1. (a) 0000   1100
       0101   1010
       0011   1001
       0110   1111

(b) (A, BCD), (B, ACD), (C, ABD), (D, ABC)
    (AB, CD), (AC, BD), (AD, BC)

3. (a) 00001   10101
       00110   10010
       01100   11000
       01011   11111

(b) (A,BCD,ABDE,CE) , (B,ACD,DE,ABCE)
    (C,ABD,BCDE,AE) , (D,ABC,BE,ACDE)
    (E,ABCDE,BD,AC) , (AB,CD,ADE,BCE)
    (AD,BC,ABE,CDE)

5. Every alias set contains main effects or two factor interactions. Thus, under the basic assumption, there is no estimate
   of Exp.Error and therefore no way of testing any hypotheses.
   If one can reasonably assume that no interactions exist, the
   following ANOVA is appropriate

Source	df
A	1
B	1
C	1
D	1
E	1
Exp.Error	2

Each main effect can be tested but with only 2 degrees of freedom for Exp.Error, these tests are not very sensitive.

7. (a)

000	102	201
012	120	210
021	111	222

   (b) (A, BC, AB^2C^2), (B, AC, AB^2C), (C, AB, ABC^2)
       (AB^2, AC^2, BC^2)

9. (a)

0000	1021	2012
0120	1111	2102
0210	1201	2222

   (b) $A = AB^2C^2 = ABCD = ABCD^2 = AD = BC = AB^2C^2D^2 = BCD^2 = D$

   $B = AB^2C = BC^2D^2 = AC^2D^2 = ABD^2 = AC = CD = ABC^2D = AB^2D^2$

   $C = ABC^2 = BC^2D = AB^2D = ACD^2 = AB = BD = AB^2CD = AC^2D$

   $AB^2 = AC^2 = ACD = AB^2CD^2 = ABD = BC^2 = ABCD = CD^2 = BD^2$

11. (a) Every main effect is aliased with a 4-factor interaction component. Two-factor interaction components are aliased with 3-factor interaction components.

   (b)

Effect	Contrast Value	Effect	Contrast Value
A	7	D	75
B	23	AD	17
AB	-27	BD	1
C	-21	CE	11
AC	-7	CD	-43
BC	9	BE	39
DE	19	AE	-33
		E	5

(c)

Source	df	SS (=MS)	Source	df	SS (=MS)
A	1	3.06	AE	1	68.06
B	1	33.06	BC	1	5.06
C	1	27.56	BD	1	0.06
D	1	351.56	BE	1	95.06
E	1	22.56	CD	1	115.56
AB	1	45.56	CE	1	7.56
AC	1	3.06	DE	1	1.56
AD	1	18.06			

(d)   There is no estimate of Exp. Error variation under the
stated assumptions. Thus, no tests of main effects and
2-factor interactions exist. An examination of the mean
squares reveals that the important treatment components
are main effect D, the BE and CD interactions.

Chapter 13

1. (a) $y_{ijg} = \mu + \tau_i + \rho_j + \epsilon_{ijg}$; $\begin{aligned} i &= 1, 2, \ldots, 5 \\ j &= 1, 2, 3, 4 \\ g &= n_{ij} \end{aligned}$

μ = overall mean

τ_i = effect of i-th treatment

ρ_j = effect of j-th block

ϵ_{ijg} = random component explaining all extraneous
variation of (i,j,g)-th observation.

(b)   The ϵ's are i.i.d. $N(0, \sigma_\epsilon^2)$

The τ's are fixed such that $\Sigma\, \tau_i = 0$

The ρ's may be fixed or random. If fixed, they sum to
zero. If random, they have a $N(0, \sigma_\rho^2)$ distribution,
independent of the ϵ's.

(c)

Source	df
Blocks(unadj.)	3
Trts(adj.)	4
Intrablock error	8

3. (a)

Block	1	2	3	4	$N_{i.}$
Trt. 1	1	1	1	1	4
2	1	0	1	1	3
3	0	1	1	1	3
4	1	1	1	0	3
5	1	1	0	1	3
$N_{.j}$	4	4	4	4	16=N..

(b) $3\hat{\tau}_1 - (3\hat{\tau}_2 + 3\hat{\tau}_3 + 3\hat{\tau}_4 + 3\hat{\tau}_5)/4 = 10.5$

$3\hat{\tau}_2 - (3\hat{\tau}_1 + 2\hat{\tau}_3 + 2\hat{\tau}_4 + 2\hat{\tau}_5)/4 = 27.75$

$3\hat{\tau}_3 - (3\hat{\tau}_1 + 2\hat{\tau}_2 + 2\hat{\tau}_4 + 2\hat{\tau}_5)/4 = -22.75$

$3\hat{\tau}_4 - (3\hat{\tau}_1 + 2\hat{\tau}_2 + 2\hat{\tau}_3 + 2\hat{\tau}_5)/4 = -16.75$

$3\hat{\tau}_5 - (3\hat{\tau}_1 + 2\hat{\tau}_2 + 2\hat{\tau}_3 + 2\hat{\tau}_4)/4 = 1.25$

5. (a) $15\hat{\tau}_1 = 42$

$7\hat{\tau}_2 - 5\hat{\tau}_1 = 111$

$7\hat{\tau}_3 - 5\hat{\tau}_1 = -91$

$7\hat{\tau}_4 - 5\hat{\tau}_1 = -67$

$7\hat{\tau}_5 - 5\hat{\tau}_1 = 5$

(b) $\hat{\tau}_1 = 2.8$; $\hat{\tau}_2 = 10.35$; $\hat{\tau}_3 = -8.02$; $\hat{\tau}_4 = -5.84$; $\hat{\tau}_5 = 0.71$

7. (a)

Source	df	SS	MS
Blocks(unadj.)	3	298.8	---
Treatments(adj.)	4	597.8	149.45
Intrablock error	8	25.2	3.14

(b) $F = 149.45/3.14 = 47.6$, $P < 0.005$

9. Randomly assign T_1 and T_4 to the EU of 10 pairs while T_2 and T_3 are randomly assigned to the EU of the remaining 10 pairs.

Chapter 14

1. (a) b = 7, t = 7, k = 4, r = 4, λ = 2

3. (a) NO   (b) Add one block: 1, 3, 4, 5, 6

5. Use 5 ovens and assign treatments as follows:

Oven	Treatments
1	1 2 4 5
2	1 3 4 5
3	1 2 3 5
4	1 2 3 4
5	2 3 4 5

7. Cannot construct a BIB design with 8 schools. One would need
   b = 15 schools to construct a BIB design for t = 6 and k = 4.

9. Numerous possibilities exist: Once a LS is available, any
   column can be omitted, with each row forming a block.

11. One of many possibilities is:

1	2	3	4	4	8	12	16	2	7	9	16
5	6	7	8	1	6	11	16	3	6	12	13
9	10	11	12	2	5	12	15	1	7	12	14
13	14	15	16	3	8	9	14	4	6	9	15
1	5	9	13	4	7	10	13	2	8	11	13
2	6	10	14	1	8	10	15	3	5	10	16
3	7	11	15	4	5	11	14				

13. One of many possibilities is:

1	2	3	2	4	6	3	5	6
1	4	5	2	5	7			
1	6	7	3	4	7			

15. (a)

Source	df	SS	MS
Blocks(unadj.)	6	7165.2	
Trts.(adj.)	6	12026.3	2004.4
Intrablock error	8	489.7	61.21

(b) F = 2004.4/61.21 = 32.75, P < 0.005
(c) MSB = 167.05, w = 0.0453
     MS*(Trts.adj.) = 2739.3, MSE* = 72.3
     F = 2739.3/72.3 = 37.9, P < 0.005.

Chapter 15

1. (a)

Treatment	First Associates	Second Associates
1	2,3,4,5,6,7	8
2	1,3,4,5,6,8	7
3	1,2,4,5,7,8	6
4	1,2,3,6,7,8	5
5	1,2,3,6,7,8	4
6	1,2,4,5,7,8	3
7	1,3,4,5,6,8	2
8	2,3,4,5,6,7	1

(b) $r = 3$, $t = 8$, $b = 6$, $k = 4$, $\lambda_1 = 1$, $\lambda_2 = 3$, $n_1 = 6$, $n_2 = 1$

$$P_1 = \begin{pmatrix} 4 & 1 \\ 1 & 0 \end{pmatrix} \qquad P_2 = \begin{pmatrix} 6 & 0 \\ 0 & 0 \end{pmatrix}$$

3. (a)

Treatment	First Associates	Second Associate
1	2,3,4,5	6
2	1,4,5,6	3
3	1,4,5,6	2
4	1,2,3,6	5
5	1,2,3,6	4
6	2,3,4,5	1

(b) $r = 2$, $t = 6$, $b = 3$, $k = 4$, $\lambda_1 = 1$, $\lambda_2 = 2$, $n_1 = 4$, $n_2 = 1$

5. The new design is:

Block	Treatments
1	4,5
2	2,3
3	1,6

which is a PBIB design with parameters of the first kind:
$r = 1$, $t = 6$, $b = 3$, $k = 2$, $\lambda_1 = 0$, $\lambda_2 = 1$, $n_1 = 4$, $n_2 = 1$

7. Yes, the parameters are $r = k = 3$, $t = b = 5$, $n_1 = n_2 = 2$, $\lambda_1 = 2$, $\lambda_2 = 1$, and

$$P_1 = \begin{pmatrix} 0 & 1 \\ 1 & 1 \end{pmatrix} \qquad P_2 = \begin{pmatrix} 1 & 1 \\ 1 & 0 \end{pmatrix}$$

9. (a)

Treatment	First Associates	Second Associates
1	2, 3, 4, 6, 7, 8	5
2	1, 3, 4, 5, 7, 8	6
3	1, 2, 4, 5, 6, 8	7
4	1, 2, 3, 5, 6, 7	8
5	2, 3, 4, 6, 7, 8	1
6	1, 3, 4, 5, 7, 8	2
7	1, 2, 4, 5, 6, 8	3
8	1, 2, 3, 5, 6, 7	4

(b) $t = b = 8$, $r = k = 3$, $n_1 = 6$, $n_2 = 1$, $\lambda_1 = 1$, $\lambda_2 = 0$

$$P_1 = \begin{pmatrix} 4 & 1 \\ 1 & 0 \end{pmatrix} \qquad P_2 = \begin{pmatrix} 6 & 0 \\ 0 & 0 \end{pmatrix}$$

11. (a)

Source	df	SS	MS
Blocks(unadj.)	7	13.45	
Treatments(adj.)	7	31.59	4.51
Intrablock error	9	3.89	0.43

(b) $F = 4.51/0.43 = 10.5$, $P < 0.005$
(c) 0.23, 0.43

13. (a)

Treatment	$\hat{\tau}_i^*$	Treatment	$\hat{\tau}_i^*$
1	0.774	5	-1.270
2	-0.478	6	0.615
3	0.644	7	2.072
4	-0.034	8	-2.320

(b) 0.58; 0.61

INDEX